Integer Linear Programming in Computational and Systems Biology

Integer linear programming (ILP) is a versatile modeling and optimization technique that is increasingly used in nontraditional ways in biology, with the potential to transform biological computation. However, few biologists know about it. This how-to and why-do text introduces ILP through the lens of computational and systems biology. It uses in-depth examples from genomics, phylogenetics, RNA and protein folding, network analysis, cancer, ecology, co-evolution, DNA sequencing and sequence analysis, pedigree and sibling inference, haplotyping, clustering, and more to establish the power of ILP. This book aims to teach the logic of modeling and solving problems with ILP, and to teach the practical "workflow" involved in using ILP in biology.

Written for a wide audience, with no biological or computational prerequisites, this book is appropriate for both entry-level and advanced courses aimed at biological and computational students, and as a source for specialists. Numerous exercises and accompanying software (in Python and Perl) demonstrate the concepts.

Dan Gusfield is Distinguished Professor of Computer Science at the University of California, Davis (UCD), and a fellow of the IEEE, the ACM, and the International Society of Computational Biology (ISCB). His previous books include *The Stable Marriage Problem* (with Robert W. Irving) and *Algorithms on Strings, Trees and Sequences*; and *ReCombinatorics*. He has served as chair of the computer science department at UCD (2000–2004), and was the founding editor-in-chief of *The IEEE/ACM Transactions of Computational Biology and Bioinformatics* until January 2009. He has been instrumental in the definition and development of the intersection between computer science and computational biology.

INTEGER LINEAR PROGRAMMING IN COMPUTATIONAL AND SYSTEMS BIOLOGY

An Entry-Level Text and Course

DAN GUSFIELD
University of California, Davis

CAMBRIDGE
UNIVERSITY PRESS

University Printing House, Cambridge CB2 8BS, United Kingdom

One Liberty Plaza, 20th Floor, New York, NY 10006, USA

477 Williamstown Road, Port Melbourne, VIC 3207, Australia

314–321, 3rd Floor, Plot 3, Splendor Forum, Jasola District Centre, New Delhi – 110025, India

79 Anson Road, #06–04/06, Singapore 079906

Cambridge University Press is part of the University of Cambridge.

It furthers the University's mission by disseminating knowledge in the pursuit of education, learning, and research at the highest international levels of excellence.

www.cambridge.org
Information on this title: www.cambridge.org/9781108421768
DOI: 10.1017/9781108377737

First published 2019

Printed and bound in Great Britain by Clays Ltd, Elcograf S.p.A.

A catalogue record for this publication is available from the British Library.

Library of Congress Cataloging-in-Publication Data
Names: Gusfield, Dan, author.
Title: Integer linear programming in computational and systems biology : an entry-level text and course / Dan Gusfield, University of California, Davis.
Description: Cambridge, United Kingdom ; New York, NY : Cambridge University Press, 2019. | Includes bibliographical references and index.
Identifiers: LCCN 2018061722 | ISBN 9781108421768 (hardback : alk. paper)
Subjects: LCSH: Computational biology–Mathematical models. | Systems biology–Mathematical models. | Linear programming.
Classification: LCC QH324.2 .G87 2019 | DDC 570.285–dc23
LC record available at https://lccn.loc.gov/2018061722

ISBN 978-1-108-42176-8 Hardback

Dedicated to Dick Karp, mentor and role model to me for 45 years and counting. Thank you Dick for all your wisdom, support, and friendship.

Contents

Introduction to the Book and Course

Quoting Dick Karp, "Quoting Dan Gusfield, 'When you have an instance of a hard computational problem, try integer programming – It might work.'"

I.1 WHY INTEGER PROGRAMMING?

Integer (linear) programming, abbreviated "ILP," is a versatile modeling and optimization technique that was developed for complex planning and operational decision-making. However, it has been increasingly used in *computational and systems biology* in *non traditional* ways, most importantly and inventively as a computational tool to *model* biological *phenomena*, to *analyze* biological *data*, and to *extract* biological *insight* from the models and the data. Integer programming is often (but not always) very effective in solving *instances* of hard computational problems on *realistic* biological data of current importance, despite the fact that many of those problems lack *general* algorithmic solutions that are efficient (in a provable, worst-case sense); and that the problem of solving integer programs also lacks a provable worst-case efficient algorithm.

Moreover, even for a problem where a general, worst-case efficient algorithm might be possible, the time and effort needed to find it, and the time and effort needed to implement it as a computer program, are typically much greater than the time and effort needed to formulate and implement an ILP solution to the problem. Further, it is often desirable to solve both the *minimization* and the *maximization* variants of a problem. For many problems, this requires finding and implementing two quite *different* algorithms, while an ILP solution for one variant can often be converted to an ILP formulation for the other variant, just by interchanging the words "minimize" and "maximize."

Highly engineered, commercial ILP solvers are available now (free to academics and researchers) to *solve* ILP formulations. Alternative open-source solvers are also available, and these can be effective in *some* biological applications. The improvement of the *best* solvers has been spectacular, with an estimate that (combined with faster computers) benchmark ILP problems can now be solved 200 *billion* times faster than 25 years ago.[1] Exploiting ILP, some biological problems of importance can be modeled in a way that allows a solution in seconds on a laptop, while solutions to the more common (statistically based) models require days, weeks, or months of computation on large clusters.[2]

The effectiveness of the best ILP solvers on many problem *instances* of importance in biology opens huge opportunities. The impact of faster and easier-to-implement computation could be truly transformative in many areas of biology. However, few biologists know about integer linear programming, and when I explain that the ILP approach might translate an instance of a simple-to-describe computational problem into hundreds of thousands, or even millions, of linear inequalities that are then solved, I am often met with incredulity that such an approach could work – but it does, and often!

So, there are challenges to effectively using the ILP tools for biological problems, and educational and outreach issues that must be addressed. This book is motivated by those successes (and some failures), opportunities, and challenges. It is my attempt to reach and educate a broad range of computational biologists and students, particularly those with biology, rather than computation, backgrounds.

I.2 ABOUT THIS BOOK AND COURSE

My previous books are heavy on abstract theoretical thinking, emphasizing theorems, proofs, and methods, and are aimed at an advanced audience. Those books explicitly state that they are *not* "how-to" books. But in addition to my theoretical work, I have also done some "practical" work, particularly in my use of integer linear programming for computational biology. So in contrast to my past books, this *is* a "how-to" and a "why-do" book, and it is aimed at an audience with little or no background in either integer linear programming or computer programming.[3] The book uses meaningful examples from a wide range of topics in computational and systems biology in order to argue for the utility of integer programming, and to teach the practical "workflow" involved in using integer programming in biology.

Although the book discusses a wide variety of integer programming applications in computational and systems biology, it is *not* intended as a *comprehensive survey* of the literature, even restricted to *molecularly oriented biology*. I direct the reader to a review by G. Lancia [118], which covers much of the literature up to 2008; to deeper treatments of a narrower range of selected applications in [117] by Lancia in 2004; and to an overview of several problems [7], written by E. Althaus, G. W. Klau,

[1] However, the difference in efficiency and reliability between the best solvers and their competitors can be *enormous*, and the use of less efficient or less reliable solvers can give a very distorted impression of what is currently possible using an ILP approach.

[2] Most of the runtimes I report in this book were obtained on a Macbook Pro laptop, 2.3 GHz i7, with four processors allowing eight threads, costing under $2,000. Most of the ILP computations were done with the Gurobi Optimizer, version 6.5, from Gurobi Optimization ®. But, during the writing of the book, Gurobi released versions 7.02, 7.5, and 8.0, and some computations were done with those versions. A few computations were done using Cplex 12.6 for additional validation of the qualitative results reported in the book, and several results were obtained on an iMac i5, four processors with four threads. That processor is slightly faster than the one on the Macbook Pro, but with only four threads, the run times were similar, with the iMac at most 20% faster than the Macbook Pro. These parameters should be contrasted with typical results for computations in biology that often run for months on a cluster with hundreds to thousands of processors.

[3] However, readers with such background should still find many of the specific topics of interest.

O. Kohlbacher, H. P. Lenhof, and K. Reinert in 2009. The paper [69] also surveys several computational problems in computational biology, but the problems are first cast as *quadratic* integer programming problems, and later converted to *linear* integer programming problems.

Intended Audience My intended audience consists of computational and systems biologists, and students in biology and/or computer science/mathematics. I am particularly interested in reaching real biologists who have had little or no exposure to integer programming. Why the emphasis on biologists?

Many biologists in computational biology are excellent computer programmers, and many are very sophisticated in methods of *statistical* modeling and computation, and in many types of computational methods. For example, there are many biologists who are highly proficient with classical statistical tests and sampling, with statistical genetics, with Markov chain Monte Carlo methods, with dynamic programming, with the use of sophisticated string algorithms, with data structures, and with databases. In contrast, very few biologists are familiar with integer programming. Overwhelmingly, the papers in computational and systems biology that use integer programming are authored by researchers from computer science, engineering, mathematics, or statistics. This book, which discusses integer programming through the lens of computational and systems biology, and is written for an audience with no background in ILP, aims to change that. However, I believe that the book will also be of interest to a more advanced audience, and a source for specialists.

As a Text for a Course Part I of the book can be used as a text for an introductory course on integer linear programming in computational and systems biology.[4] All of the instructional material on integer programming, and the use of the ILP solver Gurobi Optimizer, is in Part I of the book. Additional topics from Part II can be added to match the interests of the instructor or class. I recently designed and taught such a course, the first of its kind, using material from Part I, to undergraduate students with *no* required prerequisites. About half the students were biology students, and half were from computer science. Only a few of the biology students had any background in computer programming. I believe the material in the course was quite accessible to all of the students, as most of the students did very well, particularly in their term projects.[5]

I.3 TASKS AND SKILLS

In addition to demonstrating the power of integer linear programming to model and solve instances of computational problems in biology, this book develops *skills* for four tasks:

(A) The task of developing the *solution idea* or *solution logic* for an *already well-articulated* computational problem in biology. Of course, the solution ideas must be developed with an eye toward translating the logic into an integer linear program (task B). Often, the solution logic will be straightforward or even trivial, but in a few of the problems we discuss, the logic will be more subtle.

[4] Or even for a general-purpose course on the use of integer programming.
[5] Some of these are discussed in the book.

(B) The task of creating an *abstract* ILP *formulation* for a biological problem, translating the logic developed in task A into general integer linear (in)equalities, and a linear objective function. Task B is the main focus of this book.

(C) The task of writing a *computer program* that takes in the details of an *instance* of a biological problem, and creates the *concrete* ILP formulation specifying the linear (in)equalities (coming from the result of task B) that are used to solve the given problem instance.[6]

Computer programs that accomplish task C have been written for many of the computational problems discussed in this book, and these programs can be downloaded from the book website. You do *not* need to know any computer programming to *use* those programs. But, some computer programming can be learned by examining those programs, and additionally, the book website contains a short tutorial on programming in the computer language Python, to accomplish task C. However, the book is structured so that this material is *optional*, and can be completely *skipped*. A reader can understand the material that addresses tasks A, B, and D, without any need to engage with task C.[7]

(D) The task of *using* an ILP *solver* to find an optimal solution for a concrete ILP formulation (created by task C). In this book we mainly discuss the ILP solver developed by Gurobi Optimization ®.[8] Discussion of task D is interwoven with the discussion of tasks A and B.

Biological and Mathematical Modeling There is an additional skill that is *crucial* for effective computational and systems biology. But, it is a skill that I cannot formally teach, although I can display and critique it. It is the skill of *translating* biological phenomena into *formal, mathematical models*. In this book, I address biological phenomena that are formally modeled in a way that lead to computational problems solvable by integer linear programming.

I.4 WHY LEARN FROM ME?

Of the four tasks addressed in the book, I am most proficient in tasks A and B, which most closely involve the kind of abstract, logical thinking that forms my professional training and career. But, one of the most appealing properties of integer programming is that the logical thinking required for task A is often not that demanding. And, for task B, there is a small set of *idioms* or *modeling tricks* that are very helpful in moving from a high-level solution *idea* to a collection of linear inequalities that encode that idea.

I am much less proficient in skills C and D, and I consider this a *virtue*. In this book, skill D is developed by using the ILP solver *Gurobi Optimizer*, and skill C is

6 For small problem instances, one can write out the needed inequalities directly using a text editor or word processor, without writing a computer program. If your interest is only in learning about the power of integer programming, how to formulate ILPs, or how to use an ILP solver, small examples are sufficient. But for typical problem instances that arise in practice, you will need a computer program to *create* the concrete ILP formulations.

7 The recent course I taught on integer programming in computational biology skipped task C entirely.

8 Although, all of the concrete ILP formulations produced by the software can also be solved with Cplex ®, and an instructor who wants to use Cplex in place of Gurobi could easily supplement the Gurobi-specific materials in the book.

addressed with a tutorial on Python programming, posted on the book's website. Compared to true experts, I am a novice in both Python and Gurobi. I typically run Gurobi with its *default* parameters (there are over 50 adjustable parameters in Gurobi), and my Python programs usually sacrifice elegance and efficiency in order to reduce my programming effort. There are huge (thousand-plus page) books on Python, and the Gurobi documentation is equally massive. I have read and understand a *minuscule* fraction of those, and that is one of my strongest qualifications for writing this book – I don't know much about Gurobi or Python, but I know enough to get the job done. So, what I will teach you is just enough Gurobi and just enough Python to join me. True connoisseurs of Python will be disgusted by my crude Python programming and the way I explain it, and true aficionados of Gurobi will be appalled by my lack of sophistication. But sometimes ignorance is bliss, and helps to keep the work simple.

I.5 THE ORGANIZATION OF THE BOOK

I have divided the book into two parts. Both parts discuss real problems in computational and systems biology that have been, or could be, solved by integer linear programming. Most of the topics come from published research papers.

I.5.1 Part I

Part I is self-contained and can be the basis for a course on integer linear programming in computational and systems biology. All of the material I used in a recent undergraduate course on computational biology and integer programming comes from Part I. Part I starts with biological problems that can be solved with *basic* ILP techniques, and then incrementally develops and explains integer programming through biological problems that need additional, or more complex, ILP techniques. The ILP formulations in Part I are thoroughly explained, and many of the formulations are accompanied by software (available online) to generate or read problem instances, and to create the *concrete* ILP formulations to solve those instances. Using this software allows the reader to generate their own examples, and to test and extend their understanding of the topics in the book. I highly recommend that the reader do that.[9] The software also allows an instructor using this text to develop additional examples or test problems for use in the class. Part I also contains more course-related exercises than Part II, along with exercises on the practical use of Gurobi Optimizer.

I.5.2 Part II

Part II continues with additional topics in computational and systems biology that are solved by integer programming. Several of the chapters in Part II parallel chapters in Part I, further developing biological and ILP topics introduced in Part I. Most of the ILP formulations in Part II are explicitly described (as they are in Part I), but more

[9] In the class I recently taught, students made many interesting observations that were new to me, based on such explorations.

of the *explanations* for the formulations are developed in exercises. There are also more *extended exercises*, where most of a topic is developed through exercises.

Thus, Part II further exposes the power of ILP and its wide range of applications in biology, but relies on the discussions of ILP and Gurobi from Part I. Finally (at this moment) only a few of the topics in Part II are accompanied by software, although this may change in the future, as software written by myself, or others, is added to the webpage for this book.

I.5.3 Software, Empirical Results, and Computer Programming

As mentioned earlier, many of the topics in Part I (and some in Part II) are accompanied by computer programs that read or generate data, and then create concrete ILP formulations for that data. The programs are written in the programming language Python (version 2.7), or in PERL. The reader does *not* need to know either of those languages, or even need to have any knowledge of computer programming. The reader can simply *use* those programs without needing to look inside of them. But, for the reader who wants to develop a few skills for task C, the tutorial that can be downloaded from the book website is where you can start. The URL for the book website, maintained by Cambridge Press is:

www.cambridge.org/9781108421768

The awful 13 digit number at the end is the ISBN number for the book. It is better to bookmark the page, or download all programs in a zip file, than trying to correctly remember and enter the ISBN number each time.

Empirical Results Throughout the book, I report on computational experiments I have done using Gurobi Optimizer or Cplex to solve instances of ILP formulations developed in the book. Many of these computations were done with versions of ILP solvers that are *not* the most current. The purpose of these empirical results is to *roughly* distinguish between practical and impractical computational solutions, and to illustrate the *range of practicality* of an ILP approach. Empirical results showing that an ILP approach is practical today, for a useful range of data, will remain valid as faster ILP solvers are released. So, these empirical tests are *not* intended as comprehensive, fine-grained explorations leading to advice on the best ILP solvers or the best computer architectures or the best parameters to use today for solving particular ILP problems. Periodically, papers of that type appear in the literature, often making very fine distinctions between different approaches. But those results quickly become outdated as computers, operating systems, and solvers change.

Python Programs and Programming No computer programming is needed to read this book and use the software, and the reader need never learn any. However, for a reader who wants to see details of the whole "workflow" (i.e., all tasks A through D), the tutorial on the book website introduces elements of the Python programming language. The tutorial explains the Python programs that accomplish task C for two of the biological problems discussed in the book. These programs, and all the Python programs available on the book website, are written using the *simplest* possible elements of Python.[10] Thus, the tutorial on Python introduces just enough Python to understand those two programs, and to begin some of your own programming, but is very far from a complete tutorial on Python or computer programming.

On Figures I have drawn most of the figures in the book. Several others have been generously contributed by colleagues and students. A fair number of figures have been copied, and sometimes modified, from papers in PLoS and BMC journals, under the Creative Commons Attribution 4.0 International License (http://creativecommons.org/licenses/by/4.0/), or the Creative Commons CC0 public domain dedication. A few others have been reprinted from journal papers under licenses granted by the publishers.

I.6 NO MATH

One of the prepublication reviewers of a draft of the book said that most biologists (even the ones who already want to learn about integer programming) will see the mathematical *symbols* in the book and close it fast and hard. But actually, the book contains *no* math beyond arithmetic[11] – it just looks like scary math until you read and parse it. The mathematical notation is really just a *shorthand* or a *language* for stating what is required and desired of an integer programming solution.[12] Yes, there are occasional ideas that take some time and thought to understand, and sometimes the material requires *logical puzzling* – some of which is challenging. But nothing in the book requires knowing, understanding, or learning any math you didn't already learn by the eighth grade. Of course, there is deep and beautiful mathematics *underneath* the whole field of linear programming and integer linear programming. But this book does not get into any of that – it is only about *using* integer linear programming to *model* and *solve* certain computational problems that arise in biology.

I.7 ACKNOWLEDGMENTS

Of course in writing a book, and in doing all the reading and research before that, there are many people who helped, and many people to thank. In no particular order I want to thank Bob Bixby, Ed Rothberg, Steven An, Gunnar Klau, Roded Sharan, Yufeng Wu, Chase Maguire, Thong Le, Julia Matsieva, Yelena Frid, Yun Song, Dan Brown, Russell Schwartz, Fumei Lam, Rob Gysel, Kristian Stevens, Balaji Venkatachalam, Matthias Koeppe, David Amar, Giuseppe Lancia, Carl Kingsford, Tandy Warnow, Mohammed El-Kebir, Tanya Berger-Wolf, Dick Karp, Kent Wilken, Jim Orlin, David Shmoys, Ron Shamir, Thomas Magnanti, Paola Bonizonni, Sorin Istrail, Teresa Przytycka, Sylvia Spengler, and all of the students in the fall 2017 ECS 189 class on integer programming in computational biology at UC Davis, particularly Kat Arnott for conversations on community detection, and Jonathan Kim, Parsoa Khorsand, Jessica Au, and Jessie Vuong, whose class projects I write about in this

[10] Some of the programs accompanying this book are written in PERL, and those programs are more complex. The tutorial does *not* explain the PERL programs, although they can be easily *used* by the reader.

[11] OK, maybe some very, very elementary high school algebra.

[12] We will define each piece of mathematical notation the first time it is used, and list it in the index of the book.

book; and who have generously allowed their class-project software to be distributed with this book. I also acknowledge the support of the Simons Institute for Theoretical Computing, where I wrote some of the book while on sabbatical, and the research support from the NSF grant 1528234, on integer linear programming in computational biology.

Last, but certainly not least, I want to thank my wife Carrie (again) for her continuing forbearance. In the acknowledgments of a prior book, I swore to her that it would be my last – and I lied (again).[13]

[13] But the good news is that I only have two more books I want to write.

PART I

1

A Flyover Introduction to Integer Linear Programming

Integer linear programming (more commonly called *integer programming*) is a versatile computational method that is widely used in management, engineering, industry, banking, transportation, etc. However, it is almost unknown to biologists, although computer scientists and mathematicians are increasingly using integer programming in computational and systems biology. This chapter introduces integer linear programming (ILP), starting with the foundational topic of *linear programming* (LP).

1.1 LINEAR PROGRAMMING (LP) AND ITS USE

It is helpful to divide the discussion of linear programming (and integer linear programming) into four parts: the *problem to be solved*; the *concrete formulation* of a linear program (or model), given all the data required to specify a specific problem instance; the *solution* of a concrete formulation; and the *abstract formulation* of a linear program.

We will explain these four elements using a simplified version of a real problem (discussed in [42, 46]) from *conservation ecology*. We take liberties in describing that work in order to simplify the presentation.[1]

1.1.1 The Threatened Species Protection Problem

The initial work in [46] concerns conservation of "remnant patches of bush on the Eyre Peninsula, South Australia." The latter work in [42] extends the approach in [46], addressing conservation of endangered plant species in the *Cape of South Africa*, which is

[1] This problem is a bit atypical for this book, first because it comes from *ecology*, while most of the book involves problems in *genomics, genetics, phylogenetics, RNA, protein, networks, and disease*; second, because it translates almost directly into a linear program and integer linear program, while most of the problems in the book have less direct translations to LP and ILP. But, the ease of translation allows a simpler introduction to linear programming. Finally, the problem is a bit atypical because it involves a question of biological *management* rather than biological science.

> ... one of the most botanically species-rich areas of the world with more than 9000 species ... [42]

These species span more than 700 *genera*,[2] and at least 274 of them contain one or more species that is classified as "vulnerable, endangered, or critically endangered." And,

> [o]f the 274 threatened genera, 17 belong to the top-20 most-threatened genera of South Africa, based on the proportion of their species that are threatened. [42]

Conservation agencies have (limited) resources to preserve *some* areas of the Cape, and hence to help protect some of the threatened species. So, the general question is how to most effectively use the available resources to protect threatened species. The analogous question for threatened animal species was recently addressed in a perspective piece in *Science* magazine [76].

There are many variants, extensions, and specializations of this conservation problem. We start with a simple variant, but will return to the general problem throughout the book, extending it as we explain more about LP and ILP.

In [42], the Cape is divided into about 200 square regions, each containing about 675 square kilometers. It is assumed that in each region, we know the abundance of each of the 274 threatened species. In our telling of the story, we measure abundance of a species in a region by the area (in square kms) occupied by that species.[3] The cost to preserve *all* the land in any region is also assumed to be known, but partial preservation of a region is also possible. Then, for each threatened species, a *conservation target* is established, meaning that the species would be considered protected if the total area occupied by the species in the preserved land (over all the regions) contains more than its conservation target. With all of this data, the first general problem is

> **The Species Protection Problem** What is the *least expensive* way to protect *all* of the threatened species?

1.1.1.1 A Toy Concrete Example

To make all of this concrete, we will look at an artificially small *instance* of the species protection problem, with only five regions, and two threatened species. The regions are named *A, B, C, D*, and *E*; and the species are named α and β. Table 1.1 shows the area occupied by each species in each of the regions; the cost of preserving *all* the land in each region; and the conservation target for each species.

The Concrete Problem Instance The specific data in Table 1.1 defines a concrete instance of the species protection problem: What is the minimum cost way to preserve parts of the five regions, so that for each of the two threatened species, the total area in the preserved land is at least the conservation target for that species?

More important than the actual solution for this concrete problem instance, what is a *method* we can use to solve any concrete problem instance?

2 According to Wikipedia: "genus, pl. *genera* is a taxonomic rank used in the biological classification of living and fossil organisms in biology. In the hierarchy of biological classification, genus comes above species and below family." [216]

3 We assume that the species is distributed uniformly over the region, so for example, any half of the region will have half of the species in the region.

Table 1.1 Concrete Data for An Instance of the Species Protection Problem.

Region	Area occupied by α in square kilometers	Area occupied by β in square kilometers	Cost for preserving the region
A	24	83	$97K
B	36	11	$73K
C	0	29	$22K
D	15	0	$11K
E	40	18	$45K
Target	64	87	

The answer is to *formulate* a *linear program (LP) model* (also called an *LP formulation*) that describes (or expresses) the details of the specific problem instance. That formulation, with all the details of the problem *instance*, is called a *concrete* LP formulation. Then, to solve the concrete instance, we *solve* the concrete LP formulation using an LP *solver*. The solution will specify how much of each region to preserve, and what the total cost will be. We next discuss more about the LP formulation.

1.1.2 Creating a Concrete LP Formulation

To create an LP formulation (model) for a concrete problem instance, we begin by creating linear programming *variables*. The term "variable" in this context is the same as in high school mathematics. An LP variable can take on a *numerical value*. The LP variables for the species protection problem express the unknown values that we ultimately want to determine: variable X_A denotes the *fraction* of region A that will be preserved. Being a fraction, we restrict X_A so that it can only take on a value between 0 and 1 (inclusive). The variables X_B, X_C, X_D, and X_E have analogous meanings for regions B, C, D, and E, respectively.

The next step in formulating a concrete LP model for a problem instance is to develop *linear constraints*, which are either *inequalities* or *equalities*. The inequalities and equalities express the additional constraints on the values that are *permitted* to be assigned to the variables. A further explanation follows below.

Linear Functions and LP Constraints A *linear function* of a set of variables is formed by multiplying each variable by a specific *coefficient* (or constant number) and adding together the resulting terms. For example, suppose the set of variables is $\{X_A, X_B\}$. Then,

$$3X_A + 4X_B,$$

is a linear function of those two variables, with coefficients 3 and 4, respectively. A linear *equality* consists of a linear function followed by the equality sign ("=") and a constant; for example,

$$3X_A + 4X_B = 17.$$

A linear *inequality* consists of a linear function followed by a symbol for an inequality *relation*, ($\leq$, $\geq$, $<$, or $>$), followed by a constant number. For example,

$$3X_A + 4X_B \leq 13,$$

is a linear inequality.

The *constraints* in a linear program consist of linear inequalities (which could actually be linear equalities). Although the *definition* of a linear inequality includes the cases of *strict* inequality, i.e., < and >, those are not allowed symbols in linear programming solvers I have used; but there are ways to achieve the effect. We will see that in several examples throughout the book.

In the concrete formulation for the toy instance of the species protection problem, we have the simple, initial constraints:

$$\begin{aligned} X_A &\leq 1, \\ X_B &\leq 1, \\ X_C &\leq 1, \\ X_D &\leq 1, \\ X_E &\leq 1. \end{aligned} \tag{1.1}$$

We do *not* need to explicitly include the constraints $X_A \geq 0$, etc., because it is already *assumed* in linear programming that the value of any variable will be *nonnegative*.[4]

More Interesting Constraints The more interesting, and complex constraints come from the requirement to protect both of the threatened species. The following constraint expresses what is needed to protect species α:

$$24X_A + 36X_B + 0X_C + 15X_D + 40X_E \geq 64. \tag{1.2}$$

To understand this, note that each term in the inequality specifies the area occupied by species α in one specific region, times the value of the X variable for that region. For example, the constant number 24 in the term $24X_A$, is total area in region A that is occupied by species α. So, once the value of X_A is specified (which is a number from 0 to 1), $24X_A$ is the amount of species α that will be protected in region A. Hence, the linear function in (1.2) specifies the *total preserved* area that will be occupied by species α.[5] The right end of the inequality has "$\geq$ 64." The overall result is that inequality (1.2) states that the total preserved area occupied by species α, *must be* greater or equal to 64, the conservation target for α.

The analogous constraint for species β is:

$$83X_A + 16X_B + 19X_C + 0X_D + 18X_E \geq 87. \tag{1.3}$$

Feasible Solutions Some combinations of values for the variables X_A, X_B X_C, X_D *satisfy* (make true) all the inequalities in (1.1), (1.2), and (1.3), but some combinations of values *violate* one or more of the inequalities. A combination of values assigned to the variables that satisfies *all* of the inequalities is called a *feasible solution* to the constraints. If there are no feasible solutions, then the set of constraints is called *infeasible*.

For example, if we set all five variables to 0.66, then the inequalities in (1.1) are satisfied, and the sums in inequalities (1.2) and (1.3) are 75.9 and 89.76, respectively.

4 The default assumption in linear programming that a variable can only have nonnegative value, is for convenience; it is not limiting. There are standard ways to get around it, but in all of the problems discussed in this book, all the variables will naturally have nonnegative values.

5 Recall that we have assumed that in any region, species α is distributed equally (uniformly) throughout the region.

Hence, that assignment of values to variables satisfies all of the inequalities, and is a feasible solution. However, if we set all the variables to 0.6, then the first sum becomes 69, and the second sum becomes 81.6. In that case, the inequalities in (1.1) and (1.2) are satisfied, but inequality (1.3) is violated. Hence, that assignment of values is *not* a feasible solution.

The Objective Function So far, we have not used the *costs* required to preserve different regions, yet the stated problem is to protect both of the threatened species, spending the *minimum* (i.e., least) amount of money possible. How is that objective included in the model? When values are assigned to the five X variables, the total cost (in thousands of dollars) will be equal to:

$$97X_A + 73X_B + 22X_C + 11X_D + 25X_E. \tag{1.4}$$

So, the total cost is a *linear function* of the five X variables. Therefore, the *objective function*, which expresses the goal of spending the least money possible (while protecting both species) is stated as:

$$\text{Minimize } 97X_A + 73X_B + 22X_C + 11X_D + 25X_E. \tag{1.5}$$

1.1.2.1 The Concrete Linear Programming Formulation

The objective function (1.5) together with the inequalities in (1.1), (1.2), and (1.3) form the *concrete linear programming formulation* for the concrete problem instance of the species protection problem. Summarizing, the full concrete LP formulation for the toy problem instance is shown in Figure 1.1.

A *feasible solution* to a concrete LP formulation is an assignment of values to the variables that satisfies all of the constraints. However, a feasible solution does *not* need to be one that *minimizes* the objective function. A feasible solution the minimizes the objective function is called an *optimal solution*.[6]

Given the values of the variables in a feasible solution, the resulting value of the linear function in the objective is called the *objective value* or the *value of the solution*.

For example, when all the variables are given the value 0.66, the value of the solution is 150.48. As we will see later, this feasible solution is *not* an optimal solution, because there is a feasible solution with smaller objective value.

LP Solvers for Concrete LP Formulations A concrete LP formulation has all the information required to allow a solution to the specific problem instance. The formulation can then be input to an *LP solver* (in the proper format). If there is a feasible solution to the concrete LP formulation, the LP solver will determine and report an *optimal* solution. Notice that the phrasing allows for the possibility that there is more than one optimal solution, which is often the case.

[6] The terms *feasible solution* and *optimal solution* can be confusing, and keeping the distinction between them will often be crucial in this book. It is even more confusing when an assignment of values is just referred to as a *solution*. In general (and I hope I have been completely consistent in this book), when the term "solution" is used by itself, it is shorthand for a "feasible solution," which *might not* be an optimal solution. The word "optimal" should always be included when making the point that a feasible solution is an optimal solution.

$$\text{Minimize } 97X_A + 73X_B + 22X_C + 11X_D + 25X_E$$

Subject to the constraints:

$$24X_A + 36X_B + 0X_C + 15X_D + 40X_E \geq 64$$

$$83X_A + 16X_B + 19X_C + 0X_D + 18X_E \geq 87$$

$$X_A \leq 1$$

$$X_B \leq 1$$

$$X_C \leq 1$$

$$X_D \leq 1$$

$$X_E \leq 1$$

Figure 1.1 **The Full Concrete LP Formulation of an Instance of the Species Protection Problem. This is called a "concrete" LP formulation because it contains all of the information in this particular problem *instance*. Usually, the phrase "subject to the constraints" is abbreviated to "st," or "such that." Note that the objective function is a linear function of a subset (possibly the whole set) of the LP variables, and that each of the constraints is a linear inequality, defined on the LP variables. The last five inequalities are also called *bounds* because each one provides a bound (upper bound in this example) on a single LP variable.**

Alternatively, if there is *no* feasible solution to the concrete LP formulation, the LP solver will determine and report that; and if there is a feasible solution, but there is no bound on the value of the feasible solutions (essentially infinity for maximization problems, or negative infinity for minimization problems), the LP solver will determine and report that fact. An unbounded solution is usually an indication of an error in the general problem specification, or in the logic of the LP formulation, or in the concrete LP formulation.

The Concrete Optimal Solution In the concrete LP formulation in Figure 1.1, an optimal solution has objective value (after rounding up) of 108.7, which is achieved by setting (after rounding up) X_A to 0.831, X_D to 0.269, X_E to 1, and the other two variables to 0. So, in this solution, none of regions B and C will be preserved, all of region E will be preserved, and regions A and D will be partly preserved. Note, that there might be other optimal solutions that will assign different values to the variables, but *all* optimal solutions will have the same (rounded up) objective value, i.e., 108.7 in this example.

Exercise 1.1.1 *Use the values given to X_A, X_B, X_C, X_D, and X_E in the optimal solution detailed above, to determine the total preserved area occupied by species α, and the total preserved area occupied by species β. Do you see anything interesting?*

1.1.3 Refinements to the Model

One of the most useful features of linear programming is the ease in which *"what if"* questions can be explored, once the first LP formulation is created and solved. As an illustration, suppose there is pressure to specifically preserve some land in region B, which is not preserved at all in the optimal solution found above. But, land in region B is relatively expensive and the conservation agencies have limited resources. From the optimal solution to the concrete formulation above, we know that both of the threatened species can be protected for \$108.7K. Even with spending no more than that amount, but reducing what is spent on the other regions, it *might* still be possible to preserve *some* of region B, while protecting both species α and β. So, we can ask:

> How much of region B can be preserved, without spending more than \$108.7K, while still protecting both species α and β? And, how do we figure out the answer to this question?

The answer to the second question is to use linear programming again, modifying the concrete LP formulation in Figure 1.1. We change the objective function to:

$$\text{Maximize } X_B,$$

and add the constraint:

$$97X_A + 73X_B + 22X_C + 11X_D + 25X_E \leq 108.7.$$

Then we use the LP solver to find an optimal solution to the modified concrete LP formulation. Running the solver, we get a new optimal solution with objective value of 0.00296. That means that it is *not* possible to preserve more than a very small amount (less than one third of 1%) of region B, without increasing the total amount spent for land preservation.

Exercise 1.1.2 *In the optimal solution to the modified LP formulation, the values given to the five variables are:* $X_A = 0.830, X_B = 0.00296, X_C = 0, X_D = 0.263$ *and* $X_E = 1$. *What do you think will be the result if you plug those values into the original objective function in Figure 1.1. That is, what is the result of plugging the values of the five variables into the linear function:*

$$97X_A + 73X_B + 22X_C + 11X_D + 25X_E.$$

Try to answer this and to explain you answer without *actually plugging in the values. After that, plug in the values. What did you learn?*

Another What If Given that very little of region B can be preserved (while protecting both α and β) without increased spending, we could next ask:

> How much would we have to spend if we want to preserve at least 10% of region B? And, how do we solve this problem?

Of course, the answer to the second question is to use linear programming. But how, specifically? The answer is to start with the LP formulation in Figure 1.1, and add in the constraint:

$$X_B \geq 0.1.$$

Solving this concrete LP formulation results in an optimal solution with objective value of \$111.7K, an increase of only \$3K (a steal!).

Summary This toy example illustrates the three parts of every linear programming formulation: an objective function (either to *maximize* or *minimize*) a *linear* function of a (sub)set of the LP variables; a set of *linear* inequalities (constraints), each defined on a (sub)set of the LP variables; and a set of *bounds*, each defined on a single LP variable. Each bound is actually a constraint, and so could be considered as part of the constraints, but are historically distinguished from the other constraints.

1.1.4 Algorithms and LP Solvers

In the above example, we showed optimal solutions for concrete LP formulations, but did not say *how* they were obtained. The short answer: with algorithms and LP solvers.

Algorithms There are several *algorithms* (well-specified methods) that can take any concrete LP formulation and find an optimal solution; or determine that the formulation is infeasible; or that the solution value is unbounded. The latter case is usually not a sensible result, and usually indicates a user-created error.

The first and most famous LP algorithm is the *Simplex Algorithm*, developed by George Dantzig shortly after World War II. It is still the basis for many practical LP solvers, although additional refinements have been made to the original method. Further, other algorithms were later developed that are based on very different ideas than the simplex algorithm. Some of these later algorithms have theoretical properties that the simplex algorithm lacks. For example, some LP algorithms are *provably efficient* in a *worst-case* theoretical sense which is a property that the simplex algorithm does not have, despite its efficiency *in practice*. However, for the purposes of this book, we don't need the details of any of these algorithms, or any of their theoretical properties. What is important in this book, is the fact that highly engineered computer programs have been developed that implement LP algorithms, and these programs are very effective in practice.

LP Solvers When the details of an LP algorithm are written into an executable computer program, the program is called an *LP solver*. An LP solver takes in a concrete LP formulation (in some, usually rigid, format), and returns the value of the optimal solution, together with values assigned to the LP variables in the optimal solution.

In this book, I discuss the LP solver developed by *Gurobi Optimization* ®. In my experience, Gurobi is the fastest and most reliable of the two major LP solvers, and it has excellent documentation and support. Gurobi is a proprietary, commercially created LP solver that has been extensively tuned and engineered. Fortunately, Gurobi offers free licenses, of their full software, for academic and research users.[7]

[7] It is beyond the scope of this book to review all available LP solvers, but I should mention that the other major LP solver is *Cplex*, and it is currently owned by IBM. Cplex ®, following Gurobi's lead, currently also makes free licenses available to academic users. Two free, open-source LP solvers are COIN-LP, available from *COIN-OR*, and GLPK (*GNU linear programming kit*), available from the *GNU Project*. In my experience, *GLPK* is effective on some moderate-size LP formulations, although it generally runs much slower than Gurobi or Cplex, and for more demanding formulations, it does not find the optimal in a reasonable time, while Gurobi and Cplex do. I don't have experience using COIN-LP, although I believe it has a good reputation.

1.2 INTEGER LINEAR PROGRAMMING (ILP)

Linear programming allows the LP variables to be given *fractional*, i.e., noninteger, values, as we saw in the optimal solution to the species protection problem. *Integer* linear programming refines linear programming by *requiring* that certain variables in a formulation only be given *integer* values.[8] We will also refer to an *integer linear function* to mean that the function is linear, and that its variables are only allowed to take on integer values. Similarly, an *integer* linear inequality is an inequality whose variables are restricted to be integers. However, the *coefficients* (i.e., constants) in the integer linear functions or inequalities (in the objective function and in the constraints) are still allowed to be fractional.

For example, suppose the five variables in the LP formulation in Figure 1.1 are now required to take only integer values. That turns the LP formulation into an ILP formulation. Now, because of the inequalities in (1.1), each variable must be assigned either 0 or 1. This means that in any feasible solution, for every region, the solution must either preserve *all* of that region, or *none* of it. It is no longer possible to preserve only part of a region. Does that really affect what objective value is possible?

Yes. The integer optimal solution to the toy species preservation problem has objective value of \$122K, in contrast to the optimal LP solution, which has value of \$108.7K. Remembering that *value* in this case is actually a *cost*, the LP optimal solution is less expensive than the ILP optimal solution. In one ILP optimal solution, variables X_A and X_E are set to 1, and the other three variables are set to 0. In that solution, the total preserved area occupied by species α is 64, exactly meeting the conservation target for species α; but, the preserved area occupied by species β is 101, well above the conservation target for β.

Exercise 1.2.1 *Exercise 1.1.1 asked you to calculate the total preserved area occupied by species α and β, respectively, implied by the optimal LP solution to the toy species preservation problem. Compare those numbers to the areas of 64 and 101, stated above for the ILP optimal. Do you see a general explanation for what you observed?*

More Terminology A integer variable that is further constrained to only take on value 0 or 1 is called a *binary variable*. An ILP formulation where the variables are further constrained to only take on values of 0 or 1, is called a *binary* formulation. Binary formulations are very common. The ILP for the toy species protection problem is a binary formulation, and most of the ILP formulations in this book will be binary formulations.

When the values assigned to the variables in an ILP formulation satisfy all of the constraints, and all of the variables required to have integer values, do have integer values, the solution is called an *integer feasible solution*. An integer feasible solution with the best objective value (maximum or minimum, depending on the objective function) is called an *integer optimal* solution.

Relating LP and ILP Suppose $\mathcal{P}$ denotes an ILP formulation, and $\mathcal{P}'$ denotes the same formulation where we allow *all* of the variables to take on *fractional* values,

[8] More commonly, when some variables are required to have integer values, and some are allowed to have fractional values, the term *mixed integer linear programming (MILP, or sometimes MIP)* is used. For simplicity, in this book, I will just use the acronym "ILP," even if some of the variables are allowed to have fractional values.

although an integer value is also allowed. Then, LP $\mathcal{P}'$ is called the *LP relaxation* of $\mathcal{P}$. Now we state a simple, but very important observation:

> An optimal solution to the LP relaxation, $\mathcal{P}'$, of the ILP $\mathcal{P}$, has an optimal objective value, which is *no worse*, and is often *strictly better*, than the integer optimal solution to the ILP formulation $\mathcal{P}$.

This is true because any integer feasible solution to $\mathcal{P}$ is also a feasible solution to $\mathcal{P}'$. As an example, in the toy species protection problem, the integer optimal solution has value \$122K, while the optimal solution to the LP formulation has value \$108.7K.

Exercise 1.2.2 *Suppose $\mathcal{P}$ is an ILP, and $\mathcal{P}'$ is its LP relaxation. Suppose that there is an optimal solution to $\mathcal{P}'$ where all of the variables take on integer values. Is that optimal solution from the LP also an optimal solution to* ILP *problem $\mathcal{P}$? Explain.*

More Biological Fidelity Later The species protection problem comes from the research paper [42], although the presentation here is actually closer to what is in [46]. At several places in the book, after we explain the needed integer programming material, we will return to the species protection problem, extending and modifying it to contain more of the detailed biological model used in [42]. For example, we will include the use of *evolutionary trees* and *species diversity*, and the importance of preserving regions with neighboring boundaries, to form *connected areas* where a species is protected.

1.2.1 ILP Solvers

All of the LP solvers discussed earlier, Gurobi, Cplex, GLPK, are also ILP solvers,[9] returning an integer optimal solution, or determining that no integer feasible solution exists, or that the solution value is unbounded. To specify an ILP formulation, the user must list which variables are restricted to having only integer values, or only binary values, in addition to specifying the LP formulation. Any variable not listed is allowed to take on fractional values.

At the high level, the algorithms that ILP solvers use to find an integer optimal solution are quite different from the algorithms used to solve LP formulations. But ILP methods usually require creating and solving *many* concrete LP formulations. Thus, the time to solve an LP relaxation of an ILP formulation is generally much less than the time needed to solve the ILP. Further, unlike the case of linear programming, where worst-case efficient LP solvers have been created, no such ILP solver exists.[10] Despite this, highly tuned ILP solvers such as Gurobi and Cplex, are surprisingly effective for problem *instances* from a wide range of ILP formulations.

9 Another noncommercial ILP solver (which is not an LP solver) that has a good reputation is called SCIP, but I have not had much experience with it.

10 In fact, the problem of solving ILP formulations is known to be *NP hard*, which suggests that no theoretically efficient ILP solver is possible. If you don't know about or understand what "NP" or "NP hardness" means, don't worry about it. That knowledge is not required for the material in this book. However, if you are curious, a classic reference is [74], and most college textbooks for courses on "Algorithms" discuss NP and NP completeness, for example, [49, 109].

1.2.2 Abstract Problem Statements and Abstract Formulations

The LP and ILP examples given above are *concrete* formulations, because they completely specify a problem *instance* and (suitably formatted) can be input to an LP or ILP solver. In contrast, when we describe a problem in *general* terms, and demonstrate how the problem can be solved by integer programming, we are describing an *abstract* problem, and an *abstract* LP or ILP formulation. For example, the following is an abstract statement of the species protection problem:

> **Abstract Species Protection Problem** Given n regions, and m threatened species; and data on how much area each species occupies in each region; and the cost of preserving each region; and a conservation target for each species, determine the cheapest way to protect all of the m threatened species, by preserving some part (possibly none or all) of each region.

Abstract LP or ILP Formulations An *abstract* LP or ILP formulation for an abstract problem describes the *general* form of the formulation to solve the abstract problem. The abstract formulation must be detailed enough that when a concrete problem instance is specified, the abstract formulation can be converted into a *concrete* formulation for that problem instance.

The task of creating an abstract formulation is what is called *Task B* in the Introduction. An *abstract LP formulation* for the species protection problem is:

> For each pair (i,j) of n regions and m threatened species, let $r(i,j)$ denote the area in region i occupied by species j. For each region i, let $C(i)$ denote the cost of preserving all of region i; and for each species j, let $t(j)$ denote the conservation target, i.e., the total preserved area that must be occupied by species j in order to protect species j. All of these data, $r(i,j)$, $C(i)$, and $t(i,j)$ are constant numbers, given as input to any concrete instance. Then, for each of the m regions, let $X(i)$ denote the *fraction* of region i that will be preserved. Each $X(i)$ is a variable can will take on any value between 0 and 1 (inclusive).

The *abstract LP formulation* for the species protection problem is shown in Figure 1.2.

$$\text{Minimize} \sum_{i=1}^{n} C(i) \times X(i)$$

Such that
For each species j from 1 to m:

$$\sum_{i=1}^{n} X(i) \times r(i,j) \geq t(j)$$

And, for each region i from 1 to n:

$$X(i) \leq 1$$

Figure 1.2 The Abstract LP Formulation for the Species Protection Problem. The symbol "$\sum$" means to take the summation of the terms indicated after the $\sum$ symbol.

As discussed before, in linear programming, the default assumption is that each variable can only take on nonnegative values. The abstract ILP formulation for the species protection problem just adds the requirement that the value given to any variable $X(i)$ must be an integer.

The LP and ILP formulations for problems described in this book will be a mixture of concrete and abstract formulations.

1.3 EXPRESSIBILITY OF INTEGER LINEAR FORMULATIONS

The use of *linear* programming (allowing fractional values for variables) is essential for stating and solving *integer* linear programming problems, however, the focus of this book is not on LP but ILP problems and solutions. So it is appropriate to ask:

> What kind of problems can be solved by *integer* linear programming?

At first exposure to integer programming, it might seem to be a pretty limited set. After all, an ILP objective function can only maximize or minimize a *linear* function; and each constraint must be a linear inequality, without even a symbol for strict inequality. What do we do, for example, if we want to *minimize the maximum* of a set, or express the *absolute value* of variable? These and many other *nonlinear-looking* objects *can* actually be expressed in integer linear programming, but it often takes some cleverness to do so. We will discuss many of these throughout the book.

And, there is a theoretical answer to the question that is very positive. The fact that the problem of solving ILP formulations is *NP hard* suggests that no *worst-case, theoretically efficient* ILP solver can ever exist. That is the widely known *negative* consequence of NP hardness. But, the NP hardness of integer linear programming also establishes a more *positive* fact:

> Any problem in a huge class of natural problems can provably and efficiently be expressed by a compact ILP formulation, or a "relatively short" series of compact ILP formulations.

A discussion of this fact is outside the scope of the book, and you need not understand this technical point. The key thing is that it establishes that despite being restricted to *linear* constraints and *linear* objective functions and *integer* values, ILP formulations are actually *very expressive* in a provable way. They are able to compactly express most natural discrete optimization problems. Rarely does one encounter a discrete optimization problem that cannot be efficiently formulated as a single ILP problem, or a series of ILP problems. Sometimes the formulation is tricky to find, but in many cases it is quite simple, although that does not guarantee that its *solution* will be easy. Moreover, sometimes a less compact, mathematically equivalent ILP formulation is even simpler to design, and more efficiently solved by the ILP solver. We will see examples of this in the book. So, although integer linear programming may seem very limited, it is actually a very expressive language.

2

Biological Networks, Graphs, and High-Density Subgraphs

In this chapter I introduce a general topic that has huge number of diverse applications in computational and systems biology, and even more outside of biology. The topic is that of *graphs* and *networks*, and the problem of finding different types of *high-density subgraphs* and/or *communities* in those graphs. The extreme case, and most easily defined version is the problem of finding a *maximum clique* in a graph

I start this chapter with a discussion of *biological* graphs and networks, and then discuss a simple integer linear programming formulation for the *maximum clique* problem; and its solution using the ILP solver Gurobi Optimizer. In the next chapter, I develop an application of the maximum-clique problem in the area of *phylogenetics*, solving it with a modification of the ILP formulation discussed in this chapter. After that, I move to the related topic, in Chapter 4, of *near cliques* and more general definitions of *high-density subgraphs*. In Chapter 13, I return to the topic of community structure in graphs, and cuts in graphs, and more on finding high-density subgraphs in biological graphs.

Note to the Reader Who Is Eager to Get on with Integer Programming I chose to start this chapter with a discussion of biological graphs and networks before continuing our treatment of integer programming. It is central to my concept of this book to fully explain the biological context of every computational problem that we examine, and biological networks and graphs are particularly important. However, the biological material in this chapter is *not* required in order to understand the *technical* aspects of integer programming. So, the reader who is already familiar with biological graphs and networks, or simply wants to continue an uninterrupted focus on integer programming, can skip over the biological material in this chapter, but pay attention to the definitions about graphs and networks. That is the fastest way to get to Section 2.2, where we develop an ILP formulation for finding the largest clique in a graph (Figure 2.1).

2.1 BIOLOGICAL GRAPHS AND NETWORKS

An *undirected graph* $G = (V, E)$ consists of a set of *nodes*, V, (also called *vertices* or *points*); and a set of *edges*, E, (also called *lines*) that run between certain *pairs* of

Figure 2.1 Zebras in the Field, with a Superimposed Graph Representing Pairs of Zebras That Are in Close Contact. Image created by Tanya Berger-Wolf. Used with permission.

nodes (see Figure 2.2). An *undirected edge* between nodes u and v is denoted (u, v). A *directed graph* is a graph where each edge is given a specific direction (see Figure 2.3). The convention is that if (u, v) is an edge in a directed graph, then the direction of the edge is *from u to v*, i.e., from the first node of the pair to the second node. However, sometimes, in order to emphasize that the graph is directed, we will use the notation $< u, v >$ in place of (u, v) to denote an edge that is directed *from u to v*. The term "network" is generally used in biology to refer to a directed graph; but sometimes it also refers to an undirected graph (see Figures 2.3 and 2.4).

Graphs and networks are used extensively in biology to represent *relationships* between biological elements, or to represent *processes* that the elements participate in. There are hundreds of *types* of graphs and networks used in biology, and *thousands* of published networks and graphs that display specific biological information. We will describe a few of these here, and introduce several computational problems defined on those graphs and networks. Additional problems defined on biological networks will be discussed in later chapters.

2.1.1 Examples of Biological Networks

Food Webs One of the earliest uses of graphs and networks in biology is in the field of *ecology*. In one kind of network, called a *food web*, each node represents (and is named by) an organism in some geographic area, and there is a *directed* edge $< u, v >$ from node u to another node v, if organism v is a *predator* of organism u. In a different kind of food web, there is an *undirected* edge (u, v) between nodes u

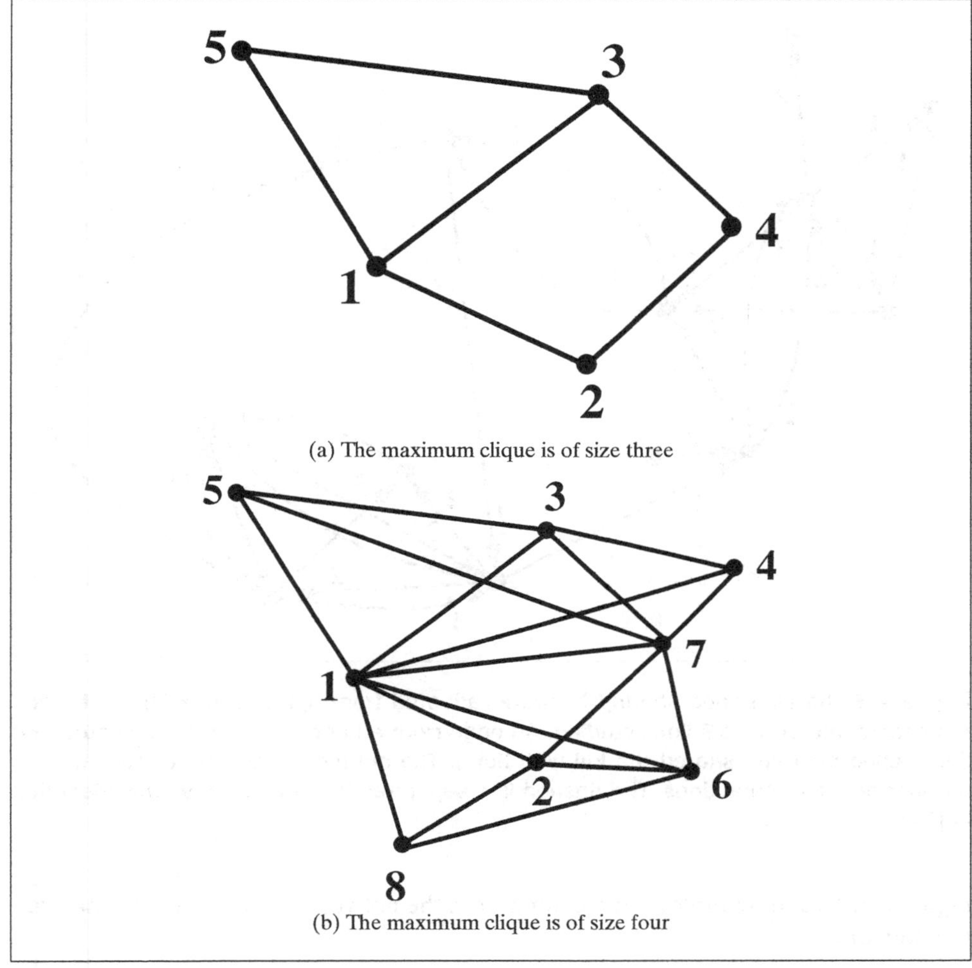

Figure 2.2 (a) An Undirected Graph with a Single Maximum Clique of Size Three. (b) An Undirected Graph with Three Maximum Cliques of Size Four.

and v if organisms u and v are *competitors* for the same food source, or *share* food, or *inhabit* the same ecosystem.

A recent article [64] involving a food web received considerable attention in the popular press, including in *The New York Times*. The feeding and co-feeding behaviors of 13 mountain lions (with GPS tracking collars) in Wyoming were observed using several hundred motion-triggered cameras placed at sites where mountain lions had previously killed and eaten large animals. Mountain lions are thought to be solitary animals that do not generally share food or cooperate with other mountain lions (except to mate). The cameras caught 48 occurrences of a single lion with a kill that was larger than the lion could eat in one meal, and the subsequent arrival of another lion. The data of interest was whether the first lion shared the carcass with the second lion, or fought with it.

The recorded data was organized as a network, where each node represented one of the 13 lions, with a *directed* edge from the node for the first lion to the node for second lion, *if* the carcass *was* shared. A representation of that network is shown in

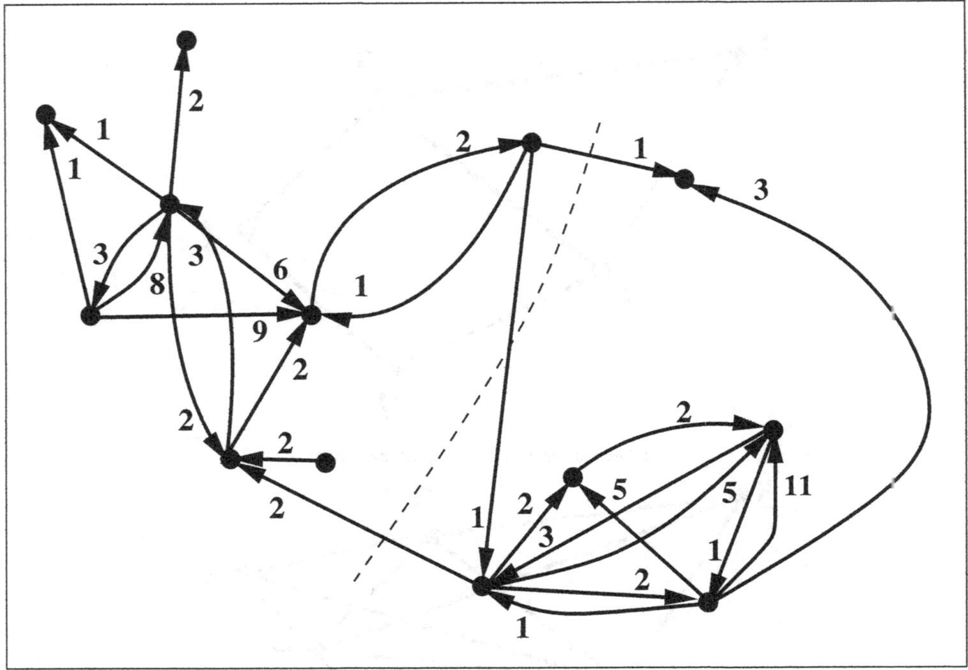

Figure 2.3 **The Lion Food-Sharing Network (adapted from figure 1 in [64]). Each node represents one of the 13 lions studied. An edge from a node *i* to a node *j* indicates that lion *i* killed prey and shared the kill with lion *j*. The number on the edge indicates the number of such interactions. The dashed line separates the two communities identified in [64].**

Figure 2.3. To extract biological meaning from the network, the researchers analyzed the network

> to quantify the extent to which the network was composed of distinct communities (also called 'clusters') within which many edges occurred, and between which few edges occurred. [64]

The result, as can be seen in Figure 2.3, is a division into two distinct food-sharing communities, separated by the dashed line. The community membership of most of the lions is clear, but the membership of the upper-right lion might be debated. Much is unknown about the communities, for example, how they were established, but knowing that they exist and which lions are members, is an important and necessary start. More details on the lions (e.g., their sex, weight, and age) is given in [64].

Social Interaction Networks Figure 2.4 shows an interaction network that records the pairwise interactions of ants over 26 seconds of observation. Such networks are compared under different conditions, and examined for distinctive patterns that lead to insight into ant behavior.

Metabolic Networks

> Metabolism is the set of life-sustaining chemical reactions in organisms. The three main purposes of metabolism are: the conversion of food/fuel to energy to run cellular processes;

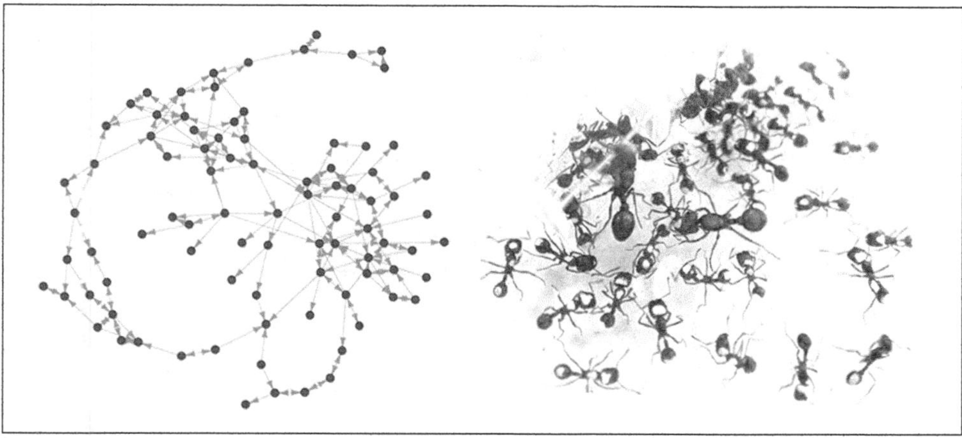

Figure 2.4 Ant Colony Interaction Network. The left figure shows a directed network of interactions between workers in a single *Phidippus californicus* colony based on 26 seconds of colony behavior. The right figure shows a typical instantaneous configuration of the ants. Each ant is painted for unique identification during the period of observations. This figure is extracted from figure 1 in [199].

> the conversion of food/fuel to building blocks for proteins, lipids, nucleic acids, and some carbohydrates; and the elimination of nitrogenous wastes.
>
> These enzyme-catalyzed reactions allow organisms to grow and reproduce, maintain their structures, and respond to their environments. The chemical reactions of metabolism are organized into metabolic pathways, in which one chemical is transformed through a series of steps into another chemical, each step being facilitated by a specific enzyme ... A striking feature of metabolism is the similarity of the basic metabolic pathways among vastly different species [217].

A huge number of large networks, called *metabolic networks*, have been developed that represent metabolic reactions, pathways, and cycles. A metabolic network depicts the biochemical reactions and energy transformations that occur in a cell. It represents all of the interacting metabolic pathways. In most metabolic networks, each node represents a *metabolite* (a molecule involved in metabolism), and there is a directed edge from the node for metabolite A to the node for metabolite B, if A is biochemically involved in the production of B (A may be involved in the production of several metabolites, with additional edges emanating from A). There are two types of "involvement": one where A *activates* (contributes to) the production of B; and one where A *inhibits* (or retards) the production of B. The edges in the network must distinguish those two types of involvement. The actions of the enzymes are usually depicted on the edges (see Figures 2.5 and 2.7). The latter network contains both *activation* edges (with an arrow) and *inhibition* edges (with a perpendicular line at the node being inhibited).

Genomic Networks More recently, with the tremendous growth of *genomic* data, many *gene interaction* or *gene influence* or *transcription* networks and graphs have been created and studied. In those networks, each node represents (and is named

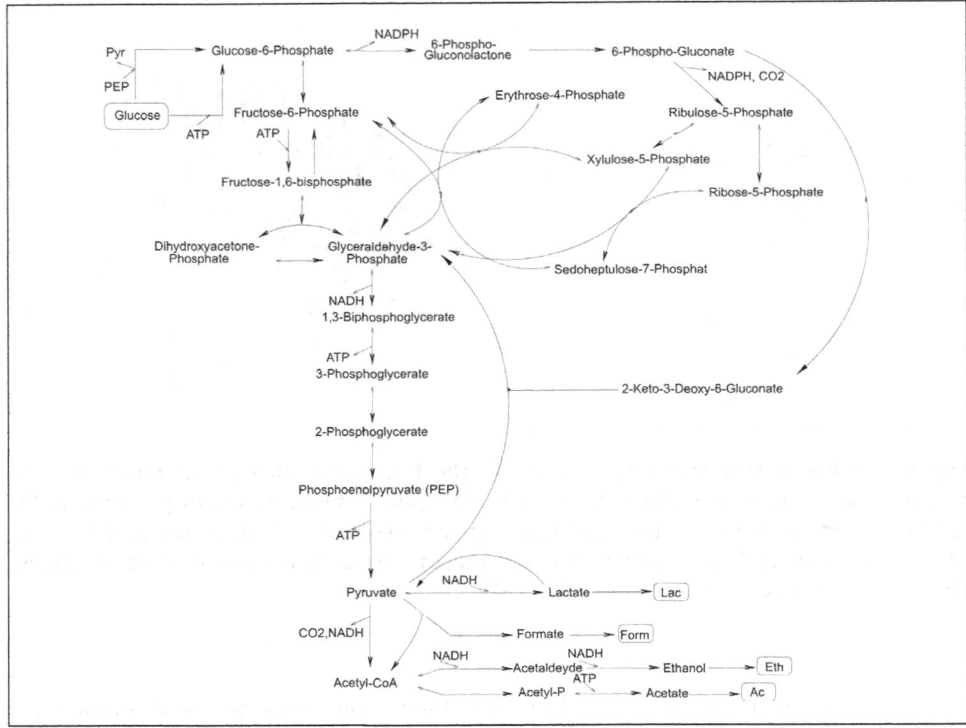

Figure 2.5 Metabolic Network Showing the Conversion of Sugar Glucose Into Fuel Ethanol By the Bacteria *Escherichia Coli*. This figure is modified from figure 1 in [196].

by) a single gene. Then, there is a *directed* edge, $< u, v >$, from a node u to a node v if the transcription[1] of gene u leads to the transcription of gene v (via a process that is not represented in the network). There can also be edges that represent *repression*, where the expression of a gene" of gene A acts to repress the expression of gene B.[2]

In an *undirected* gene-interaction graph, there is an undirected edge between node A and B if genes A and B interact in a biologically important way. There are many ways that genes interact. For example, an edge might indicate that the genes are transcribed at the same *time*, or that one gene codes for a *promoter* or *repressor* of a second gene, or that they are part of the same biological *process*, or they code for proteins in the same biochemical *pathway* (but the order in which the genes are expressed is unknown), or are in the same *location* in a cell or organ, etc. An example of a gene-interaction graph is given in Figure 2.6.

Molecular Signaling Pathways As the name suggests, *signaling pathways* transmit information between cells or inside cells, and involve interactions between genes and proteins. A signaling pathway is typically a *cascade* of successive gene-protein and protein-protein interactions (most activating with a few repressing) that result in the transmission of some information. The signaling pathway shown in Figure 2.7

[1] Stated simply, "the transcription of a gene" means "the turning on" of the gene.
[2] "Expression of a gene" means that the protein or regulatory RNA that the gene codes for is produced.

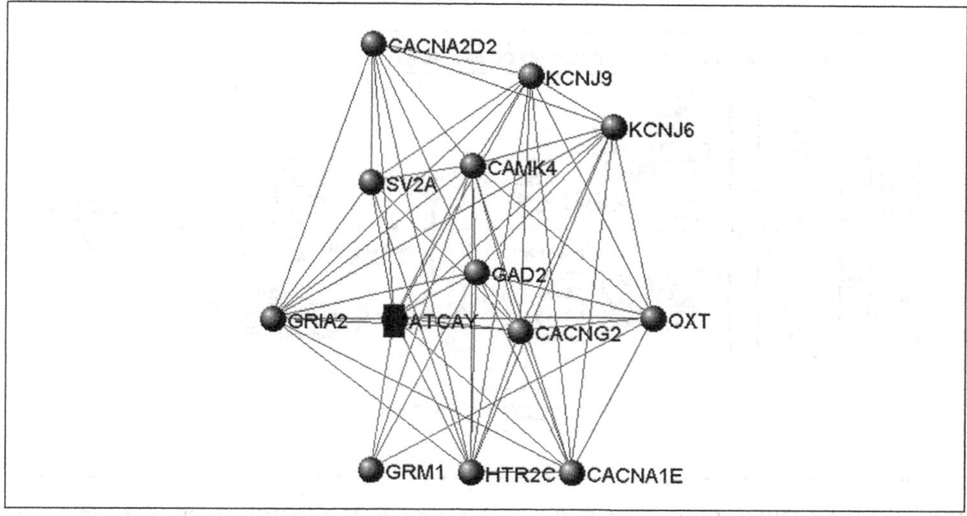

Figure 2.6 **Gene-Interaction Graph Showing the Interactions of Genes That Each Interact with the *ATCAY* Gene. The *ATCAY* gene is known to be central in *ataxia*, a neurological disease of the *cerebellum* that leads to loss of muscle control and voluntary motion. Each node represents a gene, and an edge reflects evidence of a functional interaction between two genes. This figure is modified from figure 6 in [79].**

depicts the NOTCH gene/protein signaling pathway in humans. This is an extremely important pathway that transmits information between neighboring cells.

Protein-Protein Interaction Networks

> Protein-protein interactions form the molecular basis for organismal development and function. [126]

> Protein interactions form a network whose structure drives cellular function and whose organization informs biological inquiry. [98]

Protein-protein interaction (PPI) networks (which are often undirected graphs, but can have directed edges) are extremely important networks in computational biology. They depict a wide range of information. For example, they can depict *physical* contacts between *pairs* of proteins; or depict pairs of proteins that are part of the same protein *pathway* or the same *complex*; or depict pairs of proteins that are present at the *same time in the same region* of a cell; or depict pairs of proteins that are *co-regulated*; or depict pairs of proteins that are both involved in a specific biological *function*; or depict the relationship of proteins that are expressed in a specific *organ* or in the *brain*. When edges in a PPI network are directed, the direction usually represents the *action* of one protein on another, but other kinds of information, such as chronological ordering, can be represented by edge direction. Several hundred thousand pairwise, or directed, interactions between proteins have been discovered already, and new technology is making it possible for faster and cheaper discovery of additional interactions [68]. Figure 2.8 shows a small PPI network for the proteins involved in the NOTCH signaling pathway in humans (shown in Figure 2.7). A larger PPI network, from *yeast*, is shown in Figure 2.17. We will discuss several problems

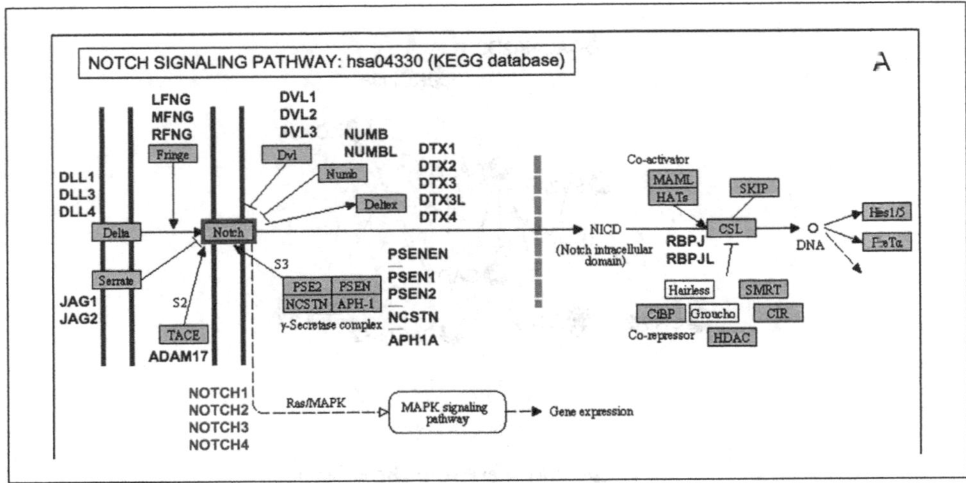

Figure 2.7 The NOTCH Signaling Pathway in Humans. The NOTCH pathway is critical in many biomolecular processes, and defects in the pathway are involved in several types of cancers and in several neurological diseases. The figure shows the pathway including nine proteins directly connected to NOTCH. Arrows (e.g., from Delta to NOTCH) show the direction of influence (activation), and edges with a perpendicular line at the end (e.g., from Serrate to NOTCH) show the direction of repression (inhibition). This figure is extracted from figure 3 in [165]. The caption here is modified from the original caption in [165].

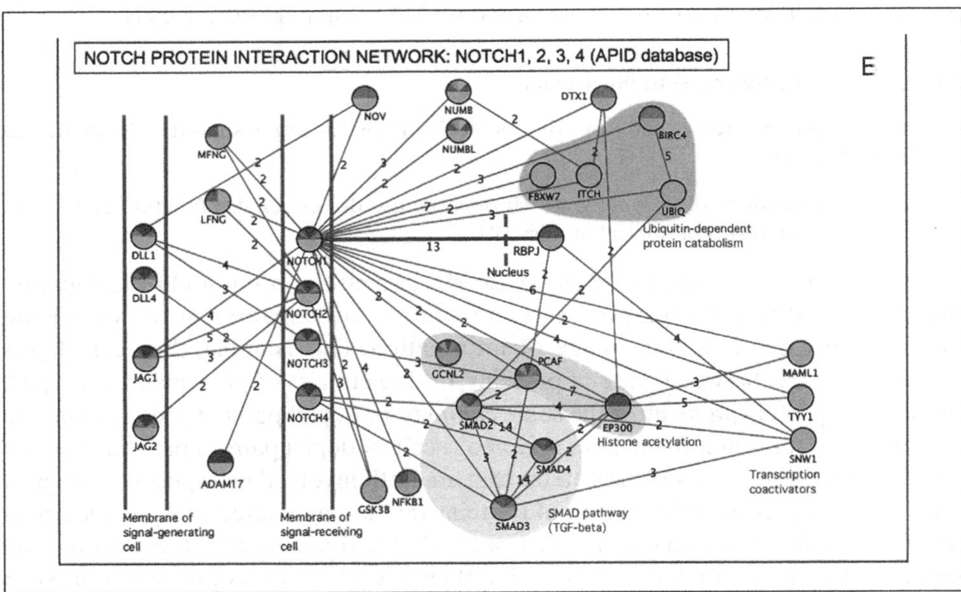

Figure 2.8 The NOTCH PPI Network, Including All Interactions Proven with at Least Two Different Experiments. The number of experiments is indicated next to each edge. The PPI network provides complementary information to the signaling pathway, revealing the particular links of each of the four NOTCH protein variants (NOTCH1, 2, 3, and 4) present in the human proteome. The biomolecular elements included in both networks are quite similar and the information that can be deduced from them is complementary. This figure is extracted from figure 3 in [165]. The caption here is modified from the original caption in [165].

involving PPI networks in later sections in the book, particularly in Chapters 7, 16, and 17.

General Correlation Networks A final common kind of general graph that is frequently used in biology represents *correlated events*. For example, *function magnetic resonance imaging (fMRI) scans* show small locations in the brain that *light up* in response to certain stimuli. High correlations in those data suggest connectivity between pairs of locations. In the brain, fMRI data is widely used, but other kinds of correlated data can be used, such as the amount of gray matter, or the cortical thickness of the brain in specific locations. In the resulting graph, each node represents one of the small locations studied, and an edge between two nodes represents the fact that the data at the two locations is highly correlated under varying conditions.

> Coordinated variations in brain morphology have been proposed as a valid measure to infer large-scale structural brain networks. This approach depends upon the assumption that positive correlations between morphometric parameters of different brain regions indicate connectivity. ...[95]

Figure 2.9 shows a correlation network in the human brain of subjects who have survived acute lymphoblastic leukemia, the most common form of childhood cancer; and in normal, control subjects. The size of a node represents its degree, i.e., the number of edges incident with the node. A "hub" is a node of very high degree, above some threshold. The hubs that appear in one of the networks, but not the other, are colored white. The figure shows some clear differences in hub structure between the two groups of subjects.

A huge National Institutes of Health (NIH) program is underway to determine the important connections and communication pathways in the human brain. That network is called the human *connectome*.

Many More There are *many* more types of graphs and networks that are currently used in biology, and several will be shown later in the book. But, this short survey should be sufficient to convince you of the importance of graphs and networks in biology. Similar kinds of networks and graphs are also widely used in the study of *social* relations, *communication or transportation* networks, and in the study of *friends, colleagues, criminal* and *terrorist* networks, etc. Many of the computational questions that arise in studying biological networks are common to the study of all of these disparate kinds of networks and graphs. [65]

Exercise 2.1.1 *Use Wikipedia to find figures of protein-protein interaction networks. Do they have directed or undirected edges, or both? Does the network have edges that are neither directed or undirected? If so, what interactions do those edges represent?*

Use Google to find figures for at least five different types of biological networks.

2.1.2 Biologically Informative Features of Graphs and Networks

There are a variety of features of biological networks and graphs that are believed to have biological meaning. The simplest feature is the *number* of edges that touch a node (technically, this is called the *degree* of a node). A node of high degree is called a "hub." Hubs are thought to represent elements (e.g., genes or proteins or predators) that have central importance in the system under study. Generally, if the

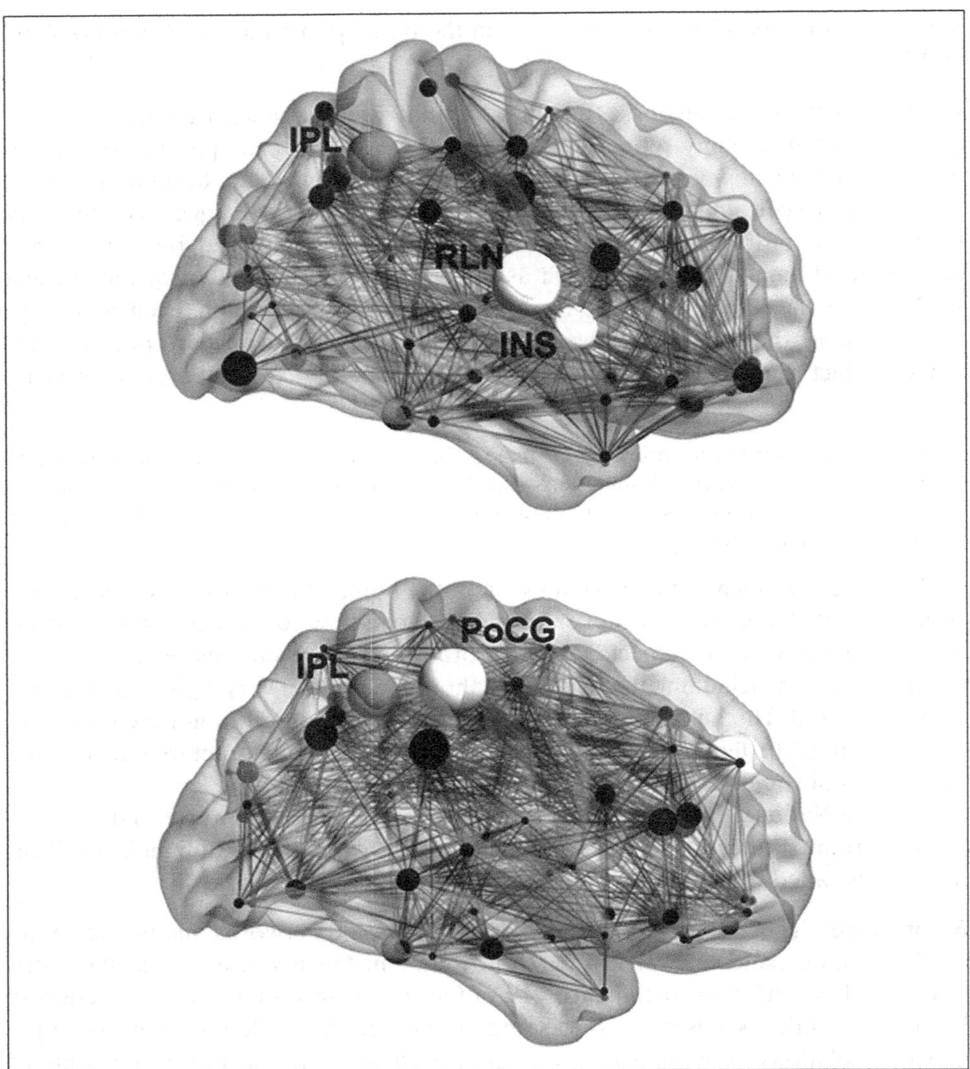

Figure 2.9 Correlation Network and Hubs in the Brain Network of Humans Who Have Survived Acute Lymphoblastic Leukemia (top), Compared to the Correlation Network in Brains of Control Subjects (bottom). The nodes are displayed in their correct locations in a two-dimensional projection of the brain. The white circles show hubs that are in one network, but not in the other. The four labeled regions are inferior parietal lobule (IPL), insula (INS), rolandic operculum (RLN), and the postcentral gyrus (PoCG). The figure is extracted from figure 8 in [95].

element represented by a hub is removed from the system, then the system will not function properly. Although conceptually simple, analysis of hubs can reveal important biological information:

> ... the distribution of hubs are altered in structural networks of patients with Schizophrenia, Alzheimers disease and multiple sclerosis. [95]

Another feature with possible biological meaning is a small *node-cut*. A node-cut is a subset of nodes in the network or graph whose removal disrupts (cuts) all paths between *at least one pair* of nodes that had previously been connected by some path(s) in the graph. We can analogously define an *edge-cut* as a set of edges whose removal disconnects at least one pair of nodes. Small cuts (node or edge) have several biological interpretations, and may be important in identifying *drug targets*, i.e., molecules that might be regulated (through drug intervention) to stop the progression of a harmful biological process, say the transformation of a normal cell to a cancerous cell. We will discuss several applications of graph cuts later in the book, particularly in Chapter 13.

Related graph and network features can be defined for *directed* graphs, and they can be more discriminating. For example, in directed graphs we can distinguish between the number of edges that are pointed *into* a node (the *in-degree*), and the number of edges pointing *out of* the node (the *out-degree*). Then, the degree of a node is the sum of its in-degree and its out-degree. Similarly, we can modify the notions of hubs and cuts to incorporate the direction of the edges. See [65] for a general discussion of network features that are believed to be informative. For more biologically oriented reviews, see [50] and [157]; and see [151] and [215] for reviews of networks and network features of importance in the study of cancer.

2.1.3 High-Density Subgraphs: A Nontrivial Feature

A more complex feature of biological graphs that is widely thought to have very significant biological importance is the existence of *high-density* subgraphs, which in the extreme case are *cliques*.

2.1.3.1 Defining Subgraphs, Density, and Cliques

Suppose that $G = (V, E)$ is an undirected graph, and that V' is a strict subset of the nodes of G. In symbols, this is written $V' \subset V$.[3]

The *induced subgraph* $G(V') = (V', E')$ of G is the graph formed by the nodes in V', and the edges E', where an edge $e = (u, v)$ is in E' if and only if *both* u and v are in V'. When $G(V')$ is an induced subgraph of a graph G, we also say that $G(V')$ is *induced by* the set of nodes V'. For example, in Figure 2.2, the subset of nodes {1,3,5} defines the induced subgraph with three edges $\{(1,3), (1,5), (3,5)\}$. Note that the two edges $\{(1,3), (1,5)\}$ also forms a subgraph involving the nodes $\{1,3,5\}$, but it is not an induced subgraph because it does not have edge $(3,5)$.

Density We will discuss density in more detail in Chapter 4, but need an informal definition here. The *density* of a graph $G = (V, E)$ is simply the number of edges, $|E|$, in G, divided by the number of nodes, $|V|$, in G. A *high-density* subgraph $G(V')$ of G is an induced subgraph of G where the density of graph $G(V')$ is above some large, fixed *threshold*, for example, 0.8.

[3] The symbol "$\subset$" denotes *strict inclusion*. For example, $V' \subset V$ means that every element in set V' is also in set V, but there is at least one element in V that is *not* in V'. When V' is a subset of V, but it is possible that $V' = V$, we use the symbol "$\subseteq$", and write $V' \subseteq V$.

Cliques in Graphs Suppose G is an undirected graph with n nodes. If there is an undirected edge between every pair of nodes in G, there will be exactly $n(n-1)/2$ edges, so every graph has that number of edges or fewer.

A *clique*, K, in an undirected graph $G = (V, E)$ consists of a *subset* of V (possibly all of V) with the property that for *every* pair of nodes (u, v) in K, there is an edge between u and v in G. So if the clique has k nodes, it will have $k(k-1)/2$ edges (see Figure 2.2).

A *maximum* (or maximum-size) clique in G, is a clique that has a number of nodes that is greater or equal to the number of nodes of any other clique in G. A *maximal* clique, K, is a clique in G such that the addition to K of any node not in K would result in a subset of nodes that is not a clique in G. That is, there is no node outside of K that is adjacent to every node in K. The maximum (i.e., largest) clique in G must be a maximal clique, but the converse does not hold. That is, graph G can have maximal cliques that are smaller than the largest-size clique in G. For example, a single edge whose end points have no neighbors in common is a maximal clique. It is not always easy to see the maximal cliques in a graph, and even harder to identify maximum cliques. Figure 2.2a shows an undirected graph with a single maximum clique of size three. You can see this at a glance.

Figure 2.2b shows an undirected graph with three maximum cliques of size four. A glance here is not sufficient – you have to examine this graph systematically to see all the cliques of size four, and to be sure it does not have any cliques of size five. I do not see any *maximal* cliques of size three in this graph (do you?), but if we add the edge $(5, 8)$, then the subset of nodes $\{1, 5, 8\}$ would form a maximal clique of size three.

Exercise 2.1.2 *Does the graph in Figure 2.2b have a maximal clique of size three?*

Would the addition of edge $(4, 6)$ *to the graph create another maximal clique of size three?*

2.1.3.2 Examples of Cliques and High-Density Subgraphs in Biological Graphs

High-density *subgraphs* (which, in the extreme are cliques) are probably the most studied nontrivial feature of biological graphs, and graphs from other application areas. There are hundreds (perhaps thousands) of publications in biology (and more outside of biology) that use high-density subgraphs (often cliques) to define features of importance in many different kinds of biological networks. We mention only a few here, to illustrate the prevalence and importance of cliques and high-density subgraphs in biology.[4]

[4] A numerical reflection of the prevalence of high-density subgraphs in biological networks is the "clustering coefficient" of a network. The clustering coefficient of a *node*, v, in a graph, is the number of edges in the subgraph induced by all of the $d(v)$ (degree of v) neighbors of v, divided by $\frac{d(v)\times(d(v)-1)}{2}$, which is the maximum possible number of edges in that subgraph. A clustering coefficient for node v that is close to one means that the neighbors of v are very cliquish. Then, the clustering coefficient of a network is the *average* of the clustering coefficients of the nodes of the network. The key fact is that clustering coefficients in PPI networks and other biological interaction networks are typically 100 times larger than the clustering coefficients in random graphs with the same number of nodes and edges.

Protein Complexes and Modules

> Many biological functions are carried out by the integrated activity of highly interacting cellular components, referred to as functional modules. [205]

Functional modules and protein complexes can often be identified in graphs and networks (such as PPI networks) as highly connected subgraphs.

> ... [c]omplexes may also be revealed from network topology alone as clusters of highly interconnected proteins. Because no prior knowledge is required, community detection algorithms may associate new proteins with known complexes and identify unknown complexes. [98]

Protein Cores The *core* of a *protein complex* is a set of proteins that are believed to be a single *functional unit*, and appear as a high-density subgraph in a protein-protein interaction network.

> Biochemists ... are now developing tools to chart more comprehensive sets of protein-protein interconnections at levels from the organellar to the organism ... Self-contained clusters of interconnected proteins that emerge from these webs may represent key complexes, and communal functions or ... provide clues to the roots of disease. [68]

The entire protein complex can have more isolated nodes hanging off of the core, and a PPI network can have several cores that are either separated or are connected to each other by low-density subgraphs.

Identifying Causal Genes in Genetically Influenced Diseases High-density subgraphs in gene-influence graphs or protein-interaction networks have been used to search for genes (and the proteins they code for) that collectively contribute to *genetically influenced disease*. The method first identifies a set of *candidate* genes that *might* be involved with the disease of interest. This is done by finding genes that are more frequently mutated in individuals who have the disease, than in individuals who don't. The problem is that this finds far too many candidates genes. So, then a *restricted* interaction graph is built for proteins encoded by the candidate genes (extracted from general protein-protein interaction databases, and other interaction and co-occurrence data); and *high-density subgraphs* are identified in that graph. Genes coding for proteins in those high-density subgraphs are then considered *prime candidates* for genes that influence the disease. The biological basis for this approach is reflected in the following quote:

> It is widely postulated that the (protein)[5] products of genes whose variants are implicated in the same disease, participate in the same biological function or process whose disruption leads to the disease. This concept is supported by examples of complex disease in which the proteins encoded by the implicated genes *interact*, form a molecular complex, or function at different steps of the same biochemical pathway. [94] (italics added)

Thus, finding high-density subgraphs in the candidate-restricted network is a central task in this approach. The results in [94], using a particular graph they call an "integrated phenotypic-linkage network," are summarized in the following quote, which is part of the caption for figure 3 in [94]. The graph in that figure is shown here in Figure 2.10.

[5] I have inserted the word "protein" here for clarity.

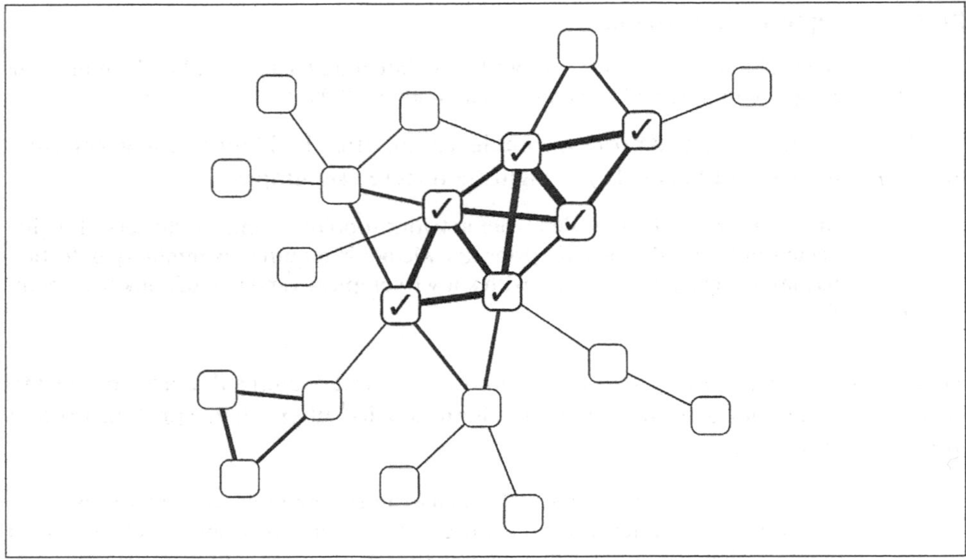

Figure 2.10 "The Genes Mutated in the Same Disease, Cluster Most Significantly." The mutated genes are depicted with check marks. The graph in extracted from figure 3 in [94].

> We have examined the network properties of whole sets of genes with ... mutations implicated by recent ... sequencing studies in autism (ASD), severe intellectual disability (ID), epilepsy or schizophrenia (S). The implicated genes are significantly more strongly interconnected with each other ... than random gene sets of the same size, ... The genes mutated in the same disease, cluster most significantly in the integrated phenotypic-linkage network, while genes mutated in healthy controls do not cluster.

Cliques in Cancer In [153], cliques are used to identify highly conserved biological relationships in the development of *colorectal cancer* in several populations with very different dietary behaviors. Collections of maximal cliques in gene co-expression data are identified and used in [181] to study the progression of *breast cancer*, and could potentially be used in cancer prognosis.

Be Afraid: Be Very Afraid The next example of the use of high-density subgraphs (actually cliques) in a biological network, comes from a study of *fear* in humans [107]. The experimenters showed subjects scary movies (e.g., *The Birds* and *Halloween*) while monitoring their brain activity using fMRI. During the viewing, the experimenters were able to monitor points in a subject's *amygdala*, widely believed to be the part of the brain most involved with fear and strong emotions.

In the experiment, a subject signaled any time that they felt fear. The data was summarized in an undirected graph F (for "fear"), where each node in F represents a small part of the brain (technically, these parts are called "voxels") whose activity was monitored. An undirected edge between nodes A and B was added to F if the two voxels associated with A and B were *both* active when the subject reported fear, and *not* active together at any other time. Then, large cliques were extracted from F, and each clique was hypothesized to represent a functional *unit* or *module* involved with the emotion of fear.

Cliques for Structural Genome Variants Massive, high-throughput DNA sequencing has now resulted in the creation of "reference genomes" for *thousands* of species. Re-sequencing methods have produced the genomes of *hundreds of thousands* of *individuals*.[6] Of course, the individual genomes differ from the reference genome in many ways and places. These differences include *mutations* at single sites, and *insertions* and *deletions* of small segments of DNA, along with larger differences such as the *duplication* of long segments of DNA, or the *inversion* of segments, etc. Common mutations at single DNA sites (called SNP sites) have been systematically cataloged and heavily studied. The other types of differences, now called *structural genome variants*, have been less studied and cataloged. In [132] the authors addressed the question of how to find structural variants that are common to many individuals. Sequences from many individuals are aligned and represented in a graph, where a *maximal clique* helps identify possible structural variants in the sequenced individuals.

Cliques in Evolutionary Biology In [93], clique finding is the central task used to clarify the controversial and "bedeviling" evolutionary history of cormorants and shags (related seabirds). We will discuss this in more detail in Section 5.4 of Chapter 5.

From Pairs to Structures More generally, in many biological systems, data from *pairwise* interactions between biological elements reveal *high-density* or *highly structured* subgraphs that correspond to functional biological subsystems.

> ... [Pairwise][7] genetic interaction screens in model organisms have shown that [pairwise] genetic interactions frequently cluster into highly structured motifs, where members of the same pathways share similar patterns of [pairwise] genetic interactions. [198]

Note that when several elements in a network have *highly similar* pairwise interactions, the subgraph induced by those elements must be dense. Hence, searching for dense subgraphs is a way of searching for elements that likely are together in some biological system.

More Biological Graphs and Networks to Come Here we have discussed a few examples of biological graphs and networks. Additional biological graphs and networks will be presented throughout the book.

2.2 THE MAXIMUM-CLIQUE PROBLEM AND ITS SOLUTION USING ILP

Having motivated biological networks and graphs, and the importance of high-density subgraphs, we now turn to the question of how integer linear programming can be used to *find* high-density subgraphs in a large graph. We begin with the special, but widely adopted, case of finding a *largest clique* in an undirected graph.

The Maximum-Clique Problem:

> Given an undirected graph G, find a *maximum-size* clique K in G. That is, find a clique K that is as large, or larger, than any other clique in G.

[6] Amazingly, it was only 20 years ago that sequencing a *single* human genome was considered a multibillion-dollar "moon-shot."

[7] The three uses of the word "pairwise" are added in this quote for clarity, and are not in the original quote.

For small graphs, one might be able to solve the maximum-clique problem by eye, but for graphs even slightly larger, this is not possible. And most applications of the maximum-clique problem that arise in biology involve graphs with tens, or hundreds or even thousands of nodes and edges.

2.2.1 An Abstract ILP Formulation for the Maximum-Clique Problem

In the introduction, we outlined four tasks involved in solving a problem by integer programming. The first one, task A, is to develop the core *logic* (or idea) of a solution, always with an eye toward implementing the idea in terms of integer linear inequalities. The logic for solving the maximum-clique problem is explained next.

The Logic Expanding on what it means for K to be a clique, it is required that *if* a node i is chosen to be in K, *and* a node j is chosen to be in K, *then* (i,j) *must* be an *edge* in G. But that is equivalent to saying that *if* (i,j) is *not* an edge in G, then we *cannot* choose *both* i and j to be in K.[8] That is the logic that we will implement with integer linear inequalities.

The ILP Variables In the abstract ILP formulation for the maximum-clique problem, we create one *binary* variable,[9] $C(i)$, for each node i of G. These are the only variables needed for the formulation. For example, if G has five nodes, then the ILP variables are $C(1), C(2), C(3), C(4)$, and $C(5)$.[10]

We use variable $C(i)$ to indicate whether or not node i will be included in a set called K^*, defined as follows:

> If $C(i)$ is set to 1 in the optimal ILP solution, then node i will be put into K^*; and if $C(i)$ is set to 0, it will not be put int K^*.

We will show that K^* *must* be a maximum-sized clique in G, so the C variables in the optimal ILP solution do specify a maximum-size clique in G.

The Inequalities For each pair of nodes (i,j) in G, we create the following inequality if (and only if) there is *no* edge in G between nodes i and j.

$$C(i) + C(j) \leq 1. \tag{2.1}$$

For example, in the graph given in Figure 2.2a, there is no edge between nodes 1 and 4, and so the ILP formation would include the inequality

$$C(1) + C(4) \leq 1.$$

Correctness To see the correctness of this formulation, note that when there is no edge between nodes i and j, inequality (2.1) would be violated if both variables

8 One can formally verify this equivalence by using *propositional logic*, but we rely instead on the reader's intuition.

9 Recall that a binary variable can only be set to value 0 or 1.

10 I often use variable names, such as these, that start with letters and then have numbers (sometimes inside parentheses), but that is just my practice. A variable name in Gurobi is just a string of characters, so you have great flexibility in choosing variable names, but I advise starting the name with a alphabetical character, rather than a number.

Maximize

$$\sum_{i=1}^{n} C(i)$$

subject to

$$C(i) + C(j) \leq 1, \tag{2.2}$$

for each pair of nodes (i,j) in G where there is *no* edge in G between nodes i and j.
All variables are binary.

Figure 2.11 The Abstract ILP Formulation for the Maximum-Clique Problem.

$C(i)$ and $C(j)$ were set to 1. Remember that K^* is *defined* as the set of nodes whose corresponding C variables are set to 1 in the optimal solution of the ILP formulation. Therefore, two nodes i, j can *both* be in K^* *only if* there *is* an edge between nodes i and j. And since this applies to *any* two nodes included in K^*, it follows that K^* must form a clique in G.

The Objective Function Because want to find a *maximum* clique, we use the following objective function:

$$\texttt{Maximize} \sum_{i=1}^{n} C(i).$$

The objective function, along with the inequalities specified in 2.1, ensure that an optimal ILP solution will specify a *maximum-size* clique in G. Hence K^* will be one. Summarizing, the abstract ILP for the maximum-clique problem is shown in Figure 2.11.

2.2.2 A Concrete Problem Instance

We illustrate the ILP solution to the maximum-clique problem with a concrete problem instance. Consider the graph, G, with five nodes, shown in Figure 2.2a.

The Matrix Representation of G We use a representation of G called an *adjacency matrix*, shown in Figure 2.12. The adjacency matrix for G is a five-by-five matrix where each cell has a value of either 0 or 1. A cell (i,j) with value 0 means that there is no edge (i,j) in G. For example, a 0 in cell $(1,4)$ means that there is no edge in G between nodes 1 and 4; an entry with value 1 in cell $(1,5)$ means that there *is* an edge in G between nodes 1 and 5. In an undirected graph, the entry in cell (i,j) must be the same as the entry for cell (j,i). That is, the adjacency matrix must be *symmetric*. In general, for a graph with n edges, the adjacency matrix is an n-by-n matrix, and every cell (i,i) on the main diagonal has value 0. The entries on the main diagonal are 0 because we have assumed that a graph does not have any *self-loops*.

The Concrete ILP Formulation for Graph G Examining the adjacency matrix for G (or looking at Figure 2.2a), we see that the edges that are *not* in G are

```
     12345
     -----
  1: 01101
  2: 10010
  3: 10011
  4: 01100
  5: 10100
```

Figure 2.12 The Adjacency Matrix for the Undirected Graph in Figure 2.2a

Maximize $C(1) + C(2) + C(3) + C(4) + C(5)$

Subject to

$$C(1) + C(4) \leq 1$$

$$C(2) + C(3) \leq 1$$

$$C(2) + C(5) \leq 1$$

$$C(4) + C(5) \leq 1$$

All variables are binary.

Figure 2.13 The Concrete ILP Formulation for the Maximum-Clique Problem, for the Graph in Figure 2.2a.

$(1,4), (2,3), (2,5)$, and $(4,5)$. The *concrete* ILP formulation for this instance (based on the ILP formulation in Figure 2.11), is shown in Figure 2.13.

Exercise 2.2.1 *Would the ILP formulation in Figure 2.13 work to find the maximum clique if the variables were allowed to be nonnegative integers, i.e., not explicitly required to be binary?*

2.2.3 Formatting for Gurobi

Above, we developed the inequalities and objective function for a specific problem instance of the maximum-clique problem. This is a concrete ILP formulation for the problem instance. To *solve* the ILP formulation, i.e., to find the values for the variables that define an *optimal* solution to this problem instance, we use an ILP solver (Gurobi Optimizer in our case). For that, we must put the formulation in a *format* that the solver can handle. Input that could be given to Gurobi for this small example is shown in Figure 2.14.

The objective function in the Gurobi-formatted formulation is the same as in the ILP formulation given in Figure 2.13. The inequalities differ, but only by the replacement of "≤" with "<=."[11] The list of variables after the word "binary" tells Gurobi which variables are restricted to have only values 0 or 1. Finally, each input file ends with the word "end" (duh). Although we work with Gurobi in this book,

[11] Although we don't use it in this example, the relation "≥" is written in Gurobi as ">=."

```
Maximize  C(1) + C(2) + C(3) + C(4) + C(5)
subject to
C(1) + C(4) <= 1
C(2) + C(3) <= 1
C(2) + C(5) <= 1
C(4) + C(5) <= 1

binary
C(1)
C(2)
C(3)
C(4)
C(5)
end
```

Figure 2.14 The Gurobi-Formatted Version of the Concrete ILP Formulation Shown in Figure 2.13.

the above formatted example would also be acceptable in Cplex, the other major ILP solver.

Annoying Details One point to note is that the Gurobi format requires a space before and after any "+" or "−" symbols. Otherwise, the symbols will be considered as part of a variable's name, leading to odd and incorrect results. Also, Gurobi requires that all *variables* be on the *left* of the relation symbol ("<=" here), and hence only a single *constant* (i.e., a number) can appear to the *right* of a relation symbol. For example, the inequality "$X \leq Y$" must be formatted for Gurobi as "$X - Y <= 0$." These are mathematically equivalent, but I find this requirement very annoying, and it sometimes obscures the logic of the inequality. Still, the requirement is easy to obey.

The ILP Formulation Must Be in a File The way I use Gurobi requires that the concrete formulation, no matter how it is generated, be saved to a file on the computer. The above Gurobi-formulated ILP is small enough that it could be generated by using a text editor (or word processor[12]) to type it into a file (and I suggest you do that in this example). But usually, the formulation is too large, and we need to use a computer program to *generate* and write the ILP formulation to a file. I discuss the topic of writing Python programs to generate concrete ILP formulations in a tutorial available on the book website.

Suppose that the above Gurobi-formatted ILP formulation has been saved to a file called *clique.lp*. We could have used any file name, provided that the extension is ".lp." That extension tells Gurobi the ILP formulation is in *LP format*. We will not

12 It is best to use a simple text editor such as TextEdit on a Windows machine, or vi on a Unix or OSX (mac) machine. That way, there will not be any invisible characters in your file that might cause trouble.

formally describe what the LP format is. In this book, LP format is taught by default. Everything we say about "Gurobi-formatted ILPs" is actually about ILPs in the LP format.[13]

2.2.4 Running Gurobi to Solve a Concrete ILP Formulation

In this book, we will use the ILP solver from Gurobi in the *simplest possible* way, i.e., by issuing *commands* on the *command line* of a *terminal window*. This is because the emphasis of the book is on learning to create ILP *formulations*, and seeing the power of integer programming in biology, rather than becoming sophisticated users of Gurobi. But, with what you learn about Gurobi in this book, you should be able to easily learn more sophisticated ways of using Gurobi (e.g., the interactive shell, or an Application Programming Interface (API)) from the extensive Gurobi documentation and tutorials.

To find an optimal solution to the concrete ILP formulation stored in the file "clique.lp," I open a *terminal* window, move to the same folder (in MS Windows) or directory (in Unix or OSX) that the file clique.lp is in, and then issue the following command[14] at the prompt:

```
gurobi_cl clique.lp
```

This tells Gurobi to read in file clique.lp and then (using default parameter settings for Gurobi) find values for the variables giving an optimal solution to the ILP problem instance. Gurobi executes this command and produces the output shown in Figure 2.15.

Much of what is output is of little interest to us now. The main thing to note, at the bottom of the output, is that Gurobi found a maximum-size clique of size three, in 0.00 seconds (reported to two decimal places). If we want to see the values that Gurobi set for the variables in the optimal solution, we would execute Gurobi with the command:

```
gurobi_cl resultfile=clique.sol clique.lp
```

The inclusion of "resultfile=clique.sol" tells Gurobi to write the values given to the variables in the optimal solution into a file called "clique.sol." The prefix "resultfile=" is a Gurobi specification that *must* be part of that command, but the user makes up the name of the file to write to. The only requirement for that file name is that it must have the extension ".sol". As a simplifying convention, the name I use for the resultfile is the same name I use for the ILP file, only changing the extension from ".lp" to ".sol." After Gurobi solves the concrete ILP formulation, the results will be in the specified .sol file, and they can be viewed with a text editor

[13] Gurobi can work with several different formats, but the LP format is the simplest and most natural. Quoting from the Gurobi Reference Manual: "The LP format captures an optimization model in a way that is easier for humans to read … and can often be more natural to produce."

[14] A common error at this point is to have opened the Gurobi *Interactive Shell*. If you see those words, or if the prompt has changed to say "Gurobi," you are in the shell and need to exit it (use control-d) before issuing the command.

```
Gurobi Optimizer version 6.5.0 build v6.5.0rc1 (mac64)
Copyright (c) 2015, Gurobi Optimization, Inc.

Read LP format model from file clique.lp
Reading time = 0.00 seconds
(null): 4 rows, 5 columns, 8 nonzero
Optimize a model with 4 rows, 5 columns and 8 non-zeros
Coefficient statistics:
  Matrix range     [1e+00, 1e+00]
  Objective range [1e+00, 1e+00]
  Bounds range     [1e+00, 1e+00]
  RHS range        [1e+00, 1e+00]
Found heuristic solution: objective 2
Presolve removed 4 rows and 5 columns
Presolve time: 0.00s
Presolve: All rows and columns removed

Explored 0 nodes (0 simplex iterations) in 0.00 seconds
Thread count was 1 (of 8 available processors)

Optimal solution found (tolerance 1.00e-04)
Best objective 3.00e+00, best bound 3.00e+00, gap 0.0%
```

Figure 2.15 The Gurobi Output (slightly edited) When Solving the ILP Formulation in the File clique.lp.

or word processor. For example, after Gurobi solves the ILP in clique.lp, file *clique.sol* contains:

```
# Objective value = 3
C(1) 1
C(2) 0
C(3) 1
C(4) 0
C(5) 1
```

So, Gurobi found a maximum-size clique consisting of nodes 1, 3, and 5. In this small example, you can easily verify by eye that this is a correct answer.

Exercise 2.2.2 *Referring to Figure 2.14, plug in the values given above for the five C variables, to verify that all of the inequalities in the ILP formulation are satisfied. That is, verify that this is a feasible solution.*

Exercise 2.2.3 Gurobi *Using an editor or word processor, type the Gurobi-formatted ILP shown in Figure 2.14 into a file called "clique.lp," and then run Gurobi to check that it gets a correct solution.*

Then modify the graph by adding new nodes and edges; change the ILP formulation to reflect the changes, and run Gurobi to verify that it again finds a maximum-size clique. Learn from experimentation! Report on what you did.

$$\text{Maximize } C(1) + C(2) + C(3) + C(4) + C(5) + C(6) + C(7) + C(8)$$

Subject to

$$C(2) + C(3) \leq 1$$

$$C(2) + C(4) \leq 1$$

$$C(2) + C(5) \leq 1$$

$$C(3) + C(6) \leq 1$$

$$C(3) + C(8) \leq 1$$

$$C(4) + C(5) \leq 1$$

$$C(4) + C(6) \leq 1$$

$$C(4) + C(8) \leq 1$$

$$C(5) + C(6) \leq 1$$

$$C(7) + C(8) \leq 1$$

All variables are binary.

Figure 2.16 The Concrete ILP Formulation for the Instance of the Maximum-Clique Problem Shown in Figure 2.2b.

A Computer Is Needed For small problem instances of the maximum-clique problem, such as the one above, you can easily see the largest clique, but for even moderately larger graphs, finding a largest clique without a computer is very difficult. For example, in the graph in Figure 2.2b, which is still quite small, I cannot spot a maximum clique at a glance. I have to systematically analyze the graph to find it and be sure it is maximum size.

Exercise 2.2.4 Gurobi *The concrete ILP formulation to solve the maximum-clique problem for the graph in Figure 2.2b is shown in Figure 2.16. Format the ILP formulation for Gurobi; run Gurobi to solve it, and print out the values of the variables in the optimal solution. How long did Gurobi take to solve this concrete formulation? Did Gurobi find a correct solution?*

2.2.5 Practicality

The maximum-clique problem is in a class of problems called *NPhard*, implying that no provably efficient, provably correct[15] algorithm is known that solves the maximum-clique problem in general. But the ILP formulation that we discussed above is correct – any optimal solution to a concrete ILP formulation will always

[15] In a way that can be more precisely defined, but we will not do it in this book.

specify a maximum-size clique. Hence, we shouldn't expect the ILP to be efficient for all problem instances, particularly as the size of the instance grows. Still, the range of practicality of the ILP approach to the maximum-clique problem is impressive, and covers a wide range of realistic problem instances in biology.

For example, in our experiments with *randomly* generated graphs with 100 nodes, where each potential edge is in the graph with probability of *one-half*, Gurobi typically finds a maximum-size clique in under one second. For randomly generated graphs with 300 nodes, where each potential edge is in the graph with probability of *one-quarter*, Gurobi finds a maximum-sized clique in under 10 seconds. But for 500 nodes with the same edge density, the typical solution time increases to around 40 minutes. Whether that is a practical time depends on the user and the application. And remember, the computation can always be sped up by using more sophisticated computers or clusters, rather than my four-core MacBook Pro laptop.

On a Real Biological Network To illustrate the practicality of maximum-clique finding in real biological networks, consider Figure 2.17. It shows part of the protein-protein interaction (PPI) network[16] from a class of *yeast*. There are 209 nodes and

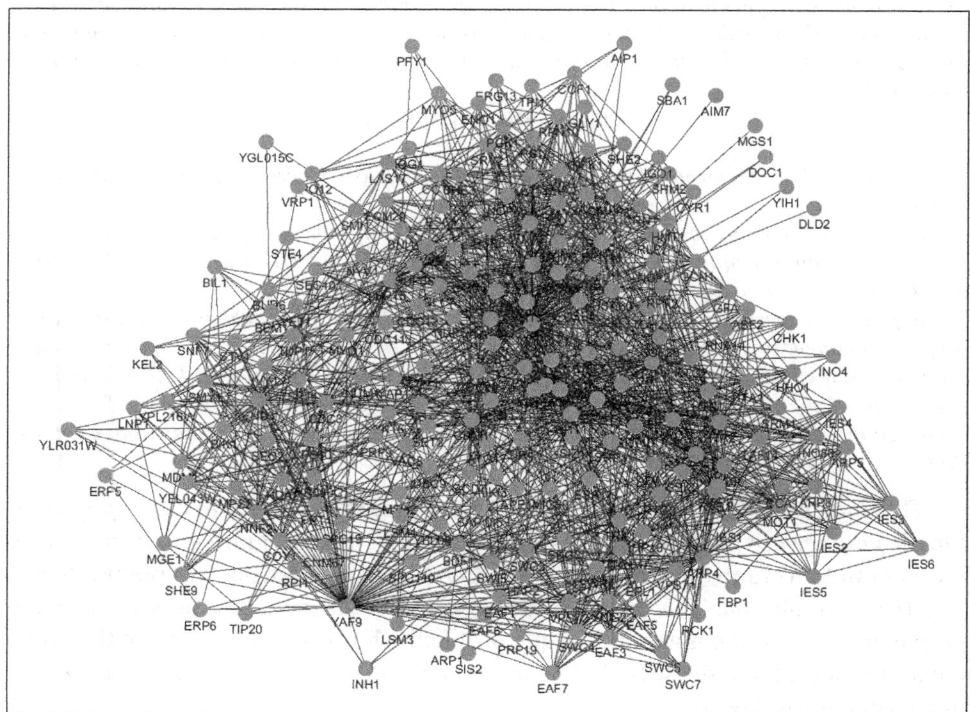

Figure 2.17 The Protein-Protein Interaction Graph for a Subset of the *Yeasts*. There are 209 nodes and 1776 edges. This is a moderate-size biological network, but already much too big for processing by hand, or by brute force computation. Figures like this are called "hairballs," and they are not very informative. The ILP approach found a maximum-sized clique (of size 12) in this hairball in under one second. (Try to match that by hand, or by eye, or even with a *brute-force* computer program.)

[16] Thanks to David Amar for producing this graphic.

1776 edges, so the edge density is about 8.2%, although the distribution of the edges is less uniform than in the random graphs discussed above. Gurobi solved the maximum-clique problem for this graph, finding a clique of size 12, in about one-half of a second. The maximum-clique problem for randomly generated graphs with 209 nodes and edge-probability of 0.082 took 2.5 to 4 seconds to solve. The faster solutions for real biological networks is likely due to the greater *structure* and *asymmetry* in real graphs compared to random graphs.

2.2.5.1 Software for the Maximum Clique Problem

The Python program *CLgraphgen.py* (available at the class website) asks the user to specify a *number* of nodes, n, and an *edge probability*, r. It then generates a *random* graph G with n nodes, where each potential edge is in G with probability r (and so the expected number of edges in G is $n \times r$). The program then constructs and outputs the concrete ILP formulation to solve the maximum-clique problem on graph G. It also outputs the adjacency matrix that represents graph G.

Exercise 2.2.5 Software *Assuming you have Python (with a variant of version 2, not 3) working on your computer, download the program CLgraphgen.py. You don't need to know Python or to understand the workings of CLgraphgen.py to run it, although the program is explained in the tutorial on the book website. Running the program (on a mac) just involves writing the command:*

```
python CLgraphgen.py
```

on a command line in a terminal window, in the same directory where program CLgraphgen.py is found.[17]

Run CLgraphgen.py to produce a random graph, G, and the concrete ILP formulation to solve the maximum-clique problem for G. Use $n = 50$ *and* $r = 0.6$*. Then, use Gurobi to solve the ILP formulation to find the maximum-size clique in G.*

Repeat this experiment, varying the number of nodes and the edge probability of the random graphs generated, to establish in more detail the range of practicality of the ILP approach to the maximum-clique problem. How long does Gurobi take to solve the ILP formulation, as a function of the size and density of the graphs generated? How does the size of the maximum clique depend on the size and density of the graphs generated? Report on what you find.

The book website also has two other Python programs related to *CLgraphgen.py*. One is *graphgen.py* and the other is *CLgraphfile.py*. The program *graphgen.py* produces a random graph the same way that *CLgraphgen.py* does, but without producing any ILP formulation. This is useful because there will be many uses for random graphs, not just for the maximum clique problem. The adjacency matrix of the graph created is output to a file specified by the user. Call the program on a command line in a terminal window as:

```
python graphgen.py adjacency-matrix-file
```

The second program, *CLgraphfile.py*, reads a file containing the adjacency-matrix of a graph, and creates the concrete ILP to solve the maximum clique problem for

[17] The instructions for a Windows machine are similar, but the reader should check with a Windows guru if this instruction does not work.

that graph. The file containing the adjacency matrix, and the output file for the concrete ILP formulation, are specified by the user. Having CLgraphfile.py in addition to CLgraphgen.py is useful because the adjacency matrices input to CLgraphfile.py can from anywhere, and need not be randomly generated. This allows the user to experiment more with the ILP formulation for maximum clique, and allows the solution of the maximum clique problem in applications that output specialized graphs. Call the program on a command line in a terminal window as:

```
CLgraphfile.py adjacency-matrix-file ILP-file.lp
```

Example data: *example-graph*

Note that if you run *graphgen.py* and then use its output as input to the program *CLgraphfile.py*, you get the same effect as running *CLgraphgen.py*.

Exercise 2.2.6 Software *Run the program* CLgraphgen.py *to generate the adjacency matrix, $A(G)$, of a random graph G, and to create the concrete ILP formulation to solve the maximum clique problem on G. Use Gurobi, with the option "resultfile," to find a maximum clique in G, and output it to a .sol file.*

Next, run CLgraphfile.py *using $A(G)$ as input, generating a concrete ILP formulation. Use Gurobi to solve that ILP formulation, and output an optimal solution to a .sol file. Does this optimal solution have the same value as the one found in the first execution of Gurobi? Should it? Are the variables set in the same way in this Gurobi execution as they are in the first execution? If they are the same, must they always be the same?*

Bigger and Better Later Several additional biological applications of clique finding will be discussed in Sections 3.4.1, 3.5, 3.5.2, and 4.6. Those applications involve finding cliques in graphs larger than the ones considered here. For example, in graphs with 2,000 nodes. An interesting observation is that the maximum clique problem for graphs that arise in those applications is solved more efficiently than for the *random* graphs considered here. I think this is a property that often arises in applied problems – real problem instances may contain *structure* that we can exploit to improve the computation; or the structure may be unknown to us, but it indirectly allows the ILP solver to solve problem instances more efficiently.

Exercise 2.2.7 *An* independent set *of nodes in a graph is a set of nodes S such that there is no edge between any pair of nodes in S. An independent set is essentially an* anti-clique. *Write out and explain an ILP formulation to find a maximum-size independent set of nodes in a graph.*

If you know Python, modify the program CLgraphfile.py to create a concrete ILP formulation to find a maximum-size independent set in a graph. Even if you don't know Python, you might be able to figure this out by looking at CLgraphfile.py and reading the comments, i.e., what is written to the right of a hash character, "#." Hint: only a single character in the program needs to be changed.

2.2.6 Adding Node Weights

Often, there is a *weight* (such as a *reliability* score) associated with each *node* in G. The weights are specified at the input of a concrete problem instance. They are constants, not variables. Given these node weights, the goal is to find a clique that maximizes the *sum of weights* of any clique in G. We will discuss such an application, in the study of cancer, in Section 3.5.2.

It is straightforward to extend the ILP formulation for the maximum-clique problem to handle weighted nodes. Suppose that $w(i)$ represents the specified weight of node i. Then, the abstract ILP formulation for this problem is the same as for the maximum-clique problem, but we change the objective function from

$$\sum_{i=1}^{n} C(i),$$

to

$$\sum_{i=1}^{n} w(i) \times C(i).$$

In Section 7.1.2 we will see how to also incorporate *edge* weights into the maximum-clique problem.

Exercise 2.2.8 *In some applications, each node in G has a* probability, *specified in the input to a concrete problem instance; and the problem is to find the most* probable *clique, where the probability of a clique is given by the* product *of the probabilities of the nodes in the clique. So, the natural objective function for this problem would involve the* product *of variable values. However, linear programming only allows linear functions, where the product of variables is not permitted.*

Still, this problem can be solved by integer linear programming. How? Hint: logarithms. *Give the full ILP formulation for the problem of finding the most probable clique. This exercise is harder than most of the exercises in the book.*

2.2.7 Finding a *Second* Largest or *Second-Largest* Clique

We have developed an abstract ILP formulation for finding *one* largest clique. But often it is of interest to find *all* of the largest cliques, so we want to find a second- (or third-, etc.) largest clique. It is also often of interest to find a clique that is largest among all cliques that are *smaller* than the largest clique, but are *not* contained in the largest clique.[18] Such a clique is called a *second-largest* clique.

There are two major variations on these problems: either we require that the additional cliques share *no* nodes with the (first) largest clique found; or we do allow nodes to be shared between the cliques.[19]

An Iterative Approach The conceptually simplest way to solve these problems is to *first* solve the ILP formulation to find a largest clique in G, and *then* use information about the solution to formulate and solve *another* concrete ILP problem.

In particular, let $\mathcal{I}(G)$ denote the ILP formulation shown in Figure 2.11, for the maximum-clique problem with input graph G, and let K denote the clique that the ILP solver finds. If we want to find a largest clique in G that shares *no* nodes with K, we can add the inequality:

$$\sum_{i \in K} X(i) = 0,$$

18 Allowing a second-largest clique to be fully contained inside a largest clique would be uninteresting – a clique of size k always has within it a clique of size $k-1$ and $k-2$, etc.

19 There are many other variations of the kinds of clique-finding tasks that are meaningful in biology, and additional variations on *second largest vs. second-largest* pun.

to $\mathcal{I}(G)$, and then solve the modified ILP formulation.[20] So, if we want to find a largest clique in G that is *not* contained in K, we can add the inequality:

$$\sum_{i \notin K} X(i) \geq 1,$$

to $\mathcal{I}(G)$, and solve that modified ILP.

Exercise 2.2.9 *Explain, for both problems, why these modifications create the correct ILPs that solve the problems.*

Let K' denote the clique found by the second ILP computation. Suppose we want to continue the process to find many optimal solutions (if there are many). How should we next proceed to find a largest clique that shares no nodes with K or K'; or proceed to find a largest clique that is not contained in either K or K', but might be contained in the set $K \cup K'$.[21]

A Single ILP It is somewhat cleaner to formulate the *second largest* and the *second-largest* clique problems as a *single* ILP, avoiding the need to successively formulate and solve two ILPs. Here, we develop an abstract ILP that finds *two* cliques in graph G, assuming that G is not itself a clique. One of the reported cliques will be a largest clique, K, in G, and the other will be a largest clique, K', that shares *no* nodes with K. Hence, if there is a clique with the same size as K that shares no nodes with K, the ILP solver will find one; and otherwise, it will find a largest clique (smaller than K) that shares no nodes with K.

We start with the abstract ILP formulation, $\mathcal{I}(G)$, shown in Figure 2.11, and add new inequalities to it. In detail, for each inequality $C(i) + C(j) \leq 1$ in (2.2), we add the inequality:

$$D(i) + D(j) \leq 1,$$

where each D variable is binary. These new variables and inequalities are essentially duplications of the original ILP formulation for maximum clique. We also add the inequalities:

$$\sum_i D(i) \leq \sum_i C(i),$$

and the inequality:

$$C(i) + D(i) \leq 1,$$

for each node i in G. Finally, we change the objective function to

$$\sum_i D(i) + \sum_i C(i).$$

The entire abstract ILP formulation is shown in Figure 2.18.

[20] Recall (from high school perhaps) that the symbol "$\sum$" denotes the *summation* of a set of values; the set is specified by the terms below (and sometimes above) the symbol "$\sum$," together with the terms that follow the "$\sum$." Recall (again from high school perhaps), that the symbol $\in$ means "is a member of," so $j \in C$ means that j is a member of set C. For the converse, $j \notin C$, means that j is *not* a member of set C.

[21] The symbol "$\cup$" denotes the *union* of the two sets specified to its left and right. So $K \cup K'$ is the set consisting of the union of the set of nodes K with the set of nodes K'.

Maximize

$$\sum_{i=1}^{n}[C(i) + D(i)]$$

subject to
For each node i :

$$\sum_i D(i) \leq \sum_i C(i)$$

$$C(i) + D(i) \leq 1$$

For each pair of nodes i, j, where (i, j) is not an edge in G :

$$C(i) + C(j) \leq 1$$

$$D(i) + D(j) \leq 1$$

All variables are binary.

Figure 2.18 The Abstract ILP Formulation for Finding the Two Largest Cliques, Where the Cliques Share No Nodes.

Exercise 2.2.10 *Explain why the $D(i)$ variables that are set to 1 in a feasible solution to the ILP shown in Figure 2.18 must define a clique in G.*

Explain why an optimal solution to this abstract ILP formulation will find two cliques as described above.

What is the purpose of the inequality $\sum_i D(i) \leq \sum_i C(i)$?

A more involved ILP formulation is needed when we allow the two reported cliques to *share* some nodes, but continue to require that the second clique *not* be contained in the first one. We will address that problem in Section 12.3, after we have developed some additional tools.

2.2.7.1 Software for the Two Largest Cliques

The Python program *CL2graphfile.py* can be downloaded from the book website. The program reads a file containing the adjacency matrix of a graph and creates the concrete ILP for the problem of finding two cliques. One is a largest clique, K in the graph; and the other is a largest clique that shares no nodes with K. Call the program on a command line in a terminal window as:

```
Python CL2graphfile.py adjacency-matrix-file output-ILP-file
```

A Few Tests Recall that the graph in Figure 2.17, with 209 nodes and 1776 edges, has a largest clique, K, of size 12. Does it have another clique of that size?

We used the program *CL2graphfile.py* to find a largest clique, K, and a largest clique that shares no nodes with K. Gurobi 7.5 found two cliques, one of size 12 and the other of size 10, in 1.85 seconds. Hence, the graph does not have a clique of size 12 or 11 made up entirely of nodes not in K. However, it might have a clique of size

12 or 11 that contains *some* node in K and at least one node *not* in K. We will resolve that question in Section 12.3.

More generally, the single-ILP approach runs well on tests of *sparse* random graphs (400 nodes with edge probability of 0.1), but is slower[22] than the iterative approach on tests with much *denser* random graphs (400 nodes with edge probability 0.4). In the end, the single-ILP approach is a good exercise in ILP formulation, and is a convenience if the alternative is to modify ILP formulations by hand. But otherwise, the iterative approach seems superior.

Exercise 2.2.11 *Compare CL2graphfile.py with the iterative approach. That is, generate random graphs over a range of sizes and densities. Use CL2graphfile.py to generate concrete ILP formulations for each of the test graphs, and then Gurobi to see how long it takes to solve the ILP formulations. Similarly, see how long it takes Gurobi to solve the two separate ILP formulations with the iterative approach. Don't count the time involved in modifying the ILP between the first and second executions.*

2.2.7.2 Gurobi Tools for Multiple Solutions

Gurobi has a way for the user to request multiple optimal solutions, and get back both the *count* of how many there are, and a full description of them. However, in the *command-line mode* (which is what we use in this book), currently the user can only get the count. The descriptions of the multiple optimals can be obtained in the more advanced *interactive* or *API* modes. In the command-line mode the following magic incantation issued on a single line will give you the number (assuming there are no more than 1 million of them) of distinct optimal solutions in the ILP formulation called *example.lp*.[23]

```
gurobi_cl poolsearchmode=2 poolsolutions=1000000

                    poolgap=0 example.lp
```

The *poolsearchmode=2* parameter setting says "compute multiple solutions." The second parameter indicates a maximum number of solutions to find (if there are that many), and the third says "Throw away any that are more than this far away from the best." So, in fact, if you want to include some *suboptimal* solutions as well as the optimal ones, you can use this approach to get their count.

Exercise 2.2.12 Software *The adjacency matrix for the yeast PPI network shown in Figure 2.17 is in the file called* yeastAdjMatrix, *which can be downloaded from the book website. Using it, and a program (call it A) previously discussed, create the ILP to find a maximum clique in the network. Then use another program (call it B) to find if there are two optimal solutions, and to output their descriptions, if there are two. Then use the gurobi command described above to find the* total number *of distinct maximum cliques there are in that network.*

Now suppose, purely hypothetically, that the largest clique is of size 12, and you find that there are five cliques of size 12 in the network. However, when using the program B, the optimal value is not 24, i.e., two times 12. Certainly, the optimal value will be less than 24 in that case. What can you deduce about the five cliques of size 12 in the network?

Also, find how many cliques there are which are one smaller *than the maximum size.*

[22] We only count the execution time of Gurobi on the two problem instances, and not the time it takes to modify the first formulation by hand. A more advanced use of Gurobi can automate those changes and reduce the time for the modification to almost nothing.

[23] Of course, when you use this, put in the name of the appropriate file in place of "example.lp."

More Cliques Later We will see additional applications of cliques in computational and systems biology in later sections, particularly in the next two chapters and in Sections 7.1.2.3 and 17.1 of Chapter 7.

2.3 BOUNDS AND GUROBI PROGRESS REPORTING

When Gurobi solves a concrete ILP formulation, it outputs a running *progress report*, which is also accumulated in (actually, appended to) the log-file called "gurobi.log." During the computation, the most important items in the report are the *Objective Bounds*, which consist of the "Incumbent" value, the "BestBd" and the "Gap." We will first discuss the Incumbent value and the BestBd.

Bounds and Guarantees To better define and explain the objective bounds, we consider the case of a *maximization* problem. We let *opt* denote the objective value of an *optimal* ILP solution. Of course, before the ILP instance is solved, we do not know what *opt* is.

At any point in the computation, the best feasible solution found so far is called the *incumbent solution* (don't ask why – I have no idea). We denote the objective value of the current incumbent solution by V^*. Since the objective is maximization, V^* will always be *less than or equal* to *opt*, and it will never decrease during the computation.

Gurobi also maintains a value, called *"BestBd"* (an abbreviation of "Best Bound"), which, for a maximization problem, has the *guaranteed* property that $opt \leq BestBd$ throughout the computation. So, we always know that an optimal solution cannot have a value larger than the current *BestBd*. As the computation proceeds, the value of *BestBd decreases*[24] So, in the case of a maximization problem, it is guaranteed that $V^* \leq opt \leq BestBd$, throughout the computation.

ILP solvers solve concrete ILP formulations, and determine the value of *opt*, by *alternately* focusing on finding better feasible solutions, i.e., increasing V^* (in the case of a maximization problem), and decreasing *BestBd*. In fact, the ILP solver only knows it has found an optimal solution when V^* equals *BestBd*, at which point, *opt* has also been determined – it is equal to those identical values. Moreover, that equality is the *proof* that the value of *opt* has actually been found.

In the case of a *minimization problem*, the incumbent value V^* *decreases* as the computation proceeds, and *BestBd increases*. Again, the value of *opt* is determined when $V^* = BestBd$.

The Progress Report Figure 2.19 shows the progress report that Gurobi produces while solving a concrete ILP formulation for the maximum clique problem for a randomly generated graph with 150 nodes and edge probability of 0.6. This progress report is more extensive than the report shown in Figure 2.15, because the ILP instance solved there was solved in 0.00 seconds (to two decimal point accuracy).

[24] We won't explain how Gurobi determines when to decrease *BestBd* as the computation proceeds, but its *first* value is easy to describe: It is the optimal objective value of the *LP-relaxation* of the ILP instance. Remembering that the problem is a *maximization* problem, and that the optimal objective value of an LP-relaxation of an ILP instance is larger or equal to the optimal objective value of the ILP instance, it follows that this first *BestBd* is larger or equal to *opt* (in the case of a maximization problem).

```
    Nodes    |    Current Node    |     Objective Bounds      |     Work
 Expl Unexpl |  Obj  Depth IntInf | Incumbent    BestBd   Gap | It/Node Time

     0     0   24.19001    0  150    9.0000   24.19001   169%     -    0s
H    0     0                       11.000000  24.19001   120%     -    0s
     0     0   23.85906    0  149   11.0000   23.85906   117%     -    0s
     0     0   23.71799    0  150   11.0000   23.71799   116%     -    0s
     0     0   23.71223    0  150   11.0000   23.71223   116%     -    0s
     0     0   23.66983    0  150   11.0000   23.66983   115%     -    0s
     0     0   23.65555    0  150   11.0000   23.65555   115%     -    0s
     0     0   23.65259    0  150   11.0000   23.65259   115%     -    0s
     0     0   23.65249    0  150   11.0000   23.65249   115%     -    0s
     0     0   23.61698    0  150   11.0000   23.61698   115%     -    0s
     0     0   23.61314    0  150   11.0000   23.61314   115%     -    0s
     0     0   23.61162    0  150   11.0000   23.61162   115%     -    0s
     0     0   23.61114    0  150   11.0000   23.61114   115%     -    0s
     0     2   23.61114    0  150   11.0000   23.61114   115%     -    0s
*  203   166                  16   12.000000  22.88016  90.7%  80.3    1s
H 1031   575                       13.000000  19.57340  50.6%  65.9    2s
  1737   535   18.62161   22   81   13.0000   19.57340  50.6%  61.4    5s

Cutting planes:
  Clique: 38

Explored 6738 nodes (307332 simplex iterations) in 8.46 seconds
Thread count was 8 (of 8 available processors)

Optimal solution found (tolerance 1.00e-04)
Best objective 1.30e+01, best bound 1.30e+01, gap 0.0%
```

Figure 2.19 Gurobi Progress Report (Slightly Edited) While Solving a Concrete ILP Formulation for the Maximum-Clique Problem. Note that as the computation proceeds, the incumbent value (which we have called V^*) increases, and *BestBd* decreases. What is shown here is the part of the report after Gurobi finished the *presolve* phase. The presolve part of the report has been omitted to make a smaller figure. The presolve phase of the computation for a different problem instance was included in Figure 2.15. Note the line "Clique: 38" below the main body of the report. The reference there to "clique" has to do with the internal workings of Gurobi, and is totally *unrelated* to the maximum-clique problem that we have been discussing. So, don't be confused by that line.

The key information is on the right half of the figure. The incumbent value and *BestBd* are written under the heading *Objective Bounds*. The time taken is written in the right-most column. We see that the first solution is a clique of size 9, and that *BestBd* is 24.19001, so we are guaranteed that no clique of size 25 is possible. The computation to that point took zero (reported) seconds. As the computation proceeds, the clique size increases, and upper bound *BestBd* decreases. After five seconds, a clique of size 13 has been found, with a guarantee that no clique of size 20 is possible. The final information reported is that matching bounds, of 13, have been found after 8.46 seconds. There is no information about what occurred between the 5 second and 8.46 second marks.

Note that in the progress report, there are three lines where the incumbent solution improved, first from 9 to 11, then 11 to 12, and finally from 12 to 13. In two of those three lines, the character "H" is written on the extreme left end of line. Character "H" there stands for "Heuristic," indicating that Gurobi applied some (probably proprietary) heuristic method that resulted in an improved incumbent solution.[25]

Mind the Gap The third piece of information in the objective bounds section of the report is the *Gap*, which measures the difference between the incumbent value, V^* and *BestBd*. The column titled "Gap" shows $(BestBd - V^*)/V^* \times 100$ (which is a percentage), for a maximization problem, and $(V^* - BestBd)/BestBd \times 100$, for a minimization problem. By definition, the gap *decreases* as the computation proceeds, and will be zero when an optimal solution is found.

2.3.1 Parameters to Affect Gurobi's Progress

Gurobi has several ways that the user can try to influence how Gurobi does its computation. Here we describe several of those.

Early Termination Based on GAPs For some classes of problems, a common observation is that ILP solvers quickly find a feasible solution whose objective value, V^*, is equal to, or very close to the optimal objective value, *opt*. Then, the majority of the computation time is used to improve *BestBd*, so that it equals V^*. Often, a precise optimal solution is not needed, and the computation can be terminated when the difference between V^* and *BestBd* is smaller than some threshold, i.e., when *GAP* is small enough.[26]

The user can instruct Gurobi to stop the solution of a concrete ILP formulation when the *GAP* falls *below* a user-specified threshold. This is very useful if you expect Gurobi to take excessive time to find an optimal solution, and a "close-to-optimal" feasible solution is almost as good as an optimal one.

The Gurobi magic word (command) to specify the threshold as an *absolute* amount, is "mipgapabs." For example, to instruct Gurobi to start solving the ILP formulation "problem.lp," but stop when the gap is smaller then the absolute value of 50, use:

```
gurobi_cl mipgapabs=50 problem.lp
```

The user can also specify the GAP threshold in terms of a *percentage*. This may be more natural than specifying an absolute value. In this case, the magic word is just "mipgap." To instruct Gurobi to stop when the reported gap falls below $x\%$, use the threshold $x/100$. For example, to instruct Gurobi to start solving "problem.lp," but stop when the reported gap falls below 20%, use:

```
gurobi_cl mipgap=0.2 problem.lp
```

Although you might not end up with an optimal solution, or you might have an optimal solution but not know it, you will have a guaranteed bound on the deviation

[25] The line where the incumbent improved from 11 to 12 is marked with a "*." Its meaning is more difficult to explain now, and is outside of the scope of this book.

[26] This is another practical advantage of integer programming over some other solution methods that only produce a useful solution when the *entire* computation is complete.

of the solution you have from the optimal solution value. This is again a advantageous property of integer programming that is missing for most other types of optimizations.

If the command also includes "resultfile=problem.sol," then the values of the variables in the best solution found so far, will be written to "problem.sol."

MIPFocus Sometimes the user is more interested in getting a good solution quickly, rather than improving *BestBd* quickly. At other times, the user wants the opposite, i.e., for *BestBd* to improve quickly, at the possible expense of V^* improving more slowly. To direct Gurobi to prioritize one improvement over the other, the user can use the *MIPFocus* parameter. Quoting from the Gurobi user manual:

> The MIPFocus parameter allows you to modify your high-level solution strategy, depending on your goals. By default, the Gurobi MIP solver strikes a balance between finding new feasible solutions and proving that the current solution is optimal. If you are more interested in finding feasible solutions quickly, you can select MIPFocus=1. If you believe the solver is having no trouble finding good quality solutions, and wish to focus more attention on proving optimality, select MIPFocus=2. If the best objective bound is moving very slowly (or not at all), you may want to try MIPFocus=3 to focus on the bound.

For example, the following command (issued at the command line) instructs Gurobi to solve the concrete ILP formulation in the file *problem.lp*, by putting more effort into finding good solutions quickly, and putting less effort into improving *BestBd* quickly:

```
gurobi_cl MIPFocus=1 problem.lp
```

It is not obvious (to me) what the difference is between setting the *MIPFocus* parameter to 2 or to 3. My understanding of the quote from the Gurobi manual is that setting the MIPFocus parameter to 2 tells Gurobi to emphasize trying to prove optimality of the incumbent solution, rather than trying to get better solutions. Setting the parameter to 3 tells Gurobi the same thing, but louder: *Now I really mean it!!*[27]

Empirical examples where the use of a MIPFocus parameter makes a significant impact are discussed in Section 4.6.1.1 and in Section 7.2.3.

Heuristics Parameter Gurobi also has a feature that allows the user to encourage more aggressive uses of *heuristics*, in order to improve the incumbent solutions more quickly, again, at the expense of improvements in *BestBd*. Similar to the motivation for setting the MIPFocus parameter to 1, if you want the best feasible solution as quickly as possible (even without knowing exactly how close to optimal it is, because the GAP might remain large) you can set the *heuristic* parameter to 0.5 or to 1, when starting an execution of Gurobi. For example, as follows:

```
gurobi_cl heuristics=0.5 problem.lp
```

Early Termination Based on Time To instruct Gurobi to stop the solution of a concrete ILP formulation when the computation time exceeds a given amount (say

[27] In correspondence with the CEO of Gurobi, I got the advice that MIPFocus = 3 is for when you want to "throw everything you've got at the bound."

100 seconds), we include an instruction to change the "timelimit" *parameter*. This is done on the command line when we call Gurobi to solve *problem.lp*, as follows:

```
gurobi_cl timelimit=100 problem.lp
```

Then, if the computation exceeds 100 seconds, the computation will be terminated, and the value of the best solution found *so far*, will be reported. If we had also included "resultfile=problem.sol," then the values of the variables in the best solution found so far, would be written to file "problem.sol."

3

Maximum Character Compatibility in Phylogenetics

In this chapter we discuss a problem that comes from both classical and modern *phylogenetics*, and is one of the *many* problems in computational biology that are essentially forms of the *maximum-clique* problem. Its ILP solution is therefore closely related to the ILP solution for the maximum-clique problem developed in the previous chapter.

3.1 BASIC DEFINITIONS

Phylogenetics is the subfield of evolutionary biology that tries to deduce the evolutionary history of a set of species or individuals or molecules, or tumor cells, etc., given information about the *states* of certain *characters* or *traits*. Originally, the term, and the field, just concerned the history of *species*, but now the term "phylogeny" is applied to the evolutionary history of any set of objects (usually living, or having once lived).

A *character* or *trait* is a *discrete* property or characteristic of an individual entity of interest (e.g., a species or individual or a molecular sequence). By "discrete," we mean that there is a *finite* number of *states* or *variants* (*alleles* in more genetic vernacular) that a character can take on. For example, the birth gender of a human is a *binary character* taking on one of *two* possible states (male, female). Another binary character is the *presence* or *absence* of a particular mutation at a specified location in a genome. As another example, the nucleotide ($A, T, C,$ or G) present at a particular site of a DNA sequence is a *four-state* character. For simplicity, we refer to an entity of interest as an "individual," no matter what it actually is.[1]

In the phylogenetic problem that we will consider, the data consists of an n by m *matrix* M whose n rows represent n individuals, and whose m columns represent m binary characters. Each cell (f, c) of M specifies the state of character c, of individual f. For example, Figure 3.2 shows a matrix M with $n = 5$ individuals and $m = 6$ binary characters.

[1] In phylogenetics, "taxon" and "OTU" are often used as the generic terms to refer to an object of interest. "OTU" stands for "Operational Taxonomic Unit" (Ugh!).

3.2 PHYLOGENETIC TREES

The most common way to represent the evolutionary history of a set of species is with a *phylogenetic tree*.[2] The root of the tree represents the earliest time included in the history, and the leaves of the tree represent current time.[3] Each node of the tree is labeled by a sequence indicating the states of the characters in M. As the tree is traversed from the root to the leaves of the tree, the states of some of the characters *mutate*. The character(s) that mutate are noted on the edges of the tree, so the sequence that labels a node v is obtained from the root sequence by applying the mutations along the unique path in the tree from the root to v. This also means that if node v' is the unique parent of node v, then the sequence labeling v is *derived* from the sequence labeling v', by changing the state of every character written on the edge (v', v). In particular, the mutations on the path from the root to a leaf f, starting at the root sequence, create the sequence labeling leaf f. A rooted tree T is said to *represent* or *derive* the data M, if the sequence at each leaf f equals the data in row f of M (see Figures 3.1 and 3.2).

3.2.1 Perfect Characters

In phylogenetics, a character c that mutates only *once* in the history of the individuals, is called a *perfect character* [91, 161] or an *ideal* character [202]. In some areas of evolutionary biology, it is called a *monophyletic* character. Another way to say this is that a perfect character can change *only once* in a phylogenetic tree.

Since a perfect character, c, mutates only once, if f is an individual who has the derived (mutated) state of character c, then every descendant of f must also have the derived state of c. All other individuals will have the ancestral (root) state of c. Hence, when the evolutionary history is represented by a tree T, if a perfect character c mutates on edge e of T, an individual g will have the mutated state of character c if and only if the leaf labeled g is in the subtree of T *below* edge e (see Figure 3.1). This is stated in the biological literature as follows:

> The ideal character contains some state that both uniquely defines a set of species and has not been reversed in evolution, so that all existing species which possess this state can be said to have descended from one species in the past that evolved the state. For every such character state that can be identified, a branch in the phylogenetic tree can be added. [202]

There are many molecular characters that have been used in phylogenetics that are believed to be perfect characters. These include *point mutations*, which are the best-known and most widely studied mutations in DNA; *insertions*, *deletions*, *inversions*, or *duplications* of whole segments of DNA; *micro-RNA* sequences, *copy number variants* (CNVs), *SINEs*, *LINEs tandem repeats*, and *transposable elements*. This list of molecular characters is certainly not exhaustive. Further, there are also

[2] A wonderful nontechnical book, [158] for the "popular audience" has recently been published that discusses the history of phylogenetics, many of the people involved in its founding, and the use of the tree as a representation of evolutionary history. It also discusses the limitations of the tree metaphor now that non-treelike evolutionary phenomena have been better understood.

[3] For some historical reasons, in computer science, we generally place the root of a tree at the *top* and extend the rest of the tree *downward*, below it. Other fields put the root at the bottom, more like a real tree, and some fields put the root at the left and grow the tree horizontally to the right.

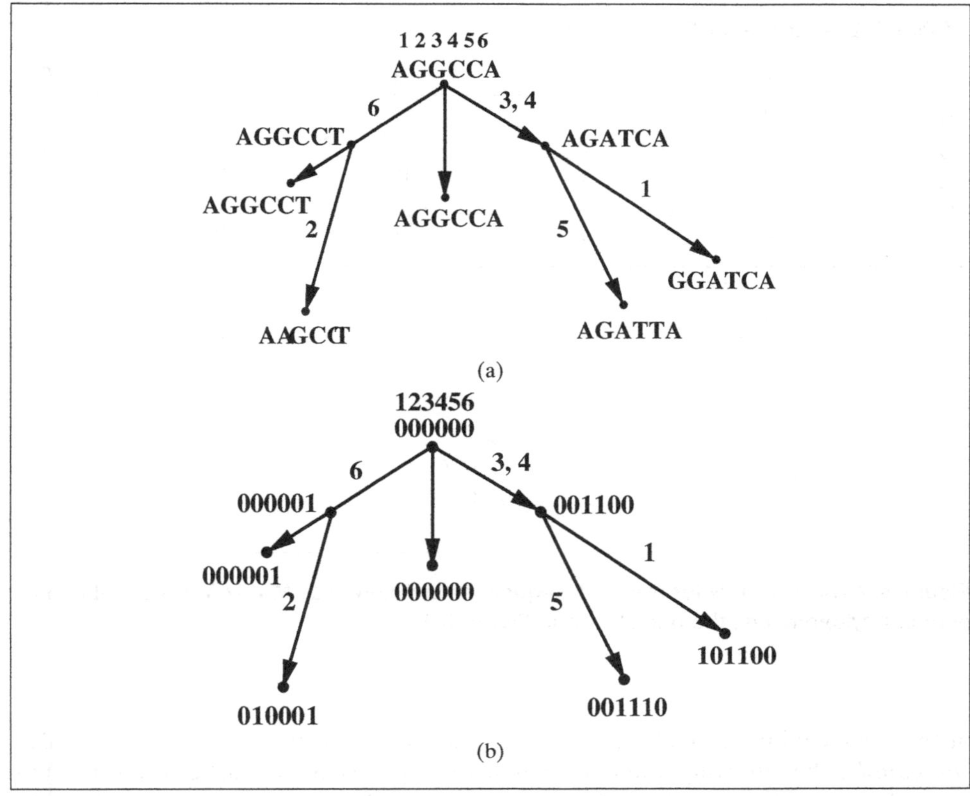

Figure 3.1 (a) The Phylogenetic Tree Presented in [138] for Five Sequences at Six Sites in the Human DTNBP1 Gene (Dysbindin) on Chromosome 6. Each of the six sites mutates exactly once in the tree. So, there are exactly two DNA variants (alleles) at each site in *M* (for example, A and G at site 1, and C, T at site 5). (b) Therefore, the Sequences Can Be Recoded Into Binary, Using 0 for the *Ancestral* State, and 1 for the *Derived* State. The sequence at the root is the *all-zero* sequence, and the number written on an edge indicates the character whose state mutates from 0 to 1 along that edge. The tree is a *perfect phylogeny*.

morphological (nonmolecular) characters that have been proposed to be perfect characters.

3.2.2 Character Compatibility

If there is a tree T with the *all-zero ancestral (root)* sequence that derives the data in matrix M, and every character is perfect, i.e., mutating only once in T, then T is called a *perfect phylogeny* for M, and M is said to be *derivable* on a perfect phylogeny. For example, the tree in Figure 3.1 is a perfect phylogeny for the data M shown in Figure 3.2. For an in-depth discussion of the biological basis for perfect characters, the perfect-phylogeny model, and applications of it in computational biology, see [86]. Not all binary data can be derived on a perfect phylogeny, with an all-zero root sequence, but a simple condition establishes when binary data can be. Two characters in M are said to be *incompatible* if their associated columns in M contain all *three*

Table 3.1 Matrix M for Exercise 3.2.2.

	c_1	c_2	c_3	c_4	c_5
r_1	1	1	0	0	0
r_2	0	0	1	0	0
r_3	1	1	0	0	1
r_4	0	0	1	1	0
r_5	0	1	0	0	0

```
000000
000001
010001
001110
101100
```

Figure 3.2 The Matrix M for the Leaf Sequences in Figure 3.1. Matrix M is derivable on a perfect phylogeny, i.e., the one shown in Figure 3.1.

of the binary pairs (0,1), (1,0), and (1,1). Otherwise, the two characters are called *compatible*. The following widely known theorem precisely establishes when a matrix M can be derived on a perfect phylogeny, with the all-zero ancestral sequence at the root.

The Perfect-Phylogeny Theorem

A binary matrix M is derivable on a perfect phylogeny *with the all-zero root sequence*, if and only if every pair of columns in M is compatible. Equivalently, when *no pair* of columns in M contain all *three* of the binary pairs (0,1), (1,0), and (1,1). Further, when M can be derived on a perfect phylogeny with the all-zero root sequence, the perfect phylogeny is *unique* – there is only one.

For example, the leaf sequences in Figure 3.1b are given in the matrix M shown in Figure 3.2.

Exercise 3.2.1 *Verify that matrix M satisfies the perfect-phylogeny theorem, and that the perfect phylogeny for M is shown in Figure 3.1.*

Exercise 3.2.2 *Is there a perfect phylogeny with the all-zero root sequence that derives the data in the matrix M in Table 3.1? Either create and show a perfect phylogeny that derives M, or explain why there is none.*

Hint: Although we have not discussed an algorithm to build perfect phylogenies, if there is a perfect phylogeny that derives M, it will be unique, so it can be found by making successive deductions about what certain parts of the tree must look like. For example, characters c_1 and c_2 must both mutate on the same path from the root, and the mutation for c_2 must be closer to the root than is the mutation for c_1 (explain why). What about characters c_1 and c_5? After making deductions such as these and others, a little fiddling with the tree should lead to the unique perfect phylogeny – if there is one for M.

3.3 PERFECT PHYLOGENY AND CANCER

Cancer is often referred to as a "disease of genes," but not in the sense of a "genetic disease" that is inherited from one or both parents.[4] Rather, cancer is a disease of genes in the sense that *mutations* in certain genes of the *cancer-affected individuals* themselves (instead of their parents) are necessary for the onset of the cancer. Additional gene mutations occur as the cancer progresses and spreads in the body.[5] In fact, one way to gauge the progression of some cancers, and to predict how they will respond to certain therapies, is to examine how severely the affected genes have mutated. For example, mutation studies of affected genes in breast cancer found an average of 67 mutations in the genes in breast tumors, compared to genes in healthy cells; and found 52 mutations in the genes in colorectal tumors [213], compared to healthy cells. The promise of *personalized medicine* for cancer is based on examining the relevant genes of the individual to see which mutations have occurred. So, understanding the roles of genes and mutations is critical to understanding the causes, treatments, and hopefully preventions, of cancer.

Recently, extensive gene mutation data has been collected in individuals with cancer, as the cancer develops and spreads. Such gene mutation data has lead to the viewpoint that the history of the mutations in certain cancers can be productively represented on a *phylogenetic tree*, and that the analyses of these trees can yield important biological insights. As one early example, see [213], and for an excellent review of this area, with a large annotated bibliography, see [172]. See [131] for an example of the use of phylogenies in the study of cancer.

Shortly after this early work, many of these trees were recognized in the cancer literature as *perfect phylogenies* (for example, see [167, 115, 172]), although sometimes with different terminology. Hence, computational problems about several cancers have, essentially, been formulated and addressed as questions about perfect phylogeny. Here, we introduce one such study, and two others will be discussed in Section 3.5.

Endometrial Cancer The study in [213] focused on *endometrial cancer*, the fourth most common cancer in women. Twenty-six mutations were studied in 30 tissue samples from an endometrial tumor in one individual. For each of the 26 mutations, the *presence or absence* of that mutation defined a *binary* character, and each of the 30 samples was then characterized by a vector of 26 binary values.

The 30 tissue samples were taken at the same time, and from the same tumor (although from different location in the tumor), but the genetic makeup (binary vectors) of those samples could differ due to the exact history deriving each cell. Remember that cells mutate, divide, and propagate (and eventually die), and a mutation that occurs in a cell, is passed on to all of its successive descendents (which are cells, not children). But unless that same mutation occurs again in another cell, only those descendent cells will have that mutation. So although the samples came from living cells, the different binary vectors of the cells reflected their differing histories. Then, those 30 binary vectors were used to build a phylogenetic tree showing the

[4] There are, however, some cancers, such as *retinoblastoma*, where genetic inheritance is central to the development of this cancer. And, there are certain genes, such as BRCA1 and BRCA2, where particular *inherited* alleles do significantly increase the likelihood of breast and ovarian cancer.

[5] The technical term for "spreading in the body" is "metastasis."

deduced evolutionary history of the mutations in the samples. In that way, the successive mutations in the replicating cancer cells in an individual can be deduced from sampled cells. To build the tree, they assumed that

> ... genetic mutations are inheritable and the likelihood of any two cancer cells acquiring the same mutation *de novo* is negligible ... [213]

These are the *perfect-phylogeny* assumptions, in different terms. The phrase "genetic mutations are inheritable" means that once a mutation occurs in a cell, all the successive descendants (cells) of that mutated cell will contain the mutation, so there are no back mutations. And, the rest of the quote means that no mutation is likely to occur more than once in the evolutionary history of the cancer cells. Hence, although they didn't explicitly observe the connection of their approach to the perfect-phylogeny model, they were nonetheless following the model, and using it to construct the phylogeny. That phylogeny gives insight into cancer, because

> [a] phylogenetic tree of a tumor can reconstruct the tumor's genetic progression and mutational distribution [213].

More Complex Perfect-Phylogeny Problems in Cancer Above, we saw that biological assumptions about mutations in cancer cells imply that the history of cancer in an individual should be reconstructable on a perfect phylogeny. However, some problems are more complex, due to the way that data is obtained, even when the biological assumptions agree with the perfect-phylogeny model. There have been several papers that have addressed this situation, trying to infer the "correct" perfect phylogeny from a *mixture* of cancer cells and noncancer cells. As one example, see [63].

3.4 CHARACTER REMOVAL IN THE PERFECT-PHYLOGENY MODEL

We have introduced the perfect-phylogeny model, which is widely adopted in computational biology, and have detailed one application of the model in the development of cancer. We next explore a common computational issue that arises in the use of the perfect phylogeny model, and show how integer linear programming can be used to address that issue.

Perfection Can Be Elusive Recall that not every binary dataset, M, can be derived on a perfect phylogeny. There are multiple reasons why a real, biological dataset that should in principle be derivable on a perfect phylogeny, actually is not. When the input data M *cannot* be derived on a perfect phylogeny, the common practice (for example, see [66, 115]) in phylogenetics is to *remove characters* so that every pair of *remaining* characters is compatible. The remaining characters can then be derived on a perfect phylogeny. The expectation in phylogenetics is that the removed characters where probably the result of data error, but even if that data were good, when a perfect phylogeny can be constructed using a large percentage of the original characters, then the resulting tree will give valid information about the evolution of the individuals. Essentially, the data represents a perfect phylogeny, but it is "hidden behind misleading artifacts [163]."[6]

6 It is rarely advised to remove *individuals* because the evolutionary history of the individuals is usually the main reason for building the phylogeny. The characters provide the raw data used to find the

But Minimization Is Helpful Of course, in order to get the most informative tree, we want to remove *as few* characters as possible. The number of characters that must be removed also gives a measure of the "tree-ness" of the original data.

Due to the perfect-phylogeny theorem, removing the fewest characters so that the remaining characters satisfy the perfect-phylogeny theorem is equivalent to finding the *largest* set of characters that are *pairwise* compatible (i.e., so that each pair in the set is compatible). One approach is to remove all of the characters that are in *at least one* incompatible pair. This approach is computationally simple, and the remaining characters *will* be pairwise compatible. But, this approach will typically remove *many more* characters than necessary. Hence, we need a method that is *guaranteed* to find the true *minimum* number of characters to remove so that the remaining characters are pairwise compatible. This leads to

The Minimum Character-Removal (MCR) Problem Given M, find a *minimum-size* set of characters to remove from M, so that the resulting matrix M' can be derived on a perfect phylogeny. Equivalently, find a *maximum-size* set of characters that are pairwise compatible.

For problem sizes of current interest, integer linear programming can solve the MCR problem very quickly. The ILP approach to the MCR problem was first discussed in [87].

3.4.1 The MCR Problem Is a Version of the Maximum-Clique Problem

The reader may already see that the MCR problem is just a form of the maximum-clique problem. Specifically, given an n by m matrix M, we could create a graph G with n nodes, one for each character in M. Next, we would put an edge between a pair of nodes (i, j), if and only if characters i and j *are compatible*. Then any maximum-size clique K in G specifies a maximum-size set of characters of M that are *pairwise compatible*, i.e., where every pair of nodes in K specifies a pair of compatible characters in M. Further, the nodes that are not in K specifies a optimal solution to the MCR problem, i.e., minimum-size set of characters to remove, so that the remaining characters are pairwise compatible.

With this approach, we would explicitly create G from M, and then create the concrete ILP for the maximum-clique problem for G. But we can short-circuit the process and go directly from M to the ILP formulation for the MCR problem.

Directly Creating the ILP Formulation for the MCR Problem We call a subset of characters K in M a *character clique* if every pair of characters in K is a compatible pair. Then the **Maximum Character-Clique Problem (MCCP)** is:

> Given a binary matrix M, find a character-clique K^* with the largest number of characters, over all character cliques in M.

The abstract ILP formulation for the MCCP is identical to the ILP formulation for the maximum-clique problem, but the meaning of the variables is different. Variable $C(i)$ now indicates whether or not character i will be *included* in K^*: a value of 0 indicates that it will *not*, and a value of 1 indicates that it *will*.

phylogeny, but if, after removing characters, enough signal remains to build the correct phylogeny, the intended goal has been achieved.

M
1010101
1100110
0101010
0001111
1001101

Figure 3.3 A Concrete Problem Instance of the Character Removal Problem.

3.4.2 A Concrete Problem Instance

Consider the matrix shown in Figure 3.3 with five rows and seven columns. The *incompatible* pairs are $(1,2), (1,4), (1,6), (1,7), (2,4), (2,5), (4,5), (4,6)$ $(4,7), (5,6)$, and $(6,7)$, so every character other than 3 is part of some incompatible pair. This example demonstrates that a solution obtained by removing every column that is in some incompatible pair would be a *terrible* solution – it would remove all but a *single* column. Instead, we should formulate and solve the MCCP for the data. As we will see, there is actually a character clique of size *three*, so we only have to remove four (i.e., $7 - 3 = 4$) columns.

Using Gurobi The Gurobi-formatted ILP formulation for the problem instance in Figure 3.3 is shown in Figure 3.4. Suppose it is file *cc.lp*. Then the command

```
gurobi_cl resultfile=cc.sol cc.lp
```

produces the Gurobi execution report shown in Figure 3.5; and the resulting content of file *cc.sol* is shown in Figure 3.6.

Exercise 3.4.1 *Verify by hand that removing columns 1, 4, 5, and 6 from the matrix in Figure 3.3 results in a matrix where all pairs of columns are compatible.*

Exercise 3.4.2 *Type into a file the Gurobi-formatted ILP instance shown in Figure 3.4, and run Gurobi to check that it finds an objective value of three. Then experiment with Gurobi by modifying the ILP formulation (add new variables, add some new inequalities, remove some, etc.) and rerun Gurobi. Learn interactively.*

Exercise 3.4.3 *Consider the matrix M shown in Figure 3.7. Find and report all of the pairs of columns that are incompatible.*

Then use Gurobi to find the minimum number of columns to remove so that the remaining data can be derived on a perfect phylogeny. Explain exactly what you did.

3.4.3 A Computer Is Still Needed

As in our earlier discussion, for small problem instances we can find a maximum-size character clique by hand or by eye, but for even moderately larger matrices, a computer, and an solution method, are necessary. For example, the following matrix M has 11 rows, 43 columns, and 83 pairs of incompatible columns. It has 28 characters that are incompatible with some other characters. If we remove all those 28 characters, what remains is a character clique of 15 characters. However, Gurobi finds that

```
Maximize
C(1) + C(2) + C(3) + C(4) + C(5) + C(6) + C(7)

subject to
C(1) + C(2) <= 1
C(1) + C(4) <= 1
C(1) + C(6) <= 1
C(1) + C(7) <= 1
C(2) + C(4) <= 1
C(2) + C(5) <= 1
C(4) + C(5) <= 1
C(4) + C(6) <= 1
C(4) + C(7) <= 1
C(5) + C(6) <= 1
C(6) + C(7) <= 1

binary
C(1)
C(2)
C(3)
C(4)
C(5)
C(6)
C(7)
end
```

Figure 3.4 The Gurobi-Formatted ILP Formulation for the Character-Clique Problem for the Data in Figure 3.3.

the largest character clique has size 34, so we can find a perfect phylogeny using 34 columns, more than twice the number (15) that would be used in the more naive solution. So, optimization can make a large difference.

```
                    M

000000001100000000110111011110000000000000
001000000000000000110111011110000000000000
000000000000000000000000000000000010000101
000000000000000011000000000000000010011000
000110001011001111000000000000000001000000
001000000000000100000000000001010111000010
001000000000000100000000000011111101000000
111110001011100100000000000011111101100000
111110001011100100000000000011111101100000
111110001011100100000000000011111101100000
111111111000010100001000100011111101000000
```

```
Gurobi Optimizer version 6.5.0 build v6.5.0rc1 (mac64)
Copyright (c) 2015, Gurobi Optimization, Inc.

Read LP format model from file cc.lp
Reading time = 0.00 seconds
(null): 11 rows, 7 columns, 22 nonzeros
Optimize a model with 11 rows, 7 columns and 22 nonzeros
Coefficient statistics:
  Matrix range     [1e+00, 1e+00]
  Objective range [1e+00, 1e+00]
  Bounds range     [1e+00, 1e+00]
  RHS range        [1e+00, 1e+00]
Found heuristic solution: objective 3
Presolve removed 11 rows and 7 columns
Presolve time: 0.00s
Presolve: All rows and columns removed

Explored 0 nodes (0 simplex iterations) in 0.00 seconds
Thread count was 1 (of 8 available processors)

Optimal solution found (tolerance 1.00e-04)
Best objective 3.00e+00, best bound 3.00e+00, gap 0.0%

Wrote result file 'cc.sol'
```

Figure 3.5 **The Gurobi Report After Solving the Concrete ILP Formulation Shown In Figure 3.4.**

```
# Objective value = 3
C(1) 0
C(2) 1
C(3) 1
C(4) 0
C(5) 0
C(6) 0
C(7) 1
```

Figure 3.6 **The Contents of File *cc.sol* After the Execution of the ILP Instance From Figure 3.4.**

The ILP formulation for this matrix is too large to show here. But in Gurobi terms, it is still very small. It only has 43 binary variables and 83 inequalities, and Gurobi found the optimal solution, of size 34, in 0.01 seconds (reported to two decimal points). Further, the Gurobi function *Presolve* removed 43 rows and 27 columns,

```
10100110
10100100
11100101
00010100
10100010
10100000
10101001
10101111
```

Figure 3.7 Matrix M for Exercise 3.4.3.

simplifying the inequalities by applying rules of linear algebra, and exploiting the fact that a feasible solution must be integral, before doing any optimization.

A Larger Illustration As another illustration, consider the following matrix M, with 10 individuals and 100 characters:

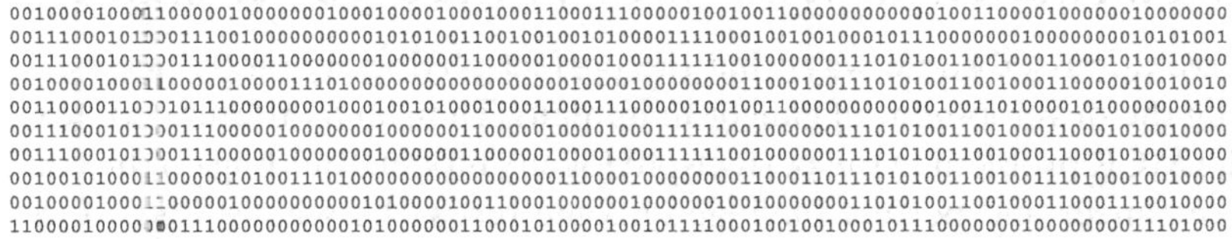

```
0010000100001100000100000000100010000100010001100011100000100100110000000000000000100110000010000000010000000
0011100010100011100100000000000010101001100100100101000001010000011110001001001000101110000000010000000001010 1001
0011100010100011100001100000000100000001100000100001000000100011111100100000011101010011001000110001010010000
0010000100001100000100001110100000000000000000000100001010000010000000001100010011101010011001000110000010010010
0011000011000101110000000001000100101000100011000111000000111000000100100110000000000000000100110100001010000000100
0011100010100011100000100000000100000001100000100001000000100011111100100000011101010011001000110001010010000
0011100010100011100000100000000100000001100000100001000000100011111100100000011101010011001000110001010010000
0010010100001100000101001110100000000000000000001100001011000010000000001100011011101010011001001110100010010000
0010000100001100000100000000000101000010011000100000001000000100000001001000000001101010011001000110001110010000
1100001000000001110000000000000101000000011000101000010010001001011110001001001000101110000000010000000011101000
```

In this M, there are 493 incompatible pairs of characters, and 57 characters that are part of at least one incompatible pair of characters. Removing those 57 characters would leave a character clique of size 43. But the maximum-size character clique actually has 74 characters, and was found by Gurobi in 0.01 seconds, on an Macbook Pro laptop computer with four processors (costing under $2,000).

If You Are Not Impressed, You Are Not Paying Attention At this point, you really should be impressed with the practicality of the ILP approach, together with Gurobi, for solving the MCCP. In the above maximum character-clique problem, we must find a largest subset of the 100 characters with a given property (of being pairwise compatible). Without a tool such as ILP, and a solver such as Gurobi, the best approach you probably would come up with (to *guarantee* that a maximum-size character clique has been found) is pure *brute force*: *enumerate* and check *all* subsets of the 100 characters. But there are 2^{100} subsets. Good luck checking all of those before the sun dies, even with the fastest available computer. You may be tempted to buy a huge cluster, put it in a heavily refrigerated room, and run thousands of CPUs in parallel, but even that won't help.

And You Want More? On the book website, you can find a problem instance of the maximum character clique problem with 40 rows and 1,000 characters. In that problem instance, there are 645 characters that are incompatible with some other character, and out of the 499,500 pairs of characters, 69,460 pairs are incompatible. Try to solve that instance by brute force! Gurobi 7.02 (with default settings) found the maximum character clique, of size 576, in 1.06 seconds on my Macbook Pro. The ILP solver Cplex 12.6 found the solution in 24.65 seconds (slower than Gurobi, but still amazing compared to brute force).

3.4.4 Software for the MCCP (and Hence MCR) Problem

The Perl program *charcliquefast.pl* can be downloaded from the book website. The program creates the concrete ILP formulation to solve the maximum character clique problem (MCCP), given an input matrix. If the optimal solution has size k, and the input matrix has n sites, then the optimal solution to the minimum character removal (MCR) problem has size $n - k$. Further, the sites in the two optimal solutions are complements of each other. Call the program as:

```
perl charcliquefast.pl data-file
```

Example data is in the file: *char-data*, which is also on the book website.

Exercise 3.4.4 Software *Download the program* charcliquefast.pl *and the data file* char-data *from the book website. Then use the program to create the concrete ILP formulation for the MCCP on this data. Solve the ILP formulation using Gurobi. How much time does Gurobi take to solve the formulation? What is the size of the optimal solution?*

3.5 MINIMUM CHARACTER REMOVAL (MCR) IN THE STUDY OF CANCER

In Section 3.3, we discussed some ways that the perfect-phylogeny model has been used in deducing the development of cancer in humans. Then, in a different context, we discussed the MCR problem in the perfect-phylogeny model. Building on the connection between cancer and perfect phylogeny, some researchers have studied the development and spread of cancer, using the MCR problem [167], and the solution of the MCR problem using integer linear programming. In this section, we introduce two such papers, [163] and [193]. We focus on [163], and later in Chapter 14, we discuss it more deeply, along with the approach in [193].

3.5.1 Metastasis in Human Cancers

The paper "Reconstructing Metastatic Seeding Patterns in Human Cancers" [163] addresses the problem of deducing how cancer spreads (metastasizes) in cancer-affected individuals. The study considers *pancreatic, ovarian*, and *prostate* cancers; but the paper mostly focuses on pancreatic cancer, which metastasizes to the lungs and liver. The key question is *where* the cancerous cells that metastasize to the lungs and liver *originate*, and what is the *temporal order* of the metastasis events. For example, cancerous cells, possibly mutated in the form of "subclones,"[7] might repeatedly move from the original tumor in the pancreas, or alternatively, successive cancerous cells might get their seeds from cancerous cells in one of the secondary tumors. More broadly, the key issue is the *pattern of evolution* of cancer:

> The principles governing this evolution are still an active area of research, particularly for metastasis, the final biological stage of cancer that is responsible for the vast majority of the deaths from the disease ... we do not yet know how malignant tumours evolve

[7] Cancer at its original location in the body is called the "primary tumor," and the original cells in that tumor are called "founder cells." Cells that remain in the primary tumor but acquire additional mutations, beyond those in the founder cells, are called "subclones." If the cancer metastasizes, a tumor at the new location is called a "secondary" tumor. A cancerous cell that moves to a new location, causing a secondary tumor, is called a "seed."

> the potential to metastasize ... Moreover, the temporal, spatial and evolutionary rules governing the seeding of metastases at spatially distinct sites distant from the primary tumours have mostly remained undetermined. [163]

This view is reenforced in a recent paper in *Nature*:

> One question that researchers are starting to tackle is the origin of metastatic cells, which emerge from the primary tumour and invade sometimes distant organs. They tend to be the hardest tumour cells to vanquish and the ones most likely to kill patients. [171]

The study in [163] looked at multiple samples from cancerous cells in 10 different locations, two from the primary tumor in the pancreas, and eight from the lung and liver; and also sampled neighboring, noncancerous, cells to help identify mutations in the cancerous cells. The sampled DNA was fully sequenced, and 90 (informative) mutations (which they call *variants*) consisting of single nucleotide substitutions (*SNPs*), or short insertions and deletions, were identified. So, for each individual, the data is represented in a 10-by-90 matrix M, where cell $M(i, j)$ has value 1 if mutation j was identified in cell-location i; and has value 0 otherwise. Each of the 90 mutations was given a *reliability* score, reflecting the believed reliability of the data for that mutation. A larger score indicates that the reported mutation is more reliable than is data with a lower score. We will not discuss how those scores are obtained.

The study assumes that cancer mutation data without errors would fit the *perfect-phylogeny* model, as defined in Section 3.2.[8] However, *real* data have *artifacts*, which are sequencing errors and uninformative characters that are believed to be *noise*, obscuring an underlying perfect phylogeny. Hence, in real data for an individual, there were *incompatible pairs* of characters, exactly as discussed in Section 3.2. In the individual, with alias *Pam03*, who is the most discussed patient in the paper, about 30% of the 90 characters were incompatible with some other character, so no perfect phylogeny for the full data is possible for patient Pam03. Quoting from [163]:

> ... [A] perfect ... tree consistent with the observed (noisy) data of Pam03 cannot be inferred. We show that such a phylogeny indeed exists but that it is hidden behind misleading artifacts ...

And,

> If a phylogenomic method does not account for sequencing artifacts, the mutation patterns of a large fraction of variants will often be inconsistent with any inferred evolutionary tree.

3.5.2 The Artifact Problem and Weighted MCR

The paper uses ILP to address the problem of artifacts in the data, *correcting* the artifacts and building a perfect phylogeny for the corrected data. In the following,

[8] Unfortunately, the authors of [163] deviate from established terminology, calling a perfect phylogeny a "perfect and persistent" phylogeny. In their usage, "perfect" means that a mutation in a character arises once in the evolutionary history, and "persistent" means that there are no *back* mutations. However, the standard definition of *perfect phylogeny* already includes both of those assumptions, so there is no need for the added word "persistent." Worse, the term "persistent phylogeny" is used in the phylogenetic literature with a very different meaning. Persistent phylogeny will be discussed in Section 14.4.

we present a simplified version of the method in [163], and later develop the full approach in Chapter 14. Interestingly, the simplified version discussed here, is exactly what was used in the first part of the method in a different study [193] to identify *subclones* in tumors. We will develop the second part of that method in Chapter 14.

> **Simplified Approach to the Artifact Issue:** Remove characters from *M* so that the remaining characters are *pairwise compatible*, and hence can be derived on a perfect phylogeny; and over all such solutions find one that *minimizes* the *sum* of the reliability scores of the removed characters.

This task can be viewed as a *weighted MCR problem*, as discussed in Section 2.2.6. So, we want to find a *maximum weighted clique*, $\mathcal{K}$, in a graph $G(M)$, where each node represents a character in *M*, and the weight on the node is the reliability score of the data for that character. There is an edge between two nodes in $G(M)$ if and only if their corresponding characters *are* compatible. The characters in the maximum weighted clique are *retained* and used to build a perfect phylogeny, and the other characters are *removed* from *M*. As discussed in Section 2.2.6, this weighted MCR problem can be solved using integer linear programming, and the ILP formulation given there is identical to the ILP formulation presented in [163] for the artifact problem.[9]

Consistent with what we saw in earlier examples, solving the MCR problem greatly reduces the number of characters that have to be removed. In the case of patient *Pam03*, although about 30% of the characters were incompatible with some other character, only 8% of the characters needed to be removed in order to make the remaining characters pairwise compatible.

Once an instance of the weighted MCR problem is solved for an individual (e.g., Pam03), the characters that were *not* removed (i.e., characters in the optimal clique) are guaranteed to be derivable on a perfect phylogeny. The expectation is that these perfect phylogenies (one for each individual in the study) will be informative for the history of character mutations, and the history of each of the tumors, as we explain next.

3.5.2.1 Results

The resulting perfect phylogenies (see Figure 3.8), are thought to have significant biological meaning, resolving open questions about the development and metastasis of pancreatic, ovarian, and prostate cancers. The major conclusion in [163] obtained from studying these perfect phylogenies, is that in pancreatic cancer, the seeding of secondary tumors come *from subclones in the primary tumor*, "... in contrast to colon cancer where liver metastases are assumed to seed lung metastases ... [163]." Further,

> ... liver metastases were seeded from genetically distinct subclones.
>
> ... spatially and genetically distinct subclones in the primary tumour have the capacity to seed metastases.

[9] However, the solution in [163] is derived through the logic of a problem called the *minimum vertex cover*, and the solution in [193] is derived through the logic of a problem called *maximum independent set*. Both the vertex cover and the independent set problems are essentially equivalent to the maximum-clique problem, in that any of these three problems can be easily transformed into either of the other two.

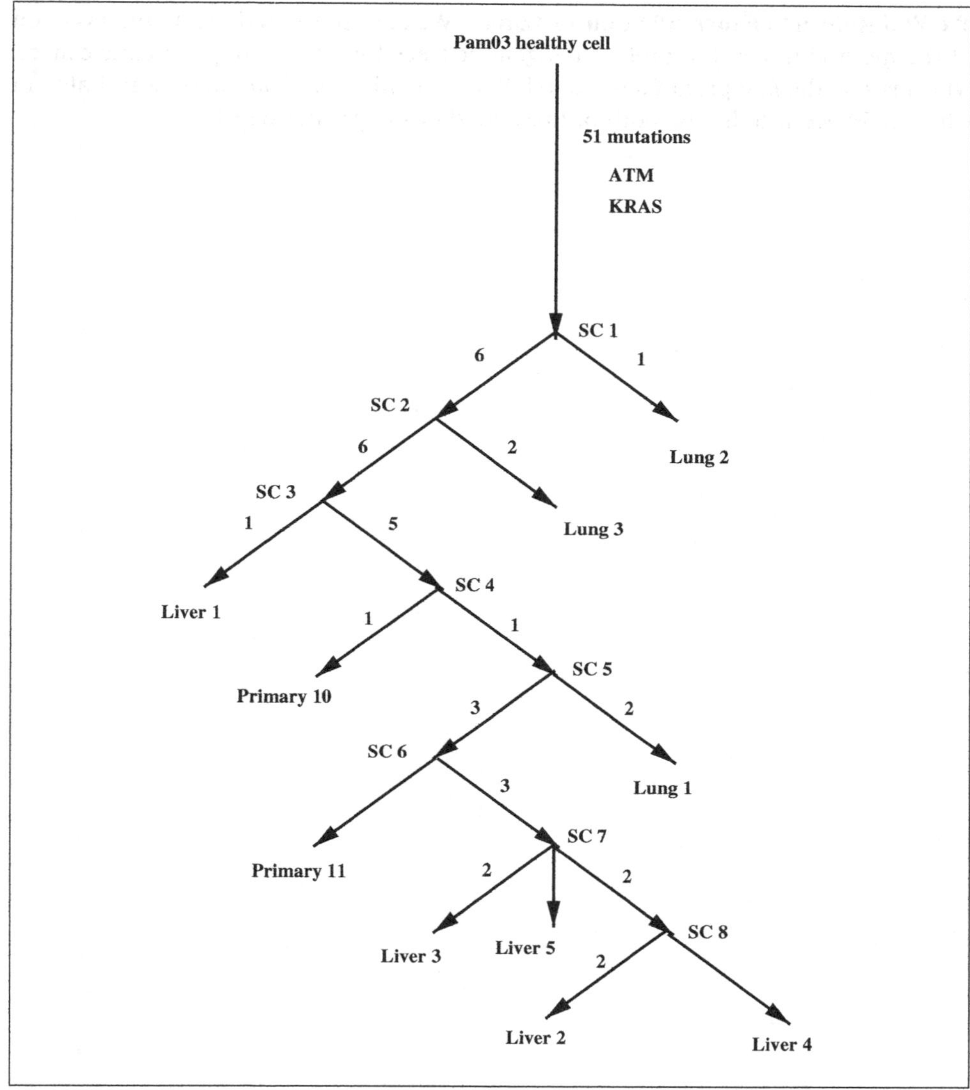

Figure 3.8 The Perfect Phylogeny in [163] Based on the Modified Data from Patient *Pam03*. Each "SC" refers to a deduced subclone of the primary tumor in the pancreas; each "PT" refers to a sample taken from the primary tumor; each "liver" refers to a sample taken from the liver; and each "lung" refers to a sample taken from the lung of patient *Pam03*. The number written on an edge of the tree is the number of mutations that occur along that edge. The edge from the healthy cell to node labeled "SC1" has 51 mutations, including mutations in genes ATM and KRAS, which are commonly mutated genes in several types of cancer.

And,

> [t]he reconstructed phylogenies also indicate that distinct subclones in the primary tumour were equally capable to seed metastases in the same and different organs. [163]

Exercise 3.5.1 *What is it about the phylogeny in Figure 3.8 that suggests the conclusion that in pancreatic cancer, the seeding of metastases comes from subclones in the primary tumor?*

We Will Return to Pancreatic Cancer Later We have presented a simplified version of the method in [163] to build phylogenetic trees for data from pancreatic cancer, which is also the *first* part of a method [193] to identify *subclones* in tumors. Later, in Chapter 14, we will discuss both of these methods in greater depth.

4

Near Cliques, Dense Subgraphs, and Motifs in Biological Networks

Many biological functions are carried out by protein complexes. During the past decade, the main strategy used to identify protein complexes from high-throughput network data has been to extract *near-cliques* or *highly dense subgraphs* from a single protein-protein interaction (PPI) network.

Many approaches have been developed to discover protein complexes, ... these tools aim to extract from a single PPI network *near-cliques* or *highly connected* clusters, which may have the potential to be considered protein complexes. [130]

In Chapters 2 and 3, we developed an ILP formulation for the maximum-clique problem and discussed applications of it in biological networks and in phylogenetics. However, those were just two of the *huge* variety of problems and models in computational and systems biology that involve finding large cliques, near-cliques, or high-density subgraphs in graphs and networks.

In this chapter, we discuss several *near-clique* and *dense subgraph* problems, develop ILP formulations for them, and discuss some of the ways that these problems arise in computational biology.[1] This discussion will also lead to our first ILP *idiom*, i.e., a formalized way to implement certain logical statements as integer linear inequalities. At the end of the chapter, we will discuss how cliques and near cliques can be used in *motif-search* problems.

4.1 NEAR CLIQUES

Recall that a clique of size m in a graph G is a subset K of m nodes, such that there is an edge in G for every pair of nodes (i,j) in K. Often however, we are interested in finding a subset of nodes with a high *percentage* of the possible edges, but the requirement of having every possible edge is too severe. For example, in graph G in Figure 2.2b, the largest clique has four nodes, but the subgraph induced[2] by the subset of the five nodes $\{1,2,6,7,8\}$ is *almost* a clique – it is only lacking edge $(7,8)$.

[1] Cliques and near cliques will also be central in a protein problem discussed in the Chapter 7.

[2] Recall from Section 2.1.3, that the *induced subgraph* $G(K) = (K, E')$ of a graph $G = (V, E)$ contains the node set K, and every edge whose two end points are in set K.

In many applications, that larger "near clique" would be as biologically informative as a true maximum-size clique. Examples of the biological use of near cliques (also called "quasi cliques") in the case of protein-interaction networks are discussed in [130] and [210].[3]

One Detailed Motivation for Near Cliques In the above discussion, we implicitly assumed that there were no errors in the data, so a pair of nodes (i,j) in G that does *not* have an edge between them represents the fact that there is *no* biological interaction between proteins i and j. Still, a near clique identifies a set of proteins, which form a complex or module, even without being a full clique.

In [210], the authors take a different view. They assert that in PPI networks, the missing edges in a large near clique (which they call a *defective clique*) are caused by *errors* in the data. In particular, they assert that the most errors in PPI networks are *false negatives*, where a true interaction between two proteins has not been observed, due to experimental errors. Hence, they are interested in finding large near cliques in order to *predict* that a missing edge in a near clique indicates a *true* protein interaction that should be sought out, and *will* eventually be found. Moreover, they assert that as the size of the maximum near clique increases, and/or the number of missing edges decreases as a percentage of the observed edges in the near clique, the reliability of those predictions should increase. Essentially, they assert that the representation of pairwise interactions in a true protein complex or module, *should* form a clique.

Near Cliques in Breast Cancer Data A different biological use of cliques and near cliques is discussed in [181], where the progression of breast cancer is studied. Each node of the graph represents a gene, and two genes are connected by an edge if their expression pattern is sufficiently similar. That is, they are both expressed (co-expressed) at a sufficient number of time points, and under a sufficient number of different conditions in the progression of breast cancer. Over time, and over the progression of the cancer, different collections of genes are co-expressed. The graph feature of interest is not a clique, but the overlap of several *maximal* cliques. Such a union of maximal cliques is called a *module*[4] in [181]. Being a collection of overlapping cliques, a module is a subgraph that would be a clique if a few additional edges (relative to the edges already there) were added. In the paper, 17 modules are identified, which have the property that

> ... the expression pattern of the modules could cluster patients into subgroups with significantly different relapse-free survival times and provided useful prognostic information that was independent of tumor grade. [181]

4.1.1 A Near-Clique Definition

Having motivated near cliques, we now turn to the question of how to *formally* define them, and to find them. There are many ways to formalize the concept of a near clique, and we will consider several. We start with the simplest one:

[3] Even the title of this paper "Predicting Interactions in Protein Networks by Completing Defective Cliques," indicates the importance of near cliques (that they call "defective cliques").

[4] Actually, there is some ambiguity in their use of the term "module." Sometimes it is used for "clique," and sometimes for a collection of overlapping cliques found by a specific iterative algorithm. The details of that algorithm are not needed here.

A *near-clique* K is a set of nodes, which would be a true clique if *one* edge was added between some pair of nodes in K.

This leads to

The Maximum Near-Clique Problem: Given an undirected graph G, find the largest near clique, K, in G.

4.1.2 An ILP Formulation to Solve the Maximum Near-Clique Problem

Recall that in the ILP formulation (in Figure 2.11) for the *maximum clique* problem in a graph G, a node i is said to be *selected* if and only if variable $C(i)$ has value 1. Then, K was defined as the set of selected nodes, and we showed that K was necessarily a clique. In that ILP formulation, there is an inequality

$$C(i) + C(j) \leq 1, \tag{4.1}$$

if and only if there is *no* edge between nodes i and j in G. That made it *impossible* to select a subset of nodes that is not a clique.

But now, we want to allow the use of *one* missing edge in a set of nodes. We do this by modifying the ILP formulation. We *remove* the inequalities that strictly *prohibit* the selection of two nodes if there is no edge between them, and instead, we introduce variables and inequalities that *record* that such a pair of nodes has been selected. Then, we introduce an inequality that *limits* the number of such selected pairs.

In particular, let $\overline{E}$ be the set containing every *pair* of nodes (i,j) in G where there is *no* edge between i and j. For every pair (i,j) in $\overline{E}$, we use the binary variable, $V(i,j)$ to *record* whether both nodes i and j have been selected, although there is no edge between them. This is implemented with the inequality:

$$C(i) + C(j) - V(i,j) \leq 1. \tag{4.2}$$

Notice that *if both* $C(i)$ and $C(j)$ are assigned the value of 1, so that their sum is two, *then* variable $V(i,j)$ *must* be set to 1 in order to satisfy inequality 4.2. Notice also, that if $C(i)$ and $C(j)$ are both 0, inequality 4.2 reduces to $V(i,j) \geq -1$. Finally, if *exactly* one of $C(i)$ or $C(j)$ has value 1, then inequality 4.2 reduces to $V(i,j) \geq 0$. These two reduced inequalities are automatically satisfied,[5] because $V(i,j)$ is a binary variable. So, the inequalities in 4.2 do *not* have any bad side effects – they only have the effect we want.

Hence, to create the abstract ILP formulation for the maximum near-clique problem, start with the ILP formulation for the maximum-clique problem (shown in Figure 2.11), and replace each inequality 4.1 with inequality 4.2. Then add the inequality

$$\sum_{(i,j)\in\overline{E}} V(i,j) \leq 1.$$

The set of selected nodes, K, will be a clique or a near clique. Summarizing, the abstract ILP for the *maximum near-clique problem* is shown in Figure 4.1.

[5] In math speak, we would more often say that the inequalities are "trivially" satisfied.

$$\text{Maximize} \sum_{i \in V} C(i)$$

subject to

$$\sum_{(i,j) \in \overline{E}} V(i,j) \leq 1, \tag{4.3}$$

and for each node pair $(i,j) \in \overline{E}$:

$$C(i) + C(j) - V(i,j) \leq 1, \tag{4.4}$$

where all variables are binary.

Figure 4.1 The Abstract ILP Formulation for the Maximum Near-Clique Problem.

Note the Asymmetry The inequality $C(i) + C(j) - V(i,j) \leq 1$ ensures that $V(i,j)$ will be set to 1 *if both* $C(i)$ and $C(j)$ are set to 1. However, it does *not force* $V(i,j)$ to be 0 when $C(i)$ and $C(j)$ are *not* both 1, and there is no inequality in the above formulation that does that.[6] This does not cause a problem, because the inequality

$$\sum_{(i,j) \in \overline{E}} V(i,j) \leq 1,$$

limits the number of V variables that can be set to value one.

4.1.2.1 A Concrete Example

As an example, the concrete ILP formulation for the graph in Figure 2.2 (formatted for Gurobi) is:

```
Maximize
C(1) + C(2) + C(3) + C(4) + C(5) + C(6) + C(7) + C(8)

subject to

V(2,3) + V(2,4) + V(2,5) + V(3,6) + V(3,8) +
V(4,5) + V(4,6) + V(4,8) + V(5,6) + V(7,8) <= 1

C(2) + C(3) - V(2,3) <= 1
C(2) + C(4) - V(2,4) <= 1
C(2) + C(5) - V(2,5) <= 1
C(3) + C(6) - V(3,6) <= 1
C(3) + C(8) - V(3,8) <= 1
C(4) + C(5) - V(4,5) <= 1
```

[6] However, there would be no harm in including such an inequality in the formulation, and we will develop one shortly.

```
C(4) + C(6) - V(4,6) <= 1
C(4) + C(8) - V(4,8) <= 1
C(5) + C(6) - V(5,6) <= 1
C(7) + C(8) - V(7,8) <= 1

binary
V(2,3)
V(2,4)
V(2,5)
V(3,6)
V(3,8)
V(4,5)
V(4,6)
V(4,8)
V(5,6)
V(7,8)
C(1)
C(2)
C(3)
C(4)
C(5)
C(6)
C(7)
C(8)
end
```

Gurobi solved this ILP (on my Macbook Pro) in 0.00 seconds (reported to two decimal places). The solution file (i.e., the ".sol" file) from this execution contains:

```
# Objective value = 5
C(1) 1
C(2) 1
C(3) 0
C(4) 0
C(5) 0
C(6) 1
C(7) 1
C(8) 1
V(2,3) 0
V(2,4) 0
V(2,5) 0
V(3,6) 0
V(3,8) 0
V(4,5) 0
V(4,6) 0
V(4,8) 0
V(5,6) 0
V(7,8) 1
```

The solution file shows that the near clique found by Gurobi consists of the nodes $\{1, 2, 6, 7, 8\}$, and that the one added edge is $(7, 8)$.

Exercise 4.1.1 *The abstract ILP formulation for the near-clique problem is shown in Figure 4.1. Suppose we change the right-hand constant in inequality (4.3) from 1 to 0. What problem does the resulting ILP solve? Do you think it would ever be sensible to make that change?*

When a concrete ILP formulation is input to Gurobi and other ILP solvers, they begin by identifying obvious mathematical simplifications or redundancies in the formulation. The part of the solver that does this work is called the preprocessor. If we did change the right-hand constant in 4.3 from 1 to 0, what do you think would be the net effect, after an ILP preprocessor examines the formulation and simplifies it?

4.1.3 Related Notions of a Near Clique

In the previous section, we used a simple definition of a "near clique" where only a single new edge is permitted, and an ILP formulation shown in Figure 4.1. An immediate generalization is to allow *more* than one new edge to be added. Suppose that $k > 1$ new edges can be added. Implementing that change in the ILP shown in Figure 4.1 only requires that we change the right-hand side of inequality 4.3 from 1 to k. For example, if we change the inequality

$$V(2,3)+V(2,4)+V(2,5)+V(3,6)+V(3,8)+V(4,5)+V(4,6)+V(4,8)+V(5,6)+V(7,8) \leq 1$$

to

$$V(2,3)+V(2,4)+V(2,5)+V(3,6)+V(3,8)+V(4,5)+V(4,6)+V(4,8)+V(5,6)+V(7,8) \leq 2,$$

the optimal solution is still of value 5, meaning that there is *no* way to add just two edges to obtain a graph with a clique of size six. However, changing the constant on the right side to 3 yields a solution (again in 0.00 seconds) with value six, meaning that the addition of three edges yields a graph with a clique of size six. In fact, there are several ways to achieve this. One is to add edges $(2, 5)$, $(5, 6)$ and $(7, 8)$ to make the clique consisting of $\{1, 2, 5, 6, 7, 8\}$.

As another illustration, we solved the near-clique problem for the *Yeast PPI* network discussed in Section 2.2.5. Recall that the largest clique in that graph is of size 12. We used Gurobi three times to determine the size of the largest clique when one, two or three additional edge are allowed. Gurobi took 10 minutes to determine that the largest clique remains of size 12 when only one new edge can be added; it took 7 minutes to find a clique of size 13, when two new edges are allowed; and took 12 minutes to determine that the largest clique remains of size 13, when three new edges are allowed.

Another Generalization Another simple generalization is to allow up to one new edge *per node* to be added. That means we now define a near clique, K, of size m as a set of m nodes, such that each node in K is adjacent with *at least* $m - 1$ of the other nodes of K. Implementing that change only requires that we replace the single inequality described in (4.3) with n inequalities, one for each node in G. For each node i, the inequality would be:

$$\sum_{j:(i,j)\in\overline{E}} V(i,j) \leq 1.^7$$

Exercise 4.1.2 A different generalization *Modify the abstract ILP formulation for the near-clique problem for a graph $G = (V, E)$, so that* each *node is allowed to be adjacent to at most one added edge, and the* total *number of added edges does not exceed 10% of the number of original edges, i.e., $|E|$, in G.*

Exercise 4.1.3 *Describe an abstract ILP formulation to solve the following problem:*

Given an undirected graph $G = (V, E)$, and a number, p, between 0 and 1, find the largest set of nodes, $K \subseteq V$, such that K becomes a clique after the addition of at most $p|K|$ edges to G. So, the number of new edges is related to the number of nodes in the near-clique K.

Note, that since K is not known when the ILP formulation is created, this is a more involved problem than, for example, limiting the number of added edges to $p|V|$ or $p|E|$.

4.1.4 Software for Near Cliques and Dense Subgraphs

The Python program *Nclique.py* can be downloaded from the book website. Given the adjacency matrix of a graph, the program generates the concrete ILP formulation for the near-clique problem (i.e., only allowing one new edge to be added). Call the program on a command line in a terminal window as:

```
python Nclique.py graph-file
```

Exercise 4.1.4 software *Use* Nclique.py *to generate the concrete ILP formulation for the near-clique problem for the* Yeast PPI *graph (also available on the book website). Save the ILP formulation in a file named* NcPPI.lp. *Then, by modifying one line in NcPPI.lp you can find the size of the largest clique created when any specific number of new edges are allowed.*

With that approach, determine the smallest number of new edges that need to be added to the Yeast PPI, so that the resulting graph has a clique of size 14. How many times did you have to run Gurobi? How long did Gurobi take at each step of this process?

The Perl program *largest-dense.pl* creates an ILP formulation to find a largest subgraph (number of nodes) that has density at least that of a specified value.

Call on a command line in a terminal file as:

```
perl largest-dense.pl graph-file-name density
```

Example data file: *example-graph* or *zacharymatrix.txt*

Note that if the density is set to 1, then the program should find a largest clique in the graph.

Exercise 4.1.5 Software *Download the Perl program* largest-dense.pl *and the datafile* zacharymatrix.txt. *The file* zacharymatrix.txt *holds the adjacency matrix for the graph shown in Figure 13.2 on page 243.*

[7] The colon symbol ":" has many different uses, so that its meaning varies by context. In the context here, the colon is shorthand for the phrase "such that." Hence "$j : (i,j) \in \overline{E}$" means "$j$ such that (i,j) is in $\overline{E}$."

Run the program with datafile zacharymatrix.txt, *using a density of 0.9. Looking at the figure, does the set of nodes selected make sense? Now try density of 1. Next run the program on the same input with density 0.5. Does the set of nodes selected make sense? Explain. Hint: It does make sense – the ILP solver did what was asked – but at first glance, the result did not make sense to me. Why? Then, I understood it. How?*

4.1.4.1 Near Character Cliques in the Phylogenetic CR Problem

We can also define and find *near character cliques* in the phylogenetic *character-removal (CR) problem*, allowing exactly one pair of selected characters to be incompatible. It should be easy to see how to do this, because we saw in Section 3.4.1 how to cast the CR problem as a maximum-clique problem in an undirected graph.

As an example, recall that the largest character clique in the 1,000 character problem instance discussed in Section 3.4.1, has 576 characters. The maximum near clique in the graph associated with that instance of the character-clique problem, has 577 nodes, so the maximum near character clique has 577 characters. This was found by Gurobi in just under 3 minutes. When we allow up to 10 pairs of incompatible characters, the largest near character-clique only increased by one, to 578, and Gurobi took 14 minutes to solve the ILP. These times show that finding the largest *near* character clique is harder than finding the largest character clique, but the time is still practical for most applications, and is amazingly faster than brute-force computation.

4.2 INVERTING THE NEAR-CLIQUE PROBLEM

Now that we have developed ILP formulations for the *near-clique* problem, and several variants of it, we can see how to modify it further to solve a problem that at first may appear very different and difficult. The problem is:

The Inverse Near-Clique Problem Given an undirected graph G whose largest clique has size k, and given a positive integer d, what is the *minimum* number of edges needed to be added to G, so that the resulting graph has a clique of size at least $k + d$?

An abstract ILP formulation for the inverse near-clique problem is shown in Figure 4.2.

A Gurobi Detail for Integer Variables We have seen that in Gurobi, all binary variables must be listed under the word "binary" or "binaries" at the end of the formulation. Similarly, variables that must take on *integral* values, but are not necessarily binary, must be be listed under the word "generals,"[8] and any variable not listed is assumed in Gurobi to be allowed to take on *fractional* values.

Exercise 4.2.1 *Explain the ILP formulation for the inverse-clique problem, and why it is correct.*

Software for Inverse Clique The Python program *inverseclique.py* reads a file holding a description of a graph; asks the user for relevant information; and then creates the concrete ILP formulation to solve the *inverse-clique* problem on the given graph.

[8] The word "generals" seems like it would identify variables that could have fractional values, but in Gurobi it means variables that must have integer values.

Minimize W

such that

$$\sum_{i \in V} C(i) \geq k + d \tag{4.5}$$

$$\sum_{(i,j) \in \overline{E}} V(i,j) \leq W \tag{4.6}$$

and for each node pair $(i,j) \in \overline{E}$:

$$C(i) + C(j) - V(i,j) \leq 1, \tag{4.7}$$

where all C(i) and V(i,j) variables are binary, and W is a variable that can be given any positive integer value.

Figure 4.2 An Abstract ILP Formulation for the Inverse Near-Clique Problem.

Call the program on a command line in a terminal window as:

```
python inverseclique.py graph-file ILP-file.lp k d
```

For example, the file *example-graph* holds the adjacency matrix for a graph with 40 nodes. The graph has a clique of size 8. So, to find the minimum number of edges to add so that the resulting graph has a clique of size 10, we would call:

```
python inverseclique.py example-graph eg.lp 8 2
```

and then run Gurobi with the ILP file *eg.lp*.

Exercise 4.2.2 *What is the minimum number of edges needed to be added to example graph so that the resulting graph has a clique of size 10. What about 11? What about 18? How does the time taken by Gurobi relate to d?*

Exercise 4.2.3 *Use program* inverseclique.py *and Gurobi to verify the results claimed in Section 4.1.3, and Exercise 4.1.4. Show the results of the Gurobi computations.*

Exercise 4.2.4 *Given the results of Exercise 2.2.12, what do you expect will be the minimum number of edges that need to be added so that the yeast PPI network has a clique of size 13?*

Now find out for sure what that minimum is. How well did you do on the fist part of this question?

4.3 A SHORT INTERRUPTION: OUR FIRST ILP IDIOMS

We have seen that the inequality 4.2 ensures that the binary variable $V(i,j)$ will be set to 1 *if* both binary variables $C(i)$ and $C(j)$ are set to 1.

But how did we know to use that particular inequality? The answer, in many cases, is *experience* and *fiddling*. We make educated guesses and then examine them to see if they work or not. However, there are certain kinds of *logical constructs* that come up frequently in mathematical modeling, and general linear integer inequalities have been developed to implement them. We use to the term "idiom" to refer to such logical constructs and their ILP implementations, but these are more commonly referred to as "modeling tricks" [28].

Many ILP idioms will be discussed in the next chapter, and later in the book, but here we more formally discuss the idiom that we have already used informally, and its converse idiom, that we will need shortly.

4.3.1 An "If-Then" Idiom for Binary Variables

Overwhelmingly, the most important ILP idiom that arises in computational biology (and elsewhere) is the *If-Then* idiom. In those terms, inequality 4.8 expresses the logical construct that

If $C(i) + C(j)$ is equal to 2, *then* we require that variable $V(i,j)$ be *set* to 1.

Restating (4.2), it is implemented as:

$$C(i) + C(j) - V(i,j) \leq 1. \tag{4.8}$$

This is a simple instance of the *If-Then* idiom for *binary* variables. More general versions of the *If-Then* idiom will be discussed later in the book.

4.3.2 An "Only-If" ILP Idiom for Binary Variables

Inequality 4.8 sets $V(i,j)$ to 1 *if* $C(i) + C(j)$ equals 2, but it does *not prevent* $V(i,j)$ from being set to 1 *even if* $C(i) + C(j)$ is *less than* 2. As noted earlier, that is fine in the ILP for the maximum near-clique problem, but it is not always what we want. In some formulations, we need to enforce the *converse* of an *If-Then* construct. In the case of inequality 4.8, the converse construct is:

$V(i,j)$ is *allowed* to be set to 1 *only if* $C(i) + C(j)$ is equal to 2.

This construct is called an *Only-If* idiom (for binary variables).[9]

Implementing the Idiom To implement this Only-If idiom, we might try the inequality:

$$C(i) + C(j) - V(i,j) \geq 1. \tag{4.9}$$

This looks promising because *if* $V(i,j)$ equals 1, $C(i)$ and $C(j)$ *must* both be set to 1 in order to satisfy inequality (4.9). That is the construct we want to implement.

However, inequality (4.9) is actually *not* good to use, because when $V(i,j)$ has value 0, inequality 4.9 reduces to $C(i) + C(j) \geq 1$, so at least one of the variables $C(i)$ and $C(j)$ will be *forced* to have value 1. But there is no reason why either node i or node j should be forced to be selected in a maximum near clique. That would

[9] Calling this idiom an *Only-If* idiom is a matter of emphasis, i.e., what point we are trying to make in the context of its use, because the idiom is *logically equivalent* to: "If $V(i,j)$ equals 1, Then $C(i) + C(j)$ must be equal to 2," which takes the form of an *If-Then* construct. In fact, it is exactly the converse of inequality (4.8).

be an undesirable *side effect* of inequality (4.9). The reader should verify again, that inequality (4.8) does not have a bad side effect of this kind.

Whenever we implement a logical statement as an inequality, or a set of inequalities, we must check for correctness under the conditions it was designed for. For example, check what happens when $V(i,j)$ has value 1. But, we also have to check that they do not have bad side effects under all the other possible conditions. For example, when $V(i,j)$ has value 0.

A Correct Only-If Inequality A correct inequality that implements the *Only-If* idiom (for binary variables) *without* bad side effects is:

$$2V(i,j) - C(i) - C(j) \leq 0. \tag{4.10}$$

As in (4.9), if $V(i,j)$ equals 1, $C(i)$ and $C(j)$ *must both* be set to 1 in order to satisfy inequality (4.10). However, unlike in (4.9), when $V(i,j)$ has value 0, inequality (4.10) reduces to $-C(i) - C(j) \leq 0$, which is *always* satisfied no matter what values are given to $C(i)$ and $C(j)$, and so (4.10) has no bad side effects.

A Different Implementation Another way to implement the Only-If idiom for binary variables is with the *two* inequalities:

$$\begin{aligned} V(i,j) &\leq C(i), \\ V(i,j) &\leq C(j). \end{aligned} \tag{4.11}$$

Exercise 4.3.1 *Explain why the inequalities in (4.11) correctly implement the Only-If idiom implemented in (4.10). Explain why they have no bad side effects.*

Exercise 4.3.2 *The single inequality in (4.10) and the two inequalities in (4.11) have the meaning, that $V(i,j)$ has value 1 only if $C(i)$ and $C(j)$ both have value 1. Since they have the same meaning, we expect that (4.10) should* formally imply *(4.11), and vice verse. Adding together the two inequalities in (4.11) gives the single inequality in (4.10), so (4.11) does imply (4.10). Now, establish the opposite direction, that (4.10) implies (4.11). Hint: Remember that all the variables are binary.*

There are many more idioms that are useful in creating abstract ILP formulations. In the next chapter, we will discuss most of these, and some additional ones will be discussed in later chapters.

4.4 RETURN TO NEAR CLIQUES

We now continue our discussion of near cliques and related structures. We will use the two idioms just discussed, to handle a more biologically meaningful definition of a near-clique K in a graph $G = (V, E)$.

Review of Near-Clique Definitions In earlier definitions of a near clique, we were allowed to add a *fixed number* of edges to G. But that definition is not completely satisfactory because the number of allowed added edges is *independent* of the size of G, and independent of the size of the near cliques we seek. We also had a definition of a near clique where we could add a fixed *percentage* of the number of edges or nodes in the *entire graph* G. We could similarly bound the number of added edges to a fixed percentage of the number of *missing* edges in G. Those definitions were easy to implement because the number of nodes, edges, and missing edges in graph G are *known* when G is *input* to the computer program creating concrete ILP formulations.

Next, in Exercise 4.1.3, we introduced the idea of limiting the number of edges to a user-specified *fraction* of the number of *nodes* in K. That definition is a bit more meaningful than the others, and a bit more interesting, because K, and hence $|K|$, is *not* known until the concrete ILP formulation is *solved*. Still, this definition, like the others, is not completely satisfactory because it does *not relate* the *percentage of new edges* to the number of *original* edges in the *near-clique* K found by solving the concrete ILP.

A More Meaningful Definition What is more meaningful, is a definition of a near-clique K where the number of edges *added* to K (to make K a true clique) is a small percentage of the *original number* of *edges*, or the *total number* of edges, in the resulting clique K. This definition is more meaningful, because with it, we can enforce the desired condition that the majority of the edges in K come from the *original* graph G.

The Resulting Problem The new definition of a near clique leads to

> **The Largest Dense-Near-Clique Problem** Given an undirected graph $G = (V, E)$, and a fraction, p, between 0 and 1, find the largest set of nodes, $K \subseteq V$, such that the induced subgraph $G(K) = (K, E')$ in G becomes a clique after the addition of at most $p|E'|$ edges to G.

Note that the limit on the number of *new* edges is $p|E'|$, not $p|E|$, or $p|K|$. The later two cases are easier, and were essentially handled in Exercises 4.1.2 and 4.1.3. Note also, that $E' \subseteq E$, so that the limit, $p|E'|$, relates the number of added edges to the number of original edges in the resulting clique.

4.4.1 An ILP Formulation for the Largest Dense-Near-Clique Problem

A Key Idea A key thing to keep in mind is that the ILP solver will determine K, when it solves a concrete ILP formulation. It indicates the nodes chosen to be in K by setting some $C(i)$ variables to 1, and the others to 0. So, we can use the values of certain C variables to to count the number of added edges, and the number of original edges, that are in the clique.

In detail, for each pair of nodes (i,j), where (i,j) is *not* an edge in E (so it is in $\overline{E}$), the following inequalities will set variable $V(i,j)$ to value 1 *if and only if* both variables $C(i)$ and $C(j)$ are set to 1:

$$\begin{aligned} C(i) + C(j) - V(i,j) &\leq 1, \\ 2V(i,j) - C(i) - C(j) &\leq 0. \end{aligned} \tag{4.12}$$

This is correct because these implement the *If-Then* and the *Only-If* idioms for binary variables that we developed earlier. Hence, the value of

$$\sum_{(i,j) \in \overline{E}} V(i,j),$$

is exactly the number of new edges that will be added.

Similarly, we need the ILP solution to count the number of *original* edges in G that are between nodes in K. This is because we need to be sure that the number of new edges in K is at most $(p \times 100)\%$ of the number of original edges in the resulting clique defined by K. We do this in essentially the same way we did for the node pairs not in E. We create one binary variable, call it $Y(i,j)$, for each edge (i,j) in E, i.e., an original edge in G. Then, for each edge (i,j) in E, we will have the inequalities:

$$\begin{aligned} C(i) + C(j) - Y(i,j) &\leq 1, \\ 2Y(i,j) - C(i) - C(j) &\leq 0, \end{aligned} \tag{4.13}$$

so that $Y(i,j)$ will be set to value 1 *if and only if* both end points of edge (i,j) have been selected to be in K. Therefore, the value of

$$\sum_{(i,j)\in E} Y(i,j),$$

is exactly the number of edges from E that are in K.

Finally, we can then limit the number of the added edges to $(p \times 100)\%$ of the number of original edges in the clique defined by K, using:

$$\sum_{(i,j)\in \overline{E}} V(i,j) \leq p \times \sum_{(i,j)\in E} Y(i,j). \tag{4.14}$$

All of the inequalities can be created by the computer program that produces the concrete ILP formulation for a concrete problem instance, because $G = (V, E)$ is given as input to that program. In summary, the abstract ILP formulation is shown in Figure 4.3.

Exercise 4.4.1 *Is it true that the* Largest Dense-Near-Clique Problem *can be solved with the ILP formulation shown in Figure 4.3, even if we remove the second inequality in (4.16). Similarly, is it true that the ILP formulation would be correct if we start from the formulation in Figure 4.3, and remove the first inequality in (4.17). Is it true that the formulation would be correct if we remove both of these inequalities? Explain all of your answers.*

Exercise 4.4.2 *Modify the ILP formulation to solve the largest high-density near-clique problem when the number of new edges added is limited by a fraction, p, of the* total *number of edges in the clique defined by K, i.e., limited by* $p \times |K|(|K|-1)/2$, *where K is the selected set of nodes. Remember that every inequality must be a linear function of the ILP variables.*

4.5 FINALLY: THE LARGEST HIGH-DENSITY SUBGRAPH PROBLEM

Chapter 2 started with a discussion of the biological importance of *dense subgraphs*, and then focused on the clearest case, of cliques. In this chapter, we moved to *near cliques* and *largest dense near cliques*. Now, we finally return to the original issue of *high-density subgraphs*, and an ILP formulation to find a largest high-density subgraph. That ILP formulation is similar to the ILP formulation for the *largest dense-near-clique problem*, but, unlike the largest dense-near-clique problem, there is no possibility of *adding* new edges. So, the ILP formulations will have some differences. Still, what we learned about the ILP for dense near cliques will help with understanding the ILP for high-density subgraphs. To start, we need some definitions.

$$\text{Maximize} \sum_{i \in V} C(i)$$

subject to

$$\sum_{(i,j) \in \overline{E}} V(i,j) \leq p \times \sum_{(i,j) \in E} Y(i,j), \tag{4.15}$$

and for each node pair $(i,j) \in \overline{E}$:

$$\begin{aligned} C(i) + C(j) - V(i,j) &\leq 1, \\ 2V(i,j) - C(i) - C(j) &\leq 0, \end{aligned} \tag{4.16}$$

and for each node pair $(i,j) \in E$:,

$$\begin{aligned} C(i) + C(j) - Y(i,j) &\leq 1, \\ 2Y(i,j) - C(i) - C(j) &\leq 0, \end{aligned} \tag{4.17}$$

where all variables are binary.

Figure 4.3 The Abstract ILP Formulation for the Largest Dense Near-Clique Problem.

We define the *density* of graph G with at least one edge, as the number of edges in G divided by $n(n-1)/2$ (i.e., the *maximum* possible number of edges in G). Note that the density of G is a number that is larger than 0 and less than or equal to 1. Note also, that a graph is a clique if and only if its density is 1.

The *density* of an *induced subgraph* $G(V') = (V', E')$ of G is $|E'|$, the number of edges in $G(V')$, divided by $|V'|(|V'|-1)/2$. For example, in Figure 2.2b, the subset of nodes $\{1,2,3,4\}$ defines an *induced subgraph* with four edges, and density $\frac{4}{6}$

The main problem we consider here is

The Largest High-Density Subgraph Problem Given an undirected graph $G = (V, E)$, and a *density threshold* d between 0 and 1, find an induced subgraph, $G' = (V', E')$, in G with the maximum number of nodes, such that the density of G' is greater or equal to d.

4.5.1 An ILP Formulation

As before, the ILP formulation for this problem will contain a binary variable, $C(i)$, for each *node* i in G. We will design the abstract ILP formulation so that for any input graph G, if $C(i)$ is set to 1 in a solution, then we will include node i in the selected node set V' of G'; and will not include node i otherwise. The subgraph G' will then be the subgraph of G *induced* by the nodes in V'. That is, the edge set E' of G' will contain every edge (i,j) from E such that *both* $C(i)$ and $C(j)$ are set to 1.

The Potential-Edge and Real-Edge Variables The ILP formulation will also have a variable $P(i,j)$ for each *pair* of nodes (i,j) in G, whether or not (i,j) is an edge in G; and will have a variable $E(i,j)$ for each *edge* (i,j) in G. All of these variables are binary.

We will design the ILP so that variable $P(i,j)$ will be set to 1 in a solution, *if* both $C(i)$ and $C(j)$ are set to 1. Similarly, for every *edge* (i,j) in G, variable $E(i,j)$ will be set to 1 *only if* both $C(i)$ and $C(j)$ are set to 1. These requirements are implemented by:

$$C(i) + C(j) - P(i,j) \leq 1, \tag{4.18}$$

and

$$2E(i,j) - C(i) - C(j) \leq 0. \tag{4.19}$$

Variables $P(i,j)$ is set to 1 to indicate that (i,j) is a *potential* edge in G', while variable $E(i,j)$ is set to 1 to indicate that it is a *real* edge in G'. Finally, note that together with the inequalities in (4.18),

$$\sum_{i \in V, j \in V, i<j} P(i,j),$$

is the *maximum* number of edges that could *possibly* be in G', i.e., $n'(n'-1)/2$, where $n' = |V'|$.

The Density Requirement Next, we need an inequality that implements the requirement that the chosen nodes identify a subgraph with density at least d, where d is the density threshold, between 0 and 1, specified by the user. That requirement is equivalent to the requirement that:

$$\frac{\sum_{i \in V, j \in V, i<j}[E(i,j)]}{\sum_{i \in V, j \in V, i<j}[P(i,j)]} \geq d. \tag{4.20}$$

That requirement does not look very linear, but it is equivalent to the linear relation:

$$\sum_{i \in V, j \in V, i<j} [E(i,j)] - d \times \sum_{i \in V, j \in V, i<j} [P(i,j)] \geq 0. \tag{4.21}$$

The objective function for the formulation is

$$\texttt{Maximize} \sum_i C(i).$$

In summary, the abstract ILP formulation for the *largest high-density subgraph* problem is shown in Figure 4.4.

Exercise 4.5.1 *Explain why the ILP formulation for the largest high-density subgraph problem does not need an inequality for the converse of (4.22). Similarly, it does not need an inequality for the converse of (4.23).*

Exercise 4.5.2 *Another proposed definition for the density of a graph $G = (V, E)$ is $|E|/|V|$, in place of the earlier definition of $2|E|/(|V|(|V|-1)$. It is claimed that this definition of density is more meaningful for PPI networks, than is the earlier definition. It is claimed that with the earlier definition, the largest high-density subgraph problem tends to find large cliques, while the latter definition encourages the discovery of dense subgraphs that are more biologically meaningful, although they are smaller. With the new definition, and a parameter d, we are again interested in finding the largest subgraph of a given graph G, having density at least d.*

Describe in full detail an abstract ILP formulation to solve this version of the largest high-density subgraph problem.

$$\text{Maximize} \sum_{i \in V} C(i)$$

subject to
For each pair of nodes i,j in G:

$$C(i) + C(j) - P(i,j) \leq 1, \tag{4.22}$$

and for each edge $(i,j) \in E$:

$$2E(i,j) - C(i) - C(j) \leq 0, \tag{4.23}$$

and

$$\sum_{i \in V, j \in V, i<j} [E(i,j)] - d \times \sum_{i \in V, j \in V, i<j} [P(i,j)] \geq 0. \tag{4.24}$$

where all variables are binary.

Figure 4.4 The Abstract ILP Formulation for the Largest High-Density Subgraph Problem.

Exercise 4.5.3 *Describe an abstract ILP formulation for the following problem: Let k be a number of nodes in the largest clique in an undirected graph G. Find the densest subgraph in G that has size larger than k.*

4.5.2 Practicality and Software

The meaning of "practical" depends on the full context of the problem instance. If it takes years to accumulate the data (which is the case for some biological data) then an ILP solution time of several days is practical. However, in this book we use a much more stringent dividing line between the practical and the impractical.

The Perl programs *denseILP.pl* and *densesub.pl* (available at the book website) can be used to test the practicality of the ILP approach to the *largest high-density subgraph problem*. The first program creates a random graph $G = (V, E)$, and the concrete ILP formulation to solve the *largest high-density subgraph problem* on input G. The program asks the user for the *number* of nodes, n, the *probability*, p, that there is an edge between any pair of nodes in G, and the *density threshold*, d, that must hold in the chosen subgraph G' of G. The second program reads the adjacency matrix for a graph G from a file specified by the user, and then creates the ILP formulation for G.

We tested the ILP formulations, and solved them with Gurobi, for a range of values for n, p, and d. The formulations take longer to solve than do the formulations for the maximum clique problems with the same n and p. Still the range of n and p where the approach is successful illustrates the power of the ILP approach compared to brute-force methods. Some illustrative statistics are shown in Table 4.1.

The graphs used for the tests described in Table 4.1 have at most 60 nodes. For a much larger test, we used the 209 node *yeast* PPI network, where the

Table 4.1 Some Gurobi 6.5 and 7.0 Solution Times for the High-Density Subgraph Problem on a 2.3 GHz Macbook Pro, with Four Cores. When Gurobi 7.5 was released, we reran the data used for n = 60. The solution time was 163 seconds.

n	p	d	seconds for 6.5	seconds for 7.0
25	0.4	0.8	0.6	0.58
25	0.6	0.9	0.66	0.82
30	0.4	0.9	1.13	0.77
35	0.5	0.8	9.03	8.73
50	0.4	0.8	166	102
50	0.2	0.5	250	268
60	0.5	0.9	320	200

maximum-size clique of size 12 was found (see Section 2.2.5) in under a second. With the *yeast* PPI graph, and $d = 0.8$, the ILP formulation for the largest high-density subgraph problem took about 14 hours (using Gurobi 7.5) to find the optimal answer of 19. I think that is an impressive result, but is 14 hours (on a laptop) a practical computation? You decide. The time would certainly be reduced by using the typical kind of clusters that many biologists use.

Another test was run on the *yeast* PPI network, with $d = 0.5$. Gurobi was allowed to run for a full day before it was stopped. At that point, it had found a subgraph of 36 nodes with a density of 0.5 or greater, and a *guarantee* that no high-density subgraph (with $d = 0.5$) of size 63 exists in the PPI network. Moreover, the execution found a subgraph of size 30 (with $d = 0.5$) in the first 21 minutes of execution, and found a subgraph of size 35 after about 7 hours of execution.

Summarizing, on the *Yeast* PPI network, a largest subgraph with density one (i.e., a clique), was found in under a second; a largest subgraph with density 0.8 was found in 14 hours; and the largest subgraph with density 0.5 could not be found in one full day of computation. So, the practicality of the method is related to the required density of the subgraph. And fortunately, higher density leads to faster solutions.

In Section 13.3, we will return to the problem of computing a largest high-density subgraph, and discuss a modification to the ILP formulation that allows the use of much larger, sparse graphs.

Exercise 4.5.4 Software *Use programs* denseILP.pl *and* densesub.pl *to explore the practicality of the ILP approach to the* largest high-density subgraph problem, *as a function of graph size and edge density.*

4.6 MOTIF SEARCHING VIA CLIQUE FINDING

As we have seen, many computational problems in biology have been cast as problems of finding large *cliques or near cliques* in graphs. In most cases, the nodes of the graph represent biologically meaningful objects (e.g., molecules, species, individuals), and the edges of the graph represent biologically meaningful *relationships* between them. Thus, in those problems, a clique, or near clique has a clear biological *meaning*. But the utility of cliques extends beyond those cases. Clique finding can be a useful *tool* for biological applications, even when the nodes and edges in the graph have no biological meaning. *Motif searching* is one such important case.

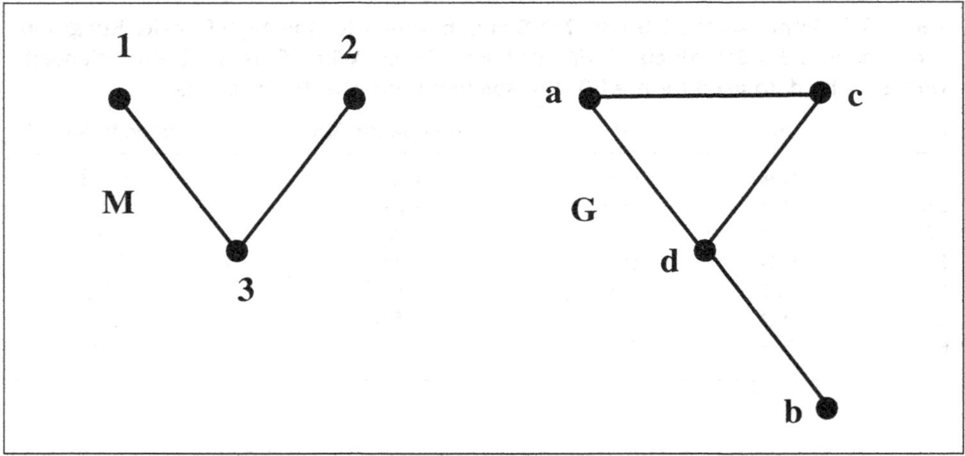

Figure 4.5 A Motif *M* and a Target Graph *G*. The adjacency matrices for *M* and *G* are shown in Figure 4.8. There are two occurrences of *M* in *G*, defined by *{b,c,d}* and *{a,b,d}*. The set of nodes *{a,c,d}* does not define an occurrence of *M* in *G*, because the subgraph induced by those nodes is a triangle rather than a path.

Modules and Motifs Recall that a *functional module* is a collection of "highly interacting cellular components" [205] that carry out some well-defined biological function(s). Looking more closely into a module,

> [m]otifs, considered to be topologically distinct interaction patterns within complex networks, may represent the simplest building blocks of such modules. [205]

Motifs are small, induced subgraphs that appear in larger biological networks.[10] In our discussion, we will assume that both the motif and the network are *undirected* graphs. For example, in Figure 4.5, the subgraph of G induced by the nodes $\{b, c, d\}$ is an occurrence of the three-node (path) motif M in Figure 4.5. Note that not every three node path in G is an occurrence of that motif. For example, the path $\{a, c, d\}$ in G does not define an occurrence of the motif because the subgraph *induced* by $\{a, c, d\}$ is a triangle, not a path.

A realistic motif of some importance is the *cycle* of five nodes (with no edges between nodes that are not neighbors on the cycle), shown in Figure 4.6. This motif appears frequently in the PPI network for *yeast* proteins. Specifically, it is stated in [205] that the cycle appears as an induced subgraph in the *yeast* PPI network vastly more times than would be expected in a random graph of the same size and edge density. Hence, it is believed that the cycle motif must have some biological significance. Evidence for this is the fact that proteins contained in such *overrepresented* motifs are more highly conserved across species, than are proteins that do not appear in overrepresented motifs [205].

More generally, there are many small motifs that occur in biological networks significantly more often than they appear in random graphs with the same number

[10] Note that the definition of "motif" given here requires that it be an *induced* subgraph, so that the *absence* of an edge between a pair of nodes in the motif is as essential to its definition as the *presence* of an edge between nodes.

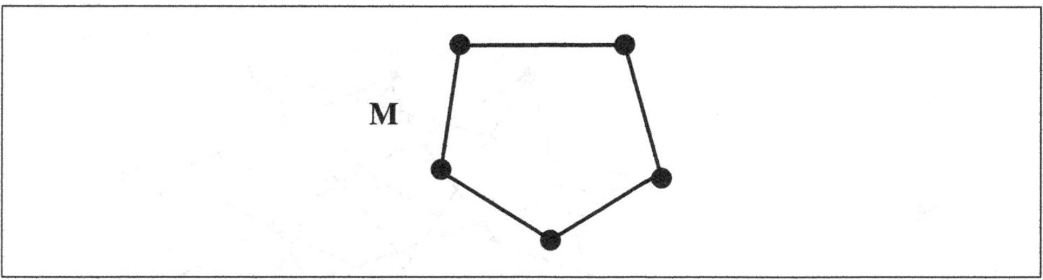

Figure 4.6 The Motif Consisting of a Cycle of Length Five. Note that the absence of any edges between nodes in the cycle is a necessary part of the definition of this motif.

of nodes and edges. Further, specific motifs are conserved across evolutionary history, so the closer two species are in evolution (say humans and chimpanzees) the more motifs they (in their biological networks) have in common. Therefore, motifs and their conservation have been widely studied for their occurrence [179], their overlaps [144], and their function [6].

4.6.1 Searching for Motifs

> Motif analysis is the identification of small network patterns (or subgraphs) that are over-represented when compared with a randomized version of the same network. Discrete biological processes such as regulatory elements are often composed of such motifs. [157]

The Motif-Search Problem Based on the above discussion, an instance of the motif-search problem is specified by a small graph M (the motif), and a larger graph G (the target graph). The problem is to determine *if*, and *where*, motif M appears in graph G as an induced subgraph of G.

A clique can be considered a motif, but not all motifs are cliques, so the motif-search problem is a generalization of the clique-search problem. Ironically, we will solve the motif-search problem by converting it to an instance of a clique-search problem, which is then solved using ILP formulations for maximum clique finding. This conversion builds on the idea of a *product graph*, discussed next.[11]

The Product Graph of M and G Given a motif, M, and a target graph G, the *product graph of M and G*, denoted MG, is a graph, which contains one node for every *pair* of nodes (i, p), where i is a node in M, and p is a node in G. So, if M has m nodes, and G has n nodes, then their product graph MG will have $m \times n$ nodes. The set of edges in MG is a bit more involved to define.

The edges of MG Let u be a node in MG, which corresponds to the node pair (i, p), where i is in M and p is in G. Similarly, let v be a node in MG, which corresponds to the node pair (j, q), where j is in M and p is in G. Then, assuming that $i \neq j$ and $p \neq q$, we create the undirected edge (u, v) in MG if and only if:

[11] I learned about this approach from a lecture by Roberto Grossi when I was at (not on) LSD, London Stringology Days, February 2018.

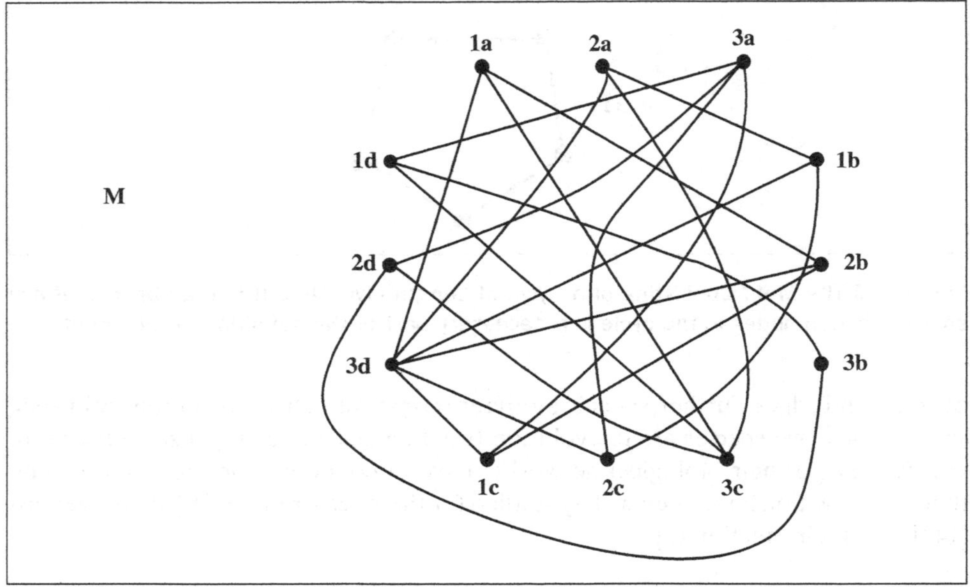

Figure 4.7 The Product Graph *MG* Created from the Motif *M* and Target Graph *G* Shown in Figure 4.5. The adjacency matrix for *MG* is shown in Figure 4.8.

i) (i, j) is an edge in M, and (p, q) is an edge in G;

Or

ii) (i, j) is *not* an edge in M, and (p, q) is *not* an edge in G.

That is, the *adjacency relation* for (i, j) is the same as for (p, q).

For example, consider again the motif M and target graph G in Figure 4.5. Their product graph, with 12 nodes and 20 edges, is shown in Figure 4.7. It is a bit messy, so the adjacency matrices for M, G and MG are shown in Figure 4.8. The node pair $(1a, 3c)$ is an edge in MG because $(1, 3)$ is an edge in M, and (a, c) is an edge in G. The node pair $(1a, 2b)$ is also an edge in MG, but this is due to the fact that $(1, 2)$ is *not* an edge in M, and (a, b) is *not* an edge in G. The pair $(1a, 2d)$ is *not* an edge in MG because $(1, 2)$ is not an edge in M, but (a, d) *is* an edge in G. Also, there is no edge between $(1, a)$ and $(1, c)$ because they have the same first element, 1.

Exercise 4.6.1 *We claim that the maximum size clique in the product graph MG for M and G can not be larger than m, the number of nodes in the motif graph M. Is this true? (hint: yes). Explain why.*

Building on this exercise, we have:

The Key Fact Graph G contains an induced subgraph that is an occurrence of motif M, *if and only if* the maximum-sized clique, K, in MG has exactly m nodes, i.e., the number of nodes in M.

Further, if K is any clique in MG with m nodes, then the M,G node pairs that label the nodes in K, identify the nodes of G that form an occurrence of M. Those node pairs also identify the mapping of nodes in M to nodes in G that defines the occurrence of M in G.

```
M| 1 2 3
--------
1| 0 0 1
2| 0 0 1
3| 1 1 0

G| a b c d
----------
a| 0 0 1 1
b| 0 0 0 1
c| 1 0 0 1
d| 1 1 1 0

MG| 1a 1b 1c 1d 2a 2b 2c 2d 3a 3b 3c 3d
----------------------------------------
1a| 0  0  0  0  0  1  0  0  0  0  1  1
1b| 0  0  0  0  1  0  1  0  0  0  0  1
1c| 0  0  0  0  0  1  0  0  1  0  0  1
1d| 0  0  0  0  0  0  0  0  1  1  1  0
2a| 0  1  0  0  0  0  0  0  0  0  1  1
2b| 1  0  1  0  0  0  0  0  0  0  0  1
2c| 0  1  0  0  0  0  0  0  1  0  0  1
2d| 0  0  0  0  0  0  0  0  1  1  1  0
3a| 0  0  1  1  0  0  1  1  0  0  0  0
3b| 0  0  0  1  0  0  0  1  0  0  0  0
3c| 1  0  0  1  1  0  0  1  0  0  0  0
3d| 1  1  1  0  1  1  1  0  0  0  0  0
```

Figure 4.8 The Adjacency Matrices for a Motif *M* and a Target *G*, and Their Product Graph *MG*.

Exercise 4.6.2 *Check the adjacency matrix for graph MG (in Figure 4.8) to see that nodes* $\{1c, 3d, 2b\}$ *form a clique in MG. That clique identifies an occurrence of the motif M as a subgraph induced by the nodes* $\{b, c, d\}$ *in G. However, there is another clique in MG that shows that* $\{b, c, d\}$ *is an occurrence of M in G. What is it, and why are there two cliques in MG for that occurrence of M in G?*

It is difficult to spot the cliques (triangles) in Figure 4.7, so some edges that are in triangles are shown as dashed edges in Figure 4.9.

Exercise 4.6.3 *Referring to Figure 4.9, what occurrences of M in G are identified by the dashed edges? What mappings between nodes in M and nodes in G do they identify? What other cliques of size three are there in MG? What occurrences of M, and what node mappings do they identify? Why is a single occurrence of M in G represented by more than one clique?*

Explaining the Key Fact The key fact is almost self-evident, once you get past the definitions. To motivate it, consider again the example in Figure 4.5. The nodes

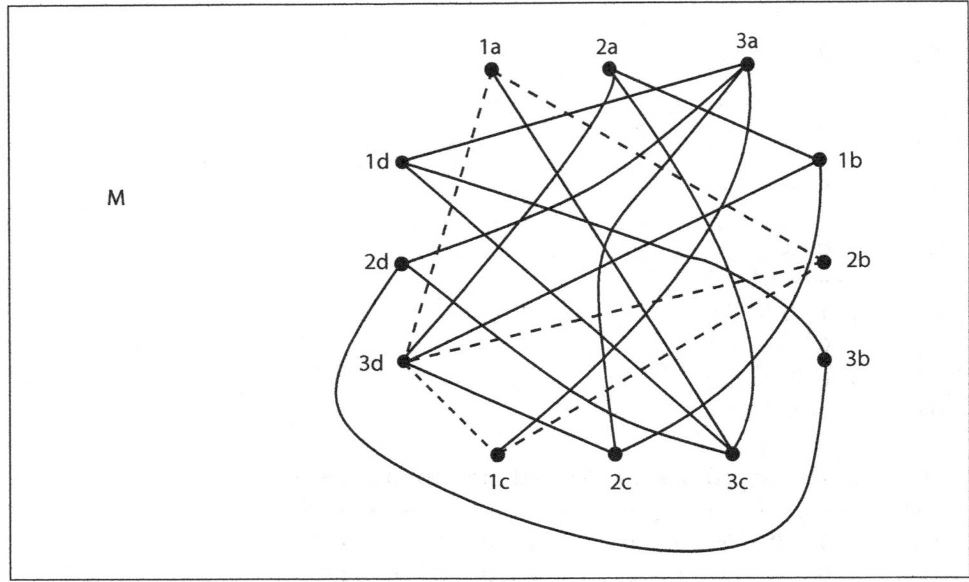

Figure 4.9 The Product Graph from Figure 4.7. The dashed lines show *some* of the edges involved in cliques of size three.

$\{b, c, d\}$ induce a subgraph in G that is identical to the three-node motif M. One way to see this identity is to map node 1 in M to node b in G; and map node 3 in M to node d in G; and map node 2 in M to node c in G. That mapping defines the nodes $\{1b, 3d, 2c\}$ in the product graph MG.

Exercise 4.6.4 *Explain the key fact in your own words.*

Exercise 4.6.5 *Suppose we relax the definition of what it means for M to occur in G, so that M only needs to be a* subgraph *of G, not a subgraph* induced *by a subset of nodes of G. For example, in Figure 4.5, the edges (a, c), (a, d) would now define an occurrence of M, even though the subgraph* induced *by the nodes {a, c, d} is a triangle.*

The motif searching problem now is to find an occurrence of M in G, under this new definition. The product-graph approach can still work to solve this version of the problem, but requires some modification. What is it? Explain.

Exercise 4.6.6 *The product graph approach can also be used to find a largest subgraph that appears as an* induced subgraph *in both of two graphs. Explain how, and then explain how to modify the ILP approach to find a largest common induced subgraph. For what size graphs do you think this ILP approach would be practical?*

4.6.1.1 Practicality

If the motif has m nodes and the target graph has n nodes, then the approach to motif searching presented here must solve a maximum clique problem on a graph with $m \times n$ nodes. For motifs in proteins, m is generally under 10, but in some biological applications n may be so large that this approach will not be practical. Our earlier tests of the use of integer programming to solve maximum clique problems (reported in Section 2.2.5) concentrated on graphs with under 500 nodes, so those tests do not by themselves resolve the range of practicality for motif searching. Therefore, we tested the product-graph and ILP approach for motif searching on a few examples to be sure that it was effective on problems of meaningful size.

The most natural example to try is the five-node cycle motif, claimed to be vastly overrepresented in *yeast* PPI networks, with the 209 node *yeast* PPI network we saw earlier. The product graph has 1,045 nodes, and a large number of edges. Still, Gurobi 7.5 running on my iMac Pro with four processors, took under 45 seconds to find an occurrence of the motif. To find additional occurrences of the motif, we could find additional maximum cliques in the product graph, which we can do by using the methods discussed in Section 2.2.7.

A smaller, four-node motif, is shown as graph G in Figure 4.5. This motif is also reported to be overrepresented in *yeast* PPI networks. Gurobi 7.5 found an occurrence of this motif in the *yeast* PPI network in just 4.4 seconds. Moving to larger motifs, I tried a cycle of 10 nodes just to see how long the ILP computation would take. Graph MG in this case had 2,090 nodes, and the ILP solution got stuck at around 14 minutes, with a remaining gap of about 10%. At that point, I thought of a simple, but likely very effective, way to speed up the computations for motif searching, and so I killed the computation.

A Huge Speedup Based on the *key fact* stated above (and made explicit in Exercise 4.6.1), the size of the maximum clique in MG can be no larger than m, the number of nodes in M. In fact, *if* there is an occurrence of M in G, then the maximum clique in MG will have size *exactly* m. So, the speedup idea is to add an equality to the ILP formulation stating that the optimal objective value (which is the size of a maximum clique in G) *must* be of size m. Note, however, that with this added equality, if there actually is *no* occurrence of M in G, then the concrete ILP formulation will be *infeasible*.

I expected that *if* there is an occurrence of M in G, the concrete ILP formulation with the new equality would solve much faster than before, since it had been given the value of the optimal solution. That knowledge helps Gurobi speed up the computation in several ways. Probably, the most important is that the computation can stop *as soon* as the *first feasible* integer solution is found. Further, when we expect (or gamble) that there will be a clique of size m in MG, we can set the parameters "Heuristics=1, and MIPfocus=1," to encourage Gurobi to try to find a feasible solution as quickly as possible.[12]

The equality could also speedup the computation when the motif does *not* appear in G. With the new equality (but without setting the heuristics and MIPfocus parameters), Gurobi will report that the problem instance is *infeasible*, as soon as it computes an upper bound (on the optimal objective function) that is *less than* m. But, without the new equality statement, Gurobi would get to that point and then continue, stopping only when it found an optimal solution to the ILP, which would objective value strictly less than m.

How to Incorporate the Upper Bound In order to incorporate the new equality into the ILP formulation, we need to make a few changes to the abstract ILP formulation,

[12] The downside of setting these parameters is that if there is no clique of size m, it might take Gurobi longer to determine that the ILP formulation is infeasible.

shown in Figure 2.11, that solves the maximum clique problem. First, we change the objective function from

$$\texttt{Maximize} \sum_{i=1}^{m\times n} C(i),$$

to

$$\texttt{Maximize } Z.$$

Then, we add the equality

$$Z = \sum_{i=1}^{m\times n} C(i),$$

and add the equality

$$Z = m.$$

It Could Work! So, how well does this work? The previously stalled search for an induced cycle of 10 nodes in the 209-node *yeast* PPI network, was now successful. The new ILP formulation found an induced cycle of 10 nodes in just 16 seconds, and most of that time was used by the preprocessor (12 seconds). In the case of the search for an induced cycle of five nodes, the time to solve the ILP formulation fell from 45 seconds, reported above, to 6 seconds.

Given those successes, I searched for induced cycles of lengths 15, 17, and 18. And, I found them – in 814 seconds, 2,952 seconds, and 2,035 seconds, respectively, without setting the *Heuristics, or MIPfocus parameters*. But, after I set those parameters and reran Gurobi 7.5, the times fell to 623, 611, and 744 seconds, respectively. It works!

Exercise 4.6.7 *What if instead of searching for full* cliques *in MG, we search for* near cliques *(under different definitions of a near clique). Are near cliques meaningful for motif searching? For example, suppose the motif of interest has m nodes, but there is no clique of size m in MG. However, there is a near clique of size m, missing one edge. What does that mean in terms of the motif? This question is open ended, and can be expanded to include all forms of high-density subgraphs we have examined. Do it.*

4.6.1.2 Final Comments

With the speedups discussed, the approach of using the maximum clique ILP for motif finding is more practical than it might have initially appeared. This is a non-trivial outcome, and it illustrates a general point I made in the introduction to the book: There are more sophisticated and more efficient algorithms (not using ILP) that search for motifs in graphs; and they may succeed on huge target graphs where the ILP approach will not be efficient enough to work. However, the ILP approach *does work* on a *meaningful range* of graph sizes that arise in parts of biology, and the ILP approach is conceptually much simpler to understand and to implement than are the more sophisticated algorithms.

5

Convergent and Maximum Parsimony Problems in Phylogenetics

In Chapter 3, we extensively examined the *perfect-phylogeny model*, which severely restricts the type and number of character mutations – a character can only mutate from state 0 to state 1, and only *once* in the evolutionary history of the character. We saw in that chapter that integer linear programming can be effectively used to solve problems concerning perfect phylogenies. Later, in Chapter 14 (particularly Section 14.1), we will examine less restricted, but still constrained, mutational models, and again see the utility of integer programming.

In this chapter, we examine two highly *unconstrained* phylogenetic models. We first consider phylogenetic models where the number of *forward* mutations is *unlimited*. We then examine the important case where *any* number of (forward or backward) mutations are allowed. In that case, the problem of building a phylogenetic tree that *minimizes* the total number of mutations is called the *maximum parsimony problem*. In other contexts, the same problem is called the *Steiner Tree Problem in Hypercubes*.

Convergent Mutations and Characters Mutations of a character that occur *more than once* in the *forward* direction, i.e., from the 0-state to the 1-state, are called a *convergent* or *parallel* mutations; and a character with a convergent mutation is called a *convergent character*. For example, the evolutionary character of *flight* is believed to be convergent – that is, to have evolved in the ancestors of birds independent of its evolution in insects, or mammals (e.g., bats). The evolutionary, convergent character of "flight," is believed to have mutated from the 0-state (no flight), to the 1-state (flight) at least *three* times in the phylogenetic *tree of life*. However, it is believed that the ability to fly is so advantageous that it has only rarely, if ever, been lost in a species that had it. A phylogeny that has convergent mutations is called a *convergent phylogeny*.

> Evolutionary convergence – the similarity of certain features of distantly related organisms – is a well recognized feature of evolution. [93]

A recently published paper shows how convergent evolution can be *deduced* (amazingly) from genomic DNA sequences, by analyzing inactive, broken, vestigial fragments of ancient DNA [180]. This work concluded that the trait of *descended testicles* is ancestral, and was *independently* lost in a group of mammals, including elephants, four separate times. In these mammals, the testes are not located *outside* of the animal torso, as in human males, but are located *inside*, near the kidneys. So, *testicle ascent* in mammals is a *convergent* trait, thought to have arisen four times.

5.1 PHYLOGENETICS VIA MAXIMUM PARSIMONY

We will consider two different models of convergent evolution: one where *no* back mutations are allowed, and one where *any* amount of back mutations are allowed. We don't know an established name for the first model – we simply call it the *forward convergent phylogeny (FCP)* model. However, the second model is well known, and is called the *maximum parsimony (MP)* model. It is a very important and highly studied phylogenetic model. We will show how integer linear programming is used to solve problems in both models.

5.1.1 The Role of Hypercubes in Phylogenetics

The ILP formulations solving problems in the FCP and MP models both choose edges from a graph called a *hypercube*. In fact, an ILP solution specifies edges forming a *rooted subtree* of a hypercube. So, before explaining the ILP formulations, we need to discuss hypercubes in detail.

Hypercubes A *hypercube* of *dimension* m is an undirected graph, H_m, with 2^m nodes, where each node is labeled with a *distinct* binary string with m bits. Hence, the labels on the nodes are the binary representations of the integers between 0 and $2^m - 1$. Two nodes in the hypercube are connected by an edge if and only if their binary labels differ by exactly one bit. We refer to any node in H_m by the binary sequence (or the equivalent integer number) that labels it.

As mentioned above, we will solve a phylogenetic problem by finding a *rooted* subtree of a hypercube, where the root represents the evolutionary origin of the tree. Hence, edges in subtree will be directed *away* from the root. Now, suppose that v_1 and v_2 are two adjacent nodes in H_m, so their labels are identical except at one position, say at position k. Suppose v_1 has a 0, and v_2 has a 1, at position k. Then, if edge (v_1, v_2) is directed from v_1 to v_2, that mutation is interpreted as a *forward* mutation of character k, i.e., from state 0 to state 1; and if the edge is directed from v_2 to v_1, it is interpreted as a *backward* mutation of character k, i.e., from state 1 to state 0.

Constructing Hypercubes To further explain hypercubes, we discuss how they are constructed. The hypercube H_m can be constructed from the hypercube H_{m-1} by first duplicating H_{m-1}; then connecting each node in the original H_{m-1} to the node in the duplicate H_{m-1} with the same label; and then adding a 0 on the left of each node label in the original H_{m-1}, and adding a 1 on the left of each node label in the duplicate H_{m-1}. This is illustrated in Figure 5.1a,b, and c, which shows the hypercubes for $m = 1, 2, 3$.

Leaves and paths in hypercubes Given the set of n taxa represented in an n-by-m matrix M, we call a node in H_m a *leaf* if it is labeled by a sequence (a taxon) in M.

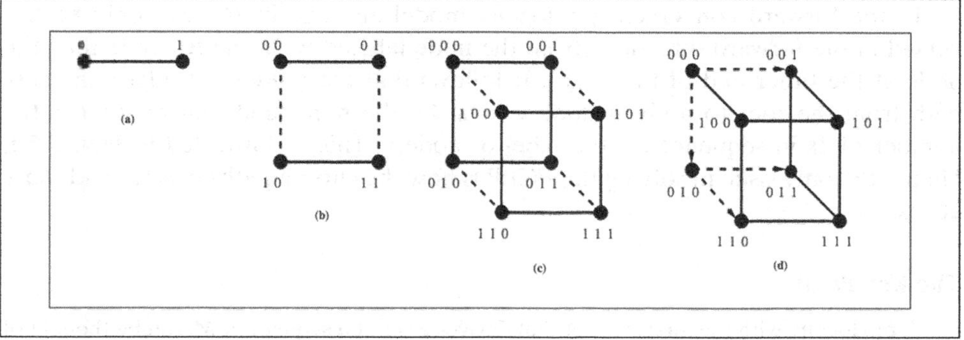

Figure 5.1 Hypercubes H_1, H_2, H_3 of dimensions one, two and three, and a convergent phylogeny. (a) The hypercube H_1 of dimension one has two nodes labeled 0 and 1. (b) The hypercube H_2 of dimension two is created from H_1, by first duplicating H_1, along with its node labels. Then new edges, shown as dashed lines, connect each pair of nodes with the same node label in H_1. Finally, the label of each node in the original H_1 is extended on the left by adding 0; and the label of each node in the duplicate H_1 is extended on the left by adding 1. (c) Using the same method, H_3 is constructed from H_2. (d) A convergent phylogeny, with dashed directed edges, for sequences $M = \{001, 010, 011, 110\}$. Note that the third character mutates twice.

Clearly, every sequence in M labels exactly one node in H_m. Generally, n is much smaller than 2^m, so the set of leaves will be a very small subset of the 2^m nodes in H_m. Notice that the edges along *any* path in H_m from the root node to a node v, specify an ordered series of mutations (either forward or backward mutations) that derive the binary sequence, S_v, labeling node v.

Unless $m = 2$, there will be many paths in H_m from the root to any given node v, and so there are many different series of mutations that derive the sequence S_v (see Figure 5.1 to see this point visually). The converse observation is also true: any ordered series of mutations that derive a particular m-length binary sequence, S, starting from the all-zero sequence, uniquely specifies a path in H_m from the root of H_m to the unique node in H_m labeled with sequence S. Further, when both forward and backward mutations are allowed, there can be paths from the root to a node v of different lengths, although the *shortest* paths from the root to node v all have length exactly equal to the number of 1s in S_v.

Having introduced hypercubes, we next return to the CP and MP models and more formally define the problems in terms of hypercubes.

5.1.2 Forward Convergent Phylogeny (FCP)

In the FCP model, any evolutionary character is allowed to mutate from the 0 state to the 1 state *any* number of times in a phylogeny, but can *never* mutate from the 1 state *back* to the 0 state. This leads to

> [t]he *forward convergent phylogeny problem (FCPP)*: Given a set of binary sequences, M, find a convergent phylogeny that derives M, using no back mutations, minimizing the total number of (forward) mutations.

In the forward convergent phylogeny model an edge in H_m can only be traversed in the forward direction (from the node labeled with one fewer 1s than the node at the other end of the edge). It follows that the *number* of edges on *every* path from the root to a given node v must be the *same*, and that number is the number of 1s in sequence S_v, the label of node v. This is illustrated in Figure 5.1. Hence, the only issue in solving the FCPP is how the chosen paths *overlap* and *share* edges.

The Key Point

A phylogeny with the all-zero root that derives a set of n sequences, M, can be thought of as the *superposition* of n paths from the root, with each path going to a different leaf node (labeled by a sequence in M). Since the length of any chosen path to a node v is fixed, the FCPP is solved by finding n paths in H_m from the root, that each go to a *distinct* leaf node using only forward edges, and that *overlap* to *minimize* the number of edges in the n paths. Note that even if an edge is in more than one of the n paths, it only contributes a count of one to the number of edges in the paths.

5.1.3 An ILP Solution of the Forward Convergent Phylogeny Problem

The characterization of a convergent phylogeny as the superposition of individual paths, so as to minimize the total *number of edges* used, leads almost directly to the ILP formulation for the FCPP. A solution to the ILP formulation will specify a set of edges, which contains, for each sequence t in M, a path from the root node to the leaf labeled t. To simplify the ILP development, we assume that the input, M, contains *no* duplicate sequences, and also does *not* contain the all-zero sequence. The first assumption certainly does not limit the generality of the solution, and neither does the second assumption, since the all zero-sequence always labels the root of the phylogeny.

The Variables Edges in H_m are defined as undirected, but it is convenient to think of an undirected edge (i,j) as two directed edges $< i,j >$ and $< j,i >$. When the label of node j has more 1s than the label of node i, then edge $< i,j >$ is called a *forward edge*, since its traversal represents a forward mutation. When the label of node i has more 1s than the label of node j, then $< i,j >$ is called a *backward edge*.

For each *forward* edge $< i,j >$ in the hypercube, we use one binary ILP variable, $F(i,j,t)$, intended to take on the value 1 if the directed edge $< i,j >$ is chosen to be on the path from the root, r, of H_m, to leaf t.

The Inequalities For each taxon t in M we have the following three types of inequalities:

$$\sum_{\substack{<r,j> \\ \text{a forward edge in } H_m}} F(r,j,t) = 1, \tag{5.1}$$

$$\sum_{\substack{<i,t> \\ \text{a forward edge in } H_m}} F(i,t,t) = 1, \tag{5.2}$$

and for every node v in H_m, other than r or t,

$$\sum_{\substack{<i,v> \\ \text{a forward edge in } H_m}} F(i,v,t), = \sum_{\substack{<v,j> \\ \text{a forward edge in } H_m}} F(v,j,t). \tag{5.3}$$

The inequalities in (5.1) state that for every taxon t, exactly one path *starts* at root r and goes to leaf t. Similarly, the inequalities in (5.2) state that exactly one path from r *ends* at leaf t. The inequalities in (5.3) state that for each t, and for each v *not* equal to r or t, exactly the same number of paths from r to t *enter* node v as *leave* node v. Together these inequalities ensure that for each leaf t in H_m, the F variables are set to specify exactly one path from the root of H_m to leaf t, so all those paths connect the root of H_m to every node labeled by a binary string in M.

Exercise 5.1.1 *Above, the inequalities in 5.3 were interpreted to say that for each leaf t, and for each node v not equal to r or t, exactly the same number of paths from r to t enter node v as leave v. However, because of other inequalities, in any feasible solution to the ILP formulation, the actual number of paths that both enter and leave v on their way to t, is either 0 or 1. Explain why this is true.*

Exercise 5.1.2 *We stated above that the inequalities (5.1), (5.2), and (5.3) ensure that: For each leaf t in H_m, the F variables are set so that there is exactly* one *path from the root of H_m to t. The correctness of this statement is a bit subtle. Write out, in your own words, why it is correct.*

Exercise 5.1.3 *What does the equality (5.3) specialize to when v is the all-one sequence, but is not t? Note that no path can traverse an edge out of v when v is the all-one sequence.*

Connecting the Leaves Inequalities (5.1), (5.2), and (5.3) set the F variables so that each of the n leaves in H_m will be connected (via a directed path from r) to root r. But, we need the chosen directed edges to form a *phylogeny*, so we need the directed edges in the n paths to specify a *tree* rooted at r, and we want that tree to have as few edges as possible. How do we achieve those two goals in an ILP formulation? It turns out that those goals are related, as we see next.

Minimizing the Number of Edges Used For each undirected edge (i,j) in H_m, we use the binary ILP variable, $X(i,j)$, to indicate whether (i,j) will be selected for the phylogeny. Variable $X(i,j)$ is intended to take on the value 1 if and only if edge $< i,j >$ is traversed on a path from the root to *some* (unspecified) leaf t. This is easily implemented with the following inequality for each i,j and t:

$$F(i,j,t) \leq X(i,j). \tag{5.4}$$

Then, the objective function for the ILP formulation is:

$$\text{Minimize} \sum_{\text{forward edge } <i,j> \text{ in } H_m} X(i,j). \tag{5.5}$$

Exercise 5.1.4 *We claimed that the overlapping paths specified in an optimal solution to the ILP formulation for the FCPP will specify a tree. But, perhaps two of the chosen paths (directed*

```
Minimize
+ X(000,001) + X(000,010) + X(000,100) + X(001,011)
+ X(001,101) + X(010,011) + X(010,110) + X(011,111)
+ X(100,101) + X(100,110) + X(101,111) + X(110,111)

subject to
.
.
.
+ F(000,001,4) + F(000,010,4) + F(000,100,4) = 1
+ F(000,001,4) - F(001,011,4) - F(001,101,4) = 0
+ F(000,010,4) - F(010,011,4) - F(010,110,4) = 0
+ F(001,011,4) + F(010,011,4) - F(011,111,4) = 0
+ F(000,100,4) - F(100,101,4) - F(100,110,4) = 0
+ F(001,101,4) + F(100,101,4) - F(101,111,4) = 0
+ F(010,110,4) + F(100,110,4) = 1
+ F(011,111,4) + F(101,111,4) + F(110,111,4) = 0
F(000,001,4) - X(000,001) <= 0
F(000,010,4) - X(000,010) <= 0
F(000,100,4) - X(000,100) <= 0
F(001,011,4) - X(001,011) <= 0
F(001,101,4) - X(001,101) <= 0
F(010,011,4) - X(010,011) <= 0
F(010,110,4) - X(010,110) <= 0
F(011,111,4) - X(011,111) <= 0
F(100,101,4) - X(100,101) <= 0
F(100,110,4) - X(100,110) <= 0
F(101,111,4) - X(101,111) <= 0
F(110,111,4) - X(110,111) <= 0
```

Figure 5.2 **A Portion of the Concrete, Gurobi-Formatted, ILP Formulation for the FCPP Instance Shown in Figure 5.1.**

from the root node) enter some node through different edges. In that case, the superposition of the paths will not specify a tree. Explain why this doesn't happen in an optimal *solution to the ILP formulation, and explain why this implies that an optimal solution for the formulation will necessarily specify a tree.*

5.1.3.1 A Concrete Example

Figure 5.1 shows an optimal solution to the FCPP for the sequences {001, 010, 011, 110}. The concrete ILP formulation for that problem instance has about 60 variables and 80 inequalities. In Figure 5.2, we show the objective function, and the inequalities for the fourth sequence, 110, in *M*. The inequalities for the other sequences in *M* are similar.

Exercise 5.1.5 *Explain why the inequalities in (5.4) set $X(i,j)$ to 1 if* any *of the n paths use the directed edge $< i,j >$. Do these inequalities allow $X(i,j)$ to be set to 1, even if $< i,j >$ is not part of any of the n paths? Hint: yes. Explain.*

Why does the ILP formulation not need inequalities to ensure that $X(i,j)$ is set to 1 only if edge $< i,j >$ is used on some r to t path?

5.1.4 The Full Maximum Parsimony Problem and Its ILP Solution

We now turn attention to the most *unrestricted* phylogenetic model. We are again given a set of n, m-length binary sequences, M, and a root sequence (assumed to be the all-zero sequence) and want to determine the minimum number of mutations needed to derive all of the sequences in M, starting from the root sequence. The history of the mutations must again be a rooted tree, but now there are *no* restrictions on the number of times a character can mutate from the 0 state to the 1 state, and *no* restrictions on the number of times a character can mutate from the 1 state to the 0 state.

A tree on which all sequences in M are derived, using the minimum number of mutations, is called a *maximum parsimony phylogeny* for M. The problem of finding a maximum-parsimony phylogeny for M is called the *maximum parsimony problem (MPP)*. In other contexts and applications, this problem is called *the Steiner-tree problem on hypercubes* [70]. We will later, in Chapters 9, 16, 17, and 19 see how other problems in computational biology are related to *Steiner-tree* problems, and how their ILP formulations are similar to those discussed here.

From the FCPP to the MPP An ILP formulation for the MPP is easily obtained by modifying the ILP formulation for the FCPP. All we need to do is allow any undirected edge in H_m to be traversed in *either* direction (but only one of the two). Again, a *forward* mutation on an edge is associated with the traversal of that edge from the node whose label has the least number of 1s to the node whose label has the most number of 1s. Traversing the edge in the opposite direction represents a *backward* mutation. Since two adjacent nodes, say i and j, in H_m have labels that differ at *exactly* one position, say k, the mutation represented on the edge (i,j) is for character k.

The conceptually easiest way to implement this modification is to again consider each undirected edge (i,j) in H_m as two directed edges, $< i,j >$ and $< j,i >$. Then, each of the inequalities (5.1), (5.2), (5.3), and (5.4) is changed to allow either a forward or a backward traversal, but only one of the two traversals is allowed. Also, to avoid confusion, the X variables (which were defined only for forward edges) will be replaced by Y variables, where $Y(i,j)$ will be set to 1 if and only if edge $< i,j >$ is traversed, regardless of whether it is a forward or a backward edge. Summarizing, the abstract ILP formulation for the *maximum parsimony* problem is shown in Figure 5.3.

Exercise 5.1.6 *Explain in detail why the ILP formulation for the MPP is correct. Note that the formulation does not have an inequality $F(i,j,t)+F(j,i,t) \leq 1$, for any pair of nodes (i,j), which would explicitly disallow the use of both directed edges $< i,j >$ and $< j,i >$. So be sure to explain how the formulation does disallow the use of any edge in both directions.*

$$\textit{Minimize} \sum_{\text{edge }(i,j)\text{ in } H_m} [Y(i,j) + Y(j,i)]. \tag{5.6}$$

Subject to
For each taxon t in M:

$$\sum_{\substack{<r,j> \\ \text{a directed edge out of} r}} F(r,j,t) = 1, \tag{5.7}$$

$$\sum_{\substack{<i,t> \\ \text{a directed edge into } t}} F(i,t,t) = 1. \tag{5.8}$$

For every node v in H_m, other than r or t:

$$\sum_{i \neq v} F(i,v,t) = \sum_{j \neq v} F(v,j,t). \tag{5.9}$$

For every undirected edge (i,j) in H_m:

$$Y(i,j) + Y(j,i) \leq 1. \tag{5.10}$$

For every undirected edge (i,j), and every leaf t in H_m:

$$\begin{aligned} F(i,j,t) &\leq Y(i,j), \\ F(j,i,t) &\leq Y(j,i). \end{aligned} \tag{5.11}$$

The X and Y variables are binary. The F variables are integer.

Figure 5.3 **The Abstract ILP Formulation for the MPP.**

5.1.5 Practicality of This ILP Approach

The abstract ILP formulation developed here is impressively practical for modest-sized data of current interest. For example, consider the following dataset with 12 taxa and 15 characters, shown in figure 5.4.

Solving the FCPP for this data by brute-force enumeration of potential solutions is completely impractical. But the ILP approach, despite creating a *huge* concrete formulation, *is* practical. The concrete ILP formulation for this data, based on the abstract formulation developed above, has over 3 *million* variables, and over 3 *million* inequalities. Yet, Gurobi 7.02 solved the formulation in only 12.47 seconds, although it took about 17 seconds to load the concrete formulation into Gurobi. Cplex 12.6 took about the same time as Gurobi, both in loading and solving. The optimal phylogeny (under the FC model) for this data has 16 edges. It is truly amazing that ILP solvers can handle such huge formulations in such little

```
000000100010110
000001000000000
100000000110110
100000010010110
100000010010111
100000010011110
100000011010110
100010010010110
100100000010110
100100010010110
101000010010110
110000010010111
```

Figure 5.4 A Modest-Size Dataset for the FCPP.

time. Further, in Section 5.2, we will see how to substantially reduce the running times.

Hitting the Wall The times mentioned above are impressive, but the limits of practicality of this abstract ILP formulation are quickly reached. The main difficulty with the formulation is that it requires $n+1$ variables for each edge of the hypercube, and the hypercube H_m has $(m-1)2^m$ edges when M has m characters. This exponential growth limits the practicality of the method to problem instances with a relatively small number of characters, say under 20. The number of taxa has less of an impact on the practicality of the formulation. Further, as m grows, the time to *create* a concrete ILP, the time to *read* it into the ILP solver, and the amount of *memory* it requires, can increase beyond the point of practicality.

In most of the ILP problems discussed in this book, we only consider the time to *solve* the ILP formulation, and not the other times or the amount of memory needed. That is because those requirements did not cause problems. But in this formulation of the FCPP, the number of variables grows so rapidly that these other requirements *can* cause problems. For example, in a typical case, when the number of characters increased from 15 to 18, and the number of taxa increased from 12 to 14, my program to generate the concrete ILP took over 2 minutes to execute; the number of inequalities in the ILP formulation increased to over *36 million*, with over *36 million* variables; Gurobi 7.02 took over 7 minutes to load the ILP, but, impressively, took less than 5 minutes to solve it. Cplex 12.6 took about half those times to load and to solve the ILP formulation. The optimal phylogeny in that case had 30 edges.

Interestingly, for instances of the FCPP, Cplex 12.6 has been consistently faster than Gurobi 7.02, both in loading the ILP formulation, and in solving it. This is in contrast to my experience with other ILP problems discussed in this book, and I have no idea what causes this deviation. But, it illustrates the suggestion made earlier, that if Gurobi is not solving fast enough, try Cplex also.

5.2 IMPROVING THE PRACTICALITY

In this section we discuss two ways to tackle problem instances of the MPP where the ILP formulation developed in Section 5.1.3 is not practical, due to an excessive number of ILP variables and inequalities. We will fully develop one method that is conceptually a small deviation from the first-developed method. It uses the same hypercube as before, but reduces the number of variables and inequalities, allowing a modest increase in the range of problem sizes that can be solved effectively. Then we will introduce a second idea, whose details are beyond the scope of this book, that allows *most* of the hypercube to be ignored, dramatically increasing the size of problem instances that can be solved.

5.2.1 A Smaller ILP Formulation for the MPP

Recall that we use n to denote the number of taxa (sequences) in M. Here we discuss an ILP formulation for the MPP where the number of variables and inequalities is reduced by a factor of n compared to the formulation developed in Section 5.1.3. This reduction in size leads to a significant reduction in the time needed to solve the formulations.

The Idea The general idea of the new abstract ILP formulation is similar to the original ILP formulation shown in Figure 5.3. As in that formulation, for a problem instance with m characters, the ILP solution will specify a subset of edges chosen from the hypercube H_m. The major change is that we will now *bundle* together all the n paths specified by a solution, rather than having variables and inequalities that apply to *each* path *separately*. As a result, for every directed edge $< i,j >$, the n binary variables $F(i,j,t)$ are replaced by a single variable $F(i,j)$ that can take on an *integer* value from 0 to n. The value of $F(i,j)$ specifies the *number* of paths from the root to leaves of H_m that traverse the directed edge $< i,j >$. We will continue to use two binary variables, $Y(i,j)$ and $Y(j,i)$, which specify whether any paths traverse the directed edges $< i,j >$ and $< j,i >$, respectively. Accordingly, the inequalities (5.7), (5.8), (5.9), (5.10), and (5.11) will be modified. The result is shown in Figure 5 5, and explained below.

Explaining the Changes The single inequality (5.13) replaces the n inequalities in (5.7). It says that n paths leave node r. We want each path to end at a distinct leaf, but unlike the n inequalities in (5.7), the destinations of the individual paths are not specified.

The change of the inequalities from (5.8) to those in (5.14) is a bit subtle. In the first formulation, where each of the n paths is treated separately, the specified path from r to a node t, ended at t. Node t could also have been on a path from r to a different node $t' \neq t$, but that would be reflected by the variable $F(r,t,t')$ being set to 1. In the new formulation, the n paths are bundled together and the inequalities in (5.14) reflect that bundling. The inequalities in (5.14) essentially say that for any sequence $t \in M$, the number of paths that enter node t must be one larger than the number of paths that leave node t, since t is a leaf, i.e., the final destination of exactly one of the n paths.

$$\textit{Minimize} \quad \sum_{\text{edge } (i,j) \text{ in } H_m} [Y(i,j) + Y(j,i)]. \tag{5.12}$$

Subject to
For each taxon t in M:

$$\sum_{<r,j> \text{ a directed edge}} F(r,j) = n. \tag{5.13}$$

For each taxon t in M:

$$\sum_{<i,v> \text{ a directed edge in } H_m} F(i,v) - \sum_{<v,j> \text{ a directed edge in } H_m} F(v,j) = 1 \tag{5.14}$$

For every node v in H_m that is *neither* the root nor a leaf of H_m:

$$\sum_{<i,t> \text{ a directed edge in } H_m} F(i,t) = \sum_{<t,j> \text{ a directed edge in } H_m} F(t,j) \tag{5.15}$$

For each undirected edge (i,j) in H_m:

$$F(i,j) \leq n \times Y(i,j),$$
$$F(j,i) \leq n \times Y(j,i), \tag{5.16}$$
$$Y(i,j) + Y(j,i) \leq 1.$$

Each Y variable is binary, and each F variable is integral.

Figure 5.5 The Abstract ILP Formulation for the MPP, Using Fewer Variables and Inequalities.

The inequalities in (5.15) replace those in (5.9), and say that for a node v that is neither the start or end of any of the n paths, the number of paths that enter v must be *equal* to the number of paths that leave v. The inequalities in (5.16) replace those in (5.11), in the obvious way. They say that if at least one path traverses the directed edge $< u,v >$ (or $< v,u >$), then $Y(u,v)$ (or $Y(v,u)$) must be set to 1, but the two variables cannot both be set to 1.

Exercise 5.2.1 *As a function of n, m and t, how many variables and inequalities are there in the two ILP formulations developed for the MPP?*

Exercise 5.2.2 *Show how to change this abstract ILP to solve the FCPP problem using fewer variables than before.*

5.2.2 The Practical Effect of the Smaller Formulation

The reduction in the number of variables and inequalities can have a significant effect on the time used to solve concrete ILP formulations. Recall the example in Figure 5.4, where the ILP formulation developed in Section 5.1.3 used over *3 million* variables and *3 million* inequalities, and took 17 seconds to load into Gurobi, and 12.47 seconds to solve. The new, smaller formulation has about 300,00 inequalities and close to 500,000 variables. It loaded into Gurobi 7.02 in 2.11 seconds, and was solved in 1.31 seconds.

Recall also the result discussed earlier, when the number of taxa increased to 14 and the number of characters increased to 18. The first ILP formulation for that problem instance had over 36 *million* inequalities and variables, and took about seven minutes just to load into Gurobi (running on my Apple Macbook Pro). The new, smaller formulation has under *3 million* inequalities and under *5 million* variables. It loaded in 27 seconds and solved in 83 seconds, using Gurobi 7.02 (Gurobi 7.5 solved in 60 seconds; Gurobi 8 loaded in 16 seconds, and solved in 55 seconds.). Cplex 12.6 took 17 seconds to load and 28 seconds to solve.

In another experiment, I tried a problem instance with 18 taxa and 20 characters. The first ILP formulation created 199,229,440 variables and 207,618,048 inequalities, and I terminated an attempt to load it into Gurobi after 2 hours. The smaller formulation created 20,971,520 variables and 12,582,910 inequalities. It loaded into Gurobi 7.02 in 79 seconds and solved in 132 seconds. Finally, I tried the smaller ILP formulation on a problem instance with 22 taxa and 20 characters. The hypercube in this case has over 1 million nodes. Gurobi 7.5 took almost 2 minutes to load, but solved the ILP in under 3 minutes. Cplex times were about half of the times for Gurobi. The maximum parsimony solution had 36 edges. A problem instance of that size could not be handled by the first ILP formulation.

Beauty Or Strength? The experience here, comparing the practicality of the two different ILP formulations (which are both correct), adds a bit of data to the general question of what makes an ILP formulation solve quickly. Twenty-five years ago the standard rule-of-thumb was "small is beautiful." That is, a smaller ILP formulation was expected to solve faster than a larger one, for the same problem instances. Our experience with the two formulations for the FCPP is consistent with this rule. However, many exceptions to that expectation have been observed,[1] and today the standard advice is a bit less applicable: the "stronger" formulation is best. However, the strength of a formulation is a somewhat technical concept having to do with the difference between the solution given by the ILP formulation, and the solution given by the same formulation when the variables are allowed to have fractional values. This is less applicable, because it is hard to determine the strength of an ILP formulation before solving it.[2]

[1] In fact, in [186], the authors discuss an ILP formulation that is provably smaller than the one developed in Section 5.1.3, but observe that it solves very poorly, taking much more time to solve than the larger formulation.

[2] More formally, the strength of a formulation is related to the difference between the LP and ILP solution spaces of the formulation. The more they differ, the weaker is the formulation.

5.3 SOFTWARE

The abstract ILP formulation for the MPP, shown in Figure 5.5 has been implemented in the Perl program *Aparsimony.pl*, which can be downloaded from the book website. That program takes a file containing sequences, and constructs the concrete ILP formulation for the MPP, with the input sequences. Call the program on a command line, in a terminal window, as:

```
Perl Aparsimony.pl sequence-file sequence-length
```

Example data appears in file: *bookpars*. File *bookpars* has 12 sequences, each of length 15.

5.3.1 Dramatic Speedups for Real Biological Data

In [186], the authors discuss several methods to reduce the time and memory needed for the ILP formulation developed in Section 5.1.3. The most important improvement is based on a surprising theorem in [14] (see also [174] for an exposition), showing that instead of using the full hypercube H_m for an m-character problem instance, M, we can use a *subgraph* of H_m. The subgraph is called the *Buneman* graph for M. We won't define the Buneman graph here, but only say that although it can (in worst case) be as large as H_m, in the real sequence data examined in [186], it is *dramatically* smaller. Further, given M, the Buneman graph can be constructed efficiently, in time *proportional* to the size of the Buneman graph. Hence, in the experiments done in [186], the memory and time reductions were dramatic, allowing a much larger range of practicality. For example, in sequence data from the Human Y chromosome with 16 sites (characters), the Buneman graph had only 510 nodes and 697 edges, instead of the roughly 32,000 nodes and a half-million edges in the hypercube H_{15}. This kind of dramatic memory reduction allowed the practical solution in [186] to problem instances with up to 98 sites. The Buneman graph is that case had only 780 nodes and 995 edges.[3]

Other speedups come from simple practice of good "data hygiene." That is, understanding when the same optimal solution can be found after reducing the size of the input. For example, in the maximum parsimony problem duplicate rows and columns can be removed from the input matrix M. Some cleanups can have a large effect on the work needed from the ILP solver. Purely mathematical redundancies might be caught and eliminated by the Gurobi preprocessor, but other kinds of cleanups need to be understood at the *logical* and *design* level by the creators of the abstract ILP formulations.

5.4 CONCERTED CONVERGENT EVOLUTION: CLIQUES AGAIN!

As we have seen, in convergent evolution, distantly related species have undergone the same mutations, not due to shared history, but due to similar evolutionary pressures. *Concerted* evolution is where *groups* of evolutionary characters (e.g. morphological or molecular characters) mutate *together*, perhaps because of some shared

[3] Note that 2^{98}, the number of nodes in H_{98}, is vastly larger than the estimated number of particles in the universe, making the use of the full hypercube impossible.

role in a function that is under evolutionary pressure. Concerted, convergent evolution is where both convergent and concerted evolution occurs. Needless to say, in such cases, deducing the correct evolutionary history is difficult.

In studying the evolution of seabirds, whose evolutionary history was contraversial, the authors of [93] conclude that these birds

> ... have been subject to similar evolutionary pressures and that their concerted response to this selection underlies many of the taxonomic problems that have bedeviled our understanding of the evolution of this group of birds.

The key tool in [93] used to clarify the role of concerted convergence is multiple *clique finding* in a graph where each node represents a character, and two nodes are connected by an edge if the characters are *compatable*. The biological basis for wanting to find cliques is that

> cliques of characters are the result of similar selective pressures and are a signature of concerted convergence. [93]

This short discussion is meant to introduce another approach to handling convergent evolution, and to again illustrate the importance of *cliques* and clique finding, in biological networks. We may never escape from cliques.

5.5 CATCHING INFEASIBILITY ERRORS USING AN IIS IN GUROBI

In discussing the FCP and the MP problems, we assumed that the root r is the *all-zero* sequence, and also that the root sequence is *not* in the input, M. Before making that assumption, I wrote a computer program (Task C discussed in the Introduction) that takes in input, M, and creates the concrete ILP formulation to solve the FCPP for M. At that point, I *did* allow the root sequence to be in M, and tried the input shown in Figure 5.1, along with the sequence 000. In detail, M was

```
000
001
010
011
110
```

But when I used Gurobi to try to solve the ILP formulation, it declared that the ILP formulation is *infeasible*, meaning that not all of the inequalities could be satisfied, no matter what permitted values were assigned to the variables. The concrete ILP formulation had *no* feasible solution. Hence, there was some internal inconsistency in the concrete ILP formulation, caused either by an error in the logic of my abstract ILP formulation, or in the computer program generating the concrete ILP formulation. I starred at both, but didn't see an error. How could I find it?

Gurobi has a very useful program, named *lp.py*, to help locate the cause of infeasibility in a concrete ILP formulation. The file containing the above-described ILP formulation was named *FCP.lp*. The following command, issued on a command line in the same directory where you run Gurobi,

```
python lp.py FCP.lp
```

executes a Python program that creates an *irreducible inconsistent subsystem (IIS)* from *FCP.lp*, and puts the IIS in a file called *model.ilp*.

An IIS is a subset of the inequalities in the ILP formulation, *FCP.lp*, such that the subset is infeasible, but the removal of *any* inequality in the IIS results in a set of inequalities that *is* feasible. So, the IIS homes in on a part of the whole ILP formulation that is causing trouble. Since the IIS is usually small, and every inequality in the IIS contributes to creating infeasibility, one can now much more easily see and fix the problem.

Of course, it is possible that the whole ILP formulation has more than one problem, and if, after fixing the first identified problem, the ILP is still infeasible, you can run *lp.py* again on the modified ILP formulation to find the next problem etc. But, I have never had to find more than one IIS, in order to find the cause of an infeasible formulation.

Specifically, when I ran *lp.py* on my infeasible ILP in the file *FCP.lp*, the program returned the following IIS, in the file called *model.ilp*:

```
Minimize

Subject To
 R0: F(000,001,1) + F(000,010,1) + F(000,100,1) = 1
 R1: F(000,001,1) - F(001,011,1) - F(001,101,1) = 0
 R2: F(000,010,1) - F(010,011,1) - F(010,110,1) = 0
 R3: F(001,011,1) + F(010,011,1) - F(011,111,1) = 0
 R4: F(000,100,1) - F(100,101,1) - F(100,110,1) = 0
 R5: F(001,101,1) + F(100,101,1) - F(101,111,1) = 0
 R6: F(010,110,1) + F(100,110,1) - F(110,111,1) = 0
 R7: F(011,111,1) + F(101,111,1) + F(110,111,1) = 0
Binaries
 F(000,001,1) F(000,010,1) F(000,100,1) F(001,011,1) F(001,101,1)
 F(010,011,1) F(010,110,1) F(011,111,1) F(100,101,1) F(100,110,1)
 F(101,111,1) F(110,111,1)
end
```

The labels R0, R1, etc. are line labels, created by Gurobi. Inequality R0 is the inequality derived from (5.1), specifying that there should be one path starting at the root, and ending at leaf 1. Such a path must start with one of the edges $(000,001)$, $(000,010)$ or $(000,100)$. Inequalities R1 through R6 are derived from (5.3), and specify that for every node v in H_3 from 001 through 110, the number of paths from r to leaf 1 that enter node v should be equal to the number of paths from r to leaf 1 that exit node v. Inequality R7 is also derived from (5.3), but it is a bit different. It says that *no* path from r to leaf 1 can enter node 111, which is correct (see Exercise 5.1.3) since a path entering 111 has nowhere to go, and leaf 1 is not labeled by 111.

Looking that these inequalities makes the error in the concrete ILP fairly simple to spot. Inequality R7 forces each of the variables $F(011,111,1)$, $F(101,111,1)$, and $F(110,111,1)$ to have value 0. Working backward from R7 to R1, we see that each of the negated variables in a line is forced by a line below it to have value 0, which in turn forces each of the nonnegated variables on that line to have value 0. Finally, from lines R7 to R1, variables $F(000,001,1)$, $F(000,010,1)$ and $F(000,100,1)$ are forced to have value 0, which means that line R0 states that $0 = 1$, which is impossible. The error is found.[4]

So, the IIS helped identify the infeasibility in the concrete formulation. And, once I saw the source of infeasibility, I realized that the logic behind the abstract ILP formulation for the FCPP (encapsulated in the inequalities in (5.1) through (5.4) and the objective function) was not quite right in the case that the root sequence is also one of the input sequences. Fixing the logic to allow the root sequence to be in M would lead to a somewhat ungainly formulation, so instead I simply required that the root sequence must *never* be in M. A similar infeasibility occurs if there are duplicate sequences in M, and this is again fixed by not allowing duplicates. As noted before, these requirements do not limit the range of FCP problems that can be solved by the ILP formulation.

Exercise 5.5.1 *Explain in detail why the logic behind the abstract ILP formulation for the FCPP (encapsulated in the inequalities in* (5.1) *through* (5.4), *and the objective function) is not quite right for the case that the root sequence is also one of the input sequences.*

Explain the similar problem if duplicate sequences occur in M.

[4] Cplex also helps find the source of infeasibility, with or without the use of an IIS. After finding the error, I tried Cplex on *FCP.lp*, just to see what Cplex would do. It also reported that the ILP formulation was infeasible, and stated that inequality R0 implies that "0 = 1." That might have been enough of a hint to let me find the error without an creating an IIS.

6

The RNA-Folding Problem

In this chapter we will develop an ILP formulation for a significant, but initially simplified, biological problem: predicting the secondary structure of an *RNA* molecule. Then we will extend the biological model and the ILP formulation to incorporate more realistic biological features of the problem.

The **RNA-folding problem** is to predict the *secondary structure* of an RNA molecule, given only its nucleotide sequence. This important, classic problem in computational biology is often solved with variants of *dynamic programming* that have been highly refined and engineered in several widely used computer programs. But here, we show how *integer linear programming* can be used to obtain the same results, with much less effort on the part of the developer or programmer, and can also be extended to model *more complex* versions of the folding problem, in ways that are difficult to model with dynamic programming.

6.1 A CRUDE FIRST MODEL OF RNA FOLDING

We start with an *extremely crude* first version of a classic RNA-folding problem.

We let s denote a string of n characters made up of the RNA alphabet $\{A, C, U, G\}$. For example, $s = ACGUGCCACGAU$. We define a *pairing* as set of *disjoint* pairs of characters in s. The word "disjoint" here means that a character cannot be in more than one pair. Note that some characters might not be in any pair in a pairing. A pair that is in a pairing is said to be a *matched pair*. A pair is called *complementary* if the two characters in the pair are *{A,U}* or *{C,G}*. A pairing is called *complementary* if all of the matched pairs in the pairing are complementary.

Folds, Secondary Structure, Pairing In our simple first model, the terms *"folding"* and *"secondary structure"* of an RNA molecule have the same meaning. And, the secondary structure of an RNA molecule is *specified* by a *pairing* in its nucleotide sequence s. In this first model of RNA folding, we require that any pairing be complementary. This constraint arises because in RNA, characters A and U represent *complementary* nucleotides that can *bond*, as can characters C and G.

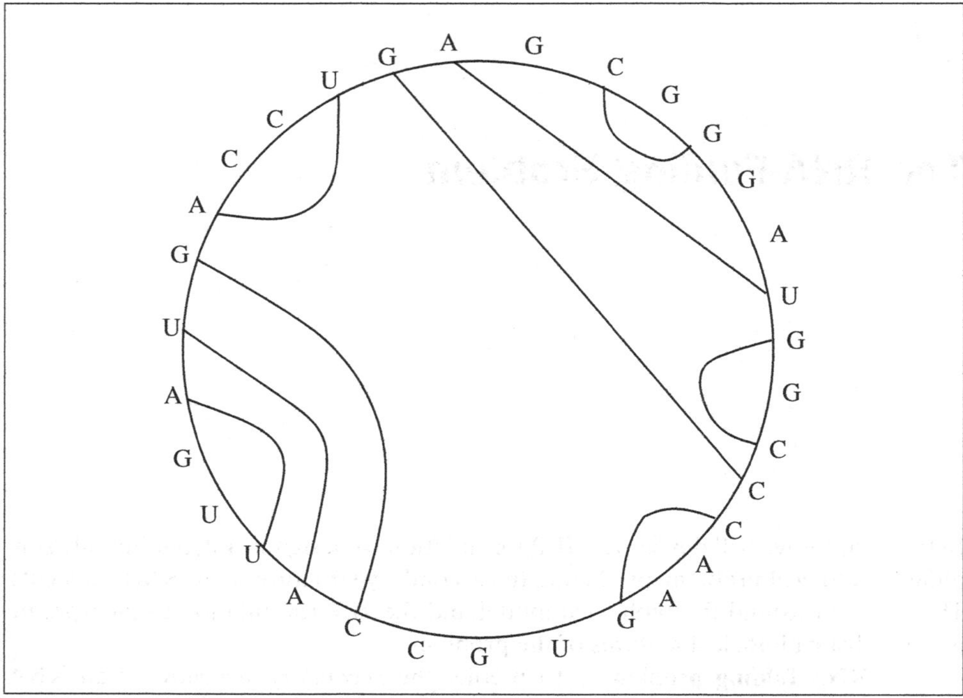

Figure 6.1 A Sequence s = *GCCAUUGAUGACCUGAGCGGGAUGGCCCAAGU* Drawn on a Circle. The sequence begins with the "G" at the bottom of the circle, and is read clockwise from there. The lines show a nested pairing, not necessarily of maximum cardinality. The sequence s is not based on a real RNA molecule, and so neither is the displayed nested pairing. Each line inside the circle represents a complementary pair in the pairing, and since it is a nested pairing, no lines cross.

Nested Pairing Displayed on a Circle If we draw the RNA string s as a *circular* string, we define a *nested pairing* (alternately called a *non-crossing* pairing) as a complementary pairing where each pair in the pairing is connected by a straight line *inside* the circle, and where *none* of the lines cross each other (see Figure 6.1).

Nested Pairing Displayed on a Line An alternative way to view a nested pairing is with the string s written, as normal, on a *straight* (horizontal) line, rather than a circle. As before, we draw a line between the two characters in each matched pair. But now the lines drawn between the two characters in a matched pair are *curved*. Again, no pair of lines are allowed to cross. A nested pairing can always be displayed by such non-crossing drawings, both with the strings on a circle, and with the strings on a straight line. Figure 6.2 shows a line representation of a nested pairing, and the corresponding RNA secondary structure.

Fold Stability From biological and chemical considerations, it is generally asserted that *as a first approximation*, the secondary structure, or the fold, of an RNA molecule is specified by a nested pairing that is the most *stable*, (more precisely, the one with *minimum free energy*).[1] We will sharpen our definition of stability

[1] So, we use the words "fold" and "pairing" interchangeably.

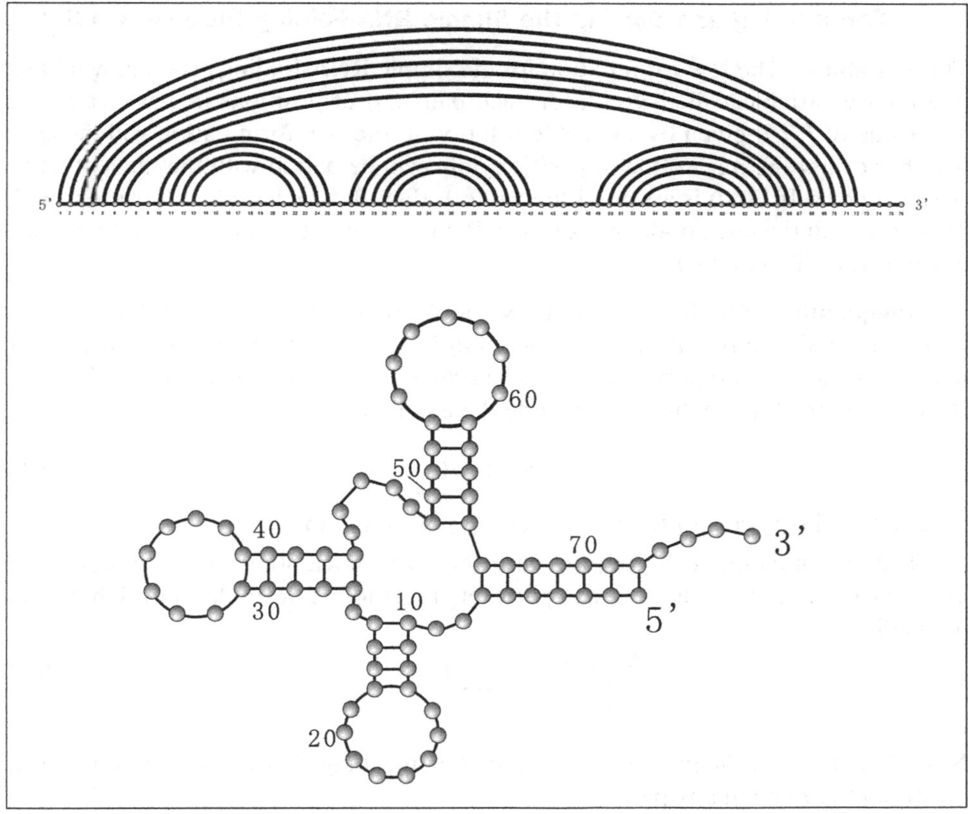

Figure 6.2 (Top) A Line Representation of a Nested Pairing. (Bottom) The Secondary Structure corresponding to the Nested Pairing. The figure is from [73].

later, but in our first simple model, the stability of a nested pairing is measured by the *number* of matched pairs it has. So, the most stable nested pairing is the one with the *largest* number of matched pairs. This leads to the following computational problem:

> **The simple RNA-folding problem** Given the nucleotide sequence s of a RNA molecule, find a *nested pairing* that pairs the *maximum* number of nucleotides, compared to any other nested pairing.

Later, we will add additional conditions that make the computational problem more accurately model the chemistry and biology of RNA folding. But first we show how the simple RNA-folding problem can be formulated and solved using integer linear programming.[2]

[2] We should point out that there is an extremely well-known dynamic programming method to solve the simple RNA-folding problem (for example, see [83]). In fact, the dynamic programming approach is very likely to be more efficient than the ILP solution, for the simple RNA-folding problem. However, as we add more conditions on the folding problem to make it more biologically meaningful, the dynamic programming methods become slower and only work for constrained versions of the problems. In contrast, the ILP approach is more versatile, allowing application on a wider range of RNA models and problems.

6.1.1 Formulating and Solving the Simple RNA-Folding Problem via ILP

The Variables The ILP formulation for the simple RNA-folding problem will have one binary variable, called $P(i,j)$, for each pair (i,j) of positions in s, where $i < j$. The value of $P(i,j)$ given by a feasible solution to the ILP formulation will indicate whether or not the nucleotide in position i of s will be paired with the nucleotide in position j of s. If $P(i,j)$ is assigned the value 1, then the nucleotide in position i *will* be paired with the nucleotide in position j. If it is assigned the value 0, then those two nucleotides will *not* be paired.

The Inequalities The first constraint we implement in the ILP formulation is the requirement that any pairing must be *complementary*. So, if the ordered pair of nucleotides in any two positions, i and j, are *not* one of (A,U) or (U,A) or (C,G) or (G,C), then the formulation will include the equality:

$$P(i,j) = 0, \tag{6.1}$$

to disallow the pairing of the nucleotides in positions i and j.

Next, we implement the requirement that each nucleotide can be paired to *at most* one other nucleotide. For each position j in s, the ILP formulation will have the inequality:

$$\sum_{k<j} P(k,j) + \sum_{k>j} P(j,k) \leq 1. \tag{6.2}$$

Note that this is an *inequality*, not an *equality*, meaning that it is permissible for a position j to *not* be in any pair.

Next, we need to implement the requirement that the pairing be *nested*, i.e., *non-crossing*. The key is to note that a pairing that is *not* nested, must contain some matched pairs (i,j) and (i',j') where $i < i' < j < j'$. So, if we disallow any such pairs, the resulting pairing will be non-crossing. We can achieve this as follows:

For every choice of four positions $i < i' < j < j'$, the ILP formulation will include the inequality:

$$P(i,j) + P(i',j') \leq 1, \tag{6.3}$$

which ensures that no pairing includes both matched pairs (i,j) and (i',j'). Remember that the variables $P(i,j)$ and $P(i',j')$ are both *binary*, and so can only be assigned value zero or one. Hence, if one of the variables is assigned the value one, the other must be assigned the value zero. So, the inequality in (6.3) enforces the requirement that the pairing be nested.[3]

The Objective Function Finally, the objective function for the ILP is:

$$\texttt{Maximize} \sum_{i<j} P(i,j), \tag{6.4}$$

which says that we want to set as many P variables as possible to the value 1.

[3] We can think of the construct in inequality (6.3) as an ILP idiom: the *NAND* ("Not And") of two *variables*. This is a particularly simple idiom, because it only involves relations between variables, rather than between inequalities.

$$\texttt{Maximize} \sum_{i<j} P(i,j)$$

Such that
For every pair (i,j) of positions in s that do not have complementary characters:

$$P(i,j) = 0$$

For each position j in s

$$\sum_{k<j} P(k,j) + \sum_{k>j} P(j,k) \leq 1$$

For every choice of four positions $i < i' < j < j'$

$$P(i,j) + P(i',j') \leq 1$$

All variables are binary.

Figure 6.3 The Abstract ILP for the Simple RNA-Folding Problem for an RNA Sequence s.

In summary, the abstract ILP formulation for the simple RNA-folding problem is shown in Figure 6.3.

6.1.2 A Toy Example

Consider the short RNA sequence $s = ACUGU$. The following is the set of equalities that guarantee that the pairing is complementary:

```
P(1,2) = 0
P(1,4) = 0
P(2,3) = 0
P(2,5) = 0
P(3,4) = 0
P(3,5) = 0
P(4,5) = 0
```

For example, characters A, C in positions one and two of s respectively are not complementary nucleotides, so they should not be allowed to pair, and this is enforced in the ILP formulation by the equality $P(1,2) = 0$.

The inequalities, formatted for Gurobi, that guarantee that each character (position) is paired to at most one other character are:

```
P(1,2) + P(1,3) + P(1,4) + P(1,5) <= 1
P(1,2) + P(2,3) + P(2,4) + P(2,5) <= 1
P(1,3) + P(2,3) + P(3,4) + P(3,5) <= 1
```

```
P(1,4) + P(2,4) + P(3,4) + P(4,5) <= 1
P(1,5) + P(2,5) + P(3,5) + P(4,5) <= 1
```

The inequalities that guarantee that the pairings are nested are:

```
P(1,3) + P(2,4) <= 1
P(1,3) + P(2,5) <= 1
P(1,4) + P(2,5) <= 1
P(1,4) + P(3,5) <= 1
P(2,4) + P(3,5) <= 1
```

Finally, the objective function is:

```
Maximize
```

$$P(1,2)+P(1,3)+P(1,4)+P(1,5)+P(2,3)+P(2,4)+P(2,5)+P(3,4)+P(3,5)-P(4,5)$$

And, for submission to Gurobi, we have to list all of the variables as binary (but we omit that here). Let's say that this input is contained in the file called *rna1.lp*. Then the command (given when in the same directory as file rna1.lp):

```
gurobi_cl rna1.lp
```

tells Gurobi to read in *rna1.lp* and find values for the variables giving the optimal solution. Gurobi responds with

```
Gurobi Optimizer version 6.0.0 build v6.0.0rc2 (mac64)
Copyright (c) 2014, Gurobi Optimization, Inc.

Read LP format model from file rna1.lp
Reading time = 0.03 seconds
(null): 17 rows, 10 columns, 37 nonzeros
Optimize a model with 17 rows, 10 columns and 37 nonzeros
Coefficient statistics:
  Matrix range     [1e+00, 1e+00]
  Objective range  [1e+00, 1e+00]
  Bounds range     [1e+00, 1e+00]
  RHS range        [1e+00, 1e+00]
Found heuristic solution: objective 1
Presolve removed 17 rows and 10 columns
Presolve time: 0.01s

Explored 0 nodes (0 simplex iterations) in 0.04 seconds
Thread count was 1 (of 8 available processors)

Optimal solution found (tolerance 1.00e-04)
Best objective 2.000000000000e+00, best bound 2.000000000000e+00, gap 0.0%
```

Gurobi 6.0 found the optimal solution, with total value 2, in 0.04 seconds; spending 0.03 seconds to read in the file, and 0.01 seconds to "presolve" the problem, meaning that it simplified the inequalities by applying standard rules of substitution and linear algebra. To see the values that Gurobi set for the variables in the optimal solution, use the Gurobi command

```
gurobi_cl resultfile=rna1.sol rna1.lp
```

After Gurobi solves the problem instance, file *rna1.sol* contains:

```
# Objective value = 2
P(1,2) 0
P(1,3) 0
P(1,4) 0
P(1,5) 1
P(2,3) 0
P(2,4) 1
P(2,5) 0
P(3,4) 0
P(3,5) 0
P(4,5) 0
```

Exercise 6.1.1 *Type the inequalities into a file called "rna1.lp." Run Gurobi to verify the results. Then change the last U in s to a G and modify the concrete ILP formulation to reflect this change. Run Gurobi to verify the result.*

Exercise 6.1.2 ***Software*** *The Python program* first-rna.py *is available at the book website. It asks the user if they want to type in an RNA string, or read one from a file called "randomstring." Once an RNA string has been input (either by typing, or from a file), the program generates the ILP formulation to solve the simple RNA folding problem for that RNA string.*

Download the program and then run it, choosing the option of typing in the RNA string. Type in the toy string $s = ACUGU$. Then, issue the appropriate Gurobi command to solve the ILP formulation and save the solution to a file. Compare the result obtained with the result stated above. The objective value must be the same, but the selected pairs might be different.

Next, run first-rna.py *again, but this time type in the RNA string shown in Figure 6.1. The linear string is given in the caption. Then, run Gurobi to solve the ILP formulation. Is the nested pairing shown in Figure 6.1 optimal?*

Exercise 6.1.3 ***Software*** *To test the option in program* first-rna.py *to read in an RNA sequence from the file "randomstring," first run the Perl program* randomrna.pl, *available on the book website. That program will ask you to specify a length for the random RNA sequence, and then it will generate a random RNA sequence using realistic frequencies for each of the characters $\{A, U, C, G\}$. The generated string is put into the file "randomstring."*

Generate a random string of length 50 and use first-rna.py *to generate the ILP formulation. Then use Gurobi to solve the ILP formulation. Record the time used by* first-rna.py *to generate the ILP formulation, and the time used by Gurobi to solve the ILP formulation.*

Repeat the experiment with differing string lengths to see how the times and objective values vary with changing string lengths.

6.1.3 Simplifications of the ILP Formulation

The above ILP formulation is correct, and Gurobi will find values for the variables which optimally solve the instance of the simple RNA-folding problem. However, there are some obvious simplifications to the ILP formulation that are possible. The simplest is that we do not need to include a $P(i,j)$ variable for any pair (i,j) that is *not* complementary. The formulation above, for $s = ACUGU$, includes an equality $P(i,j) = 0$ for each of the (i,j) pairs $P(1,2), P(1,4), P(2,3), P(2,5), P(3,4), P(3,5)$, and $P(4,5)$ because these are the pairs that are not complementary. But the effect of each of these equalities is the same as if those seven variables were *removed* from the objective function and from any inequalities that they appear in. So, the computer program that creates the concrete ILP formulation, when given a concrete problem instance, can omit any $P(i,j)$ variable for a pair (i,j) that is not complementary. Then, the concrete formulation for the above example (formatted for Gurobi) would be:

```
Maximize
P(1,3) + P(1,5) + P(2,4)

subject to

 P(1,3) + P(1,5) <= 1
 P(2,4) <= 1
 P(1,3) <= 1
 P(2,4) <= 1
 P(1,5) <= 1
P(1,3) + P(2,4) <= 1
binary
P(1,3)
P(1,5)
P(2,4)
end
```

This formulation is more compact than the first one, but Gurobi will find the same optimal solution with either formulation (you should try out Gurobi with the more compact formulation), and there is no harm in using the first version. Gurobi starts with a *presolver* that uses substitution and linear algebra to simplify the input ILP formulation, creating a more compact equivalent formulation. In particular, the presolver will see the variables that are set to zero in the formulation, and remove them everywhere, as we have done above.

So, the question of whether we create a formulation with extraneous variables and inequalities, or create a more compact formulation that omits them, comes down to the question of how much work you want to do when writing the computer program that creates concrete ILP formulations. I usually opt for the least programming effort, as long as the extraneous parts of the formulation will be detected and removed by the Gurobi presolver.

6.1.4 It's Deja Vu All Over Again!

The simple RNA-folding problem can be viewed as a problem of finding a *maximum clique* in a graph.[4] We create a graph G with one node for each pair of positions, (i,j), in the RNA string, that have complementary nucleotides. That is, each pair that is allowed to be in a nested pairing. Then, two nodes in G are connected by an edge, if and only if, the two pairs of positions, corresponding to those two nodes, are non-crossing. Then, a clique in G specifies a nested pairing. Conversely, every nested pairing specifies a set of pairs, which must form a clique in G. Hence, the largest clique in G specifies a nested pairing of maximum size in G.

[4] Did you really think you were finished with cliques?

6.1.5 Simple Biological Enhancements

Minimum Distance Constraint More realistic RNA-folding models include the chemical constraint that a position (nucleotide) cannot pair with with any position that is "too close." This is due to the fact that the RNA molecule must have a *bend* or a *loop* between any two positions that are paired. A typical constraint is that a nucleotide may not pair with any nucleotide that is less than four positions away from it on the RNA sequence. In general, we will use *"minD"* to represent the *minimum* number of nucleotides *between* any paired positions.

It is easy to incorporate this additional constraint in the ILP for RNA folding. We simply change the definition of a *complementary pairing* to require that every pair (i, j) in the pairing be a complementary pair (as before), *and* that $j - i > minD$. For example, if $s = ACCAGAGCCU$ and $minD$ is set to two, the ILP would contain several additional equalities, such as $P(3,5) = 0$ (or, as above, we would omit $P(3,5)$ from the ILP formulation), even though the nucleotides, C and G, at positions 3 and 5 are complementary. In this example, when $minD$ is set to two, the optimal nested pairing only contains two matched pairs, but when $minD$ is set to zero (as it essentially was before we introduced distance constraints) a nested pairing containing three matched pairs is possible: (1,10), (3,5), (7,8).

Differential Binding Strengths Until now, the objective function only maximized the *number* of matched pairs in a nested pairing, meaning that there was no distinction between an $\{A, U\}$ pair and a $\{C, G\}$ pair. Each contributed a score of one to the value of the objective function. However, the binding strength of a $\{C, G\}$ pair is larger than the binding strength of an $\{A, U\}$ pair, since a $\{C, G\}$ pair has three hydrogen bonds, while an $\{A, U\}$ pair only has two bonds. So, to find the most *stable* nested pairing, meaning a nested pairing of maximum binding strength, we need to enhance the ILP formulation to reflect the differential binding strengths of $\{C, G\}$ pairs and $\{A, U\}$ pairs. The enhancement is easy to implement. We simply multiply each term $P(i, j)$ in the objective function by some weight greater than one (often, in the range 1.33 to 1.5), if the nucleotide pair in positions (i, j) is (C, G) or (G, C).

The use of weights in the objective function can change the optimal solution to the ILP formulation. For example, if $s = ACCUGAGCCU$ and $minD = 1$, and we don't distinguish between the binding strengths of the matched pairs, but only count the *number* of matched pairs in a nested pairing, then there is an optimal nested pairing with value three, that has two matched $\{A, U\}$ pairs and one matched $\{G, C\}$ pair. There is also a different optimal nested pairing that has one matched $\{A, U\}$ pair and two matched $\{C, G\}$ pairs. Gurobi, could report either of these as the optimal solution it finds. But, if we multiply each $P(i, j)$ in the objective function by 1.33 for each (C, G) or (G, C) pair at positions (i, j), the objective function becomes:[5]

$$P(1,4) + P(1,10) + 1.33P(2,5) + 1.33P(2,7) + 1.33P(3,5)$$

$$+1.33P(3,7) + P(4,6) + 1.33P(5,8) + 1.33P(5,9) + P(6,10) + 1.33P(7,9)$$

[5] Note that in Gurobi format, there is a space between the constant 1.33 and the variable it is multiplying. Without that space, the constant might be considered to be part of a variable name, causing errors.

and the optimal score is now 3.66, obtained from the second nested pairing, with one matched $\{A, U\}$ pair and two matched $\{G, C\}$ pairs. The first nested pairing will only have a score of 3.33.

Allowing Some Noncomplementary Matched Pairs In some models of RNA folding, certain *noncomplementary* pairs of characters are allowed to form a matching pair, as long as appropriate weights, or multipliers, are used in the objective function for each allowed pair. Or, we could allow some noncomplementary pairing as long as "most" of the matched pairs are complementary. The most commonly allowed noncomplementary pair is $\{G, U\}$.

Exercise 6.1.4 *Modify the ILP formulations for the simple RNA-folding problem, to allow the matching pair $\{G, U\}$, but only up to some fixed percent (say 3%) of the total number of pairs in the nested pairing.*

6.2 MORE COMPLEX BIOLOGICAL ENHANCEMENTS

With the model of RNA folding used above, we were able to develop fairly simple linear inequalities to encode the RNA-folding model. This was true even when we included distance constraints, differential binding strengths, and noncomplementary pairs. Next, we extend the RNA-folding model in a chemically important way that is more subtle to encode with linear inequalities. The ILP formulations we will develop give further examples of the use of *If-Then* and *Only-If* idioms for binary variables.

6.2.1 Base Stacking

As discussed earlier, the central assumption in RNA-folding prediction is that the most *stable* fold is the most *likely* fold. Paired complementary nucleotides ($\{A, U\}$ and $\{C, G\}$) contribute to the stability of a folded RNA molecule, and were the basis for the first ILP objective function (in 6.4). However, other features of a pairing contribute significantly to RNA stability. The most important is the feature of *base stacking*.

Stacked Pairs A matched pair (i,j) in a nested pairing is called a *stacked pair* if either $(i+1, j-1)$ or $(i-1, j+1)$ is also a matched pair in the nested pairing. A *stack* in a nested pairing consists of *two equal-length* intervals, each containing at least two *consecutive* positions in the RNA, such that the characters in the first interval are paired to the characters in the second interval. So, every matched pair in a stack is a stacked pair. If (i,j) and $(i+1, j-1)$ are stacked pairs, the four positions $(i < i+1 < j-1 < j)$ is called a *stacked quartet*. In a stack, the number of stacked quartets is one less than the number of stacked pairs.

For example, the nested pairing in Figure 6.1 has one stack with three stacked pairs that contains two stacked quartets. One of the two intervals in the stack contains the substring *CAU*, and the other interval contains the substring *AUG*. As another example, the nested pairing in Figure 6.2 has four stacks.

Stacks are important in RNA folding generally, but are particularly evident in the folding of *transfer RNA (tRNA)*, because tRNA molecules have a distinctive secondary structure called a *cloverleaf*. In a cloverleaf, the RNA string is divided into alternating *stems* and *loops* (see Figure 6.2), where each stem consists of a stack. Stacks and stems are also referred to as *helixes*, since the nucleotides in a stack are known to form a helix structure in three dimensions.

Stacks and Stability It is has been shown by laboratory experiments that a stack with t matched pairs contributes more to stability than t individual (isolated) matched pairs that are not in any stacks. So, stacking contributes significantly to the stability of an RNA fold. It follows that a more realistic model of RNA folding, based on maximum stability, and a more realistic ILP formulation for RNA folding, should *encourage* paired nucleotides to be organized into (long-ish) stacks as much as possible.

6.2.2 How to Model Stacking in the ILP Formulation

As a simple first step, we will extend the objective function of the ILP by including a count of the *number* of stacked *quartets* in the nested pairing. Hence, we need the ILP formulation to calculate that number, based on the values given to the pairing variables $P(i,j)$.

We create the binary ILP variable $Q(i,j)$ to indicate whether the pair (i,j) is the *first* pair in a stacked quartet, i.e., whether (i,j) and $(i+1,j-1)$ are both in the nested pairing given by the ILP solution. This is accomplished with the following inequalities for each pair (i,j), where $j > i$:

$$P(i,j) + P(i+1,j-1) - Q(i,j) \leq 1, \tag{6.5}$$

$$2Q(i,j) - P(i,j) - P(i+1,j-1) \leq 0. \tag{6.6}$$

The first inequality, (6.5), enforces the condition that *if both* (i,j) and $(i+1,j-1)$ are in the nested pairing (i.e., $P(i,j) = P(i+1,j-1) = 1$), *then* the value of variable $Q(i,j)$ *must* be set to 1. To see this, note that when $P(i,j) = P(i+1,j-1) = 1$, $P(i,j) + P(i+1,j-1) = 2$, so the inequality can be satisfied only when $Q(i,j)$ is set to 1. The inequality in (6.5) is an instance of the simple *If-Then* idiom for binary variables.

The second inequality, (6.6), enforces the converse condition, that *if* $Q(i,j)$ is set to 1, *then both* $P(i,j)$ and $P(i+1,j-1)$ *must* be set to 1. Another way to say this is that $Q(i,j)$ can be set to 1 *only if* $P(i,j)$ and $P(i+1,j-1)$ are both set to 1. To see this, note that when $Q(i,j) = 1, 2Q(i,j) = 2$, so the inequality can only be satisfied if both $P(i,j)$ and $P(i+1,j-1)$ are set to 1. The inequality in (6.6) is an instance of the simple *Only-If* idiom for binary variables. Summarizing, inequalities (6.5) and (6.6) together ensure that $Q(i,j)$ will be set to 1, *if and only if* both (i,j) and $(i+1,j-1)$ are in the nested pairing, so $(i,i+1,j-1,j)$ is a stacked *quartet*, and i is the smallest position in the quartet.

The Objective Function To incorporate a count of the number of stacked quartets in the fold, we change the objective function from:

$$\texttt{Maximize} \sum_{i<j} P(i,j), \tag{6.7}$$

to:

$$\texttt{Maximize} \sum_{i<j} [P(i,j) + Q(i,j)]. \tag{6.8}$$

Note that by using the objective function in (6.8), the number of paired nucleotides might decrease in the optimal RNA fold, compared to the optimal RNA fold when using the objective function in (6.7), but the number of stacked quartets will increase. We can also optimize with the objective function

$$\texttt{Maximize} \sum_{i<j} Q(i,j), \tag{6.9}$$

which is consistent with the objectives used by the RNA-fold predictors that are most widely used, and are considered the most accurate.

What a Difference a P Makes What is fascinating is that these three objectives functions solve at *very* different speeds, and yet produce folds that are *not* dramatically different. For example, we implemented the above abstract ILP formulation and tested it on the RNA sequence:

GUGUCGUUAUUUGUGCAACAACAUUGAAAAACAAGCAGAAAUAAAACAUU
GACGGAACGGAUACGACCGCAGAUUGCGGGAGCUGUCCUCUCAAUCAGCC

of length 100, with the minimum distance between paired nucleotides set at five. Using the objective function in (6.7), Gurobi 7.5 took about 13 minutes and produced a nested pairing with 33 pairs and 14 quartets. Then, with the objective function in (6.8), Gurobi took 51 seconds and produced a nested pairing with 32 pairs and 20 quartets. But, with the objective function in (6.9), Gurobi took only 2.16 seconds and produced a nested pairing with 29 pairs and 20 quartets. These differences in running times and folds are typical of the *many* other test cases we did.

Pushing Gurobi a bit, we also used the objective function in (6.9) on an RNA sequence of length 250. Gurobi 7.5 solved it in *1 hour and 5 minutes*. The ILP formulation had over *10 million* inequalities, but the Gurobi presolver reduced that to only 46,516 inequalities before beginning to solve the ILP. Illustrating how much faster folding ILPs solve with the objective function in (6.9) compared to the other two, we used the objective function in (6.8) on the same 250-length RNA sequence. We let it run for about 10 hours, but terminated it then, since it had made no progress for about 4 hours, stuck on a gap of 31.5%.

The empirical fact that the most biologically favored objective function (including only stacks) is also the one that leads to the fastest solutions (by far), is a rare and wonderful result. I had no *a priori* expectation of this – it illustrates the value of experimentation in addition to theory.

The Effect of MinD To test how the solution time is influenced by the minimum allowed distance between paired nucleotides, we used the first objective function on the RNA sequence:

GUGUCGUUAUUUGUGCAACAACAUUGAAAAACAAGCAGAAAUAAAACAUUGACGGAACGG
AUACGACCGCAGAUUGCGGGAGCUGUCCUCUCAAUCAGCCCGGCAGCGGCGGCCCACAAU

of length 120, with the minimum distance between paired nucleotides first set at three. Gurobi 7.5 took about 25 minutes (on my Macbook Pro laptop), and produced a nested pairing with 40 pairs, and 25 stacked quartets. When the minimum distance between paired nucleotides was increased to six, the optimal solution had 37 nested pairs, and 23 stacked quartets, and the Gurobi running time decreased to just under 6 minutes.

Exercise 6.2.1 *The Python program* fourth-rnaf.py, *available on the book website, reads an RNA sequence from a file, and produces an ILP formulation that solves the RNA folding problem, taking note of both the number of matched pairs, recorded in variable P, and the number of stacked quartets, recorded in variable Q, in the nested matching. The user is asked what objective function to use: P only; or Q only; or P + Q.*

Program fourth-rnaf.py *is called with three arguments. The first is the name of the file holding the RNA sequence; the next is the name of the file that will hold the ILP formulation; and the third argument is an integer, called MinD, specifying the minimum number of nucleotides needed between any matched pair in a nested pairing. If you set MinD to 0, then you are allowing adjacent nucleotides to pair, if they are complementary, as in program* first-rna.py. *To run the program in a terminal window, issue the command:*

```
python fourth-rna.py RNA-file ILP-file.lp min-distance
```

After that, you can use Gurobi to find the optimal solution to the ILP formulation in the .lp file.

Do nine experiments that vary the objective function in the three ways indicated above; and vary the choice of MinD as 0, 3, and 5. For each experiment, record the time Gurobi uses; the values of P and Q in the obtained solution; and the number of stacks in the obtained solution. To get the number of stacks, as opposed to the value of Q, you will have to scan through the result file to identify the stacks. The Python program cleanres.py, *available on the book website, can help you keep your sanity. Suppose you use the name "RNA.sol" as your result file. Most of the values in the result file will be 0, and are not useful. Then run:*

```
python cleanres.py RNA.sol
```

The program removes every line that reports a variable with value 0, and outputs the cleaned file in cRNA.sol. In general, when the argument is X.sol for some name X, the cleaned file is output to cX.sol, i.e., cleanres.py *puts a "c" in front of the name of the result file.*

What is your conclusion from running these experiments?

6.2.3 Weighting Stacked Quartets in a Nested Pairing

The next, and perhaps the most important, extension of the chemical model is to incorporate *weights* into the objective function for each stacked quartet in a stack. Each weight represents the contribution that a stacked quartet makes to the stability of the RNA fold. The simplest way to do this is to multiply each P term in the objective function by a constant number, and each Q term by a different constant number, to reflect the importance of a single pair in a nested pairing, compared to the importance of a stacked quartet. More generally, the objective function is given as:

$$\texttt{Maximize} \sum_{i=1}^{i=n} [W(i,j) \times Q(i,j)], \tag{6.10}$$

	A U	C G	G C	U A
A U	9	21	24	13
C G	22	33	34	24
G C	21	24	33	21
U A	11	21	22	9

Figure 6.4 Weights for the Stacked Quartets Consisting of Complementary Matched Pairs. Note the asymmetry in the entries, due to the fact that an RNA molecule has a chemically defined direction. For example, the sequence 5' AGGGCU 3' is chemically distinct from 5' UCGGGA 3'. If AG... CU forms a stacked quartet in the first sequence, it has weight 24, while the stacked quartet UC... GA in the second sequence has weight 21.

where $W(i,j)$ is a positive constant that depends on which four nucleotides are in the stacked quartet $(i, i+1, j, j-1)$.[6] Because of that dependence, the contribution of each single pair (i,j) and $(i+1, j-1)$ can be included in the term $W(i,j)$, and hence the $P(i,j)$ variables are usually omitted from the objective function.

Note that in a stack with more than two stacked pairs, two consecutive stacked quartets, $(i, i+1, j-1, j)$ and $(i+1, i+2, j-2, j-1)$, overlap, so the matched pair $(i+1, j-1)$ influences both $W(i,j)$ and $W(i+1, j-1)$. This is appropriate since both stacked quartets contribute to the stability of the fold.

Extensive chemical studies have been done to determine good weights for stacked quartets, based on the specific nucleotides that the stacked quartet contains. For example, Figure 6.4 shows the weights for stacked quartets used in a program called Fold-Align [100]. The first matched pair in a stacked quartet is written on the left of the table, and the second matched pair is written along the top of the table. For example, the stacked quartet

```
AG
UC
```

has weight 24.

[6] We emphasize that $W(i,j)$ is a constant value, *not* a variable. The specific value of $W(i,j)$ is inserted into the concrete ILP formulation for a problem instance, by the computer program that creates the concrete ILP formulation for the instance.

6.2.3.1 A Subtle, But Common, ILP Simplification

We claim that any *optimal* solution to the RNA-folding problem will be unchanged, if we remove inequality (6.5), for each i and $j > i$, from the ILP formulation developed above. Inequality (6.5) enforces the condition that if $P(i,j) = P(i+1,j-1) = 1$, then $Q(i,j)$ must be set to 1. But, $W(i,j) \times Q(i,j)$ is now a term in the objective function, $W(i,j)$ is a positive term, and the goal is to *maximize* the value of the objective function, so the *optimal* ILP will set $Q(i,j)$ equal to 1 *whenever possible*. It is inequality (6.6) that limits when $Q(i,j)$ can be set to 1, i.e., *only when* both $P(i,j)$ and $P(i+1,j-1)$ are set to one. Hence, if we remove inequality (6.5), but retain inequality (6.6) for each i and $j > i$, the resulting ILP formulation will have the same optimal solution value as before.

This change simplifies and shortens the ILP formulation, and sometimes a smaller ILP will solve faster than an equivalent larger one. However, sometimes the larger formulation solves faster (for reasons that I don't often understand), so, in such cases, I usually try out both formulations to see which is fastest.

6.2.4 More Elaborate Models of RNA Folding

We have discussed how to formulate the RNA-folding problem as an ILP problem, using the three most important features of RNA folds – complementary, nested pairing, and base stacking. Those three features, along with appropriate weights for matched pairs and stacked quartets, were adequate to explain both the central ideas in RNA-fold prediction, and the formulation of ILPs for RNA fold prediction. However, the literature on RNA fold prediction is vast, and many additional features and refinements of RNA folds have been incorporated into fold prediction methods. In one software package for RNA folding, there are close to 200 parameter choices that the user can specify to guide the folding program. Since the point of this chapter is to further explain ILP formulations and their utility in computational biology, and not RNA folding *per se*, we only mention here that the common chemical features studied in the literature on RNA-fold prediction can also be incorporated into an ILP formulation. The exercises below explore a few of these.

Exercise 6.2.2 *In some models of RNA folding, the weight given to stacked quartet depends both on the specific nucleotides in the quartet (as in the formulation already discussed), and on* where *the stacked quartet is in a stack. The main distinction is whether the stacked quartet is the first quartet, the last quartet, or a middle quartet in a stack.*[7] *Hence, we need to extend the ILP formulation to recognize where a stacked quartet appears in a stack. Let $F(i,j)$ be an ILP variable that will be set to 1 if and only if the stacked quartet $(i, i+1, j-1, j)$ is the* first *stacked quartet in a stack. Consider the inequalities:*

$$Q(i,j) - Q(i-1,j+1) - F(i,j) \leq 0$$

and

$$2F(i,j) - Q(i,j) + Q(i-1,j+1) \leq 1.$$

Do these inequalities properly implement the definition of variable $F(i,j)$? Do they have any bad side effects? Explain.

[7] The concept of "first" and "last" is meaningful since an RNA molecule has chemically distinct ends, labeled 3′ and 5′. The first quartet in the stack is the quartet closest to the 3′ end of the RNA molecule.

Exercise 6.2.3 *Show how to extend the abstract ILP formulation for RNA folding so that some variable will be set to the* number *of stacks in the fold found. Can you think of any biological use for that number, and if so, how it might be included in the ILP formulation to get more biologically valid solutions? For example, if two optimal ILP solutions have the same number of stacked quartets, but in a different number of stacks, is there any reason to prefer one solution over the other, and if so, which one?*

Exercise 6.2.4 *Let $L(i,j)$ be an ILP variable that will be set to 1 if and only if the stacked quartet $(i, i+1, j-1, j)$ is the last stacked quartet in a stack, i.e., closest to the 5′ end of the molecule. Develop and explain inequalities that implement the definition of variable $L(i,j)$. Explain why they have no bad side effects.*

Exercise 6.2.5 *Fully explain how to modify the objective function to weight a stacked quartet differently depending on the values of $Q(i,j)$, $F(i,j)$, and $L(i,j)$.*

Exercise 6.2.6 Allowing Some Crossed Pairs *Up until now, we have only discussed pairings that are* nested, non-crossing, *since they are thought to model the secondary structure of most tRNA molecules. However, a limited number of* crossing *matched pairs are sometimes observed in RNA secondary structure, particularly for RNA molecules that are* not *tRNA molecules.*

Suppose we now change the definition of a pairing to allow some *crossing matched pairs. The first step in changing the ILP formulation is to remove the inequalities defined in* (6.3). *Next we introduce additional variables. Let $C(i,i',j,j')$ be an ILP variable that is set to value 1 if and only if $i < i' < j < j'$, and (i,j) and (i',j') are matched pairs in the pairing. That is, $C(i,i',j,j')$ should be set to 1 if and only if $P(i,j)$ and $P(i',j')$ are both set to 1, so that the matched pairs (i,j) and (i',j') cross each other.*

Develop ILP inequalities to properly implement the definition of $C(i,i',j,j')$. Once the C variables are implemented, we can add an inequality that limits the number of crossing matched pairs to a fixed number, or to a fixed percentage *of the number of matched pairs, etc. The point of allowing a few crossing matched pairs, we might be able to increase the value of the objective function (based on pairings and stacks) substantially, and to better model true RNA folding.*

Fully explain these approaches, and suggest other ways to use the C variables (perhaps in the objective function) to model the phenomenon of limited non-nested pairing.

6.2.4.1 Pseudo-Knots

Other than the features already discussed, perhaps the most important feature of many RNA molecules is a *pseudo-knot*. Pseudo-knots generalize crossing pairs, discussed in the prior exercise. The focus of the prior exercise is on detecting and limiting the *number* of crossing matched pairs, *without* regard to how they were *organized* in the pairing. But, simply counting the *number* of crossing matched pairs is often *not* a good way to model RNA chemistry and folding. When two sets of crossing matched pairs are organized into two *stacks*, they form what is called a *pseudo-knot*, which is considered to be a *single* feature of the pairing (see Figure 6.5). But, how do we incorporate this feature into the ILP formulation? We will develop an ILP approach below. Other ILP approaches appear in [152, 169].

ILP for RNA Folding with Pseudo-Knots Given a fully specified RNA pairing, let $PS(i,i')$, where $i < i'$, be a binary ILP variable, which is set to value 1 if and only if there are two stacks, $S(i)$ and $S(i')$, starting at positions i and i' respectively, such that every pair in $S(i)$ crosses every pair in $S(i')$. That is, $PS(i,i')$ will be set to 1 if and only if there is a pseudo-knot in the fold, whose two stacks begin at positions i and i' respectively.

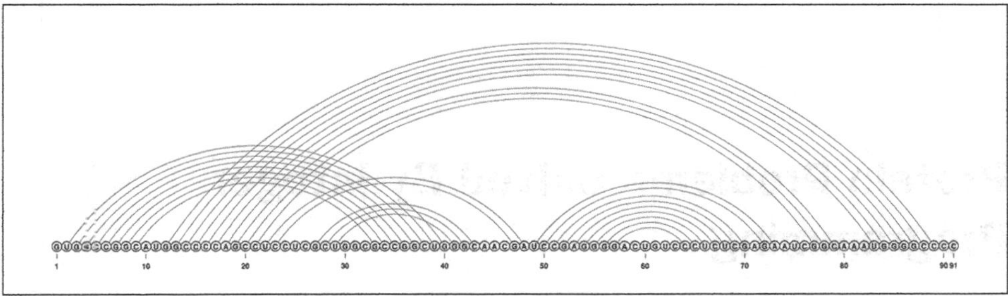

Figure 6.5 Predicted RNA Fold with Pseudo-Knots. This fold has seven stacks and three pseudo-knots. This figure is extracted from figure 7 in [99].

Exercise 6.2.7 Pseudo-Knots *Convince yourself that every pair in stack $S(i)$ will cross every pair in stack $S(i')$, if and only if the first pair in $S(i)$ crosses the first pair in $S(i')$. Are you convinced? Hint: Yes!*

Using this fact, develop and explain ILP inequalities that correctly implement the definition of the variables $PS(i, i')$, for every $i < i'$. Then, develop and explain inequalities that allow a limited number of pseudo-knots. The point again is that by allowing a few pseudo-knots, the number of pairs in the fold, and/or the number of stacked pairs in the fold, might increase, resulting in a more biologically valid RNA fold.

Exercise 6.2.8 *In an earlier draft of this section, I wrote that the RNA fold in Figure 6.5 has* six *stacks. But then I changed the number to* seven. *Which is correct? Why is it easy to get the count wrong in this figure?*

6.3 FOLD PREDICTION USING A KNOWN RNA STRUCTURE

RNA sequences (similar to the cases of DNA and amino acid sequences) can be grouped into *families* and *superfamilies* defined by similarity in sequence, structure in two or three dimensions, function, or inferred evolutionary history. Then, what is known about one *well-studied* RNA sequence in a family is often true for other *less-studied* RNA sequences in the family. Exploiting this fact allows greater leveraging of hard-to-obtain data and biological insight. This paradigm is probably the best explanation for the importance and success of computational biology and bioinformatics.

The general paradigm of using well-studied RNA to help in understanding less-studied RNA is illustrated in the use of *known* two-dimensional RNA structures to help determine the two-dimensional structure of lesser-studied RNA sequences. Similarly, researchers have used known *three-dimensional* structures in *proteins* to help deduce the three-dimensional structure in lesser-studied proteins. Integer linear programming has been a key element in some of those protein efforts. This will be discussed in Section 21.6. After understanding those integer programs for protein problems, the reader should be able to modify them for application to RNA structure prediction. So, we defer further discussion of RNA folding until Section 21.6.

A related problem of RNA-RNA alignment has also been addressed with the use of ILP. For example, see [102].

7

Protein Problems Solved By Integer Programming

In this chapter we examine three problems involving proteins and how they have been addressed using integer linear programming.

7.1 THE PROTEIN SIDE-CHAIN POSITIONING PROBLEM

Proteins are macromolecules that are the workhorses of cellular biology, performing a huge range of diverse and vital functions. Each protein consists of a chain of individual *amino acids*, and each amino acid consists of a *main chain* and a *side* chain. The main chains of the amino acids are identical, and (in common living organisms) each side chain is one of 20 possible molecules, called *residues*. It is the main chains of the amino acids that link together to form the *backbone* of a protein. Thus, we can consider a protein to be a chain of (identical) amino acid main chains, where each amino acid main chain has an attached side chain residue. When we speak of the amino acid *sequence* of a protein, we are referring to the sequence of its side chain residues.

The functions that a particular protein performs are largely determined by its *three-dimensional* structure, and the way that its structure changes in different biological and chemical environments. Thus, determining the three-dimensional structure(s) of proteins is a major focus of biochemistry and computational biology. The biochemical approach generally uses imaging, X-ray crystallography, or NMR images, for example, while the computational biology approach tries to determine the three-dimensional structure of a protein from its amino acid sequence, and any structural information that is already known about functionally related proteins, or proteins with similar sequences.

Side-Chain Positioning In the computational approach, a central *subproblem* is to determine the three-dimensional positions of the side chains of a protein, assuming one already knows the three dimensional positions of the protein's backbone. This is called the *side-chain positioning problem*. In this section, we explain an integer

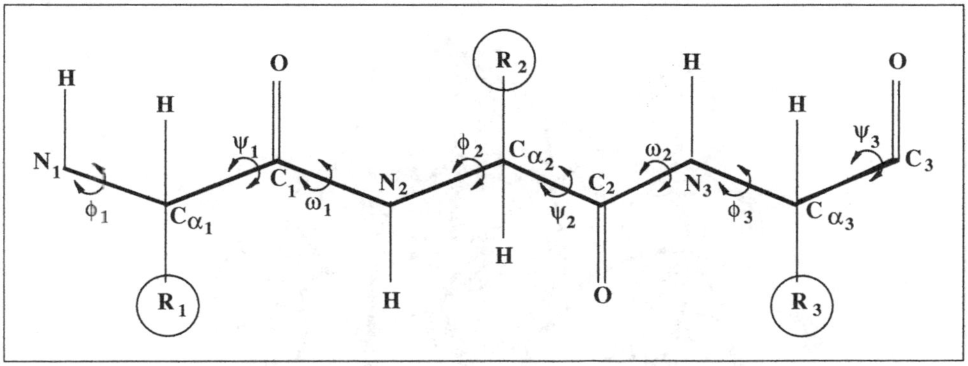

Figure 7.1 A Chain of Three Amino Acids. Reading left to right, the i'th amino acid main chain starts with the nitrogen atom, N_i, and ends with the carbon atom, C_i. Amino acids i and $i+1$ are attached by a peptide bond between atoms C_i and N_{i+1}. The i'th amino acid has a single side chain attached to the carbon atom denoted C_{α_i}. The side chain of the i'th amino acid contains a residue, denoted R_i, which is one of 20 standard amino acid residues. The figure does not specify which residues are represented by R_1, R_2, and R_3. The backbone of the amino acid chain is shown with thick lines. Points of rotation along the backbone are denoted ϕ_i, ψ_i, ω_i. Oxygen atoms are represented with an O, and hydrogen atoms are represented with an H. Each oxygen is attached by a double bond, and each hydrogen is attached by a single bond.

programming approach, developed in [106], for solving the side-chain positioning problem.[1]

7.1.1 The Biochemical Model in More Detail

We have assumed that the three-dimensional structure of a given protein *backbone* is already known, but it is helpful to understand a bit more about what determines that structure. In every amino acid main chain, there are three places where the chemical has some freedom to *rotate*, and the observed rotations at those places tend to occur in discrete amounts. The rotations in the main chain of an amino acid are at locations denoted by the Greek letters ϕ, ψ, and ω. There are three possible rotations at ϕ and ψ, while ω can only rotate in two ways. In a protein consisting of numerous amino acids (typically in the hundreds) the combined rotations of the amino acid main chains determine the overall three-dimensional structure of the protein backbone (see Figures 7.1 and 7.2).

Rotamers: Side-Chain Rotations In addition to the rotations possible in the main chain of an amino acid, the *side chain* of every amino acid other than glycine has some flexibility. Each of the 19 common residues that have flexibility, can rotate at a number of places on the side chain. That number varies from one (for *proline* for example) to four (for *arginine*). The points of rotation on a side chain are denoted χ_i, where i can range from 1 to 4 depending on the number of points of rotation. At each rotation point, the amount of rotation is chosen from a set of *three*

1 We deviate from the exposition in [106], by framing a *maximization* problem rather than a *minimization* problem, and we use somewhat different inequalities in the ILP formulation.

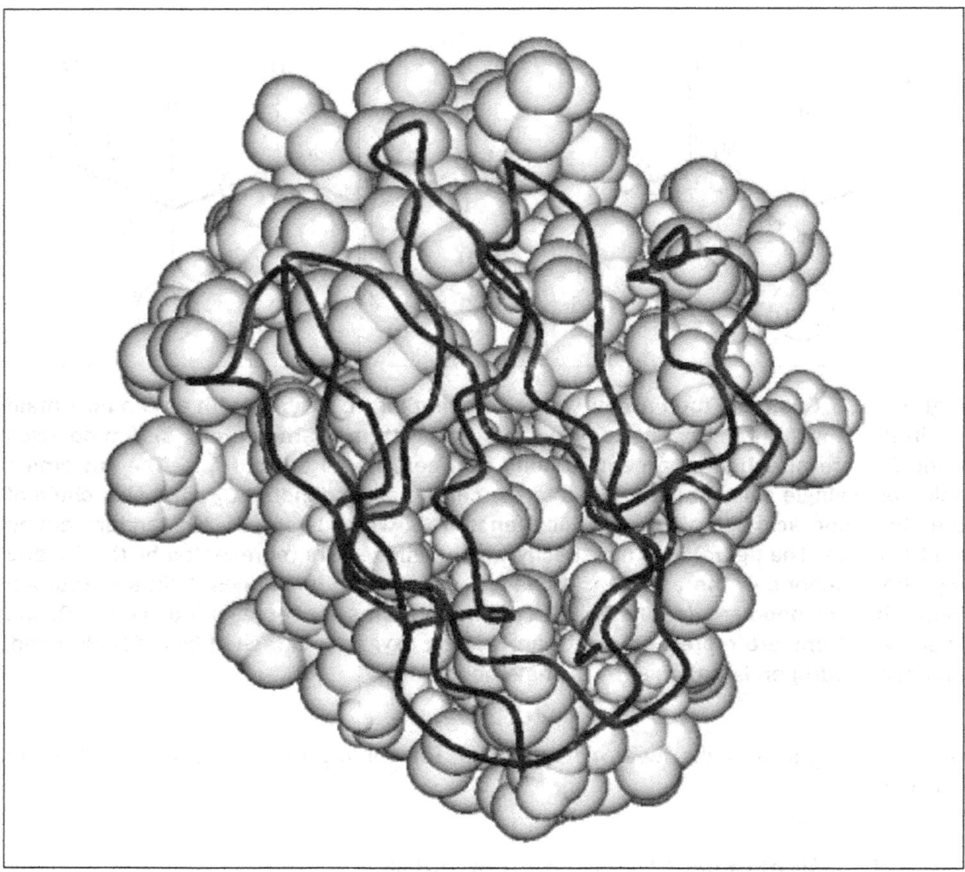

Figure 7.2 A Three-Dimensional Atomic Representation of a Protein. The protein backbone is shown as a dark curve. The atoms are shown as spheres. The three-dimensional structure of the backbone is created by the rotations in the amino acid main chains. (Image created by Carl Kingsford. Used with permission.)

specified angles. The specific rotation(s) that occur in the side chain determine the three-dimensional structure of the side chain. Of course, different combinations of rotations are observed with differing frequencies.

A permitted three-dimensional structure of a side chain, determined by the angles at its point(s) of rotation, is called a *rotamer.* Since *proline* only has a single point of rotation, it has only three rotamers, while *Arginine*, with four points of rotation, has $3^4 = 81$ rotamers. Experimentally determined libraries reflect the frequencies of each of the possible rotamers, for each of the 19 standard, flexible amino acids. More detailed libraries [61], called *backbone-dependent* libraries reflect the frequencies of each possible rotamer, for each possible *combination* of ϕ, ψ angles in the main chain of the amino acid.

Protein Structure Stability The entire three-dimensional structure of the protein is determined by the rotations in its backbone, together with the rotations in its side chains. It is generally accepted that a protein will fold into a structure that is the most *stable* – more precisely, the structure that has the *minimum free energy*. For our purposes, we don't need to understand much about free energy. The biochemical literature uses the concept of free energy, and the goal of *minimizing* free energy, but it is more natural and intuitive for non-chemists to talk about *maximizing* stability. That is the approach we take here. Chemical stability of a protein comes from four sources: the internal stability of the three-dimensional structure of its backbone; the internal stability of each rotamer; the interaction of each rotamer with the main chain of the amino acid it is attached to; and the chemical interactions of the rotamers that are positioned close to each other, either on the protein sequence, or in three-dimensional space. Note that only *pairwise* interactions of rotamers are included in this model of stability.

Recall that we have assumed that the three-dimensional structure of the protein backbone is known. Then, given the chosen rotamers, the *stability* of the protein is quantified as:

$$E_0 + \sum_{i} E_{i_r} + \sum_{i<j} E_{i_r,j_s}. \tag{7.1}$$

The term E_0 is a measure of the stability of the protein backbone; i_r is the chosen rotamer for i'th amino acid in the protein, and E_{i_r} is a measure of stability due to the interaction between i_r and the main chain of amino acid i, plus a measure of the internal stability of rotamer i_r; and E_{i_r,j_s} is a measure of the stability due to the interaction between two chosen rotamers, i_r and j_s. The value of E_{i_r,j_s} is based both on what the two chosen rotamers are, and on how close they are to each other in three dimensions. That proximity is known because we know the three-dimensional structure of the backbone, and the dimensions of the rotamers.

In practice, the value of E_0 can be determined from chemical energetics and the three-dimensional structure of the protein backbone. The stability measure, E_{i_r}, of each rotamer, i_r, has been established experimentally and is available from published rotamer libraries. The interaction between the main chain of amino acid i and the rotamer i_r is influenced by the angles ϕ_i, ψ_i and ω_i, which are assumed known. Experimentally established backbone-dependent libraries hold the values of E_{i_r} based on those angles and the specific rotamer i_r. Finally, in establishing the value of E_{i_r,j_s}, the closeness of rotamers i_r and j_s can be determined by the particular rotamers and the three-dimensional structure of the backbone.

All of the values discussed, obtained either from experimental or theoretical studies, are critical to accurate three-dimensional structure modeling, but for our purposes the specific values are not important. We just assume that the values are known, and are *nonnegative*. Our focus is on the ILP formulation, which uses them to solve the side-chain positioning problem.

Side-Chain Positioning Problem, also called the **Rotamer Prediction Problem**: Given the three-dimensional structure of the protein backbone, *choose* one allowed rotamer for each side chain to *maximize* the stability of the resulting three-dimensional protein structure.

7.1.2 The Abstract ILP Formulation for the Rotamer Prediction Problem

First, we take a small detour to establish a technical ILP point.

7.1.2.1 Allowing Edge Weights in Clique Problems

In Section 2.2.6 we discussed the problem of finding a *maximum-weighted* clique when the *nodes* of the graph have weights. Now we extend the model to allow each *edge* (i,j) to have a weight, denoted $w(i,j)$. Then, the weight of a clique is the sum of the weights of the edges in the clique, together with the weights (if any) of the nodes in the clique. Let $w(i)$ denote the weight of a node i, and for each edge (i,j) in the graph, let $W(i,j)$ be a binary variable used to record whether both nodes i and j have been chosen to be in the clique. Recall that $C(i)$ is a binary variable that indicates whether or not node i will be chosen to be in the clique. Then, for each edge (i,j) in the graph, we include the inequalities:

$$\begin{aligned} C(i) + C(j) - W(i,j) &\leq 1, \\ W(i,j) &\leq C(i), \\ W(i,j) &\leq C(j). \end{aligned} \tag{7.2}$$

The modified objective function used to find a maximum weighted clique is now:

$$\sum_{(i,j)\in E} W(i,j) \times w(i,j) + \sum_i C(i) \times w(i).$$

7.1.2.2 Back to the ILP Formulation for the Rotamer Prediction Problem

Given a protein sequence of length p, let R_i (for i from 1 to p) denote the set of allowed rotamers for the i'th amino acid in the protein. The ILP formulation will have one binary variable x_{i_r} (called a *selection* variable) for each allowed rotamer i_r in R_i. Each variable x_{i_r} will be set to 1 to specify that rotamer i_r will be chosen, and will be set to 0 to specify that it will not be chosen. This is implemented in the abstract ILP formulation with the following inequalities:

For each i from 1 to p,

$$\sum_{i_r \in R_i} x_{i_r} = 1. \tag{7.3}$$

Rotamer Interactions The next part of the ILP formulation addresses possible *interactions* between chosen rotamers: If rotamers i_r and j_s are both chosen (for the i'th and j'th amino acids respectively), then the objective value should include a contribution of E_{i_r,j_s}; but if either of those rotamers are *not* chosen, the objective value should not include a contribution of E_{i_r,j_s}. This can be achieved using a binary variable x_{i_r,j_s} (called an *interaction* variable) for each pair of rotamers (i_r, j_s), where $i \neq j$. To connect the selection and interaction variables, we require:

> **Interaction Requirements** An interaction variable x_{i_r,j_s} should be set to 1 if and only if the selection variables x_{i_r} and x_{j_s} are both set to 1.

Certainly, the *if-then* and *only-if* idioms discussed in Section 4.3, can be used to implement the interaction requirements.

Exercise 7.1.1 *Using if-then and only-if idioms, state inequalities that implement the interaction requirements for the values of variable* x_{i_r,j_s}.

Then, the objective function of the ILP formulation for the rotamer prediction problem is:

$$Maximize\ E_0 + \sum_i x_{i_r} \times E_{i_r} + \sum_{i<j} x_{i_r,j_s} \times E_{i_r,j_s}. \tag{7.4}$$

7.1.2.3 Viewing Rotamer Prediction As an Edge-Weighted Maximum-Weighted Clique Problem

The rotamer prediction problem can be viewed as a maximum-weighted clique problem, but the ILP formulation given in the prior section does not make this clear. Here we make the connection explicit.

Given a protein with p amino acids and a known backbone, create a graph G with one node for each possible rotamer, and add an edge between two nodes representing rotamers i_r and j_s respectively, if and only if $i \neq j$. That is, if and only if the rotamers are for different amino acid positions on the protein. Weight the edge between the nodes for i_r and j_s with the value w_{i_r,j_s}. That value might be zero, for example, if rotamers i_r and j_s are very far from each other on the protein backbone. Since there is no edge between any pair of nodes representing rotamers in the same amino acid position, the largest clique in G has exactly p nodes. Further, since none of the node and edge weights are negative, there is a maximum-weight clique in G containing exactly p nodes. Thus we have

> [a]ny instance of the rotamer prediction problem can be cast as, and solved as, an instance of the maximum-weight clique problem in graph G.

This is another illustration of utility of the maximum clique problem in computational and systems biology.

7.1.3 Empirical Results

The ILP solution times reported in [106] are now very dated, and would be much faster today. But even with relatively slow computers and slow ILP solvers, the results in [106] show that the ILP approach to the rotamer prediction problem is very practical. The authors conducted empirical studies with proteins of lengths between 60 and 256 amino acids, where the number of rotamers to choose from ranged between 900 and 4,000. The times to optimally solve the ILP formulations were all under 2 minutes, and most were considerably faster than that.[2] The paper reported that the accuracy of the rotamer predictions given by the ILP approach was comparable to the accuracy of the best alternative methods available at that time.

One point of great interest is that the authors report that for most of the computations, the LP relaxation of the ILP formulation gave optimal *integer* solutions. As discussed in Section 1.2, LP relaxations can be solved much faster than

[2] Although, they also used some "cleanup" methods to reduce the sizes of the problem instances before creating any ILP formulations. We recently reexamined some of their data with modern ILP solvers. Using only a laptop for the computation, we were able to avoid using the cleanup methods, and still achieve faster ILP solution times. This again demonstrates one of the premises of the book, that the best current ILP solvers often allow one to focus on ILP *formulations* without needing to get involved with the challenge of solving the ILPs.

the corresponding ILPs, and when an optimal solution to the LP relaxation has only integer values, it is guaranteed that it is an optimal solution to the ILP formulation.

7.1.4 Homology Modeling and Protein Design

So far, in our discussion of the rotamer prediction problem we have assumed that the three-dimensional structure of a protein backbone is given as input. That is a reasonable assumption because protein structure prediction methods typically first predict the backbone structure, and then solve the rotamer prediction problem. Further, often the three-dimensional backbone structure is *known* for a protein whose amino acid sequence is highly similar (more than 60% identical in an alignment of the two sequences) to that of the input protein. It is known that backbone structure is highly conserved across similar protein sequences, but the structures of the side chains are more variable. Therefore, researchers have used known backbone structure in similar proteins as a proxy for the *unknown* backbone structure of the protein under study. It is reported in [106] that this approach works well in the empirical tests that were performed. The general idea of using the structure of similar proteins as a proxy for the unknown structure is called *homology modeling.*

A more ambitious problem than homology modeling is to create (design) an amino acid sequence, which is likely to fold into a specified three-dimensional structure. This *protein design* problem is discussed in [106], where an ILP approach is again used. But in contrast to the rotamer prediction problem, ILP formulations for design problems typically took an hour or more to solve, and LP relaxations did not give integer solutions. The interested reader is directed to [106] for a detailed discussion.

7.2 PROTEIN FOLDING VIA THE HP MODEL

As discussed in Section 7.1, determining the three-dimensional structure of proteins, or learning some parts of the structure, are critical tasks in biochemistry, biophysics, computational biology, and systems biology. Many computational tools have been developed for this purpose. The most ambitious of these are *atomic-level simulators* that attempt to completely model atomic-level physics, using precise geometric and energy parameters during the entire protein folding process. The simulations hopefully end with the protein fold that minimizes free energy, or equivalently, maximizes the stability of the folded protein. These physical simulations of the full folding process are generally not able to reach the final protein fold because a prohibitive amount of computer time (and wall-clock time) is required to simulate even a fraction of the full folding event.

An alternative approach is to avoid atomic-level modeling and instead use a much *simpler model* of the *end result* of the folding process, i.e., the folded protein. A particularly attractive simple model, the *HP model*, was proposed by K. Dill [54] in 1995 and has been extensively explored since then.

The HP Model The HP model of protein folding is based on three simplifications, leading to a characterization of protein folds, and an objective function that the fold optimizes. The model *first* simplifies the 20 standard amino acids by dividing them into *two* groups: the *hydrophobic* and the *hydrophilic*; that is, water-fearful and water-loving. Somewhat more precisely, the hydrophobic amino acids are *nonpolar*, and are repelled from (or by) water, and the hydrophilic amino acids are *polar* and bind well to water. With this simplification, each hydrophobic amino acid in a protein sequence is replaced by the letter "H"; and each hydrophilic (polar) amino acid is replaced by "P." Hence a protein sequence, based on an alphabet of size 20, is reduced to a binary string. "The folding code is mainly binary" [54].

For the *second* simplification, the HP model uses the fact that proteins in a cell are surrounded by water, so the hydrophobic amino acids are generally forced into the interior of the folded protein, and the hydrophilic amino acids are generally on the exterior of the folded protein. This is particularly true of *globular* proteins, which fold into roundish, globlike structures. This is a simplification, and most folded proteins have some hydrophobic amino acids on the surface and some hydrophilic amino acids deep in the core of the protein.

The *third* simplification in the HP model is that the dominant feature of a folded protein, and the one that determines its stability, is the *number of bonds* between hydrophobic amino acids. These are called *H-H bonds*. This simplification is related to the second one, since when hydrophobics are closely packed in the interior of the fold, there are many bonds that are possible.

It may seem that these three simplifications would result in a severe loss of information, but as stated in [54]:

> Simple exact models are crude low-resolution representations of proteins. But while they sacrifice geometric accuracy, simple exact models often adequately characterize the collection of all possible sequences of amino acids (sequence space) and the collection of all possible chain conformations (conformational space) of a given sequence. For many questions of folding, we believe that complete and unbiased characterizations of conformational and sequence spaces are more important than atomic detail and geometric accuracy.

Finally, given the three simplifications, the HP model of protein folding asserts that the most stable fold for a globular protein is the fold that *maximizes* the number of H-H bonds. We next formalize the HP model in an optimization problem.

7.2.1 The HP Prototein Model and Folding Problem

A *prototein*[3] is a binary string that we embed on a *two-dimensional grid*. We can think of the grid as a graph, where each node of the grid is called a "point."

A *legal embedding* of a prototein on the grid must satisfy the following rules:

1. Each character in the string gets placed on *exactly one* point of the grid.
2. Each point of the grid gets *at most* one character of the string placed on it.
3. Two adjacent characters in the string must be placed on two points that are *neighbors* on the grid in either the horizontal or vertical direction, but not both. That is, two points across a diagonal are not neighbors. (Note that an interior point on the grid has four neighbors on the grid; that a point on a border of the grid has at most three neighbors; and that a corner point has only two neighbors).

These three rules mean that the string must be embedded into the grid as a *self-avoiding walk*, without deforming the string. In an embedding of the string, we

[3] Yes, the word is "prototein" not "protein."

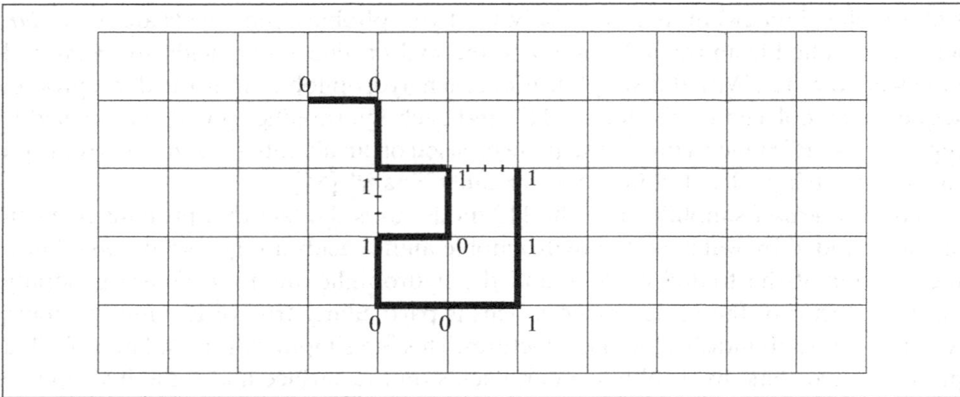

Figure 7.3 One Possible Embedding of the String 00110100111. The hatched lines show the two contacts in this embedding. Can you find an embedding with more contacts? For the answer, see Figure 7.5.

define a *contact* as a pair of 1s that are *not* adjacent in the string, but are placed on neighboring points on the grid. Note that each contact is associated with an *edge* of the grid (see Figure 7.3).

The central assumption in the HP model is that the most stable fold is one that maximizes the number of contacts (H-H bonds). This leads to

The Prototein Problem: Given a binary string, find a legal embedding of the string on the two-dimensional grid to maximize the number of contacts.

7.2.2 An ILP Solution

Here we develop an abstract ILP formulation for the prototein problem. The basic approach was first detailed in [37], which was later extended to three dimensions, along with further empirical studies, in [141].

There are two parts to the abstract ILP formulation. The first part consists of inequalities that ensure a *legal embedding*; and the second part consists of inequalities that determine, for each edge of the grid (i.e., for each neighboring pair of points on the grid), whether or not the chosen legal embedding creates a contact between those points.

7.2.2.1 Variables and Inequalities to Create a Legal Embedding

Suppose the input string is of length n, and the grid is n-by-n (so it has exactly n^2 points), although real proteins would be embedded in smaller grids. We give each point on the grid a unique integer *identifier* from 1 to n^2. For concreteness, we give the points in the top row the identifiers 1 to n, from left to right; and give the points in the next row the identifiers $n+1$ to $2n$, etc., so that the lower right-hand point gets identifier n^2.

For each position i on the input string, and each point p on the grid, let $X(i,p)$ be a binary variable that will be set to 1 to specify that the character at position i

in the string is assigned to point p on the grid. Note that there are about n^3 such variables.

To ensure that each position on the string is assigned to some point on the grid, the abstract ILP formulation will have the following inequality for each position i, from 1 to n:

$$\sum_{p=1}^{n^2} X(i,p) = 1. \tag{7.5}$$

Remember that in a concrete ILP formulation, n is a constant number, not a variable. The value of n is the length of the input string, so its value is known when the computer program creates the concrete ILP formulation for a specified input string.

Next, we need inequalities to ensure that no point on the grid is assigned more than one position in the string. For each pair of positions i,j in the string, and each point p on the grid, the ILP formulation will have:

$$X(i,p) + X(j,p) \leq 1. \tag{7.6}$$

Exercise 7.2.1 *Could the requirement that no point on the grid be assigned more than one position in the string be implemented as:*
For each point p on the grid: $\sum_{i=1}^{i=n} X(i,p) \leq 1$? *Explain.*

Inequalities to Ensure Connectedness We next need to ensure that adjacent positions on the string are assigned to neighboring points on the grid. For a point p, let N_p be the set of its neighboring points on the grid. The abstract ILP formulation needs to express the logic that for each position i (i from 1 to $n-1$) in the string, and each point p on the grid:

$$\text{If } X(i,p) = 1 \text{ then } \sum_{q \in N_p} X(i+1,q) = 1. \tag{7.7}$$

That logic can be implemented, for a fixed i and p, as follows:

$$X(i,p) - \sum_{q \in N_p} X(i+1,q) \leq 0. \tag{7.8}$$

In general, the set N_p consists of four points, two on the same row as p, and two on the same column as p. However, a corner point only has two neighbors, and a non-corner point on the top or bottom rows, or the first or last columns, only has three neighbors. For a given input string, the concrete ILP formulation for the prototein problem must reflect these different cases, but the abstract inequality specified in (7.8) includes all of the cases.

The inequalities specified in (7.5), (7.6), and (7.8) ensure that the assignment of string positions to grid points specifies a legal embedding of the string.

Exercise 7.2.2 *Verify that the inequalities do specify a legal embedding of the input string.*

Exercise 7.2.3 *In terms of n and p, and an n-by-n grid, calculate the explicit integer identifiers of the points in* N_p*, for all the locations of point p. That is, when p is an interior point with four neighbors; or p is a corner point; or p is a non-corner point in the first or last row, or column. These are precisely the calculations that must be done in the program that creates a concrete ILP formulation when an input string of length n is specified.*

7.2.2.2 Inequalities to Detect Contacts

Recall that a *contact* in a prototein is an edge (p, q) on the grid where neighboring points p and q have been assigned *nonadjacent* characters in the string, each with value 1. Here we develop inequalities to record whether or not an edge is a contact, based on the values of the $X(i, p)$ variables.

For each point p on the grid, let $I(p)$ be a binary variable, which will be assigned the value 1 if and only if some position i in the string is assigned to point p, where the character at position i is 1. Note that $I(p)$ will have value 0 if *either no* position in the string is assigned to point p; or if some position i in the string is assigned to point p, but the character at position i is 0.

Let I be the set of all positions in the string where the character is 1. The program that creates the concrete formulation for an input string knows I. Then, the following inequalities correctly set the $I(p)$ variables.

For each point p on the grid:

$$I(p) = \sum_{i \in I} X(i, p). \tag{7.9}$$

Next, for each edge (p, q) in the grid, we create the binary variable $C(p, q)$, which will be set to 1 *only if* $I(p) = I(q) = 1$. The following inequalities correctly set those variables.

For each edge (p, q) in the grid:

$$I(p) + I(q) - 2C(p, q) \geq 0. \tag{7.10}$$

Exercise 7.2.4 *Could we replace the inequality in (7.10) with*

$$I(p) + I(q) - C(p, q) \geq 1?$$

Clearly, $C(p, q)$ can be set to 1 only if both $I(p)$ and $I(q)$ are set to 1. So, what could be the harm?

Using the inequalities in (7.10), $C(p, q)$ will be set to 1 for every contact in the embedding (Figure 7.4). But, the sum

$$\sum_{(p,q) \text{ an edge on the grid}} C(p, q).$$

will be *larger* than the *true* number of contacts in the embedding, if the string has any adjacent positions where both positions have character 1. In fact, the amount of the overcount is *exactly* the number of adjacent pairs of positions in the string, where both positions have character 1. For example, the string 1110101100 would have an overcount of 3. The computer program that creates the concrete formulation for an input string can trivially count the number of such pairs, and create the equality:

```
offset = overcount,
```

where "offset" is an integer ILP variable, and "overcount" is the specific number determined by the computer program, given the input string.

$$\texttt{Maximize} \left[\sum_{(p,q)\in E} C(p,q) \right] - \texttt{offset}. \tag{7.12}$$

Such that
For each position i, from 1 to n:

$$\sum_{p=1}^{p=n^2} X(i,p) = 1. \tag{7.13}$$

For each pair of positions i, j, and each point p:

$$X(i,p) + X(j,p) \leq 1. \tag{7.14}$$

For each position i from 1 to $n-1$, and each point p:

$$X(i,p) - \sum_{q\in N_p} X(i+1,q) \leq 0. \tag{7.15}$$

For each point p on the grid:

$$I(p) = \sum_{i\in I} X(i,p). \tag{7.16}$$

For each (p,q) in E:

$$I(p) + I(q) - 2C(p,q) \geq 0. \tag{7.17}$$

All variables are binary.

Figure 7.4 The Abstract ILP Formulation for the Prototein Problem. *E* is the set of edges in the grid. All the other symbols and terms in the formulation were introduced earlier.

The Objective Function Finally, the objective function is

$$\text{Maximize} \left[\sum_{(p,q)\in E} C(p,q) \right] - \texttt{offset}. \tag{7.11}$$

In summary, the abstract ILP to solve the prototein problem is shown in Figure 7.4.

Exercise 7.2.5 Software *The Perl program,* HPb.pl *(available on the book website), takes in a binary string and produces a concrete ILP formulation, based on the abstract ILP formulation in Figure 7.4, to solve the prototein problem. To run program* HPb.pl, *create a binary string and put it on the first line of a file, say "HPstring." Then issue the following command in a terminal window.*

```
perl HPb.pl HPstring
```

The concrete ILP formulation will be in file "HPstring.lp."

Use HPb.pl *to create the concrete ILP formulation for the toy input string 1101. Clearly, the optimal value of this instance of the prototein problem is one, and we don't need an ILP computation to know that. But, looking at the concrete inequalities helps in understanding the logic of the ILP formulation.*

Look through the concrete formulation for string 1101 to separate the inequalities into blocks, where the inequalities in a single block are created by the same inequality-generating statement in the abstract ILP formulation shown in Figure 7.4.

Exercise 7.2.6 *There are two immediate generalizations of importance to the HP model. The first is to generalize the two-dimensional model to three dimensions. Fully detail that generalization.*[4]

Exercise 7.2.7 *A second generalization is to put different values on H-H contacts, depending on which particular amino acids are in contact. Contact values derived from known protein structures are developed in [136]. Then, instead of maximizing the number of contacts, an optimal solution should maximize the total value of the contacts. Fully detail this generalization.*

If you can program in Perl, modify the program HPb.pl *to implement this generalization. Use the contact values in [136]. Then run the modified program to compare the folds obtained with and without contact values.*

7.2.3 Empirical Results

I have implemented and tested the ILP formulation shown in Figure 7.4. The Perl program, *HPb.pl*, that takes in a binary string and produces a concrete ILP formulation is available on the book website. Those concrete formulations are then solved by Gurobi 6.5 on an iMac Pro, with default parameter settings. This allows practical solution of the prototein problem for strings of length up to 25 characters. Those instances *typically*, but not always, solve in under 10 minutes. However, using the ILP formulation described above (without speedups, to be discussed), length-48 sequences from [211] did not solve in 10 hours of computation, although they quickly found embeddings that seem highly compact. In one case, I allowed the computation to run for 20 hours before terminating it – at that point, the solver found an embedding with 21 contacts, and an upper bound showing that no embedding with more than 26 contacts is possible.

In another length-48 example from [211], where the optimal is known to be 18 contacts, I let Gurobi run for about 2 days, and it found a solution with 17 contacts and an upper bound of 27. It found that solution after 20 hours, but then made no improvement; and it found the upper bound of 27 after 19 hours. I then reran that ILP formulation with the added parameter *MIPFocus=1*, which instructs Gurobi to emphasize finding better feasible solutions quickly, at the expense of reductions in the upper bound (see page 47). In that rerun, Gurobi found an optimal solution, with 18 contacts, in under 40 minutes, but at that point the upper bound was 32. In contrast, running Gurobi with the default parameters at that time mark (about 40 minutes), it had found a solution with only 15 contacts, but had an upper bound of 30. So, if one is interested in getting good feasible solutions quickly, but without a proof of optimality, the time Gurobi needs can be considerably reduced.

4 An integer programming formulation for the three-dimensional model and empirical results are reported in [208].

These results suggest that the prototein problem may be one where a simple ILP formulation, together with the default use of an ILP solver running on a laptop, is not adequate to quickly find optimal solutions, *and* prove their optimality, for instances as large as typical proteins. Part of the reason is that there are a huge number of optimal, *symmetric*, embeddings that differ only by *rotation* and *translation* in the grid. Without the use of additional speedups, the ILP solvers spend a huge amount of time exploring equally good, and essentially identical solutions. We note however that optimal solutions to problem instances of length 100 are reported in [208]. They do not explicitly state the time used to find those solutions, only stating that the solution times are practical.

7.2.4 Speeding Up the Computation

Several ways to make the ILP formulation more efficient were suggested in [37, 69]. I first tried two of these: reducing the grid size, and fixing the embedding of the first two characters in the string. In the exposition above, and in the empirical results discussed above, the default grid is of size n-by-n, which is always large enough for any embedding of a string of length n. However, since we expect the prototein to embed *compactly* in the grid (which is in fact the objective), burying the hydrophobic residues in the center of the embedding, it is highly unlikely that n rows or n columns would be needed for a prototein derived from a real globular protein. So, a smaller grid should be adequate and will reduce the number of symmetric optimal embeddings that are explored by an ILP solver.

I tried using an $n/4$-by-$n/4$ size grid for sequences of length 24 and longer.[5] Empirically, reducing each side of the grid from n to $n/4$ (for strings with $n \geq 24$) had a *very significant* effect on the running time of the computation. A related speedup is to assign the first character in the string to the middle point of the grid, and to arbitrarily choose one of its neighbors for the second character of the string. These assignments are made by setting the values of $X(1,p)$ and $X(2,q)$ to 1, for two neighboring points p and q, where p is the (or "a" depending on n) midpoint of the grid. This again reduces the number of symmetric embeddings, and empirically has a large effect in reducing the time Gurobi needed to solve the ILP formulations. These speedups are implemented in the Perl program *HPb1mid.pl*, available on the book website.

The following results illustrate the utility of these speedups. Using the program *HPb.pl* to generate the concrete ILP formulation, and using Gurobi 7.5 to solve it, a difficult length-24 sequence ran for more than 10 hours when terminated, and still had a 63% gap (lower bound 8 and upper bound around 14), which had not moved for several hours. But, using the program *HPb1mid.pl*, i.e., with both of the improvements suggested here, the solution time reduced to 167 seconds, and found an optimal value of 9.[6] I then implemented an additional speedup idea, suggested in [69], which reduces the number of variables used and should reduce the number of possible solutions examined, but the solution time for Gurobi 7.5 increased to 212 seconds. Well, at least, it ran in a reasonable time and found an optimal solution. So, these speedups can help dramatically (reducing 10 hours without an optimal solution,

[5] For short sequences, say of size 11, the resulting grid is too small, so a larger grid is needed.

[6] When Gurobi 8 was released, I reran this ILP. It solved in 82 seconds. Then, using MIPFocus=1, it solved in 56 seconds.

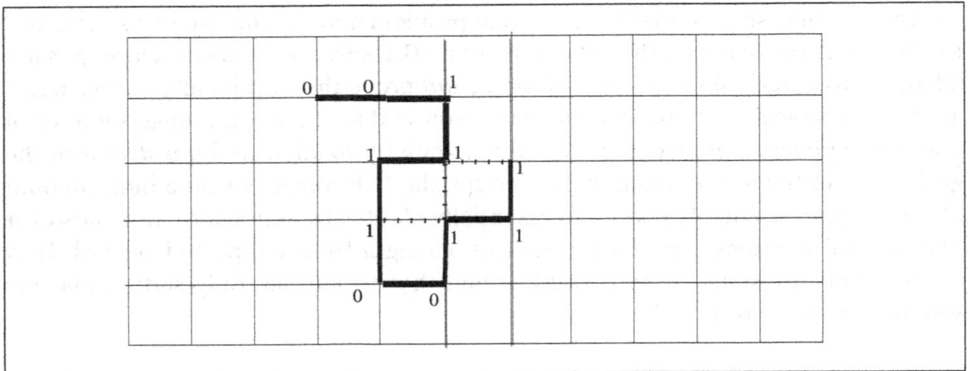

Figure 7.5 An Optimal Embedding of the String 00110100111, with Three Contacts, Shown By Dashed Lines. An embedding of this string with only two contacts was shown in Figure 7.3. This problem instance was solved by Gurobi 6.5 in about 15 seconds, and by Gurobi 7.5 in 10 seconds, and in 0.19 seconds using the two speedups discussed here.

to under 1 minute, with an optimal!), but it is hard to predict and exactly understand their effects.

Exercise 7.2.8 *The optimal number of contacts found when using HPb.pl should be the same as when using HPb1mid.pl, unless the smaller grid is too small for an optimal layout, such as produced using HPb.pl. In that case the optimal value using HPb.pl will be larger than the optimal value using HPb1mid.pl. Run the two programs on strings you create, and then solve the ILP formulations, to try to find an example where this phenomenon happens.*

Exercise 7.2.9 *The two Perl programs* HPb.pl *and* HPb1mid.pl *were discussed above. An additional Perl program,* HPb1.pl, *available on the book website, creates a concrete ILP formulation for the prototein problem on an n/4-by-n/4 grid, but it allows the first character of the input string to be put anywhere in the grid. So, it produces an ILP formulation that is a compromise between the ones produced by the two other programs.*

Run the three Perl programs on binary strings of lengths between 15 and 30, and then use Gurobi to solve each of the ILP formulations. Compare the running times of Gurobi on the three different ILP formulations, and how the times vary as a function of the length of the input strings. Note any strings where the three optimal solutions do not have the same number of contacts. What conclusions do you draw from these experiments?

Exercise 7.2.10 *The Python program* HPb1-3D.py,[7] *available on the book website, creates concrete ILP formulations for the prototein problem on a* three-dimensional *grid. Since proteins fold in three dimensions, rather than two dimensions, it seems likely that an optimal solution on a three-dimensional grid will be more chemically accurate than on a two-dimensional grid. Without checking for chemical accuracy, we should at least observe that an optimal three-dimensional solution finds more contacts than an optimal two-dimensional solution, for the same string. However, we would expect that the time to find an optimal three-dimensional solution would be greater than the time for an optimal two-dimensional solution.*

Run the programs HPb1-3D.py *and* HPb1mid.pl *on binary strings of lengths between 15 and 30, to compare running times of Gurobi on the different ILP formulations, and to compare the*

[7] The program was created by Jonathan Kim, as a project in a class on integer programming in computational biology, in fall 2017. Thanks to Jonathan for permitting this program to be distributed with the book.

number of contacts found in the optimal solutions. Summarize your overall conclusions from these experiments.

7.3 PREDICTING DOMAIN-DOMAIN INTERACTION IN PROTEINS

Proteins are often built in a *modular* form, composed of *domains*, which are units of the protein that have distinct functions or distinct folds that are *independent* of the rest of the protein. Many proteins are "built of different combinations of protein domains that have been selected from a relatively small repertoire" [44], so, the same domains are seen in many different proteins, although in different orders and combinations.

Protein interaction data, as in a PPI network, show which pairs of *proteins* physically interact. But,

> [i]nteraction between two proteins typically involves binding between specific domains ... [80]

Since, the interaction between two proteins typically occurs at the level of protein *domains*, knowing which domains interact is very valuable information. But domain-domain interaction data is very limited, while protein-protein interaction data is much more available. So, a natural question is whether we can *deduce* domain interaction information from coarser, protein interaction data.

We can typically identify, from the protein sequences, the domains contained in a protein, so when two interacting proteins each have *only one* domain, the domain-domain interaction is known. But, about half of all the known proteins have more than a single domain, and when two multi-domain proteins interact, it is unlikely that every domain in one of the proteins interacts with a domain in the other protein. So, this leads to the following

The Domain-Domain Interaction Problem (DDIP): We are given a PPI network $\mathcal{N}$, and for each protein P in the network, we are given the names of the *domains* that occur in this P. Then, deduce for each pair of interacting proteins (P_1, P_2) in $\mathcal{N}$, which *pairs* of distinct domains, one from P_1 and one from P_2, account for the protein-protein interaction. The DDP problem is the problem that is addressed using linear and integer programming in [80, 81].This section is loosely based on those papers, emphasizing the combinatorial core of their approach.

7.3.1 The High-Level Idea for Solving the DDIP

If we only examine a *single* pair of multi-domain proteins that interact, there would not be enough information to solve the DDI problem. But, with *all* of the protein interactions detailed in a PPI, we can take a *global, optimization approach* to the problem, using the biological assumption that

> [d]omain-domain interactions are *conserved* across the protein interactions in the PPI network. That is, the same domain-domain interactions show up repeatedly in a wide range of protein-protein interactions.

Thus, while two proteins may each have several domains, the domain pairs that actually interact will likely be among those domain pairs that are involved in many protein-protein interactions. Then, the *high-level* approach to solving the DDIP is suggested in the following quotes:

> ... we consider the problem of predicting interacting domain pairs as an optimization problem, in which the objective is to *minimize* the number of domain-domain interactions necessary to justify the underlying protein-protein interaction *network*.
>
> ... we want to justify each protein-protein interaction, using a *minimum* number of domain-domain interactions possible *overall*. [80] (italics added)

7.3.1.1 More Formally

We let $\mathcal{P}$ denote the set of proteins in a given PPI network $\mathcal{N}$, and $\mathcal{D}$ denote the set of known domains contained in the proteins in $\mathcal{P}$. Then, we let B be an undirected *bipartite* graph,[8] with one side of B containing a node for each known protein-protein interaction in the PPI (essentially for each *edge* in the PPI). The other side of B contains nodes that represent *pairs* of *domains*. There will be a node for domain pair (D_1, D_2) if and only if there is an interacting pair of proteins (P_1, P_2), where D_1 is a domain in P_1 and D_2 is a domain in P_2. Then there will be an edge in B between the node representing the pair (P_1, P_2) and the node representing the pair (D_1, D_2). We call the two sides of B the $\mathcal{D}^2$ side and the $\mathcal{P}^2$ side (see Figure 7.6b).

A DDI Cover of B A set of nodes S on the $\mathcal{D}^2$ side of B is called a **DDI cover** of B, if each node on the $\mathcal{P}^2$ side of B is adjacent to at least one node in S. With these definitions, the core, high-level approach to the DDI problem in [80, 81] is to solve the following problem:

DDI Cover Problem: Given a bipartite graph B derived from $\mathcal{P}^2$ and $\mathcal{D}^2$, find the *minimum-sized* DDI cover of B.

For example, in Figure 7.6, the smallest DDI cover of B has size two, and there are several DDI covers of that size. One is $\{(D_1, D_4), (D_2, D_3)\}$, and another is $\{(D_1, D_4), (D_1, D_3)\}$. That toy problem illustrates the definitions, but is too small to illustrate that finding a smallest DDI cover is a hard computational problem.

The set S that solves the DDI cover problem identifies a set of domain-domain pairs that "explain" the protein-protein interactions in $\mathcal{P}^2$, and are conjectured to be the (unknown) domain interactions that are responsible for the (known) protein interactions. The argument that the deduced set S should be small comes from the biological facts that the same domain-domain interactions appear across a wide range of proteins, and that many proteins are "built of different combinations of protein domains that have been selected from a relatively small repertoire."

7.3.1.2 An ILP Solution to the DDI Cover Problem

For each node v on the $\mathcal{P}^2$ side of B, let $\mathcal{D}(v)$ be the set of nodes on the $\mathcal{D}^2$ side of B that are adjacent to node v. For example, in Figure 7.6 if v is the node representing the pair (P_1, P_5) then $\mathcal{D}(v) = \{(D_1, D_3), (D_2, D_3)\}$.

[8] A *bipartite* graph is one where the nodes can be divided into two groups (sides), so that the only edges in the graph run between a node in one group and a node in the other group (see Figure 7.6b).

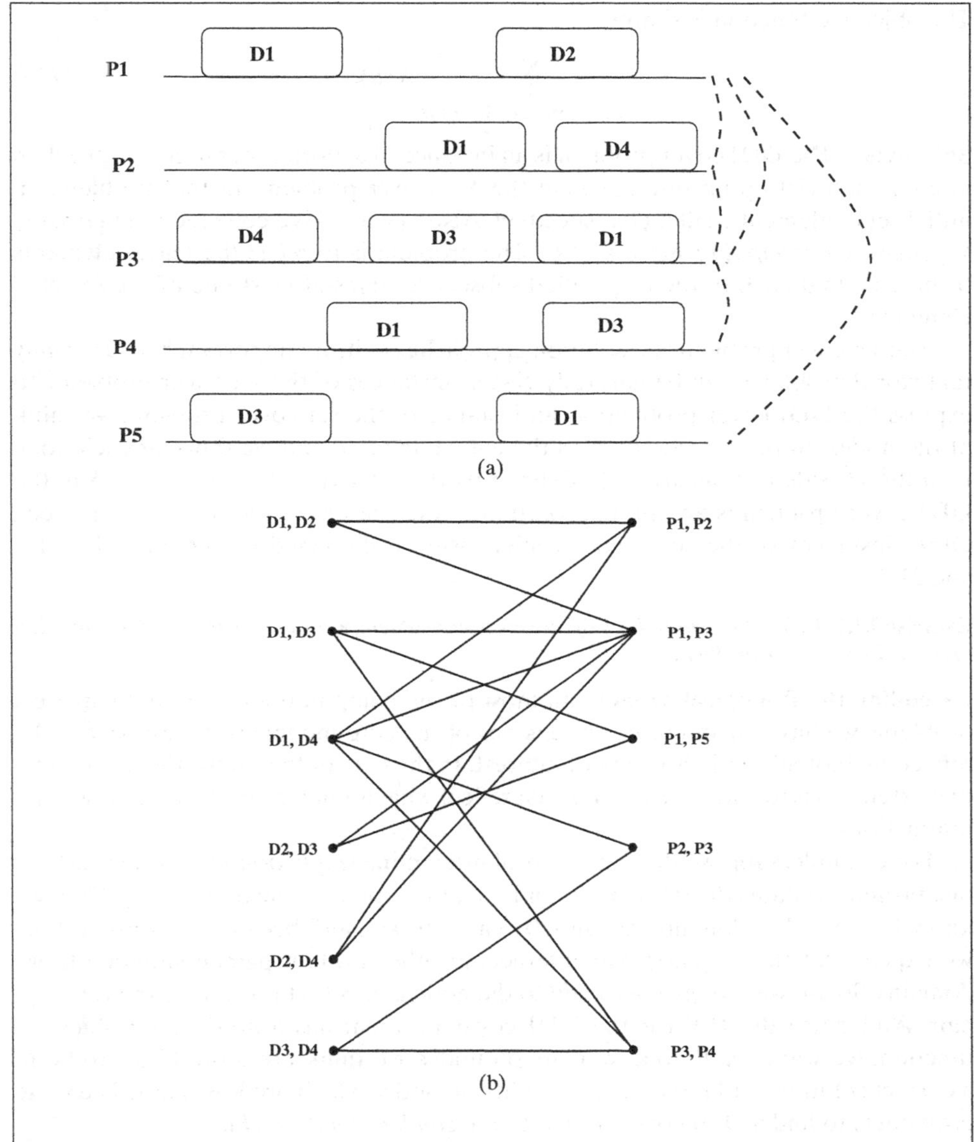

Figure 7.6 (a) Cartoon of Five Proteins P_1 ... P_5, and the Four Distinct Domains D_1, ..., D_4 They Contain. The dashed curves show which pairs of proteins interact. (b) The Bipartite Graph B Derived from the Protein Interaction and Domain Data.

In the ILP formulation for the DDI cover problem, we have one binary variable $X(u)$ for each node u on the $\mathcal{D}^2$ side of B. And, for each node v on the $\mathcal{P}^2$ side of B we have the inequality:

$$\sum_{u \in \mathcal{D}(v)} X(u) \geq 1. \tag{7.18}$$

The objective function is simply:

$$\sum_{u \text{ is on the } \mathcal{D}^2 \text{ side of } B} X(u). \tag{7.19}$$

Set Cover The DDI cover problem is an instance of a more general, classic problem in combinatorial optimization, called the *Set Cover* problem. In that problem, an initial set of elements, call it Q, is specified. Also, specified is a collection of many, say k, *generally overlapping* subsets of Q. The problem is to *select* the *fewest* elements from Q so that each of the k specified subsets contains at least one of the selected elements.

The set cover problem and solution approaches to it, are important because many real-world problems can be naturally cast as instances of the set cover problem. To express the DDI cover problem as an instance of the set cover problem, we think of the nodes on the $\mathcal{D}^2$ side of B as the elements of Q; and we think of each node v on the $\mathcal{P}^2$ side as a *subset* of Q, defined by the set $\mathcal{D}(v)$. The optimal set S in the DDI cover problem is an optimal solution to this variant of the set cover problem. Other instances of the set cover problem will be discussed in Sections 17.3, 21.4, and 21.7.

Exercise 7.3.1 *Using the terminology of the set cover problem, write an abstract ILP formulation to solve the set cover problem.*

Extending the Biological Model In most of the computational and mathematical problems we have encountered, extensions of the core model are necessary to fully reflect the biological phenomena of interest. In the case of the DDIP, the most effective extension is to add *node weights* that reflect what is known about domain-domain interactions.

For example, suppose there is a pair of interacting *single-domain* proteins, where one protein contains the domain D_1 and the other contains the domain D_2. Then we know for sure that domains D_1 and D_2 can interact, and because of conservation, we expect that the (D_1, D_2) pair interacts in other protein pairs containing those domains. So, we want to give a *weight* to the node u in B that represents the (D_1, D_2) pair. And, since the ILP for the DDI cover problem has a *minimizing* objective function, we want *small* node weights on nodes we think are more likely to be in a correct solution, and *larger* node weights on nodes which are less likely. Then, the problem is to find a DDI cover of B with the *smallest total weight.*

Another possible indicator of which domain pairs correctly account for interacting proteins comes from the *relative number* of protein pairs that contain specific domain pairs. That is, the *degrees* of the nodes on the $\mathcal{D}^2$ side of graph B. A large degree for a node u, relative to other nodes on the $\mathcal{D}^2$ side, suggests that the domain pair represented by u is more likely to reflect an actual domain interaction. Of course, a large node degree encourages the inclusion of u in a minimum-size DDI cover of B, but a more explicit encouragement may lead to more biologically meaningful choices. For example, in Figure 7.6, there are two smallest DDI covers, but the DDI cover $\{(D_1, D_4), (D_2, D_3)\}$ probably has less biological meaning than $\{(D_1, D_4), (D_1, D_3)\}$. This is because $(D_1, D_3)\}$ is a domain pair in three protein pairs, while $(D_2, D_3)\}$ only occurs in two protein pairs. Also, the first cover involves four distinct domain types, while the second only involves three distinct types. A smaller total number of distinct domain types is another indicator of biological fidelity.

Exercise 7.3.2 *How can the considerations discussed in the paragraph above be incorporated into the DDI cover problem? Think about node weights, or the objective function. For each model you suggest, write out the ILP formulation that implements that model.*

Exercise 7.3.3 *Another sensible approach to finding a set of domains that explain the known protein-protein interactions in a PPI $\mathcal{N}$, is to find the* smallest *set, S^*, of domains in $\mathcal{D}$, such that for every pair of proteins, (P_1, P_2), that interact, at least one of the domains in S^* is in P_1, and at least one different domain in S^* is in P_2. The set of domains in S^* then can "explain" the protein-protein interactions.*

Write out the abstract ILP formulation for the problem of finding S^* given $\mathcal{D}$ and $\mathcal{N}$.

Numbers The example in Figure 7.6 is a toy. The actual data used in [80] is based on 26,032 protein-protein interactions involving 11,403 proteins from 69 organisms, and 3,074 domain-domain interactions. From those, the data was trimmed to 783 unique domain-domain pairs.

More Protein Problems Later Later in the book, we will see additional ILP formulations for problems that concern proteins. Most of those will be in Chapter 17.

8

Tanglegrams and Coevolution

In this chapter we discuss several problems that arise in the study of *coevolution*. We first develop the *logic* of how to solve these problems using integer linear programming, and then show how to translate that logic into an abstract ILP formulation. The logical issues in these solutions are a bit more subtle than for the problems considered so far in this book. Up until now, the solution logic has been fairly straightforward – the challenge has been to translate that logic into an ILP formulation. Now the logical work will be a bit more challenging. The abstract ILP formulation will use ILP idioms we have already learned, and introduce two new ILP idioms.

8.1 INTRODUCTION TO COEVOLUTION

Coevolution is the process in which two organisms evolve together due to a synergistic or competitive relationship between them. The most obvious example is the coevolution of a parasite with its host organism, but there are many other examples. One common way [45, 143] to determine the existence, or extent, of coevolution of two organisms, say A and B, is to determine the "similarity" of the two evolutionary trees (phylogenies), for A and B, respectively. Identical trees support the hypothesis that A and B evolved together, and massively different trees do *not* support the hypothesis. But in between, we need some way to define and *measure* the *similarity* of two evolutionary trees. The *tanglegram* and the *tanglegram problem* derive from that effort.

8.1.1 What Is a Tanglegram?

Consider the two trees displayed in Figure 8.1. These trees are placed opposite each other, with their respective leaf sets arranged vertically, parallel to each other. The leaves of the two trees are labeled by the same set of labels (1 through 14), but some identically labeled leaves are in different positions on their respective parallel arrangements. For example, in the left tree of Figure 8.1 leaf 6 is in the sixth position from the top, but in the right tree leaf 6 is in the tenth position from the top. A dashed, straight line connects each leaf in one tree with the identically labeled leaf in the other tree. A drawing of this kind is called a *tanglegram*. The tanglegram problem, defined below, concerns the question of *how* to *draw* a tanglegram for two trees, in order to display the similarity of the trees in an informative way.

8.1.2 The Backstory

I first came upon the tanglegram problem in a seminar on molecular evolution where we read and presented sections of the book *Genes in Conflict* [32] by Austin Burt and Robert Trivers. Chapter 6 of that book had a figure (6.6B), showing the phylogenetic history of a set of 14 types of *yeast*, and another phylogenetic history of 14 forms of the element ω, a *parasitic* or *selfish* element that inserts into yeast genomes. Each of the 14 forms of ω inserts into a *different* type of the 14 yeasts; the form of ω and the type of yeast it infects were given the same label. The figure from [32] is essentially redrawn, but with numbers replacing names, in Figure 8.1 below.

The figure in the Burt and Trivers book was used to support the conclusion that there has been *little* coevolution of the *yeasts* and the ωs shown in the two trees. The caption of the figure states "The phylogeny of ω and its host are significantly different, ... " But how do we measure the "difference" between two trees? One obvious measure, suggested by the figures, and made more explicit in [143], is the number of *crosses* in the set of straight lines drawn between identically labeled leaves. A large number of crosses, or a large percentage of possible crosses, gives the *visual* impression that the two trees are very different.

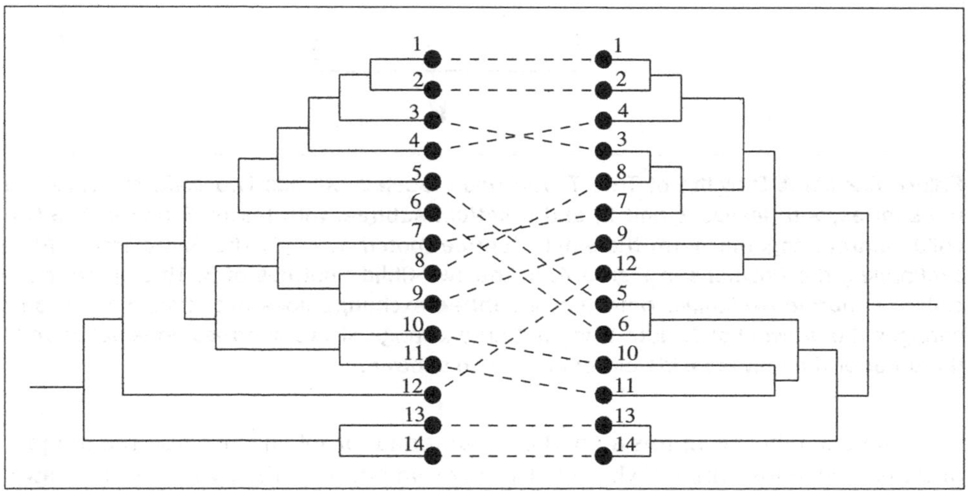

Figure 8.1 A Tanglegram, Essentially the Same as in Chapter 6 of the Book by A. Burt and R. Trivers [32]. The phylogeny for *yeast* is shown on the left, and the phylogeny for the parasite ω is shown on the right. Dashed lines connect each pair of leaves with the same label. These lines have 12 crosses in this tanglegram. The high number of crosses, and their tangled appearance, give the visual impression that the two phylogenies are very different. However, a different drawing of the two trees, shown in Figure 8.4, gives a tanglegram with only *nine* crosses and a less tangled appearance.

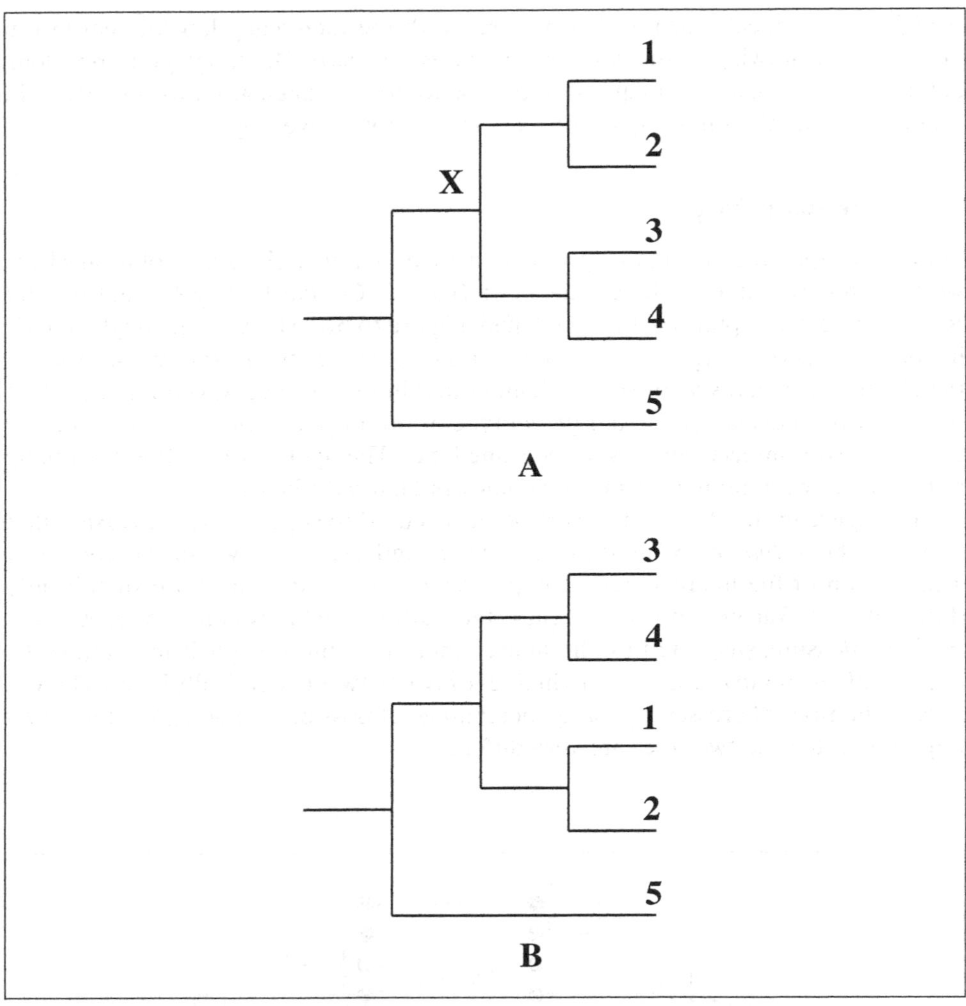

Figure 8.2 (A) A Drawing of Tree *T*. The node labeled "*X*" has two *child subtrees* – a *top* subtree, with leaves 1 and 2; and a *bottom* subtree, with leaves 3 and 4. The two child subtrees together form the single subtree *rooted* at *X*. (B) The Same Tree *T* After *Exchanging* the Positions in Figure *A*, of the Two Child Subtrees of *X*. This operation is called a *subtree exchange*. Note that a subtree exchange does not change *T* – it only changes the *drawing* of *T*. Note also, that an exchange of two subtrees does not change the order of the leaves *inside* either of the two subtree.

I immediately saw in this a good computational problem,[1] because the number of crosses depends both on what the trees are, and on *how* the two trees are drawn. And, there are many reasonable ways to draw the trees. For example, Figure 8.2 shows two drawings of the same tree, which differ by *exchanging* the positions of the two subtrees of the node marked "*X*."

[1] It turned out, however, that I was not the first to spot this problem, and there were already several papers and a book on the topic. These appeared in the biological literature, but also in the area of computer graphics, where the trees have a different use than in biological applications.

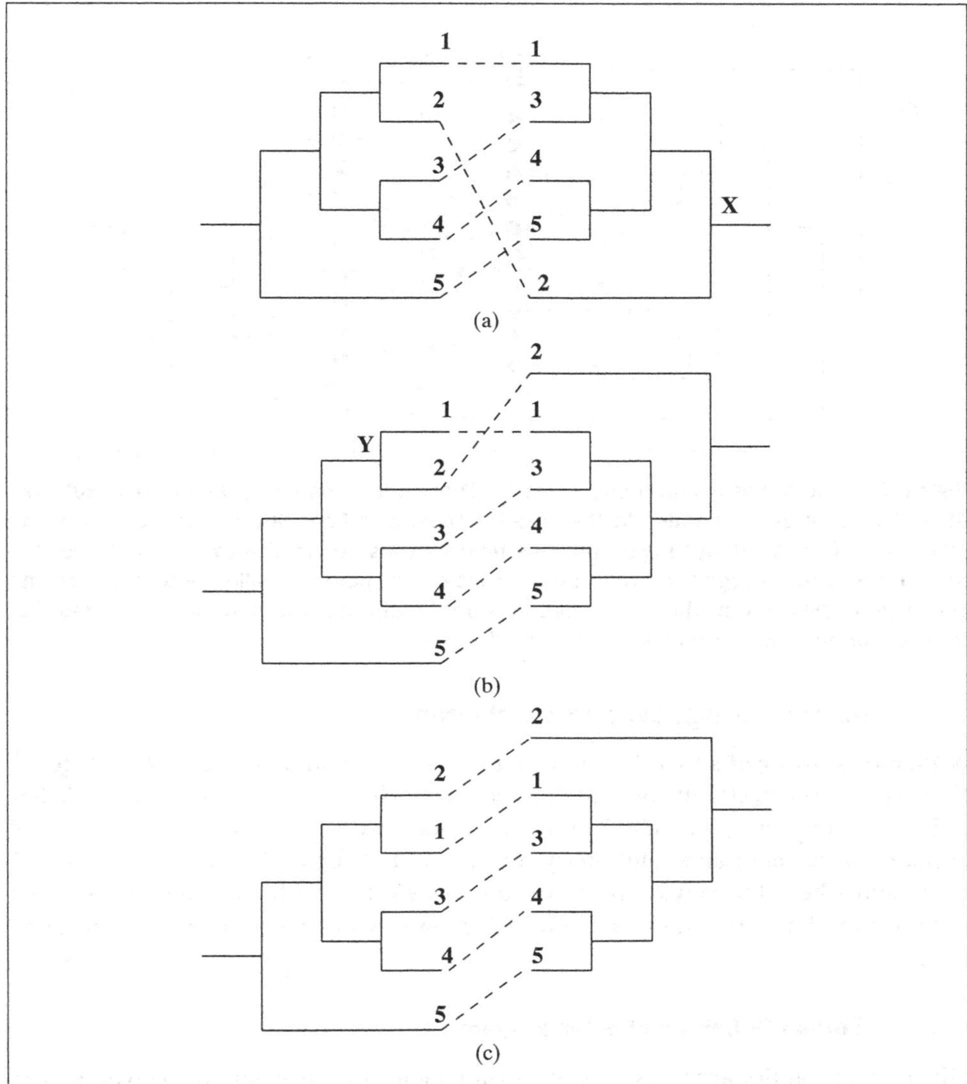

Figure 8.3 (a) Tanglegram with Three Crosses. Tree T_1 is on the left and tree T_2 is on the right. The positions of the two child subtrees of X in T_2 will be exchanged. (b) The Result of Exchanging the Positions of the Two Child Subtrees of X. The positions of the two child subtrees of Y in T_1 will next be exchanged. (c) The Two Trees Now with No Crosses.

8.2 TREE DRAWINGS, SUBTREE EXCHANGES, AND THE TANGLEGRAM PROBLEM

Figure 8.3a shows a tanglegram with three crosses, but a different tanglegram for the same two trees, in Figure 8.3c, has no crosses. Since no particular tree drawing is "canonical," it is not meaningful to define the similarity of the trees from the number of crosses in an *arbitrary* drawing of the two trees. Rather, we need a clearer definition of what drawings are allowed and are meaningful. We develop these definitions next.

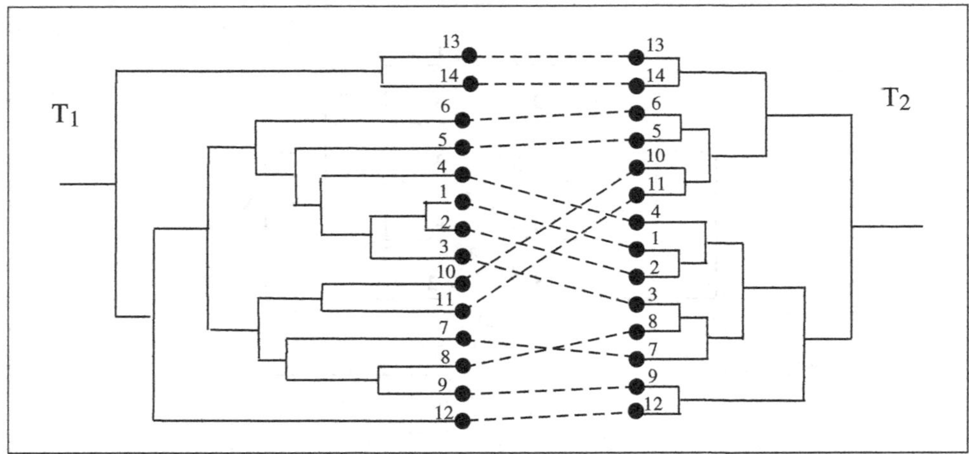

Figure 8.4 The Same Evolutionary Trees for *Yeast* and ω Shown in Figure 8.1, But Now After Five Subtree Exchanges in the Tree for *Yeast* and Four Subtree Exchanges in the Tree for ω. The resulting tanglegram has nine crosses, reduced from 12 in Figure 8.1. And, it has a more organized and modular look to it, perhaps calling into question the conclusions drawn from the earlier figure. Gurobi found the solution to the concrete ILP formulation for this instance in 0.02 seconds.

8.2.1 Planar Drawings Lead to Tanglegrams

A Planar Drawing of an Evolutionary Tree In an evolutionary tree (or phylogeny) there is one *root* node; all edges are directed away from the root; and each node has either zero or *exactly two* child subtrees. A node with no subtrees is called a *leaf*. A *planar drawing* of an evolutionary tree T is a drawing of T where *no* edges of T *cross*, and where the leaves are drawn on a single line with each leaf at a distinct point on the line. For example, Figure 8.2 shows two planar drawings of the same tree.

8.2.1.1 Formal Definition of a Tanglegram

Given two evolutionary trees T_1 and T_2 with n leaves each, where the leaves in each leaf set are distinctly numbered 1 to n, a *tanglegram* $\mathcal{T}$ for T_1, T_2 consists of a *planar* drawing of T_1 and a planar drawing of T_2, where the leaves of T_1 and T_2 are drawn on two parallel, *vertical* lines, as shown in Figure 8.1.[2] Then, each pair of identically labeled leaves are connected by a straight line. In a tanglegram for trees T_1 and T_2, the *number of crosses* of lines between the two leaf sets is denoted $C_{1,2}$.

8.2.1.2 Transforming Tanglegrams

Subtree Exchange A tanglegram can be transformed to a different tanglegram through the operation of a *subtree exchange* at a node v in tree T, where the position of the two child subtrees of v are exchanged. That is, the top and bottom child subtrees of v change positions – the top becomes bottom, and the bottom becomes top. A subtree exchange at a node v in T changes the order (on the vertical leaf line)

[2] We specify the lines to be vertical for concreteness, but other orientations are equally valid.

of two leaves in T, if and only if i and j are both in the subtree rooted at v, but in different child subtrees of v (see Figures 8.2 and 8.3). Note that a subtree exchange does not change the order of the leaves *inside* either of the child subtrees.

8.2.2 The Tanglegram Problem

Given evolutionary trees T_1 and T_2, each with n leaves labeled 1 to n, create a tanglegram for T_1 and T_2 that *minimizes* $C_{1,2}$, i.e., the number of line crossings in the tanglegram.

Input for an instance of the tanglegram problem consists of initial (arbitrary) drawings of trees T_1, T_2 in a tanglegram; and must specify which leaves are at the *top* of the drawing. An *optimal* solution to the tanglegram problem, one with the smallest number of line crosses, is specified by a set, S (possibly empty), of non-leaf nodes where subtree exchanges will be performed on the original tanglegram. Let $\overline{T_1}$ and $\overline{T_2}$ denote the final drawings of T_1 and T_2 after all the node exchanges have been made.

Since there are n leaves in T_1, there are $n-1$ non-leaf nodes (every non-leaf node has two children), so there are 2^{n-1} ways that the subtree exchanges can be chosen in T_1. Similarly, there are 2^{n-1} ways to choose subtree exchanges in T_2, so there are 2^{2n-2} different tanglegrams that can be created. Obviously, a brute-force method that examines each one, in order to find an optimal tanglegram, is not efficient. But the ILP approach we will detail next, along with the Gurobi ILP solver, found the optimal solution to the 14-node example in Figure 8.1 in a fraction of a second.

8.3 LOGIC FOR SOLVING THE TANGLEGRAM PROBLEM

In this section we develop several observations about tanglegrams and the tanglegram problem that form the core of the logic used in its ILP solution.

8.3.1 Determining Crosses in a Tanglegram

The tanglegram problem is to find a tanglegram for the two input trees that minimizes the number of line crosses. So, we have to explain the conditions that determine if two lines cross in a tanglegram. Since leaf labels are numbers from 1 to n, and the leaves are arranged on a vertical line, we say that two leaves i and $j > i$ in a tree T are *in order* if i appears *above* j in the drawing of T. Otherwise, leaves i and j are *out of order*. For example in Figure 8.1, leaves 3 and 4 are in-order in the left tree, but are out of order in the right tree.

Now, with a little playing around and reflection you should be able to discover what determines whether two lines cross in a tanglegram. But, to cut to the chase, the simple condition is:

The Ordering Lemma

In a tanglegram, the line connecting the two leaves labeled i crosses the line connecting the two leaves labeled j, if and only if leaves i and j are *in order* in one tree, but *out of order* in the other tree.

For example, in Figure 8.3a, reading from the top of the leaf sets downward, the pairs $(2,3)$, $(2,4)$ and $(2,5)$ appear *in order* in tree T_1, but appear *out of order* in tree T_2.

As prescribed by the condition above, the line connecting the two leaves labeled 2 crosses the three lines connecting the two leaves labeled 3, 4 and 5 respectively. All other leaf pairs appear in order in both trees, and hence there are no other line crossings. In Figure 8.3c, leaves 1 and 2 appear *out of order* in *both* trees, and hence their lines do *not* cross.

8.3.2 Determining Order in a Single Tree

Given the ordering lemma, the next question is what determines whether a pair of leaves (i,j) in a *single* tree, say $\overline{T_1}$, are in order or out of order. Clearly, the ordering of (i,j) in $\overline{T_1}$ depends on their original order in the input T_1, and on which subtree exchanges were made at nodes of T_1. But how exactly do subtree exchanges affect the order of the leaves?

Least Common Ancestor In Figure 8.3a, consider the unique path from leaf 1 to the root of tree T_2; and the unique path from leaf 2 to the root of T_2. The node labeled "X" is the first node where those two paths intersect. Node X is called the *least common ancestor* of leaves 1 and 2 in T_2; it is denoted $lca(1,2)$ in T_2.

Recall that a subtree exchange at a node v, exchanges the positions of the top and bottom child subtrees of v. Then, you should be able to verify the following fact with a little playing and reflection:

> A subtree exchange at a node v in T changes the order (from in order to out of order, or vice versa) of a pair of leaves (i,j) in T, if and only if node v is the least common ancestor of leaves i and j in T. That is, if and only if $lca(i,j) = v$ in T.

For example, in Figure 8.3a, the node labeled "X" is the lca in T_2 of the pairs $(1,2), (3,2), (4,2)$, and $(5,2)$, and of no other pair of nodes. A subtree exchange at node "X" changes the order of those pairs and no others.

Exercise 8.3.1 *Let S be a subset of non-leaf nodes in a tanglegram. Is it true that no matter what order we perform subtree exchanges at the nodes in S, the resulting tanglegram will be the same? Explain in detail. Hint: yes.*

Exercise 8.3.2 *Suppose tanglegram $\mathcal{T}$ is transformed to tanglegram $\overline{\mathcal{T}}$ by performing subtree exchanges at a subset of nodes S. Is it true that if we perform the subtree exchanges at the nodes in S, but on the tanglegram $\overline{\mathcal{T}}$, the result will be tanglegram $\mathcal{T}$?*

Where the Logic Has Led Us Given the logic developed above, we now see that the tanglegram problem can be solved by choosing *where* to perform subtree exchanges in the two input trees, to minimize the resulting number of leaf pairs that are *in* order in one tree but *out of* order in the other tree. So, we have completed Task A discussed in the Introduction. We now must complete Task B, translating that logic into an abstract ILP formulation to solve the tanglegram problem. That ILP formulation was developed in [194].

8.4 AN ILP FORMULATION

Any concrete ILP formulation for a specific problem instance will specify an initial drawing of the two input trees. So, in the *abstract* ILP formulation we will also assume that an initial ordering of the leaves is known.

8.4.1 The ILP Variables and Objective Function

We create one binary ILP variable $X1(v)$ for each non-leaf node v in T_1 and T_2, with the interpretation that $X1(v)$ will be set to 1 if and only if a subtree exchange is made at node v. Similarly, we create on binary variable $X2(v)$ for each non-leaf node v in T_2. So, given the initial tangelgram, the values of the $X1$ and $X2$ variables determine the ordering of the leaves in the tanglegram for $\overline{T_1}$ and $\overline{T_2}$.

Next, we create a binary ILP variable $R1(i,j)$ for each *pair* of leaves (i,j) in tree T_1, where $i < j$. We want a solution of the ILP formulation to set $R1(i,j)$ to 1, if and only if leaves i and j appear *in* order in $\overline{T_1}$, i.e., after all the subtree exchanges have been made in T_1. Equivalently, $R1(i,j)$ should be set to 0 if and only if i and j are *out* of order in $\overline{T_1}$. This will be accomplished through the inequalities to be explained below.

Similarly, we create a binary ILP variable $R2(i,j)$ for each pair of leaves (i,j) in T_2, where $i < j$. Again, any solution to the ILP formulation should set $R2(i,j)$ to 1, if and only if the leaves i and j are in order in $\overline{T_2}$.

Finally, we create a binary ILP variable, $C(i,j)$, for each pair of leaves (i,j), where $i < j$. We will develop inequalities that use these variables, to ensure that $C(i,j)$ is set to 1 in an ILP solution, if and only if the line between the two i leaves (in $\overline{T_1}$ and $\overline{T_2}$) crosses the line between the two j leaves.

These are the only variables needed in the *objective function*, which is:

$$\texttt{minimize} \sum_{i<j} C(i,j).$$

8.4.2 The ILP Inequalities

Inequalities for R variables The inequalities we use to set variable $R1(i,j)$ depend on whether the leaf pair (i,j) is in order, or out of order in the input tree T_1. Let v denote the node $lca(i,j)$ in T_1. If (i,j) are *in order* in T_1, then we create the inequality:

$$R1(i,j) = 1 - X1(v), \tag{8.1}$$

so that the value of $R1(i,j)$ is set to 1 if there is no subtree exchange at node v. Alternatively, the value of $R1(i,j)$ is set to 0 if there is a subtree exchange at node v. Similarly, If (i,j) are *out of order* in the input tree T_1, then we create the inequality:

$$R1(i,j) = X1(v), \tag{8.2}$$

so that the value of $R1(i,j)$ is set to 0 if there is no subtree exchange at node v; and is set to 1 if there is a subtree exchange at v. The inequalities used to set the $R2(i,j)$ variables are analogous to those in 8.1 and 8.2.

Exercise 8.4.1 *Verify that inequalities* (8.1) *and* (8.2) *correctly record the resulting relative order of leaves i and j in $\overline{T_1}$, after all subtree exchanges have been made, based on the input drawing of T_1 and the values of the $X1$ variables.*

Inequalities for C variables We want the binary variable $C(i,j)$ to be set to 1 if the line between the two i leaves crosses the line between the two j leaves. Hence, by the logical analysis done earlier, $C(i,j)$ should be set to 1 if the value of $R1(i,j)$ is *not* equal to the value of $R2(i,j)$ in a feasible ILP solution. This is achieved with the following two inequalities:

$$C(i,j) - R1(i,j) + R2(i,j) \geq 0, \tag{8.3}$$

and

$$C(i,j) - R2(i,j) + R1(i,j) \geq 0, \tag{8.4}$$

Inequality (8.3) forces $C(i,j)$ to be set to 1 if the leaf pair (i,j) is *in* order in $\overline{T_1}$, but *out of* order in $\overline{T_2}$. Inequality (8.4) does the symmetric thing, forcing $C(i,j)$ to have value 1 if leaf pair (i,j) is out of order in $\overline{T_1}$, but in order in $\overline{T_2}$.

Note that when the leaf pairs (i,j) are both in order or both out of order in $\overline{T_1}$ and $\overline{T_2}$, so that $R1(i,j) = R2(i,j)$, then both inequalities (8.3) and (8.4) reduce to

$$C(i,j) \geq 0,$$

which has no consequence, since $C(i,j)$ is already restricted to have value 0 or 1. Note also that we do not include inequalities to enforce that $C(i,j)$ is set to 1 *only if* $R1(i,j) \neq R2(i,j)$. This is because the objective function seeks to *minimize* the number of C variables that are set to 1. So, in an *optimal* solution to the ILP formulation, variable $C(i,j)$ will be set to 1 only if *forced* to by inequality (8.3) or (8.4). A similar issue was discussed in Section 6.2.3.1.

8.4.3 A Concrete Formulation

To illustrate the ILP formulation for the tanglegram problem, Figure 8.5 shows the concrete ILP formulation, in Gurobi format, for the two input trees in Figure 8.3a.

Exercise 8.4.2 *Using the node labeling convention explained in the caption of Figure 8.5, verify that the tanglegram in Figure 8.1 can be converted to the tanglegram in Figure 8.4 with* nine *subtree exchanges at the following nodes in tree* T_1*:* $X1(1,4), X1(1,5), X1(1,6), X1(7,11), X1(1,14)$*; and at nodes* $X2(1,4), X2(5,6), X2(5,14), X2(1,14)$ *in* T_2.

Exercise 8.4.3 *Unlike most of the ILP formulations discussed in this book, we have not provided a figure for the abstract ILP formulation to solve the tanglegram problem. Collect the inequalities introduced above, and write out the abstract ILP formulation for the tanglegram problem.*

8.4.4 One More Before We Leave

There are many additional examples of published tanglegrams in the biological literature where the number of line crossings can be reduced, and where this reduction might affect the conclusions drawn from the drawing. Without the kind of techniques discussed in this chapter, it is understandable that drawings "by eye" would likely have more crossings than necessary. The success of the ILP approach on real tanglegrams found in the biological literature fills me with joy, so I can't resist showing another one.[3] Figure 8.6 shows the published tanglegram in [185], with 29 crosses, and

[3] This one, and others, were produced by Parsoa Khorsand for a class project he did in the fall 2017 course I taught on integer programming in computational biology.

```
Minimize
+ C(1,2) + C(1,3) + C(1,4) + C(1,5) + C(2,3)
C(2,4) + C(2,5) + C(3,4) + C(3,5) + C(4,5)
subject to
C(1,2) - R1(1,2) + R2(1,2) >= 0
C(1,2) - R2(1,2) + R1(1,2) >= 0
C(1,3) - R1(1,3) + R2(1,3) >= 0
C(1,3) - R2(1,3) + R1(1,3) >= 0
C(1,4) - R1(1,4) + R2(1,4) >= 0
C(1,4) - R2(1,4) + R1(1,4) >= 0
C(1,5) - R1(1,5) + R2(1,5) >= 0
C(1,5) - R2(1,5) + R1(1,5) >= 0
C(2,3) - R1(2,3) + R2(2,3) >= 0
C(2,3) - R2(2,3) + R1(2,3) >= 0
C(2,4) - R1(2,4) + R2(2,4) >= 0
C(2,4) - R2(2,4) + R1(2,4) >= 0
C(2,5) - R1(2,5) + R2(2,5) >= 0
C(2,5) - R2(2,5) + R1(2,5) >= 0
C(3,4) - R1(3,4) + R2(3,4) >= 0
C(3,4) - R2(3,4) + R1(3,4) >= 0
C(3,5) - R1(3,5) + R2(3,5) >= 0
C(3,5) - R2(3,5) + R1(3,5) >= 0
C(4,5) - R1(4,5) + R2(4,5) >= 0
C(4,5) - R2(4,5) + R1(4,5) >= 0
R1(1,2) + X1(1,2) = 1
R1(3,4) + X1(3,4) = 1
R1(1,3) + X1(1,4) = 1
R1(1,4) + X1(1,4) = 1
R1(2,3) + X1(1,4) = 1
R1(2,4) + X1(1,4) = 1
R1(1,5) + X1(1,5) = 1
R1(2,5) + X1(1,5) = 1
R1(3,5) + X1(1,5) = 1
R1(4,5) + X1(1,5) = 1
R2(1,3) + X2(1,3) = 1
R2(4,5) + X2(4,5) = 1
R2(1,4) + X2(1,5) = 1
R2(1,5) + X2(1,5) = 1
R2(3,4) + X2(1,5) = 1
R2(3,5) + X2(1,5) = 1
R2(1,2) + X2(1,2) = 1
R2(2,3) - X2(1,2) = 0
R2(2,4) - X2(1,2) = 0
R2(2,5) - X2(1,2) = 0
```

Figure 8.5 The Concrete ILP Formulation for the Tanglegram Problem for the Input Trees in Figure 8.3a. All variables are binary, but the explicit list of variables (needed for Gurobi) is omitted. We use a node-labeling convention that labels each non-leaf node *v* with a pair of numbers *(p,q)*, where *p* is the top (first) leaf in the subtree rooted at *v*, and *q* is the bottom (last) leaf in the subtree rooted at *v*, in the input drawing of T_1 in Figure 8.3a. The same convention is used for the non-leaf nodes of T_2.

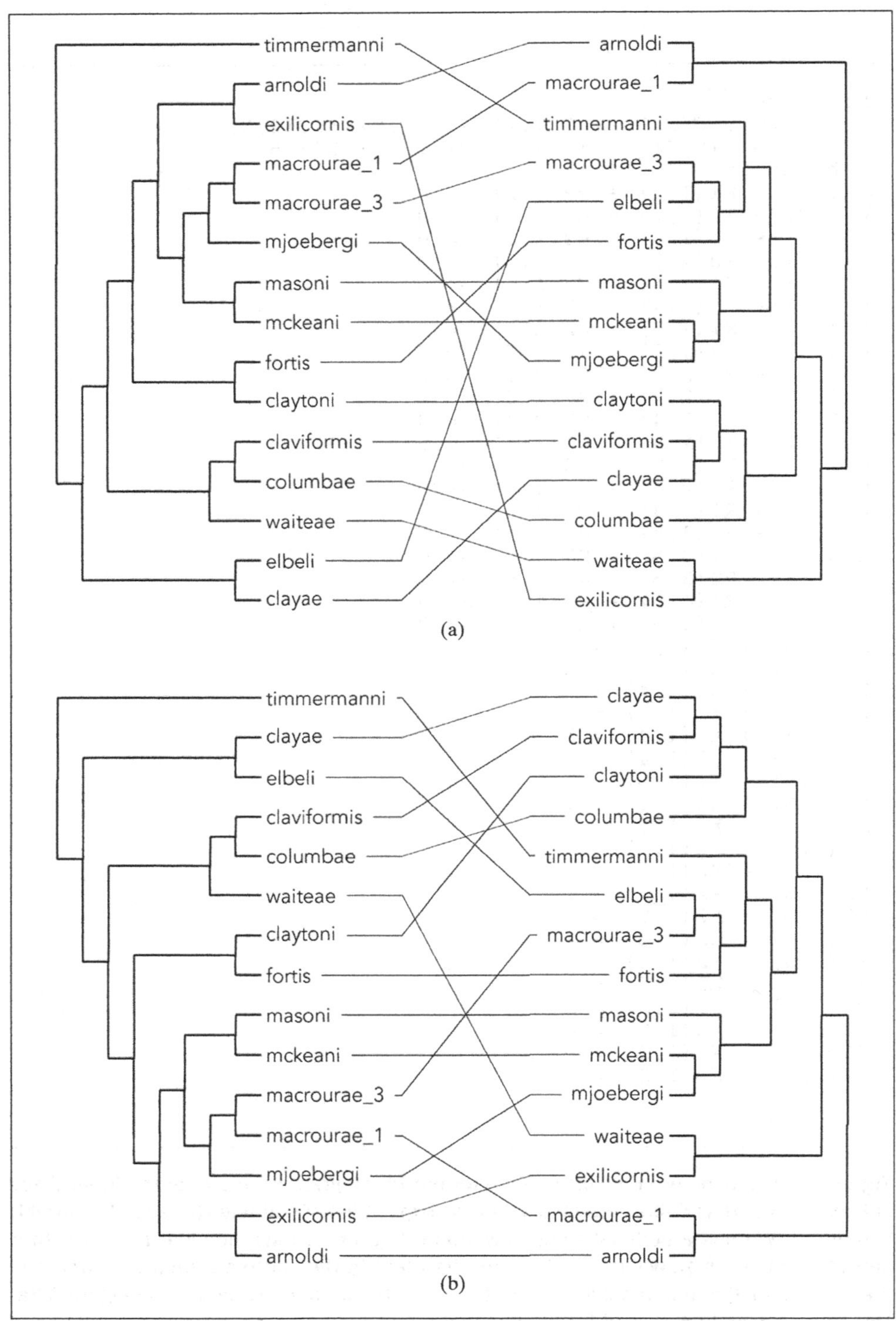

Figure 8.6 (a) The Tanglegram Shown in [185], with 29 Line Crossings. (b) An Optimal Tanglegram for the Same Trees, with Only 19 Line Crossings.

an optimal tanglegram, found by ILP, with only 19 crosses. A much larger example, with 34 species of lice and 34 species of bacteria that are parasites of lice, is is shown on page 30 of [45]. It has 129 crosses, while the optimal, found by ILP, has only 110 crosses.

8.5 SOFTWARE FOR THE TANGLEGRAM PROBLEM

A Perl program, called *tangILP.pl*, can be downloaded from the book website. Input to the program consists of three lines. The first line specifies the number of leaves in each tree, and the next two lines describe an initial tanglegram, i.e., drawings of the two input trees. Program *tangILP.pl* is executed in a terminal window using the command:

```
perl tangILP.pl file-name
```

Test date for the program is in the file *tangtest*, which can be downloaded from the book website. In order to explain the use of the program, we have to explain how a drawing of a tree can be specified in a way that a computer program can use it.

There is a well-known way to encode any evolutionary tree T with a *string*. This encoding is called the *Newick* format. We will specialize that encoding to encode tree *drawings* in a tanglegram. We need such an encoding so that the initial drawing of a tree can be input, as a string, to the program that creates concrete ILP formulations for the tanglegram problem.

We will explain our modification of the Newick format for trees with n leaves, where each leaf has a unique label from 1 to n, and each non-leaf, non-root node has exactly two children. The root has only one child. Remember that the leaves of the tree are arranged on a vertical line. To start, we look at some examples. The drawing of Tree A shown in Figure 8.2 is encoded as: (((1,2),(3,4)),5), and Tree B in Figure 8.2 is encoded as: (((3,4),(1,2)),5).

Exercise 8.5.1 *Just looking at the two examples above, and the tree drawings that they encode, try to figure out the rules to create a Newick encoding of a tree drawing.*

The rules The Newick encoding of a tree drawing can be accumulated during a *walk* around the drawing, starting at its root, and walking along each edge of the tree, first in the direction away from the root, and then toward the root. In more detail, the encoding is a string that is accumulated by adding symbols at its *right end* during the walk. The *first* time a non-leaf, non-root node v is visited, append a *left parenthesis* to the right end of the growing string, and traverse the edge to the *upper* subtree of v. When a *leaf* is visited, append its label to the right end of the growing string, and back up along the edge to the parent node of the leaf. The first time the walk *backs up* to a node v, append a comma to the right end of the string, and traverse the edge to the *lower* subtree of v. Finally, when the traversal backs up to a non-root node v for the *second* time, append a *right parenthesis*, and traverse the edge to the parent of node v. Note that we don't add anything to the string when the root node is visited, either at the beginning or end of the walk.

To better understand the Newick format, define T_v as the subtree of a non-leaf, non-root node v in T. Then note that in the Newick encoding for T, there is a pair of left and right parentheses that enclose the leaves of T_v, and do not enclose any leaves not in T_v. Further, the left and right parentheses in a Newick encoding are properly nested, so we can recreate the drawing of T from its Newick encoding by finding the matching left and right parentheses. Note that the leaves in a Newick encoding are listed in exactly the same order that they appear (top to bottom) in the drawing of the encoded tree.

Exercise 8.5.2 *Using the idea of a walk around the tree, verify that the Newick encodings given above are correct.*

Exercise 8.5.3 *Verify that the Newick encodings of the drawings of the two trees in Figure 8.1 are*
$((((((((1,2),3),4),5),6)((7,(8,9))(10,11))),12)(13,14))$ *and*
$(((((1,2),4)((3,8),7))(9,12))(((5,6)(10,11))(13,14)))$.

Exercise 8.5.4 Software *Use program* tangILP.pl *to generate the concrete ILP formulation for the tanglegram problem, using the Newick encodings (shown in Exercise 8.5.3) of the trees in Figure 8.1. Then use Gurobi to solve the formulation, to verify that the optimal solution has only nine crosses.*

Exercise 8.5.5 Software *Create the Newick encodings of the two trees drawn in Figure 8.6a. Then use* tangILP.pl *and Gurobi to verify that there is a tanglegram for those trees with only 19 crosses.*

Exercise 8.5.6 *Figure out and state clearly how to recreate the drawing of tree T from its Newick encoding. This is a classic kind of parsing problem in computer science, and would be a good exercise in several classic computer science classes.*

8.6 THE IF-XOR IDIOM FOR BINARY VARIABLES

In Section 4.3 we discussed *If-Then* and *Only-If* ILP idioms for binary variables. Now, in discussing tanglegrams, we saw that inequalities (8.3) and (8.4) ensure that binary *variable* $C(i,j)$ will be set to 1 *if exactly* one of the binary variables $R1(i,j)$, $R2(i,j)$ has the value 1, and the other has the value 0. Those inequalities implement another type of ILP idiom. Those inequalities implement the *If-XOR* idiom for *binary variables.*

Abstractly, the *If-XOR* idiom for binary variables X, Y, with the result assigned to Z is:

> Variable Z must have value 1 *if exactly one* of the binary variables X or Y has value 1:

The If-XOR idiom for binary variables can be implemented as:

$$Z - X + Y \geq 0, \tag{8.5}$$

$$Z - Y + X \geq 0. \tag{8.6}$$

In any feasible ILP solution, Z will be set to 1 *if* the values of X and Y are different. This is true because when the values of X and Y differ, one of the inequalities 8.5 or 8.6 becomes $Z \geq -1$, and the other becomes $Z \geq 1$. Further, the values of X and Y are the same, both inequalities become $Z \geq 0$, which causes no bad side effects since Z is a binary variable, and hence already limited to be 0 or 1.

Note that inequalities (8.5) and (8.6) don't enforce the property that Z will be set to 1 *only if* the values of X and Y are different. Whether or not that causes trouble in the ILP formulation depends on the other details of the formulation. As discussed earlier, these inequalities are sufficient in the ILP formulation for the tanglegram problem.

8.6.1 The-Only-If XOR Idiom

If we need to implement the XOR idiom so that Z is set to 1 *only if* the values of X and Y are different, then we can use the following inequalities:

$$Z - X - Y \leq 0, \tag{8.7}$$

$$Z + X + Y \leq 2. \tag{8.8}$$

Inequality (8.7) has the consequence that if Z has value 1, then at *least one* of X or Y must also have value 1. Inequality (8.8) has the consequence that if Z has value 1, then *at most* one of X or Y can have value 1. Together, these two inequalities implement the requirement that Z has value 1, only if *exactly* one of X and Y has value 1 (and hence the other must have value 0). So, they implement the logic for the idiom "$Z = 1$ *only if* (X XOR Y) = 1," when all the variables are binary.

We must also verify that these inequalities implement the XOR idiom for binary variables without any bad side effects. This can be done by enumerating four cases, i.e., assigning X and Y the values of 0 and/or 1 respectively, and checking that none of the four inequalities causes a bad side effect.

Exercise 8.6.1 *Show that inequalities (8.7) and (8.8) have no bad side effects.*

The If-and-Only-If XOR Idiom for Binary Variables Clearly, if we want Z to be set to 1 *if and only if* the values of X and Y differ, we can combine the inequalities (8.5), (8.6), (8.7), and (8.8). But, we still need to check that no bad side effects can occur.

Exercise 8.6.2 *Verify that inequalities (8.5), (8.6), (8.7), and (8.8) implement the if-and-only-if XOR idiom for variables, with no bad side effects.*

Exercise 8.6.3 *Verify that if any one of the inequalities (8.5), (8.6), (8.7), or (8.8) were omitted, then the remaining inequalities would not implement the if-and-only-if XOR idiom with no bad side effects.*

More Tanglegrams and Idioms Later In Chapter 15 we will return to the topic of tanglegrams and see how integer programming can be used to formulate and solve additional problems defined on tanglegrams. Also, generalizations of the tanglegram problem where a leaf of one tree may have edges to more than one leaf in the other tree are considered in [195]. Tanglegram problems of those types occur when a parasite inhabits more than a single species, or a species is inhabited by more than one parasite [185].

In Chapter 12 we will discuss several additional ILP idioms that are often used in ILP formulations. These will include an *exclusive OR* (XOR) idiom for *inequalities* and an XOR idiom that doesn't *test if* the XOR relation holds, but *requires* that it does.

9

Traveling Salesman Problems in Genomics

9.1 THE TRAVELING SALESMAN PROBLEM (TSP) IN GENOMICS

With current genomic technology, many tasks involving DNA maps, sequences, and sequencing lead to computational problems that concern the *ordering* or *permutation* of DNA sequence *fragments*. Most problems come from experimental techniques in genomics that give information (often partial, or containing errors) about (possibly overlapping) fragments of a linear molecule (e.g., a chromosome). However, the experiments don't explicitly determine how those fragments are ordered on the chromosome. Other experimental techniques give information (often partial) about the *relative* positions of molecular *markers* (sites) in a linear molecule, but again do not fully specify the complete order of those markers. The computational problems must use such information to try to *deduce* the correct, complete *order* of the DNA fragments or the molecular markers.

The specific details of the experimental techniques generating fragment data determine specific details of the computational problems. However, there are many diverse experimental techniques and they change rapidly, so we will only give a *high-level* description of one common *type* of experimental protocol. That example will serve to motivate many fragment ordering problems, and illustrate how ordering problems can be addressed using integer programming. But regardless of the experimental basis for the data, most of the current (and almost surely, future) ordering problems come down to trying to determine a best *permutation* of some set of elements, *subject to* various *constraints* on what permutations are permitted. Often, these permutation problems in genomics are naturally modeled as variants of the famous, and notorious, *Traveling Salesman Problem* (TSP). Hence, we will discuss in some detail the solution of the TSP using integer linear programming.

In this chapter, we will look in detail at a simple, *compact* ILP formulation for a TSP variant, and discuss its practical efficiency, when solved by a *single* execution of an ILP solver. We then discuss an ILP formulation for the TSP that is *not* compact, but can be solved much faster in practice, and for larger problem instances, using a *series* of ILP formulations and computations. Then, in the exercises, we introduce other simple, compact ILP formulations for TS problems,[1] and explore an extended example of the use of TSP in modeling a computational problem in genomics.

9.2 THE TRAVELING SALESMAN PROBLEM (TSP)

The classical Traveling Salesman Problem (TSP) requires finding a *minimum-cost* route for a salesman to *visit* each of a given set of cities *exactly once*. Input to a problem instance is a map (or graph) showing which pairs of cities (nodes) are directly connected by a road (edge), and the cost of traversing each road. The cost of a route that visits each city once is the *sum* of the costs of the roads traversed on the route.

Tours and Paths The TSP comes in two high-level variants. In the TS *Tour* (cycle) version, the salesman is required to start and end at the same city, hence traveling around a tour, entering and leaving each city exactly once. In the TS *Path* version, the salesman is required to start and end at *different* cities, hence traveling a path. In the path variant, the salesman must leave (but not enter) the start city once; and must enter (but not leave) the end city once; and enter and leave each of the other cities exactly once.[2] Further, there are two variants of the TS path problem, one where the start and end cities are *specified* as part of the problem input, and one where they are not.

Exercise 9.2.1 *Why don't we specify two variants of the TS Tour problem, one where the start/end city is specified, and one where it is not?*

More formally, a TS problem is defined on a *graph*, where each edge has a cost, and where the edges (for simplicity of exposition) are all (for now) *undirected*. The salesman can traverse an undirected edge in either direction, but at most once, and so cannot traverse the edge in both directions. The cost of traversing an edge is the same in either direction.

9.2.1 Converting the TS Path Problem to a Tour Problem

Most of the TS problems that arise from genomics lead to TS *path* problems, rather than TS *tour* problems. However, it is technically convenient to convert a TS *path* problem on a graph G with n nodes to a TS *tour* problem on a related graph G' with $n+1$ nodes. As a consequence, when we discuss ILP formulations for TS problems, we will focus on TS tour problems.

To explain the conversion, assume that the original problem is to find a minimum-cost TS *path* in an undirected, connected graph G, where the start and end nodes of the path are *not* specified in advance. To convert to a tour problem, we add a new node, labeled 0, to the n original nodes $\{1, \ldots, n\}$ in the input graph G, and add a new undirected edge, with zero cost, between each of the original nodes and node 0. The augmented graph is called G'. Then, a TS tour in G', starting and ending at node 0 defines a TS path in G. To get the path from the tour, just remove node 0 and the two

[1] Terminology note: When the use of the acronym "TSP" is not grammatical, as in this instance, we use "TS" as an abbreviations for "traveling salesman," which is then followed by the appropriate modifier.

[2] Informally, we say that the salesman "visits" each city once, knowing that this is shorthand for the more complete description give here.

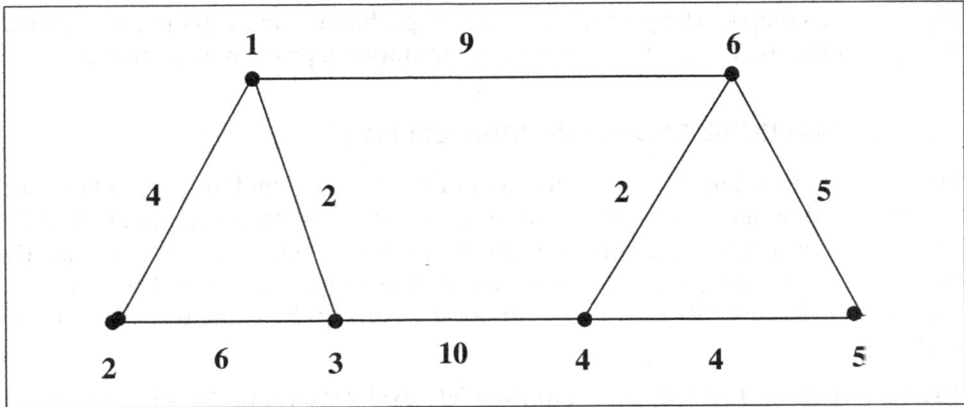

Figure 9.1 Graph G. We Want to Find an Optimal TS Path on G, Where the Start and End Nodes of the Path *Are* Not Specified As Part of the Problem Input.

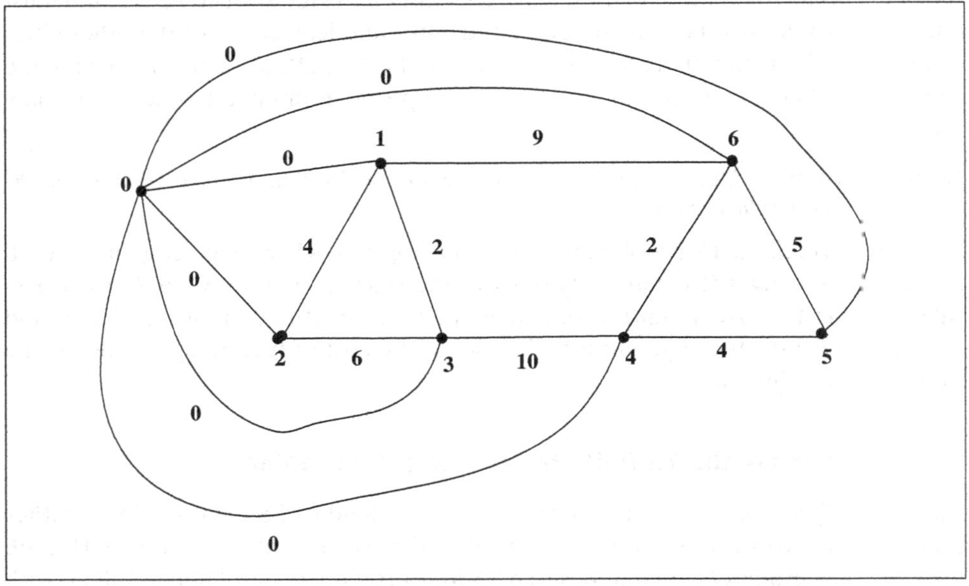

Figure 9.2 Graph G' Derived from Graph G in Figure 9.1. An optimal TS tour in G' is {0, 2, 1, 3, 4, 6, 5, 0}, with total cost of 23. This identifies a optimal TS path {2,1,3,4 6,5} in G.

edges incident with node 0 on the tour. Since the edges incident with node 0 have cost zero, the cost of the TS tour in G' will be the same as the cost of the resulting TS path in G. It follows that the TS path problem on G can be solved by solving the TS tour problem on G' (see Figures 9.1 and 9.2).

Exercise 9.2.2 *Suppose we want to solve the TS path problem on G when the start and end nodes of the path are specified as part of the problem input. Again, we want to convert this problem to a TS tour problem on a related graph, call it G''. Describe the appropriate graph G'' for this conversion. Now suppose the problem specifies the start node, but not the end node. What is G'' in that case?*

Exercise 9.2.3 *So far in our treatment of TS problems, all edges were assumed to be undirected. However, some applications are modeled by a* directed *graph, where there might be an edge from a node i to a node j, but not edge from j to i; or both edges might exist, but possibly with different costs. With directed edges, the salesman can only traverse an edge in the direction indicated on the edge.*

Show how to modify the construction of graph G' so that a TS path *problem on a graph G with some directed edges, is converted to a TS* tour *problem on a graph G' with some directed edges.*

Applications Before Solutions Before discussing how to solve TS tour problems using integer linear programming, we discuss several applications of TS problems in computational biology. Other applications of TS and TS-like problems (in data *clustering*) will be discussed in Sections 21.1 and 21.2. Applications of TSP to problems of *genome rearrangements* will be discussed in Section 18.2.6.1.[3] In particular, we will discuss in some detail a very important problem in DNA sequencing, the *DNA assembly problem*, and how it can be modeled as a TS problem. We will then discuss in detail a problem called the *marker-ordering* problem.[4]

9.3 TS PROBLEMS IN DNA SEQUENCING AND ASSEMBLY

As most readers of this book know, a DNA molecule can be considered as a sequence (or string) of nucleotides (or bases), each represented by a letter in the four-letter alphabet *A, T, C, G*. *De novo* DNA sequencing of a species refers to sequencing the genome of a species without making reference to a DNA sequence of any other individual (or mix of individuals) in that species. Once the first genome of an individual in a species has been sequenced, that sequence becomes a "reference genome sequence" for that species, and it can be used to simplify the sequencing of additional individuals in that species.

Today, *de novo* DNA sequencing of a species is almost always achieved using some variation of *shotgun* sequencing, in which the *sequence assembly* problem is a central *subtask*. DNA sequence assembly illustrates a typical way that TS problems arise in genomics.

9.3.1 The DNA Sequencing Problem

The essential technological problem in sequencing a "long" DNA molecule is that with existing DNA methods, the longest single *intact* DNA molecule that can be sequenced is relatively short. The length has varied over time, from around 300

[3] Additional genomic applications of TS problems that we will *not* discuss appear in [2, 92], where TSP is used to model *radiation-hybrid mapping* (which is a technology that is no longer current, but the structure of the problem is similar to the fragment-ordering problem in Section 9.4, and additional problems with similar structures will likely arise in future applications); in [72], where an optimal TS tour reconstructs a putative ordering of genes expressed during the cyclic *cell cycle*; in [101], where an optimal TS path is used to order *protein interaction data* in order to reveal common function; in [111], where a near-optimal *multiple sequence alignment* (under the sum-of-pairs objective) is built from a TS tour through the sequences; and in a related paper [112], where an optimal TS tour is computed as a first step in a method to build phylogenetic trees. Those trees solve the *weighted maximum-parsimony problem* (exactly as defined in Chapter 5), when the edge scores satisfy certain technical conditions that the authors believe are biologically meaningful.

[4] Some of this material is adapted from a chapter in [83].

bases in the first sequencing methods, to only 20 to 30 bases in more recent faster and cheaper methods, to a current length of a few thousand bases. But, a chromosome is a single molecule, and in humans it has several *hundred million* bases, so the chromosome must be broken up into many short pieces that can be individually sequenced. However, when breaking up the chromosome, the *original order* of the pieces becomes scrambled. No one has figured out a cheap way to avoid that. We can sequence the short, individual pieces (called "reads"), but then how do we put the reads back together in the correct order? *Shotgun sequencing* is an idea that solves this
problem.

In shotgun sequencing, we first make many copies of the DNA of interest (which is cheap and easy to do), and then cut each copy at many *random* points. The random cutting produces short *fragments* from the original DNA, but the process is such that we don't know which DNA copy produced which fragment; and we don't know the original order of the fragments on the full DNA. Instead, we have a *set* (or *soup*) of fragments without knowing where in the full DNA, or from which copy, a fragment came from.

The next step in shotgun sequencing is to sample from the soup, to select and individually sequence many fragments whose lengths fall in the range that allows the fragment to be sequenced as a single string. Each sequenced fragment is called a "read." Although we only sequence a *subset* of the fragments, the fact that many copies of the original DNA are made, and that the cutting points are random, means that the sequenced fragments from *different DNA copies* might *overlap* each other sufficiently to allow deduction of the original DNA sequence. The crucial information used to deduce that two fragments overlap, is a shared pattern at the end of one fragment, and the beginning of the other (remember that the fragments are roughly the same length, so containment of one fragment in the other is unlikely, and if it does happen, the shorter fragment can be removed). If enough overlapping fragments are sequenced, the common patterns of overlapping fragments can be used to assemble the sequence of the full, original DNA string (see Figure 9.3).

9.3.2 Sequence Assembly

Having selected and sequenced a subset of overlapping fragments, the problem is to deduce the sequence of the full string. This is the algorithmic problem of shotgun *sequence assembly*. The basic idea is to compare each *ordered* pair of fragments, (α, β) to see if there is a "long" *suffix* of α that is highly similar to a *prefix* of β. In the ideal case, there is a long, *identical* suffix-prefix match, as shown in several of the pairwise overlaps in the assembly of fragments in Figure 9.3. More typically, due to errors in sequencing (such as inserted or deleted bases or incorrect misread-read bases), the suffix-prefix match is not perfect, but is strong enough to identify a pair of fragments (α, β) that likely do overlap in the original DNA. Less than perfect overlaps are also shown in Figure 9.3.

Pairwise Overlap Scores With pairwise comparisons, we can assign a score, $S(\alpha, \beta)$ to each ordered pair (α, β) of fragments. A score reflects our level of *uncertainty* that a suffix of α overlaps a prefix of β in the original DNA sequence. A low score indicates high confidence (i.e., low uncertainty) of an ordered overlap, and a high

```
    A  T  C  G       G  C  T  A  A
          C  G  G  A  C     T  A  C  T  A          Substring layout
 A  A  T  T        G  A  G  C  T  T  C  T  A  G

 A  A  T  C  G  G  A  C  C  T  A  C  T  A  G       Contig
```

Figure 9.3 An Assembly of Seven Sequenced Fragments Into One Contig.

score indicates low confidence. For example, a score of infinity indicates that there is no evidence of overlap. The score can reflect both the *quality* of the observed overlap (e.g., how many mismatches there are) and the *length* of the overlap (e.g, longer overlap suggest a higher confidence, hence lower uncertainty). There are many ways that have been suggested for defining such scores. The point here is that given pairwise scores, the problem of finding the "best" assembly (layout) of the fragments can be modeled as a *traveling salesman path* problem.

Fragment Assembly via TS Path We model the assembly problem as a *TS path* problem on a *directed* graph G, called the *overlap graph*. Each DNA fragment is represented by a node in G, where we name the node by the fragment it represents. For any ordered pair of nodes (α, β), there is a directed edge from node α to node β, with *cost* equal to $S(\alpha, \beta)$. Then, any ordering of the fragments specifies a TS path in G (remember that no fragment is strictly contained in another fragment). The *cost* of the path is the sum of the costs on the directed edges contained in the path.

Conversely, any TS path in G specifies an ordering of the fragments. That ordering of fragments can be used to create a single sequence, and a location of each fragment in the sequence. This is done by moving through the TS path, noting the overlaps between successive fragments. When the match between a fragment and its successor is not exact, i.e., there is some disagreement on some character(s), we can use additional information, such as other overlaps to choose the character. For example, see Figure 9.3 and the contig created from the fragments. In that figure, we chose the most frequent character at each position for the character in the contig. In summary, a TS path gives an ordering of the fragments; the ordering of the fragments, and the overlaps, gives the placement of the fragments; and the placements of the fragments leads to the contig sequence. So, the choice of TS path is critical to the process of creating a contig sequence. Which TS path is best?

Selection Criteria Although not a perfect criterion, it is generally thought that the correct DNA assembly should correspond to a TS path in G with a relatively small total cost. Then, the DNA assembly problem is stated as the problem of finding a *minimum cost* TS path in the overlap graph G.

Exercise 9.3.1 *Fully flesh out the explanation of how a TS path determines an assembly of the DNA fragments, and why a minimum cost TS path is likely a good estimate of the correct assembly. What are the deficiencies of this model?*

Global Influence Note that by solving the assembly problem as a TS path problem, *all* of the data contributes *globally* to the overall assembly. This is in contrast to simpler methods (see [83] for a discussion) that built up a sequence assembly using *greedy* algorithms, where one fragment is added at a time, based on the best suffix-prefix overlap at one of the ends of the growing assembly.

In the past, the TS path approach to the assembly problem was not extensively used, because the time to compute the optimal TS path was too great. But given the huge improvement in ILP solvers generally, the increase in computer speed and parallelism, and with specialized TSP solvers (such as the program *Concorde*), the TS path model for sequence assembly is now much more practical.

9.4 MARKER ORDERING: A DIFFERENT FRAGMENT LAYOUT PROBLEM

In the sequence assembly problem, we are given the sequences of many DNA fragments. So although we don't know how the fragments are arranged in the genome, we do know how many fragments there are, and we know the DNA sequence that each fragment contains. Now we will consider a different fragment layout problem in genomics, where in addition to not knowing how the fragments are arranged in the genome, we don't know the DNA sequence of any fragment, or even how many fragments there are. We will again show that this fragment layout problem can be modeled and solved as a TS problem.

The Experimental Protocol Without going into the full details of any particular experimental protocol (which change rapidly), the *abstract* situation is the following: We have n copies of a DNA molecule, and we do n experiments, each one using a different DNA molecule. In each experiment, some fragments of one copy of the DNA are *randomly* obtained from the DNA. The extracted fragments from the experiment are *nonoverlapping* and might only cover a small percentage of the full DNA sequence (see Figure 9.4). Each of the n experiments determine which of m *molecular markers* are contained (somewhere) in the fragments extracted from the DNA copy used in the experiment. Abstractly, we can think of a molecular markers as some feature of the DNA that occurs in some fixed location(s) in the genome. The result of the n experiments can be summarized in a matrix, M, of size n-by-m, where each row represents an experiment and each column represents a marker. A row i in M has entry 1 in column j, if and only if marker j was found during experiment i, meaning that marker j is contained in the DNA fragments used in experiment i. (see Figure 9.5).

Note that the experimental data does not show the fragments, or their arrangement in the genome; and similarly, we don't know the order of the markers in the genome. Matrix M simply reports which markers were detected in which experiments. And, as always in biological data, some of the data is in error or missing. Still, the collection of fragments obtained for an experiment is sufficiently random so that two experiments will likely contain a different set of fragments, which generally cover different but possibly overlapping portions of the DNA (see Figure 9.4).

9.4.1 The Computational Problem

The Marker-Ordering Problem: Use the binary data in M to *deduce* the true order of the markers; and determine the number, composition, and placement of the fragments in the original DNA. Of course, a solution must agree with the fragment-marker data given in M.

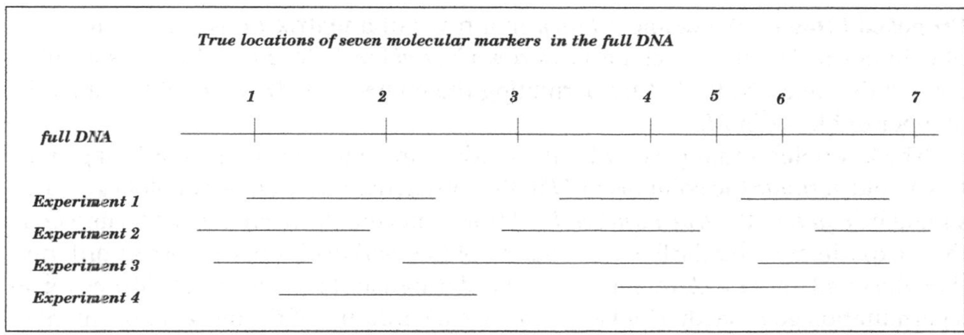

Figure 9.4 A Schematic of Human DNA Fragments That Are Created by Four Experiments. The first line represents a full DNA sequence, and shows the true (but unknown) locations of seven molecular markers. The four succeeding rows show the true (but unknown) locations of DNA fragments created in the experiments. The experiments produce the data shown in Figure 9.5.

M	2	4	5	3	1	7	6
1	1	1	0	0	1	0	1
2	0	1	0	1	1	1	0
3	0	1	0	1	1	0	1
4	1	1	1	0	0	0	1

Figure 9.5 The Fragment-Marker Data *M* from Figure 9.4. Each row in *M* corresponds to one experiment, and each column corresponds to one marker. A column labeled *j* has a 1 in row *i* if marker *j* is detected in experiment *i*, and hence is located in one of the (unknown) fragments created in experiment *i*. Since the true order of the markers is unknown and must be deduced, the column labels are shown in an arbitrary order.

The Ideas Behind the Solution The intuition behind all approaches to solving the marker-ordering problem (and many similar problems) is that the frequency with which two markers are present together, or are absent together, in an experiment depends on the physical distance between the two markers in the DNA. That follows because the closer they are the more likely it is that *both* will be contained together in some fragment created in an experiment, whenever *either* one is contained in a fragment. The experimental protocol essentially takes many random samples of the DNA, where each sample reports the markers found in a random set of nonoverlapping intervals covering only a portion of the DNA.

The true experimental protocols are more complex than described above, and are additionally complicated by errors. But the combinatorial elements are clear and we will describe an ILP approach to reconstructing the marker and fragment orders.

Proposed Model We define a *block* in a row i of a matrix M as a *consecutive* set of columns in M where each entry in row i has value 1. So, each block has a run of consecutive ones. Note that by permuting the columns of M, we might change the number of blocks in M.

The key point is that *if* we order the markers in their true, original ordering in the DNA, and *permute* the columns of M in that order, then the number of blocks created should be *equal* to the *true number*, b, of fragments that were created and sampled in the n experiments. Similarly, if an ordering of the markers is *not* the correct ordering, then the number of blocks created in that ordering should be *larger* tn b. Moreover, as a permutation gets "farther" away from the true ordering of the markers, the number of blocks should generally increase. Hence, one proposed way to solve the *Marker-Ordering Problem* is to solve the following:

> **Column-Permutation Problem** Find a permutation of the columns of M that creates the *minimum total number* of blocks of consecutive ones.

Given a solution to the column-permutation problem, each block created in a row i of the permuted M is *conjectured* to be a fragment created in experiment i. Note that since we only permute columns, the fragment-marker data in the permuted M will be identical to those in the original M.

We will see, in Section 9.4.3, how the column-permutation problem can be cast and solved as a traveling salesman problem on an undirected graph. But, first we discuss a variant of the column-permutation problem that also arises in genomics.

9.4.2 The Consecutive-Ones Problem

When the columns of M can be permuted so that in each row, *all* of the 1s are organized into a *single* consecutive block, then M is said to have the *consecutive-ones property*. Clearly, the problem of testing whether a binary matrix M has the consecutive-ones property is just a special case of the column-permutation problem. The number of blocks in an optimal solution to the column-permutation problem will be equal to the number of rows of M, if and only if M has the consecutive-ones property. So, a method that solves the column-permutation problem can be used to solve the consecutive-ones problem.[5]

Consecutive Ones in Paleo-Genetics In [39], the authors show how the goal of reconstructing the genomic sequences of *extinct* mammals, using data from highly degraded, fragmented DNA, together with DNA sequences from related living mammals, can be expressed and solved as a *consecutive-ones* problem. With that approach, if the DNA sequence for the extinct animal is correctly deduced, DNA of a living relative can be modified to become a *copy* of the DNA of the extinct animal. Then, the hope (*or fear*) is that the new (well, actually old) DNA can be incorporated into the cells of the living relative, so that the relative can give birth to living clones of a long-extinct animal.

[5] Actually, there is an efficient algorithm (in the provable, worst-case sense) that solves the consecutive-ones problem, but it is very complex to understand and to implement as a computer program. The ILP approach to this problem is valuable for its simplicity, even though the specialized algorithm is much faster.

	2	4	3	1
1	1	1	0	1
2	0	1	1	1
3	0	1	1	1
4	1	1	0	0

Figure 9.6 An Extract of Marker Data from Matrix *M* in Figure 9.5. The undirected graph for this data is shown in the Figure 9.7.

9.4.3 Solving the Column-Permutation Problem As a TS Tour Problem

Given a binary matrix M, create an undirected graph $G(M)$ with one node for each column in M, and one undirected edge between each pair of nodes. Then add an extra node s, with an edge to each other node in $G(M)$. Next, assign a cost to each edge in the graph: The cost of edge (s, v), for any node v, is the total number of cells in column v with value 1. The cost of any other edge (u, v) is the number of rows where the entries in the column for u and the column for v *differ*. This is called the *Hamming distance* between columns u and v. For example, consider the matrix M shown in Figure 9.6. The graph $G(M)$ created from M is shown in Figure 9.7.

A traveling salesman tour in $G(M)$ is a cycle that visits each node exactly once. Starting at node s, the order that the other nodes are visited creates a permutation of the columns of $G(M)$ (see Figures 9.6 and 9.7). We claim that a *minimum cost TS tour* in $G(M)$ specifies a solution to the column-permutation problem, which is a plausible solution to the marker-ordering problem. Later, in Sections 9.6 and 9.8, we will discuss ILP formulations for TS problems.

Explaining the Solution We examine three cases.

Case (1) Suppose first that column u has a 1 in row i, and that column v has a 0 in row i, and that edge (u, v) is traversed in a TS tour, from node u to node v. This means that marker u appears in cell i, but marker v does not. Therefore, the ordering of markers corresponding to the column ordering implies that a fragment in cell i must *end* after marker u, but *before* marker v. It follows that the column ordering given by the tour creates a block of ones in row i that ends with column u. The cost of edge (u, v) thus counts the *number* of blocks of ones that will end with column u (if edge (u, v) is traversed from u to v), and hence also counts the number of fragments that end after marker u in the resulting fragment reconstruction.

Case (2) Now suppose the column for u has a 0 while the column for v has a 1 in row i. Then a traversal of edge (u, v) implies that a fragment in cell i *begins* with marker v, and also that a new block of 1s in row i begins with the column for v.

Case (3) If a tour traverses an edge from start node s to any other node v, then the edge cost on (s, v) counts the number of blocks of 1s that *start* in the leftmost column of the permuted matrix. Similarly, if a tour traverses an edge from v to s, the edge cost on (s, v) counts the number of blocks that *end* at the rightmost column of the permuted matrix. Thus, with the given edge costs, each block (and implied fragment) created from the tour is counted exactly twice.

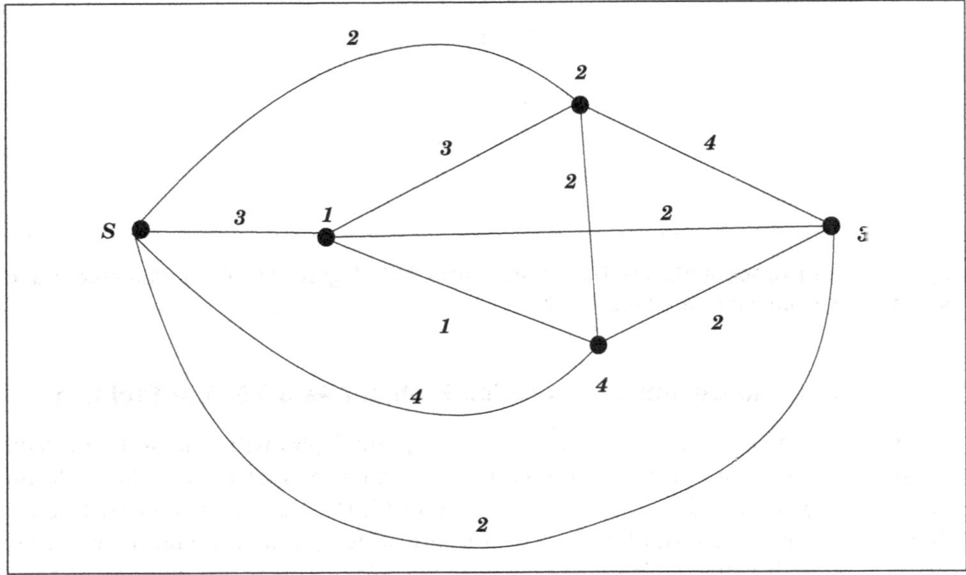

Figure 9.7 Graph *G(M)* for Matrix *M* Shown in Figure 9.6. A minimum cost TS tour, starting and ending at node s, is {s, 3,1,4,2, s}.

In summary:

> A traveling salesman tour in $G(M)$ that has total edge cost of w specifies a permutation of columns of M that creates exactly $w/2$ blocks of consecutive ones. Hence, a traveling salesman tour of minimum total cost optimally solves the column-permutation problem.

So *if* the true marker ordering minimizes the total number of blocks of consecutive ones (as has been proposed), *then* casting and solving the marker ordering problem as a column-permutation problem, which is solved as a TS problem, is an attractive approach. But, an optimal solution to the column-permutation problem will not always establish the correct marker order and identify the original blocks. Sometimes there just isn't enough data to pin down the order; and sometimes the optimal TSP solution is not unique. However, in our simulations (where we generate the blocks of the data, so we know the desired column permutation), solutions to instances of the column-permutation problems actually *did* recapture the original blocks and the correct marker orderings with high reliability, *as long* as there was "enough" data. That is, as the number of sites increases, and the lengths of the blocks increase, and the number of rows increases relative to the number of columns, the optimal ILP solution to the column-permutation problem *tends* to correctly recapture the original blocks and the marker order. A formal analysis showing that this is to be expected is given in [4]. That analysis is beyond the scope of this book, and here we can only appeal to intuition.

An Example To illustrate the above discussion, and to explain how we simulate the marker-ordering problem to test the accuracy of the column-permutation and TSP approach to it, see Figures 9.8 and 9.9. Figure 9.8a shows two rows of a simulated dataset M with 10 rows and 30 markers, where the blocks (both of zeros and ones)

```
111100000111110000111100001111
000001111000011111000011111000
              (a).

111001000011100110101101100111
000110110100011111010010011000
              (b)

010000000000011111111111111110
111111111111100000000000000000
              (c)
```

Figure 9.8 Two Rows from An Example of the Column-Permutation Problem Where the Input Matrix Has 10 rows and 30 Markers. Part (a) shows the original two rows. Part (b) shows the rows after random permutation of the columns of the 10-by-30 matrix, used for input to the column-permutation problem. Part (c) shows the two rows given by an optimal solution to the column-permutation problem. The experiment is disappointing, since the row in (c) do not recreate the original marker order, or even provide a good approximation to it.

were assumed to have lengths between four and five. This is the assumed correct (but unknown) reality in the simulated genome. Figure b shows the same rows after the columns of M were randomly reordered, reflecting the way real data would be collected, since the true order of the markers is unknown. Figure c shows the same two rows as displayed in the optimal solution to the column-permutation problem, when given the randomly ordered matrix as input. Those two rows are consistent with the result for other rows in the optimal solution, and they certainly do *not* recreate the original data. In fact, the original M had 34 blocks, while the optimal solution to the column-permutation problem with this M has only 16 blocks. This failure of the approach is due to the fact that the data is too small, too limited. In particular, it has too few rows.

In contrast, when we generated larger problem instances with 70 rows and 100 markers, and blocks with lengths between 4 and 12, the optimal ILP solution to the column-permutation problem *almost always* recreated the original blocks of the generated data. Figure 9.9 shows one row of a test case with 70 rows. The first line shows the true (but assumed unknown) first row of the 70-by-100 matrix M'. The second line shows the same row after the columns of M' were randomly reordered. The third line shows the order given by the optimal solution to the ILP for the column-permutation problem, using the randomly reordered M' as input. Notice that the first and third lines are *identical*.

Of course, the ILP solution could have generated the exact *reversal* of the original row. That result would also be counted as a success.[6]

[6] The astute reader might also ask why we *permuted* the columns in the tests we did. The graph $G(M)$ is not affected by the order of the columns in matrix M, so it must be the same whether it is constructed from the original matrix M, or from a matrix resulting from permuting the columns of M. The test of the efficacy of the column-permutation approach to the marker-ordering problem could be done using the original M. So why permute? The answer: So that the empirical results will be more convincing to

```
00001111111110000000001111111000001111111111000000000011111111111000011111100000000000111111111100000000
0010011111111100101001111110000100001110101000000001011111011010000010111001111001100110110011011010001
00001111111110000000001111111000001111111111000000000011111111111000011111100000000000111111111100000000
```

Figure 9.9 One Row of a High-Quality Solution to the Column-Permutation Problem. The first line is the original, but assumed unknown, row. The second line is the reordered first row, used in the input to the column-permutation problem. The third line is the deduced order of the row, obtained from an optimal solution to the column-permutation problem (of course, using all of the 70 rows). Notice that the first (original) and third (deduced) lines are identical, so the optimal solution perfectly recaptured the correct blocks – which it did for all 70 rows.

Exercise 9.4.1 *Explain in more detail why the graph $G(M)$ will be the same whether it is built from the original matrix M, or from a column-permuted matrix obtained from M.*

9.4.4 Software for the Marker-Ordering Problem

At the website for this book, there are several Perl programs that allow the exploration of the marker-ordering problem and the column-permutation problem, solved as a TS tour problem. These programs are: *marker-pipeline.pl*, *markerTSP.pl*, *M1graphTSPfile.pl*, *cleanres.pl*, and *pmarkersolcheck.pl*. The program *marker-pipeline.pl* is the master program that calls the others. Using it, you can choose the number of rows, sites, and the maximum permitted length of a block (the minimum is four). The programs then generate random data with those parameters; randomly permute the columns; use the data to create a concrete ILP formulation for the column-permutation problem; call Gurobi to solve the concrete ILP formulation; and return the ILP solution, compared line-by-line with the original data Use the following template to execute marker-pipeline.pl on one command line in a terminal window:

```
perl marker-pipeline.pl new-data-file-name number-of-rows
                  number-of-cols max-block-length
```

Exercise 9.4.2 Software *Download the programs named above, and use program* marker-pipeline.pl, *varying the choice of parameters, to verify the claim that given "sufficient data," the optimal solution to the column-permutation problem correctly re-creates the original matrix M, and hence solves the marker-ordering problem, correctly recreating the blocks and marker positions in M.*

anyone who thinks that using the original M might allow some kind of subtle cheating. But, in fact, we learned something surprising when comparing the results of using the original M and a column-permuted M. The optimal solutions were the same – as they must be – but Gurobi 6.5 ran much faster when solving concrete ILP formulations created from the original M, compared to solving formulations created from column-permuted matrices. This is due to some arcane implementation details deep in Gurobi 6.5 (we didn't experience this with Gurobi 8.0). So, permuting the columns can help to drive out bias, at least when testing for speed of the solver. This illustrates why I always advocate randomly permuting as much of any simulated data as possible, without changing the core of a problem, when doing empirical testing. This is also why I am often skeptical of empirical results – subtle and unanticipated biases are often possible.

9.5 FINDING SIGNALING PATHWAYS IN CERVICAL CANCER

We end our discussion of biological applications of the TS model with an example from *signaling pathways* and cancer. In [128], the authors first discuss a method for selecting a set of genes that *may* be involved in the development of *cervical cancer*. After selecting those genes, the main goal is to determine a plausible signaling pathway containing the selected genes.

Without going deeply into the biology, it is believed that a series of gene activations – each gene in the series activating the next – is involved in the onset of cervical cancer. Such a series is called a *signaling pathway*. More realistically, the signaling pathway is not a simple path, but consists of an ordering of *subsets* of the selected genes, where each subset activates the next. Signaling pathways are very common in many different biological phenomena. So, having identified a set of genes that are likely involved in the onset of cervical cancer, the issue is how to *organize* them into a plausible signaling pathway.

The proposed method in [128] uses *co-occurrence* data of *gene products* (proteins or mRNAs, usually) expressed by the identified genes. Samples of gene products are obtained at selected times, showing which gene products *coexist* at each time point. Ideally, such coexistence data of gene products should reveal the order that the underlying genes were active, and hence identify the signaling pathway. However, different gene products are expressed in different quantities, and decay at different rates, so an mRNA or protein can remain detectable *long after* the gene that expressed is active. Hence, co-occurrence data only roughly indicates how frequently each pair of genes is active at the same, or close, time points, making it challenging to deduce the correct co-occurrence data. What is needed is a method where the *global* structure of the data influences the solution. So, the authors in [128] model the problem of finding a plausible signaling pathway as a *TS path* problem on a graph G, as follows.

Each gene product is represented by a distinct node in G, and each pair of nodes, (i,j), is connected by an edge with a weight derived from the co-occurrence data of gene products i and j. The weight function is defined so that a *more frequent* co-occurrence of gene products leads to a *smaller* edge weight. Then, a *minimum-weight TS path*[7] in G gives an ordering of the gene products, and hence an ordering of the genes, which is a proposed signaling pathway. A signaling pathway containing 28 genes involved in cervical cancer was obtained in this way. The ILP formulation they used will be developed next, in Section 9.6. Note that by solving the signaling pathway problem as a TS path problem, *all* of the data contributes *globally* to the overall solution, in contrast to simpler, greedy methods where each successive node would be added to a growing path by choosing the minimum weight way to extend the path.

9.6 AN ILP SOLUTION TO THE TS TOUR PROBLEM ON G'

Having defined the TS path and TS tour problems, and given several examples of genomic problems that can be cast as TS problems, we now turn to how TS tour problems can be *formulated* and *solved* as integer linear programming problems. The

[7] They actually compute a TS tour, but I don't understand the biological basis for this choice.

ILP formulation we will develop in this section was used in [128], based on an ILP formulation given in [3, 75].

9.6.1 Initial ILP Variables and Inequalities

Recall that graph G' is input graph G augmented with a new node 0, so G' has $n-1$ nodes. Recall also, that a TS path problem on G can be solved as a TS tour problem on G'.

First Variables For each undirected edge (i,j) in the graph G', create binary variables $F(i,j)$ and $F(j,i)$. Setting a variable $F(i,j)$ to value 1 indicates that edge (i,j) will be traversed *from* node i *to* node j; and setting $F(j,i)$ to value 1 indicates that the traversal is from node j to node i.

Clearly, any specific TS tour on G' can be specified by setting each $F(i,j)$ to 1 if and only if edge (i,j) is traversed in the tour from node i to node j. For example, an optimal TS tour for G' in Figure 9.2 is specified by setting $F(0,2)$, $F(2,1)$, $F(1,3)$, $F(3,4)$, $F(4,6)$, $F(6,5)$, and $F(5,0)$ to 1, and setting the other F variables to 0.

So any TS tour specifies values for the F variables in a simple way. But the converse task requires some thinking. That is, we want an ILP formulation where the F variables set to value 1 in a feasible ILP solution specify a TS tour on G'. We will develop that ILP formulation in several steps.

The Assignment Inequalities The most basic and universal inequalities in almost all TSP formulations are called the *assignment* inequalities. These inequalities state conditions that *must* be satisfied in any feasible solution to a TS problem:

$$\text{For each edge } (i,j) \text{ in } G': \quad F(i,j) + F(j,i) \leq 1. \tag{9.1}$$

And,

$$\text{For each node } i: \quad \sum_{j \neq i} F(j,i) = 1, \quad \sum_{j \neq i} F(i,j) = 1. \tag{9.2}$$

The inequalities in (9.1) say that an edge can only be traversed in one direction. The inequalities in (9.2) say that any node i must be entered exactly once, and must be exited exactly once. Together, the inequalities say that each node must be entered and exited *exactly* once, and the entering and exiting edges must be different Note that inequalities (9.1) and (9.2) are satisfied by the way we set the values of the F variables in the example from Figure 9.2.

The objective function of the formulation is:

$$\text{Minimize} \sum_{\textit{edge } (i,j) \textit{ in } G'} [D(i,j) \times (F(i,j) + F(j,i))], \tag{9.3}$$

where $D(i,j)$ is the cost of traversing edge (i,j), in either direction.

The ILP formulation created by the objective function, together with the inequalities in (9.1) and (9.2), is called the *Assignment ILP* for G'. This name comes from the view of *assigning* each node in G' to *exactly* one other node in G'. Specifically, any node i is assigned to node j, if and only if the value of $F(i,j)$ is 1. See Figure 9.10 for a Gurobi-formatted concrete *Assignment* ILP, based on the graph in Figure 9.1.

9.6.2 Necessary But Not Sufficient

We stated above that *any* TS tour leads to values for the F variables that satisfy the inequalities in (9.1) and (9.2) (the inequalities of the Assignment ILP). However, the converse is *not* always true. For example, setting variables $F(0,2), F(2,1), F(1,3)$, $F(3,0)$ $F(4,6), F(6,5)$, and $F(5,4)$ to 1, and all other F variables to 0, satisfies the Assignment ILP (in Figure 9.10), for the graph G' in Figure 9.2. But, those F values do *not* specify a TS *tour* of G', because they do not specify just a *single* cycle. Instead, the subset of variables $F(0,2), F(2,1), F(1,3), F(3,0)$, all set to 1, describes the *cycle* $\{0,2,1,3,0\}$, which does not contain *all* of the nodes. Similarly, the remaining variables $F(4,6), F(6,5)$, and $F(5,4)$, all set to 1, describe a *different* cycle, which contains *no* nodes in the first cycle. So this feasible solution to the Assignment ILP describes two *node-disjoint* cycles.[8] Moreover, using the objective function in 9.3, those F values define an *optimal* solution to the Assignment ILP for G'.

Exercise 9.6.1 *In the example above, the edges whose corresponding F variables have value 1, are divided into* cycles. *Explain why, in* any *feasible solution to the Assignment ILP, the edges whose corresponding F variable have value 1 must specify one or more* directed cycles *in G'.*

Exercise 9.6.2 *In the example above, the cycles specified by the values of the F variables, have* no *nodes in common. Explain why this will always be true for any feasible solution to the Assignment ILP that specifies more than one cycle in G'.*

Forcing an Assignment To Be a Tour In order to create an ILP formulation whose optimal solution is *guaranteed* to be a TS *tour*, we need additional inequalities beyond those in the Assignment ILP formulation. We need inequalities to *force* the ILP solution to describe a *single* cycle containing *all* of the nodes of G'. There are many ways that have been developed to do this, and we will discuss two of them in detail. The first one is called a *compact* ILP formulation, and the second is a more involved *non-compact* formulation.

9.6.3 A Complete, Compact ILP Formulation for the TS Tour Problem

A *complete, compact ILP formulation* for the TS tour problem is shown in Figure 9.11. This abstract ILP formulation is called the *GG* TSP formulation, named after the two authors, B. Gavish and S. Graves, who described it in an unpublished working paper [75] from MIT. Although the working paper was never published, the formulation became better known after its description in [3].

The GG formulation is called *complete* because, in contrast to an approach we will discuss later, its solution completely solves the instance of the TS Tour Problem

[8] A cycle that does not contain all of the nodes of a graph is called a *subtour*.

```
Minimize
+ 4 F1,2 + 4 F2,1 + 2 F1,3 + 2 F3,1 + 9 F1,6 + 9 F6,1
+ 6 F2,3 + 6 F3,2 + 10 F3,4 + 10 F4,3 + 4 F4,5 + 4 F5,4
+ 2 F4,6 + 2 F6,4 + 5 F5,6 + 5 F6,5

such that

F1,2 + F2,1 <= 1
F1,3 + F3,1 <= 1
F1,6 + F6,1 <= 1
F2,3 + F3,2 <= 1
F3,4 + F4,3 <= 1
F4,5 + F5,4 <= 1
F4,6 + F6,4 <= 1
F5,6 + F6,5 <= 1

+ F0,1 + F0,2 + F0,3 + F0,4 + F0,5 + F0,6 = 1
+ F1,0 + F2,0 + F3,0 + F4,0 + F5,0 + F6,0 = 1

+ F1,0 + F1,2 + F1,3 + F1,6  = 1
+ F0,1 + F2,1 + F3,1 + F6,1  = 1

+ F2,0 + F2,1 + F2,3  = 1
+ F0,2 + F1,2 + F3,2  = 1

+ F3,0 + F3,1 + F3,2 + F3,4  = 1
+ F0,3 + F1,3 + F2,3 + F4,3  = 1

+ F4,0 + F4,3 + F4,5 + F4,6  = 1
+ F0,4 + F3,4 + F5,4 + F6,4  = 1

+ F5,0 + F5,4 + F5,6  = 1
+ F0,5 + F4,5 + F6,5  = 1

+ F6,0 + F6,1 + F6,4 + F6,5 = 1
+ F0,6 + F1,6 + F4,6 + F5,6 = 1

end
```

Figure 9.10 The Concrete *Assignment* ILP for the Graph in Figure 9.2. All of the variables are binary, but the Gurobi-required list of binary variables is omitted.

it is based on. The GG formulation is considered *compact* because the number of variables and inequalities is relatively small – proportional to the number of edges of G'. This is in contrast to a much larger number of inequalities used in a different ILP formulation we will discuss in Section 9.8.

$$\text{Minimize} \sum_{edge\ (i,j)\ in\ G'} D(i,j) \times F(i,j) + D(i,j) \times F(j,i), \tag{9.4}$$

such that
For each edge (i,j):

$$F(i,j) + F(j,i) \leq 1. \tag{9.5}$$

For each node j, from 0 to n:

$$\sum_{i \neq j} F(i,j) = 1, \tag{9.6}$$

$$\sum_{i \neq j} F(j,i) = 1. \tag{9.7}$$

For each node j from 1 to n:

$$b(0,j) = n \times F(0,j). \tag{9.8}$$

For each edge (i,j):

$$\begin{aligned} b(i,j) &\leq n \times F(i,j), \\ b(j,i) &\leq n \times F(j,i). \end{aligned} \tag{9.9}$$

For each node j from 1 to n:

$$\sum_{\text{edge}(i,j)} b(i,j) - \sum_{\text{edge}(i,j)} b(j,i) = 1. \tag{9.10}$$

All F variables are binary, and all b variables are integral with values between 1 and n.

Figure 9.11 **A Complete, Compact Abstract ILP Formulation for the Traveling Salesman Tour Problem on Graph *G′*. This formulation is called the *GG TSP* formulation, after the authors on [75]. Each *D(i,j)* is a constant specifying the cost of traversing edge *(i,j)* in *either* direction. Inequalities (9.5), (9.6), and (9.7) are the inequalities of the Assignment ILP; inequality (9.8) sets variable *b(0,j)* to *n*, for the unique node *j* where *F(0,j)* has value 1; the inequalities in (9.9) force the value of *b(i,j)* to be 0, *unless* the value of *F(i,j)* is 1. The *b(i,j)* variables in (9.10) are central to the correctness of the formulation, and will be explained shortly. For now, note that each *b(i,j)* variable is associated with a potential traversal of the edge *(i,j)*, from node *i* to node *j*. Note that in (9.10), index *i* can have value 0, although *j* must be in the range 1 to *n*.**

New Inequalities Inequalities (9.5), (9.6), and (9.7) are the inequalities in the Assignment ILP, so we know (from Exercises 9.6.1 and 9.6.2), that in *any* feasible solution to the GG TSP formulation, the values of the F variables specify one *or more* directed cycles containing all of the nodes. But, the GG formulation has additional variables, the b variables, and additional inequalities, beyond what are in the Assignment ILP. How do these added variables and inequalities force a feasible solution to the Assignment ILP to define a TS *tour*, i.e., with all the nodes on a *single* directed cycle? We explain that next.

The GG TSP Formulation Is Correct Suppose, for a contradiction argument, that in a feasible solution to a concrete GG formulation, the values of the F variables specify two (or more) disjoint cycles in G'. No node can be in more than one cycle, since every node is entered and exited exactly once. So, one of those cycles, call it C, must *not* contain node 0. Now pick any node j on C. By inequalities (9.5), (9.6), and (9.7), $F(i,j)$ has value 1 for exactly one node i in G'; and $F(j,k)$ has value 1 for exactly one node k in G'; and both nodes i and k are on C. Combined with the inequalities in (9.9), this implies that $b(h,j)$ has value 0 for any node h other than i; and that $b(j,h)$ has value 0 for any node h other than k. Then, inequalities (9.10) imply that

$$b(i,j) = b(j,k) + 1,$$

so $b(i,j) > b(j,k)$.

Now consider a walk around cycle C that *starts* at node j, and traverses edge (j,k) from node j to node k. From what we have deduced in the previous paragraph, and since j was a arbitrary node on C, the values of the b variables must *decline* for *each* consecutive edge on this walk. But that implies that when the walk returns to node j, $b(i,j)$ must be the smallest b value of any edge on C. Therefore, it must be that $b(i,j) < b(j,k)$, contradicting the earlier deduction that $b(i,j) > b(j,k)$. Hence, no feasible solution to the GG TSP formulation can specify two or more cycles, so the feasible GG solution must specify a TS *Tour* on G'. Then, with the objective function in (9.3), the GG TSP formulation correctly solves the TS tour problem.

9.6.4 Imagining a Meaning for the *b* Variables

In the previous section, we gave a fairly formal argument for the correctness of the GG TSP formulation. In that argument, the b variables played a very central role, so we want to have a more intuitive understanding of what they are doing. It might help to imagine a meaning for the b variables.

Imagine that the salesman has n items to sell (say n bottles of snake oil) and during his travels to the n nodes in G, the salesman must deliver *exactly one* bottle to *each* node in G. Imagine that the bottle is delivered at the time that the salesman *enters* the node. We can then interpret the value of variable $b(i,j)$ as the *number* of bottles of snake oil the salesman carries on the edge (i,j) *from* node i *to* node j. The connection between the F variables and b variables is shown in inequalities (9.8), (9.9), and (9.10). They establish that the salesman leaves node 0 with n bottles; that he does *not* carry any bottles from node i to node j *unless* he traverses edge (i,j) in that direction; and that he must deliver exactly one bottle at each node. In particular, it is the inequalities in (9.10) that say that the salesman must enter a node j with one more bottle than he has when he leaves node j. So he delivers exactly one bottle to node j. By (9.6), the salesman leaves node 0 exactly once, so all the bottles must be delivered in a single cycle, i.e., a TS *tour*. Therefore, a setting of the b and F variables that satisfy inequalities (9.5) to (9.10) in the GG formulation must define a TS *tour*. Then, the objective function enforces that the tour will be optimal, i.e., have the smallest cost possible.

Exercise 9.6.3 *Note that even though a TS tour specifies a direction for each edge in the tour, the edges themselves were assumed to be* undirected. *Modify the GG formulation in Figure 9.11 to handle the case of graphs where some or all of the edges are* directed edges. *Explain why your formulation is correct.*

Exercise 9.6.4 *The GG TSP formulation in Figure 9.11 solves the TS path problem in a graph G by solving at TS tour problem in a graph G'. Why isn't it a general ILP formulation for any TS* tour *problem?*

So, we have a general ILP formulation for the TS path *problem, but do not have a* general *ILP formulation for the TS* Tour *problem. Show how to modify the GG formulation to solve the general TS* Tour *problem.*

Exercise 9.6.5 *We claim that in the GG TSP formulation, the inequalities in (9.5) are* redundant *– the original exposition of this formulation, in [75] does not include them. Explain why they are not needed for the* correctness *of the formulation.*

Although the inequalities in (9.5) are redundant, their inclusion does speed up the solution of the ILP formulation. For example, in a test using random graphs with 200 nodes and an edge probability of 0.3, the average solution time (over 10 graphs, using Gurobi 6.5) when inequalities (9.5) were included was 34 seconds; but the average solution time without those inequalities was 436 seconds. In another trial with the same parameters, one graph required 90 seconds using the inequalities, and 3,681 seconds without them. So, the (mathematically redundant) inequalities in (9.5) really do help!

Exercise 9.6.6 Software *Download the Perl programs* randomgraph.pl *and* MgraphTSPfile.pl *from the book website. Program* randomgraph.pl *generates a random undirected, edge-weighted graph, given the name for the output file, the number of nodes, and the probability of any edge. It outputs a matrix with the pairwise edge weights. It uses the weight of 200 for a cell (i,j) if and only if edge (i,j) is not in the generated graph. Call the program on a command line in a terminal window with:*

```
perl randomgraph.pl outfile-name number-of-nodes edge-probability
```

Program MgraphTSPfile.pl *takes in the edge-weight matrix for graph G produced by* randomgraph.pl, *and creates the concrete GG TS* Path *formulation for that data,*[9] *with no designated start or end nodes. Call it with:*

```
perl MgraphTSPfile.pl name-of-edge-weight-file
                 upper-bound-on-real-weight
```

The "upper-bound-on-real-weight" is a number such that all weights less than it are for real edges; and any weight equal or larger than it should be interpreted as denoting edges that are not in the graph. It's value should be 200 for edge-weight matrices that are created by the program randomgraph.pl.

Use these two programs, together with Gurobi, to explore the efficiency of the GG TS Path formulation, and verify (or refute) the claims made in this chapter about the efficiency of the GG formulation. Report your observations.

[9] More correctly, it first converts G to G', as explained earlier, and then creates the GG TS *tour* formulation for G'.

Exercise 9.6.7 Software *Exercise 9.4.2 asked you to download program* M1graphTSPfile.pl, *which generates a GG TSP formulation to solve the TS* Tour *problem. It is executed with the command:*

```
perl M1graphTSPfile.pl  name-of-edge-weight-file
          upper-bound-on-real-weight
```

Use randomgraph.pl *and* M1graphTSPfile.pl, *together with Gurobi, to explore the efficiency of the GG Tour formulation. Report.*

Exercise 9.6.8 Software *You do not have to use* randomgraph.pl *to generate graphs to input to* MgraphTSPfile.pl *or* M1graphTSPfile.pl. *You can create your own graphs, and their edge-weight matrices with any method you want, or by hand, and then input that matrix to either of the programs that create GG formulations. Try that, and report.*

9.7 EFFICIENCY AND ALTERNATIVE FORMULATIONS

The TS path and tour problems are considered notoriously hard to solve, not only in a theoretical, worst-case sense, but in practice.[10] Improvements in TSP solving have required some of the hardest and deepest thinking in the field of optimization [48]. However, the GG TSP formulation given in Figure 9.11 leads to surprisingly *efficient* solutions of problem instances in practice, and for much larger problem sizes than I expected. Tests were run using the GG TSP formulation on *random* graphs where each potential edge was selected to be in the graph with a high probability, 0.25, and edge costs were assigned randomly (uniformly) in a range from 1 to 100. For $n = 20$, the ILP formulation solves in a fraction of a second, using Gurobi 6.5 on my Macbook Pro; for $n = 200$, the ILP solves in under 40 seconds; and for $n = 500$, it solves in under 25 minutes. These numbers illustrate the tremendous efficiency of integer programming, compared to brute-force enumeration and examination of all $n!$ (n-factorial) potential TS tours in a graph with n nodes. For example, according to Wikipedia $n!$ is 2,432,902,008,176,640,000 (a huge number) for n equal to 20 (a small number). For $n = 100, n!$ is larger than 10^{157}, a truly gargantuan number [218].

The results on some of the classic *benchmark* (nonbiological) problem instances (e.g., city data from TSPLIB [162]) were more varied, but still demonstrate the practicality of simple, *compact* ILP formulations for instances of meaningful size. For example, problem instance *wi29* with 29 cities in the western Sahara solved in under 1 second; *berlin52* with 52 locations in Berlin, solved in 3.2 seconds; problem instance ch130, with 130 cities in China solved in under 7 minutes (with Gurobi 7.5). However, problem *a280*, with 280 drilling locations took *44 hours* to solve. This is long, but still practical for many applications; and the execution took under 2 hours to reduce the gap between the best solution value (*ub*) and the best lower bound (*lb*) to 5.98%. Further, the feasible solution that was found after 3 hours was in fact the optimal solution, although it took another 41 hours to get a matching lower bound, *lb*.[11] In

[10] This is in contrast to some other NP-hard problems, such as the max-clique problem, where the common experience is that they can be solved fairly efficiently in practice.

[11] After Gurobi 7.5 was released, I reran the concrete ILP formulation for *a280*, and it solved in only 28 hours, although this time it did not obtain the optimal solution until the 27-hour mark. Then it took another hour or so to obtain a matching lower bound. Later I reran with Gurobi 8. It took 38 hours to solve.

Section 9.8, we will discuss a more advanced ILP solution method for TS problems, that solved the *a280* benchmark instance in 19 *seconds*!

I also tested simulated data that roughly mimic variants of realistic *DNA sequence-assembly* and *marker-ordering* problems, which translate into instances of TS problems with 500 nodes. Using the GG TSP formulation, Gurobi 6.5 solved each of the generated problem instances in at most 7 minutes. This illustrates the observation that TS problem instances that arise in biology are *easier* to solve than are the benchmark TSP instances with the same number of nodes. This is because many problems in biology have more structure, and have much lower density (the ratio of the number of edges to the number of nodes) than do the benchmark problem instances.

So, the ILP approach is practical for TS problems in many meaningful applications in biology. Further, the benchmark problems that took the longest to solve, for example, problem *a280* from TSPLIB, are harder to solve than the biological problems because they allow *any* pair of cities to be adjacent in the TS path, and the inter-city distance function is Euclidean, which results in many *highly similar* distances, and many near-optimal solutions. When a concrete ILP formulation has many optimal or near-optimal solutions, it takes longer for the ILP solver to find an optimal solution and establish that it is optimal.

Exercise 9.7.1 The MTZ formulation *The Wikipedia article on the traveling salesman problem [219] describes a* compact *ILP formulation, shown in Figure 9.12, that is different from the compact GG formulation developed here. This formulation was actually the first compact ILP formulation developed for the TS problem [134]. It is generally referred to as the MTZ formulation, after the three authors, C. Miller, R. Tucker, and R. Zemlin.*

In the MTZ formulation, $D(i,j)$ is again the cost of traversing edge (i,j), and $F(i,j) = 1$ again has the meaning that edge (i,j) is traversed from node i to node j. But instead of using b variables, the MTZ formulation uses U variables, where $U(i)$ can be interpreted *as the* position *(an integer from 1 to $n + 1$) of node i in the tour of G'. The tour has $n + 1$ nodes because we added node 0 to G when creating graph G'.*

Explain why the MTZ formulation correctly solves the TS tour problem on graph G'. Explain, in particular, how the last inequality works. Seeing this may be a bit subtle, since you might expect that it implements the logic: If $F(i,j) = 1$, then $U(j)$ must be equal to $U(i) + 1$. However, the last inequality does not say that, but does say something related.

Exercise 9.7.2 *Would the MTZ formulation correctly solve the TS tour problem on G' if the* inequality

$$U(i) - U(j) + n \times F(i,j) \leq n - 1,$$

is changed to an equality, *i.e., to*

$$U(i) - U(j) + n \times F(i,j) = n - 1?$$

This looks attractive because when $F(i,j)$ is set to 1, we want $U(j)$ to be exactly equal to $U(i) + 1$. But check for bad side effects.

Exercise 9.7.3 *In my experiments, when concrete instances of the MTZ formulation can be solved, the ILP solver takes much longer to finish, as compared to the GG TSP formulation. Further, the MTZ formulation does not terminate on large problem instances, where the GG formulation does. As one of many examples, we created a random instance of the consecutive ones problem discussed in Section 9.4.2, with 300 rows, 300 columns, and 30 entries of value 1 in each row. Using the GG formulation, Gurobi 6.5 solved the problem instance on my laptop in 100 seconds (the newer Gurobi 7.5 took 186 seconds – go figure!). However, using the MTZ*

$$\text{Minimize} \quad \sum_{\text{edge } (i,j) \text{ in } G'} D(i,j) \times F(i,j) + D(i,j) \times F(j,i)$$

such that:

$$U(0) = n + 1$$

For each node i, from 0 to n:

$$\sum_{j \neq i} F(i,j) = 1$$

For each node i, from 0 to n:

$$\sum_{j \neq i} F(j,i) = 1$$

For each node $i \neq 0$:

$$1 \leq U(i) \leq n$$

For each ordered pair of nodes (i,j) where i and j each range from 1 to n:

$$F(i,j) + F(j,i) \leq 1$$

$$U(i) - U(j) + n \times F(i,j) \leq n - 1$$

All F variables are binary, and all U variables are integral.

Figure 9.12 The MTZ Formulation (Featured in *Wikipedia*) [134] for the TS Tour Problem on G' [219].

formulation, Gurobi 7.5 took more than 6.5 hours to finish. It arrived at the optimal solution in about 1.5 hours, but still didn't have a matching lower bound for another 5 hours – a gap of 1.16% remained for those 5 hours. Other experiments comparing the GG and MTZ formulations gave comparable results.

Do you have any intuition for why the MTZ formulation is so inefficient compared to the GG formulation?

Exercise 9.7.4 Software *In Exercise 9.6.6, you used the GG TSP formulation to solve TS* Path *problems on graphs generated by program* randomgraph.pl*. In this exercise you will use the MTZ formulation. The Perl program to generate the MTZ formulation for the TS* Path *problem is* WgraphTSPfile.pl*. Call it on a command line with:*

```
perl WgraphTSPfile.pl name-of-edge-weight-file
                      upper-bound-on-real-weight
```

Use randomgraph.pl *to generate a random graph G, but then use G twice: once as input to* MgraphTSPfile.pl*, and once as input to* WgraphTSPfile.pl*. Then use Gurobi twice to verify* *that the same optimal TS path value is obtained in both formulations. The actual paths might be different, but the cost must be the same. Compare the times that Gurobi takes with the two formulations. Try graphs with up to 200 nodes, or more, in order to see a significant difference. Report your observations.*

Exercise 9.7.5 *Another compact TSP formulation, called the FGG formulation [71], was vastly slower than the GG formulation. It could not solve even modest-size problem instances, and was much slower on the small instances it could solve. For example, for an instance of the consecutive-ones problem in a 20-by-20 matrix with five 1s per row, Gurobi 7.5 solved the GG formulation of TSP in 0.07 seconds, but took 256 seconds to solve the FGG formulation of the problem instance. Cplex 12.6 solved the GG formulation in 0.08 seconds, but with the FGG formulation, it ran for 2 hours without even finding a first feasible solution, and was then terminated (for cause!). On the benchmark TSP instance called qa194, the GG formulation found the optimal in under* 15 *minutes, but the FGG formulation ran for 10 hours, when I killed it, without finding an integer feasible solution.*

What is the take-home lesson from these experiments?

9.7.1 One More Point

Another empirical result obtained with benchmark TSP problems illustrates the difficulty of predicting which ILP formulations will solve well. In discussing the inefficiency of MTZ compared to another formulation, the author of [147] points to the benchmark TSP problem called *p43* as a *particularly hard* TSP instance to solve, and states that it was unsolvable using the MTZ formulation. Another paper [170] reports that two variants of the MTZ formulation

> ... failed to prove optimality for the most challenging problem p43.atsp.

The larger point made in [147], based on a theoretical argument, is that *compact* TSP formulations are not viable for any but trivial instances of TS problems. And, in fact, when I tried to solve *p43* using the MTZ formulation, Gurobi 7.5 quickly stalled with a gap of about 54% – it stayed there for 9 hours before I killed the execution. So, that empirical result agrees with the point made in [147], and the experience in [170]. However, when I used the GG TSP formulation for *p43*, Gurobi 7.5 on my laptop solved the ILP in *20 seconds*![12] Moreover, that huge difference in behaviors was seen in other tests of MTZ and GG (see [82]). So, the conclusion one might draw from using MTZ alone, that compact ILP formulations are essentially useless, is not justified.[13]

Exercise 9.7.6 Software *The edge-weight matrix for TSPLIB problem* p43 *can be downloaded from the book website. Generate the GG path formulation for p43, and also the MTZ path formulation for p43. Then use Gurobi to solve the two concrete ILP formulations. Report on the time take by each. Is your experience consistent with what is written here?*

[12] And, under 10 seconds with Gurobi 8.

[13] The compact TSP formulations GG and FGG provide another illustration of how difficult it is to have reliable *intuition* about which ILP formulations will solve well. As noted earlier, the FGG formulation published in [71] was vastly inferior to the GG formulation on all of the moderate- and large-sized tests that I performed. But, the GG formulation was only written in an unpublished working paper [75], while the FGG formulation was published in a prominent journal. Further, the GG formulation is related to the FGG formulation, and since the two authors who developed GG are the authors, along with K. Fox, who later developed FGG, the GG formulation could have naturally been included in the published paper. This suggests that the authors, or the reviewers, did not appreciate how superior GG was to FGG. Good intuition and/or sufficient testing was lacking.

Summarizing Overall, the empirical results obtained with the compact GG formulation are quite amazing. They contradict what I was taught and believed until recently, that more sophisticated solution methods (of the type to be discussed in Section 9.8) would always be *required* to solve any but the smallest TSP instances. For example, in 2003, it was generally accepted [148] that any *compact* ILP formulation would only be able to solve TS problems for 50 cities or less. Now we know that this is no longer true, and the consequences for computational and systems biology can be truly transformative.[14] These empirical results show that a compact ILP approach can solve a wide range of TS problem instances that model real phenomena in biology (particularly genomics). However, this knowledge comes with two caveats.

Caveats First, the choice of ILP solver *really* does matter. In my experiments, Cplex 12.6 was able to solve TS tour instances on random graphs of size up to $n = 200$ as efficiently as Gurobi 6.5, and on some problem instances it was faster than Gurobi. But it was unable to solve instances of size $n = 500$ that Gurobi solved in under 25 minutes, even allowing Cplex to run for more than a day. And, when terminated, the gap between the best solution obtained and the lower bound was still large (in the 50% range). Also, the newer Gurobi 7.5 was sometimes slower than Gurobi 6.5 (taking about twice as much time), and when it was faster, it was only by a small amount. So, the solver matters, and the most recent release is not always the best.

Second, as noted above, you have to use the *right* compact, ILP formulation. Even though all the known, compact TSP formulation are mathematically correct,[15] in our genomic-oriented experiments, the GG formulation consistently solved faster, and sometimes *vastly* faster, than the five other (representative) compact formulations we examined in detail [82]. So, the ILP approach can be practical on problem instances of meaningful size and structure in biology, but getting that result may require art as well as science. And, sometimes you have to experiment with several different ILP formulations, and different ILP solvers. Alternatively, one can use the more sophisticated ILP solution technique to be discussed in Section 9.8; or simply run the *Concorde* program.

9.8 THE SUBTOUR-ELIMINATION APPROACH TO SOLVING TS PROBLEMS

As we have seen, the abstract GG formulation in Figure 9.11 can be solved surprisingly quickly for small to moderate-sized TSP instances, and this is sufficient for many biological applications. In that approach, any specific TSP instance is solved

[14] Of course, the compact ILP formulation given here, together with the default use of ILP solvers on a laptop, cannot solve instances that sophisticated TSP solvers have solved. Those solvers use many tools and ideas beyond compact ILP formulations, beyond integer programming, and they use large parallel computers. They have been successful at finding exact solutions for TS problem instances with 29,979 cities using 84.4 years of computing time (in 2005); and 85,900 cities using 13[illegible] years of computing time over a period of more than a year (2005–2006). They have also been successful at getting close approximations for problem instances of size 1,904,711 (world cities catalog), and 526,280,881 (celestial objects catalog). These results were accomplished with a computer program called *Concorde*, which is publicly available [9]. See [48] for details on these problems and computations, and on the development of Concorde.

[15] There are about 25 of them.

by the solution of a *single* concrete ILP formulation. But, for much larger instances, and for faster solution of moderate-sized TSP instances, there is an *ILP solution method* that creates and solves a *series* of concrete ILP formulations. Each successive concrete formulation builds on the previous one. This approach, developed by G. B. Dantzig, D. R. Fulkerson and S. M. Johnson in [52], is called the *Subtour Elimination* method in the context of TS problems, and more generally it is called the *relaxation/separation* method. The relaxation/separation method does not work for all ILP formulations, but it works particularly well for TS problems. In this section, we discuss the subtour-elimination method and introduce the general relaxation/separation method.

9.8.1 Maybe We Can Get Lucky

As discussed earlier, since a TS tour must enter and exit each of the nodes in G', exactly once, every TS tour *must* satisfy the inequalities in (9.1) and (9.2), i.e., the inequalities for the *Assignment* ILP. In fact, those inequalities are part of the abstract GG formulation shown in Figure 9.11.

A key empirical fact is that concrete formulations for the Assignment ILP solve *extremely* fast, compared to the time needed to solve a complete TSP formulation. For example, recall the problem instance called *a280* discussed in Section 9.7. Gurobi 6.5 took 44 hours, and Gurobi 7.5 took 28 hours (on my laptop) to solve the full TS tour with the GG formulation, but the *Assignment* ILP for *a280* solved in just 0.97 seconds!

Further, *if*, in the *optimal* solution to the Assignment ILP for a graph G', the set of F variables with value 1 describes a single cycle, i.e., a *tour* of the n nodes, *then* that tour is an optimal TS tour.[16] So, the following approach is very attractive:

> When given a TS problem instance (a graph G'), before we let the ILP solver try to solve the complete, concrete TSP formulation for G', we use the solver to solve only the *Assignment* ILP for that instance.

If we are lucky, the optimal solution for the Assignment ILP will define a single cycle, and we will have very quickly found an optimal TS *tour* for G'. And, since a concrete Assignment ILP solves so quickly, there is little lost if its optimal solution does not define a TS tour.

Not Likely Yes, it would be wonderful if, for *any* problem instance G', an optimal solution to the Assignment ILP on G' did describe a TS *tour* on G'.[17] However, that is *not* generally going to happen. In fact, we have already seen an example of this for graph G' in Figure 9.2. There, one optimal solution to the concrete Assignment ILP sets variables $F(0,2), F(2,1), F(1,3)$, and $F(3,0)$ to 1, describing one cycle; and also sets variables $F(4,6), F(6,5)$, and $F(5,4)$ to 1, describing another cycle. So, an optimal solution to the *Assignment* ILP does not necessarily solve the TS *tour* problem on G'. That is why the ILP formulation in Figure 9.11 has inequalities beyond (9.1) and (9.2).

16 More generally, an *optimal* solution to an Assignment ILP will have a value that is less than or equal to the value of an *optimal* TS tour, on the same graph.

17 And, it would be wonderful if I won the lottery today.

9.8.2 The Idea of Subtour Elimination

To begin, we make the following key observation:

> If we add inequalities to the Assignment ILP that *must* be satisfied by *any* TS tour, and the resulting optimal ILP solution describes a *single* cycle, then that cycle must be an *optimal TS tour*.

Following this observation, the subtour-elimination method first solves the *Assignment ILP* for G', and if we are very lucky, the ILP solution will describe only a single cycle. In that happy case, we will have very quickly found an optimal TS tour. But if the ILP solution describes more than one cycle, i.e., has a subtour, inequalities will be added to the ILP formulation to *prohibit* one or more of those subtours.

A Continuing Example We will explain the subtour-elimination method by continuing the example based on graph G' in Figure 9.2. After solving the concrete Assignment ILP in Figure 9.10, we will *prohibit* one of the subtours described by the solution. For example, to prohibit subtour $\{4,6,5,4\}$, we could add the inequality:

$$F(4,6) + F(6,5) + F(5,4) \leq 2, \tag{9.11}$$

to the ILP formulation. Since no TS tour can have a subtour, *any* TS tour *must* satisfy inequality (9.11), as well as the original inequalities for the Assignment ILP. So, by the above observation, *if* the optimal solution to this augmented ILP formulation describes a *single* cycle, that cycle must be an optimal TS tour, and we will have solved the TS tour problem for G' (and the TS *path* problem for G).

So, we next solve the Assignment ILP with inequality (9.11) added, with the expectation that augmented ILP formulation will again solve very quickly.[18] In our example, the augmented ILP formulation solved quickly, but the answer was a surprise because Gurobi was more awake and literal than I was. It set variables $F(0,2), F(2,1), F(1,3)$, and $F(3,0)$ to 1, as before; but now set variables $F(6,4), F(5,6)$, and $F(4,5)$ to 1, describing the *reverse* of the subtour we prohibited. So, adding inequality (9.11) did do what we *asked* for, but we didn't ask for the right thing. Instead, we should have added the inequality:

$$F(4,6) + F(6,5) + F(5,4) + F(6,4) + F(5,6) + F(4,5) \leq 2, \tag{9.12}$$

to the TS Assignment ILP. That inequality prohibits a subtour consisting of nodes $\{4,5,6\}$ in either direction.

Next, we solve the new extended ILP formulation, getting an optimal solution with value 18. But again, the solution describes two subtours: $\{1,3,2,1\}$ and $\{0,6,4,5,0\}$. So, we seem to be playing "whack-a-mole." We prohibited one subtour only to create another – although the value of the solution increased from 17 to 18, which is a sign of progress. So, we continue, now adding, say:

$$F(1,3) + F(3,2) + F(2,1) + F(3,1) + F(2,3) + F(1,2) \leq 2, \tag{9.13}$$

to prohibit a subtour using all the nodes $\{1,3,2\}$. This time, the ILP solution describes the single cycle, $\{0,2,1,3,4,6,5,0\}$, with objective value of 23. Since, the solution is

[18] In fact, the empirical experience is that augmented Assignment ILPs do continue to solve quickly.

a *single* cycle, and we only added inequalities to the Assignment ILP that *must* be satisfied by any TS tour, the cycle obtained *must* be an optimal TS tour of G'.

Exercise 9.8.1 Software *Recall the program* randomgraph.pl *discussed in Exercise 9.6.6. Now download the program* assignment.pl *from the book website.*

Program assignment.pl *creates the ILP formulation for the assignment ILP, given the name of a file containing an edge-weight matrix. (The edge-weight matrix could have been manually created, or created by program* randomgraph.pl.*) Call* assignment.pl *on a command line in a terminal window with:*

```
perl assignment.pl name-of-edge-weight-file upper-bound-on-real-weight
```

Use these two programs to generate a random graph and then generate the assignment ILP for that graph. Use Gurobi to solve the ILP, and note the time taken by Gurobi. Repeat, varying the number of nodes and the edge probability, and make a table of Gurobi's execution times, based on the number of nodes and edge probability. Report on your results, and conclude that the assignment ILP *is solved amazingly fast, especially compared to the time to solve a TS tour or path problem on the graph.*

9.8.3 It Works!

In the little example of the graph in Figure 9.1, with only *six* nodes, it is not clear that the subtour-elimination method, requiring the solution of a *series* of related ILP formulations, is superior to a single-solution approach such as the GG formulation. But, for much larger problem instances of the TS tour problem, the subtour-elimination approach works amazingly fast compared to any-known single-formulation approach, and greatly extends the range of problem sizes that can be solved in practice. For example, recall that the GG TSP formulation for problem instance *a280* (with only 280 nodes) required 28 hours to solve (with Gurobi 7.5). The subtour-elimination approach (automated as described below) solved that problem instance in *19 seconds*, on the same machine! And, it added under 80 inequalities to the Assignment ILP.[19]

9.8.4 It's Not As Tedious As It Seems

The subtour-elimination approach to the TS tour problem can be implemented in a computer program so that all of the steps are done automatically. Hence, one does not have to *manually* choose which subtour to prohibit, and which inequality to add, after each execution of the ILP solver. The program automatically runs the ILP solver after each change in the formulation and examines the solution after each execution of the ILP solver. It automatically determines either that an optimal tour has been found, or that an additional cycle must be prohibited.

[19] The astute reader may wonder about these numbers. We said earlier that the Assignment ILP for this problem instance takes about 1 second to solve, so a series of 80 formulations that start with the Assignment ILP and adds an inequality each time, would likely take much more than 19 seconds to solve. The explanation is that no formulation in the series is solved "from scratch," but rather every formulation is solved by *continuing* the computation done in solving the previous formulation. All of the computations done for the previous formulation are still valid for the new one. The extra time needed to solve each formulation in the series is therefore much less than would be needed by a from-scratch solution, with the result that a total of only 19 seconds is used, rather than something around 80 seconds.

9.8.4.1 Software

Gurobi has written and posted computer programs (written in different computer languages) that generate a edge-weighted *random* graph (of size chosen by the user), and then solve the TS tour problem using the subtour-elimination method. The version that is written in Python is called *tsp.py*, and it can be downloaded from the Gurobi website. Julia Matsieva has modified *tsp.py* to read a matrix of user-supplied edge weights from a file, instead of generating a random edge-weight matrix. That version, called *tspjulia.py*, can be downloaded from the book website.

9.8.5 The General Relaxation/Separation Approach

The subtour-elimination approach to solving TS tour problems is an example of a general method that we will call *relaxation/separation*.[20] This approach can be used for any ILP formulation, provided two requirements are met. First, the ILP formulation needs to contain a sub-formulation for a *subproblem* that solves *extremely* fast, and where augmentations to the sub-formulation continue to solve fast. Second, there has to be a fast way that optimal solutions to a subproblem can be checked to find any *violations* of the full ILP formulation. When such a violation is found, a new inequality is added to the formulation to prohibit that violation. That addition is called a *separation*.

The above requirements are not easily met for every optimization problem and every ILP formulation, but they are for TS problems. In fact, TS problems seem uniquely (almost *perfectly*) suited to the relaxation/separation approach.

In the TS tour problem, each subproblem consists of the *Assignment* problem with the addition of inequalities to exclude specific subtours. The task of finding a violation of the complete ILP formulation only requires finding a subtour (i.e., a cycle with fewer than n nodes) described by the current ILP solution. Prohibiting that violation just requires adding an inequality that prohibits that subtour. The empirical facts are that for TS problems, these ILP sub-formulations solve very fast, and only a few iterations of subproblems are needed. I don't fully understand why the relaxation/separation approach is so effective for TS tour problems. I can make a "hand-waving" argument for it, but ultimately its effectiveness is an empirical fact.[21]

Why Not Prohibit All Subtours from the Start? At this point, an astute reader might ask why we don't solve TS problems with an ILP formulation consisting of the Assignment ILP along with inequalities that prohibit *every* subtour, i.e., every cycle with fewer than n nodes. With this approach, for each subset of nodes, S, (other than the full set), we would include the inequality:

$$\sum_{i,j \in S} F(i,j) + F(j,i) \leq |S| - 1.$$

[20] There actually seems to be no standard term for this general method, but each application involves some *relaxation* of the full problem to an easier subproblem, and the use of some *separation* method to identify a violated constraint. So, we call the general approach the "relaxation/separation method." In Gurobi literature, the violated constraint that is added is called a "lazy constraint."

[21] The ILP literature often uses the *strength* of a formulation to explain the effectiveness of relaxation/separation in solving TSP problems. However, I do not find that explanation compelling or well supported by empirical evidence (see [82]).

We call this the *fully enumerative* ILP formulation for TS problems, and an optimal solution to this ILP will describe an optimal TS tour in G'. So why not take this approach?

The answer is that we *can* use the fully enumerative ILP formulation when the number of nodes, n, is "small." But, the number of proper subsets of n nodes is $2^n - 2$, and so as n grows, this fully enumerative approach will quickly require *far too many* inequalities. In essence, the subtour-elimination method to the TS tour problem implements this fully enumerative approach, but in a more efficient manner. Instead of prohibiting all subtours from the start, it only prohibits specific subtours that emerge in the current or past optimal solutions, and only adds inequalities to the ILP formulation as they are needed. Empirically, only a few added inequalities are required, as in the *a280* example, with 80 added inequalities, compared to $2^{280} - 2$ proper subsets of 280 nodes.

9.9 EXTENDED MODELING EXERCISES

Now that we have discussed how to model fragment assembly problems as TS problems, and how to formulate and solve the TS tour problem as an ILP problem, we can ask about ILP formulations for related problems.

Exercise 9.9.1 *In the sequence assembly problem of Section 9.3.2, we would like to include additional constraints that can increase the accuracy of a sequence assembly. For example, we know the number of bases in the original DNA sequence (this is easy to determine even without sequencing it), and we can incorporate a constraint into the ILP formulation to say that the assembly solution must (approximately) span that number of bases. Note that physical distance (number of bases) is different from the distances used to model the assembly problem as a TS path problem.*

State the abstract ILP inequalities that can be used to determine the length of the reconstructed sequence (number of bases spanned by the path), and to assure that it is approximately equal to the number of bases in the original DNA sequence. Explain your reasoning.

Exercise 9.9.2 *Another constraint that can be added to increase the accuracy of sequence assembly addresses the major problem in higher organisms, that of the frequent occurrence of repetitive (essentially identical) substrings throughout the genome. The problem is particularly severe when the repeat is longer than the typical read length. A bad assembly might pile the repeated sequences on top of each other, producing an assembly that is incorrect, and is too short.*

To increase the accuracy of a sequence assembly, we can add constraints to the ILP formulation to limit the number of fragments that span any single base. Since the sequenced fragments essentially correspond to a random sample from the fragments created by copying and breaking the original DNA, the number of times a base will be spanned is a random variable. Statistical formulas have been developed to calculate the likely number of times a random base is spanned (i.e., its "coverage"). But even without precise formulas, we can use some "rules of thumb" to try to avoid bad assemblies. For example, if the total length of all the sequenced fragments is about 10 times the length of the original DNA sequence, we might include a constraint to limit any site from being spanned more than twenty times.

State the abstract ILP inequalities that can be used to determine and limit the coverage of any base in an assembly.

10

Integer Programming in Molecular Sequence Analysis

In this chapter we discuss several problems in molecular sequence analysis and how integer linear programming has been used to address those problems.[1] Some related problems were examined earlier when we discussed the *Sequence Assembly* problem, in Section 9.3.

10.1 THE IMPORTANCE OF SEQUENCE ANALYSIS

Algorithms that operate on molecular *sequence* data (strings) are at the heart of bioinformatics and computational molecular biology. The success of computerized sequence analysis is the basis for the growth of computational biology and bioinformatics. Sequence comparison, particularly when combined with the systematic collection, curation, and search of databases containing biomolecular sequences and knowledge about them, has become essential in modern biochemistry, molecular biology, genetics, genomics, and many other subfields in biology. The key reason that sequence analysis has great biological value is the following fact:

> In biomolecular sequences (DNA, RNA, or amino acid sequences), high *sequence similarity* usually implies significant *functional* or *structural* similarity or common evolutionary history.

And, sequence similarity is not a rarity. Evolution reuses, builds on, duplicates and modifies those structures (genes, proteins, exons, DNA regulatory sequences, morphological features, metabolic pathways, gene networks, etc.) that have been "successful" (left as a vague concept). Life is based on a repertoire of structured and interrelated molecular building blocks that are shared and passed around. The same and related molecular structures and mechanisms show up repeatedly in the genome of a single species, and across a wide spectrum of divergent species.

> We should not be surprised that so many new sequences resemble already known sequences. [59]

Thus, when we study a newly sequenced biomolecule and find that its sequence is similar to another molecule that has already been studied, the new molecule will very often have properties similar to the previously studied molecule.

[1] Part of the chapter has been drawn from [83], and some of the ILP examples are drawn from [21].

Moreover, *identical*, or nearly identical *regions* in a *set* of related molecular sequences are generally found to be critical to the proper function of the molecules, while regions that have greater variation are less critical. Regions that are identical, or nearly so, in a set of related sequences, are called *conserved*. The explanation for sequence conservation in critical regions is *natural selection*: mutations in critical regions make the molecule, and the organism with it, less fit, and organisms with such mutations become less common in a population.

Two String vs. Multiple Strings When searching in molecular-sequence databases, the comparison of *two* strings at a time finds pairs of sequences that are *highly* similar or have common subpatterns (substrings or subsequences), but may *not* have been *known* to be biologically related. Indeed, the greatest value of database searching comes from using *apparent* string similarities to identify *unsuspected* biological relationships. *Multiple* string comparison is used for purposes that are somewhat *inverse* to the purposes of two-string comparison. The inverse problem is to move from *known* biological relationships to *unknown* substring or subsequence similarities. Unfortunately,

> [e]volutionarily and functionally related molecular sequences can *differ significantly* throughout much of the string and yet preserve the same three dimensional structure(s), or the same two dimensional substructure(s) (motifs, domains), or the same active sites, or the same or related dispersed residues (DNA or amino acid). [83]

Two sequences specifying the "same" protein in different species may be so different that the few observed similarities may just be due to chance. *Multiple string comparison* (often via *multiple alignment*) is a natural response to this problem. Often, biologically important patterns that cannot be revealed by comparison of *two* strings alone become clear when *many* related strings are simultaneously compared.

> "One or two ... sequences whisper ... a full multiple alignment shouts out loud." [96]

In summary, computational problems concerning the comparison of sequences, either two at a time, or in a larger *set* of sequences, are critical to the success of computational biology.

10.1.1 String Problems and ILP

Many different computational approaches have been suggested to *reveal, test, illustrate, or exploit* the overall similarity of a *set* of strings. In this chapter we discuss several such problems and the ways that ILP has been used to solve them. We emphasize problems that *lack* worst-case efficient algorithms, or only have efficient algorithms that are very complex and hard to implement. These problems illustrate the great utility of the ILP approach.

10.2 THE STRING SITE-REMOVAL PROBLEM (SSRP)

Given a set of strings of equal length, we define the *removal of site i* as the removal of the character at position *i*, from each of the strings. For example, the removal of site 1 from the pair of strings:

```
ATTCGCCAGAGCT
TTTCACCAGGGCT
```

results in the strings:

```
TTCGCCAGAGCT
TTCACCAGGGCT
```

With that definition, we can state

The String Site-Removal Problem Given a set of n strings, $\{s_1, s_2, \ldots, s_n\}$ of equal length, and a target k, find the largest subset, call it S^*, of the n strings that become *identical* after the removal of *at most k* sites.

For example, consider four strings organized into a matrix M:

```
      M
ATTCGCCAGAGCT
TTTCACCAGGGCT
TTTCAGCAGAGCT
ATACGCGAGAGCC
```

There are five columns with no variation, so seven columns, which have some variation. For example, columns 1 and 3 both contain character "A" and "T." So *seven* sites would have to be removed in order to make the remaining parts of *all* four strings identical.

But removing only *four* sites can make the remaining parts of *three* strings identical. By removing sites 1, 5, 6, and 10, the first three strings become identical – they become TTCCAGGCT. Therefore if k is set at four, the optimal solution to the string site-removal problem has value three. So, if we want $|S^*|$ equal to four, then k would have to be seven, but for $|S^*|$ equal to three, k equal to three is enough.

Exercise 10.2.1 *Without looking at the strings, but using the information in the previous paragraph, what value does the optimal solution have if k is six?*

The biological background to this problem is the assumption that many (but not all) of the strings in the input set have a *common origin*, and differ only in a few sites where mutations have occurred. The problem limits, to k, the number of sites we are allowed to remove, with the goal of finding the largest subset of strings which then become identical. Of course, there is a trade-off between k and the number of resulting identical strings, and it is not obvious which k leads to the most informative biological result. In practice, a biologist might solve the site-removal problem several times, varying k to see how those changes affect the size of the optimal solution.

10.2.1 An ILP Formulation

The Variables We use the binary ILP variable $X(i)$ to indicate whether or not string s_i will be chosen to be in S^*: $X(i)$ will be set to value 1 if string s_i is chosen, and will be set to value 0, if not chosen. We also use the binary ILP variable $C(c)$ to indicate whether or not site c will be removed from the strings (equivalently, whether column c is removed from matrix M). $C(c)$ will be set to 1 if site c is chosen for removal, and is set to 0 if it is not chosen.

The Inequalities To understand the idea behind the ILP formulation, suppose two strings s_1 and s_2 have *different* first characters. Perhaps s_1 starts with character "A" and s_2 starts with character "G". Then, certainly, *if both* s_1 and s_2 are chosen to be in S^*, site 1 *must* be removed from the strings. We can implement this logic with linear inequalities using the *If-Then* idiom for binary variables:

$$X(1) + X(2) - C(1) \leq 1. \tag{10.1}$$

Now, we generalize from this example. For every pair of the n strings, s_i and s_j, that *differ* at position c, the ILP formulation will have the inequality:

$$X(i) + X(j) - C(c) \leq 1. \tag{10.2}$$

The requirement that *at most* k sites can be removed is implemented as:

$$\sum_{c=1}^{c=m} C(c) \leq k, \tag{10.3}$$

where m is the common length of the strings. And the *objective function* is:

$$\text{Maximize} \sum_{i=1}^{i=n} X(i).$$

The concrete ILP implementation for the example shown earlier, with $k = 4$, is shown in Figure 10.1.

Exercise 10.2.2 *We have not written here a summary showing the complete abstract ILP formulation for the* SSRP. *Write it.*

Exercise 10.2.3 *The inequalities in (10.2) implement an* If-Then *idiom for binary variables, but do not implement an* Only-If *relation. So, a variable $C(c)$ might be set to 1 even though the characters at site c are identical, in the chosen strings. These are called* unneeded *removals. Explain why the ILP formulation finds an optimal S^* despite any unneeded removals.*

Exercise 10.2.4 *We would like to change the ILP formulation to avoid unneeded removals, and of course, to still find the optimal S^* using at most k removals. One approach is to add inequalities to the ILP formulation, which implement the* Only-If *relation. Do it.*

But suppose we don't add those inequalities. Would all unneeded removals be avoided if we change the objective function to

$$\textit{Maximize} \sum_{i=1}^{i=n} X(i) - \sum_{c=1}^{c=m} C(c),$$

but make no changes to the other inequalities? If not, explain why, and how you could modify the objective function to correct the problem. (This is related to material we will later discuss in Section 15.1.)

Exercise 10.2.5 *The string site-removal problem is discussed in [21],[2] where an ILP formulation is presented that is different from the one given here. We will outline that formulation. You must interpret it and explain why it is correct.*

[2] They call it "The most strings with a few bad columns problem."

```
Maximize X(1) + X(2) + X(3) + X(4)

such that

X(1) + X(2) -  C(1) <= 1
X(1) + X(2) -  C(5) <= 1
X(1) + X(2) -  C(10) <= 1

X(1) + X(3) - C(1) <= 1
X(1) + X(3) - C(5) <= 1
X(1) + X(3) - C(6) <= 1

X(1) + X(4) - C(3) <= 1
X(1) + X(4) - C(7) <= 1
X(1) + X(4) - C(13) <= 1

X(2) + X(3) - C(6) <= 1
X(2) + X(3) - C(10) <= 1

X(2) + X(4) - C(1) <= 1
X(2) + X(4) - C(3) <= 1
X(2) + X(4) - C(5) <= 1
X(2) + X(4) - C(7) <= 1
X(2) + X(4) - C(10) <= 1
X(2) + X(4) - C(13) <= 1

X(3) + X(4) - C(1) <= 1
X(3) + X(4) - C(3) <= 1
X(3) + X(4) - C(5) <= 1
X(3) + X(4) - C(6) <= 1
X(3) + X(4) - C(7) <= 1
X(3) + X(4) - C(11) <= 1
X(3) + X(4) - C(13) <= 1

C(1) + C(2) + C(3) + C(4) + C(5) + C(6) +  C(7) + C(8) + C(9)
+ C(10) + C(11) + C(12) + C(13) <= 4

All variables binary.
```

Figure 10.1 **The Concrete ILP Formulation for the Site-Removal Problem, for the Four Strings Shown Earlier.**

As before, $X(i)$ is a binary ILP variable, which will be set to 1 if and only if string i is selected to be in S^*; and $C(c)$ is a binary variable that will be set to 1 if and only if site c is selected to be removed. We denote the alphabet used in the strings by $\mathcal{A}$, and for each character q in $\mathcal{A}$, and each site c in the strings, we define the binary ILP variable $z(c, q)$, whose use is explained next.

For each string s_i, and each character q in $\mathcal{A}$, and each site c where string s_i has character q at site c, the ILP formulation has the following inequality:

$$x_i \leq z(c, q),$$

which says that if *any of the n input strings with character q at site c are chosen to be in S^*, then variable $z(c, q)$ must be set to 1. Essentially, variable $z(c, q)$ records the fact that* some *string with character q at site c has been chosen to be in S^*.*

Next, for each site c, the formulation has the following inequality:

$$\sum_q z(c, q) \leq 1 + |\mathcal{A}| \times C(c), \tag{10.4}$$

where $|\mathcal{A}|$ denotes the number of symbols in the alphabet $\mathcal{A}$.

Now for your problems:

(a) *Give an interpretation for the inequalities in 10.4. What do they accomplish?*

(b) *What additional inequalities are needed to complete this abstract ILP formulation for the string site-removal problem? Write up the full abstract ILP formulation of this problem.*

(c) *Explain why the full ILP formulation correctly solves the string site-removal problem.*

(d) *Compared to the first formulation given for the string site-removal problem, do you see any advantages or disadvantages for this formulation? If you can, write a computer program to create the two concrete ILP formulations. Implement both ILP formulations and test them with data to see if one of the formulations solves faster than the other.*

10.3 REPRESENTATIVE SEQUENCES

As discussed in the introduction, biological sequences arise through the process of evolution, and related sequences can maintain considerable similarity. Therefore, biological sequences are often organized into *families* of similar sequences with similar functions. This is particularly true for proteins, for example, the *globin* family, counting about 4,800 members, from many different organisms.

Frequently, given a protein family, or any set of similar, but *not* identical molecular sequences, we want to create a *single* sequence that *represents* the full set. This is often called a *consensus* sequence, but we will use the more neutral term of "representative sequence."

A representative sequence is useful as an illustration of the set, and is commonly used in sequence-database searching. When we have a new molecular sequence, we want to find sequences in the database that are similar to it. We could separately compare the new sequence to each sequence in the database, but if the sequences are organized into families, it is more efficient to first compare the new sequence to the representative sequences of the families. Then if the new sequence is seen to be sufficiently similar to a representative of a family, it can next be compared to all the individual members of the family.

There have been several suggestions for how to define and find a representative sequence. The most obvious one is the *plurality sequence*: Suppose the sequences in a family are aligned (possibly with some gaps in some of the sequences), so that each resulting string in the alignment has the same length. Then, we could create a representative sequence by choosing the *most frequent* character (the *plurality* character) at each position in the alignment. The resulting string is called the *plurality sequence*.

The plurality sequence is not always the most useful or informative representative for a set of sequences. Another suggestion is to use the most *central string* as the representative.

10.3.1 Central Strings

For two strings (s_1, s_2) of the same length, we define the *Hamming distance*, $D(s_1, s_2)$, between s_1 and s_2 as the *number* of positions where the letters in the two strings *differ*. For example, for

$$s_1 = ATCG,$$

and

$$s_2 = CTGG,$$

$D(s_1, s_2)$ is two.

Given a set of n strings $\mathcal{S} = \{s_1, s_2, \ldots s_n\}$, each of length m, and another string s of that length, we define the *distance*, $D(\mathcal{S}, s)$, between s and the *set* $\mathcal{S}$, as the *largest* Hamming distance $D(s_i, s)$ between s and any string s_i in $\mathcal{S}$. That is,

$$D(\mathcal{S}, s) = \text{Max}_{s_i \in \mathcal{S}} D(s_i, s).$$

Note that string s need *not* be in $\mathcal{S}$. For example, for $\mathcal{S} = \{ATCG, CTGG\}$, $D(\mathcal{S}, AACG)$ is three, because $D(ATCG, AACG) = 1$, and $D(CTGG, AACG) = 3$.

Central Strings Finally, we define a *central string* for set $\mathcal{S}$ as a string, s^*, such that $D(\mathcal{S}, s^*) \leq D(\mathcal{S}, s)$ for any string s of length m. Note that s^* need not be in $\mathcal{S}$, and generally will not be. The problem of finding a central string for $\mathcal{S}$ is called *the Central-String Problem*. A central string is also called a *closest* string.

Central strings have been widely suggested as good representatives of families of related biological sequences (DNA, RNA, and amino acid).[3] However, there is currently no efficient (in the worst-case sense) algorithm to find a central string of a set of strings, and there are theoretical reasons to doubt that one exists. Instead, we will see how to use ILP to find a central string when given a set of strings, $\mathcal{S}$.

Before giving an ILP formulation, we note that the plurality sequence for a set $\mathcal{S}$ is not always the closest string to $\mathcal{S}$ (see Figure 10.2). This suggests why it may be difficult to find a central string.

Exercise 10.3.1 *The problem of finding a central string is actually a hard problem, but it may at first seem easy, because it is easily* misunderstood *for the very simple problem of finding a string s^* to* minimize

$$\sum_{s_i \in \mathcal{S}} D(s_i, s^*).$$

Does the plurality sequence solve this problem?

[3] I am not convinced that central strings, as defined, make great representatives. But, many people are supporters of central strings, and at very least, we need the ability to find central strings efficiently in order to test whether or not they make good representatives. So, it is of interest to develop an ILP formulation for the central-string problem.

Hamming distances between sequence P (or A) and the sequences in $\mathcal{S}$

sequences		P	A
	ATGTCT	1	2
	TCACGT	5	4
	AGGTCG	1	4
	AGGTCC	1	4
	CGTCTA	5	4
Plurality sequence (P)	AGGTCT		
Alternative sequence (A)	ATTCCT		

Figure 10.2 Example Showing That the Plurality Sequence Is Not Necessarily a Closest String. The plurality sequence s is AGGTCT, and $D(\mathcal{S},s)$ is five. An alternative sequence s' is ATTCCT, where $D(\mathcal{S},s')$ is four. Is this a closest string?

10.3.2 An ILP Formulation for the Central-String Problem

We first need some notation. We use $\mathcal{A}$ to denote the alphabet used for the strings in $\mathcal{S}$, and the string s^*. We also will use the symbol "f" to denote a character in alphabet $\mathcal{A}$. For any position p between 1 and m, we use $s_i(p)$ to denote the character at position p of string s_i.

The ILP Variables For each letter, f, in alphabet $\mathcal{A}$, and each position $p \leq m$, we create the binary ILP variable $X(f,p)$. The meaning of $X(f,p)$ is that the p'th position in s^* will be set to the character f, if and only if $X(f,p)$ is set to 1 in the optimal solution to the ILP formulation. Thus, the values of the $X(f,p)$ variables in the optimal ILP solution, specify s^*.

We also will have a single, *integer* variable d, which will have a value equal to the *largest* Hamming distance between the solution string s^* and any of the input strings.

The Inequalities The formulation will have two types of inequalities. The first inequalities say that every position in s^* must be assigned a single letter from $\mathcal{A}$:

For each position p from 1 to m,

$$\sum_{f \in \mathcal{A}} X(f,p) = 1, \tag{10.5}$$

The second inequalities essentially compute and bound the Hamming distance between each input string s_i and the string specified by the values given to the $X(f,p)$ variables. For each input string s_i, we have:

$$\sum_{p=1}^{p=m} \left[\sum_{f \neq s_i(p)} X(f,p) \right] \leq d. \tag{10.6}$$

To better understand the inequalities in 10.6, consider the term

$$\sum_{f \neq s_i(p)} X(f,p),$$

inside of the square brackets. The values of i and p are fixed, so $s_i(p)$ identifies a specific character. The summation considers every character f (in the alphabet) that is *not* equal to character $s_i(p)$, and adds one to the summation if and only if $X(f,p)$ is set to 1, i.e., character f is chosen for position p. The result is that $\sum_{f \neq s_i(p)} X(f,p)$, will equal 1, if and only if, any character other than $s_i(p)$ is chosen for position p. The sum will equal 0 if character $s_i(p)$ *is* chosen for position p. For example, if string s_i is

$$ATCGC,$$

then $s_i(1)$ is "A," and $\sum_{f \neq s_i(1)} X(f,p)$ is:

$$X(T,1) + X(C,1) + X(G,1).$$

Therefore, summing over all of the positions (values of p),

$$\sum_{p=1}^{p=m} \left[\sum_{f \neq s_i(p)} X(f,p) \right],$$

will be equal to the Hamming distance between s_i and the string specified by the values $X(f,p)$ that are set to 1.

The Objective Function The objective function for the formulation is:

$$\text{minimize } d. \tag{10.7}$$

String s^* is the string specified by the settings of the $X(f,p)$ variables in the optimal solution to this ILP formulation. This may be a bit subtle. We want to create a central string s^* that minimizes the maximum Hamming distance between it and any string in $\mathcal{S}$. How do we *minimize* a *maximum*? An objective function can either minimize or maximize, but not both. The answer is in the *interplay* between the value of d in the inequalities in 10.6, and the value of d in the objective function. The inequalities in 10.6 ensure that d will be *larger than or equal to* $D(\mathcal{S}, s^*)$. The objective function tries to minimize d, so the result is that the value of d will be exactly equal to $D(\mathcal{S}, s^*)$, and s^* will be a central string for $\mathcal{S}$.

Summarizing, the abstract ILP formulation for the central-string problem is shown in Figure 10.3.

Exercise 10.3.2 *There are several problems related to the central-string problem that have been studied (for example, see [122]), and integer programming has been used to solve some of those. Here, we describe four such problems. Find, describe and explain an abstract ILP formulation for each one.*

(a) The Farthest-String Problem *We are given a set of strings $\mathcal{S}$, each of length m, written with an alphabet $\mathcal{A}$. We assume that for each position p, and each character f in $\mathcal{A}$, there is some string in $\mathcal{S}$ that has character f in position p. We define*

$$Min(\mathcal{S}, s') = \texttt{Minimum}_{s \in \mathcal{S}} D(s, s').$$

Minimize d

Such that:

For each position p from 1 to m, $\sum_{f \in \mathcal{A}} X(f,p) = 1$

For each input string s_i, $\sum_{p=1}^{p=m} [\sum_{f \neq s_i(p)} X(f,p)] \leq d$.

Every variable $X(f,p)$ is binary, and d is an integer variable.

Figure 10.3 An Abstract ILP Formulation for the Central-String Problem.

Then the farthest-string problem is: Find a string s^*, written with the alphabet $\mathcal{A}$, such that

$$Min(\mathcal{S}, s^*) \geq Min(\mathcal{S}, s)$$

for any string s of length m, written with alphabet $\mathcal{A}$.

Why isn't this problem solved by choosing the least frequent character in each position? Describe an ILP formulation for the farthest-string problem. Can you think of a biological application of the problem?

(b) The Close-to-Most Problem *We are given a set of strings $\mathcal{S}$, each of length m, and a target $d > 0$. Find a string s^* of length m, maximizing the number of strings in $\mathcal{S}$ which have Hamming distance to s^* that is at most d.*

Explain exactly how this problem differs from the central-string problem, and then describe an abstract ILP formulation for it.

(c) At least *k* close strings *We are given a set of strings $\mathcal{S}$, each of length m, and a target $k > 0$. Find a minimum value d, and a string s^* of length m, such that at least k strings in $\mathcal{S}$ have Hamming distance to s^* that is less than or equal to d.*

Explain exactly how this problem differs from the central-string problem, and then describe an abstract ILP formulation for it. When is the problem the same as the central-string problem?

(d) Separating Sets of Strings *This problem is* motivated *by the problem of designing sequence-based drugs that can target pathogens, but not harm the hosts. For example, targeting (perhaps cutting) the DNA sequences of a family of viruses that are harmful to humans, without harming any human DNA sequences. The specific problem stated here is actually too far from the realistic problem of drug design to be of immediate use, but it introduces the central issues.*

We are given a set of strings $\mathcal{S}_1$, each of length m, and a target $d_1 > 0$; and given a second set of strings $\mathcal{S}_2$, each of length m, and a target $d_2 > 0$. Find a string s^* if one exists *of length m such that*

$$D(\mathcal{S}_1, s^*) \leq d_1,$$

and

$$Min(\mathcal{S}_2, s^*) \geq d_2.$$

That is, $D(s, s^*) \leq d_1$ for each sequence s in S_1; and $D(s, s^*) \geq d_2$ for each sequence s in S_2.

Describe an abstract ILP formulation for this computational problem. Could a specific instance of this problem ever be infeasible?

Note that above, d_1 and d_2 were constants given by the user. Now, consider d_1 and d_2 as variables. Extend the ILP formulation to maximize *$d_2 - d_1$, subject to the above constraints on $\mathcal{S}_1$ and $\mathcal{S}_2$. Could a specific instance of this problem ever be infeasible?*

10.4 THE LONGEST COMMON-SUBSEQUENCE (LCS) PROBLEM

Another classic way to measure the similarity of two (or more) biological sequences is by calculating the length of their *longest common subsequence (LCS)*. In this section we discuss how to solve this problem for two strings, using integer linear programming. But rather than developing an ILP formulation for the LCS problem from scratch, we relate it to a different problem already discussed in the book. That problem is the *simple RNA-folding* problem, introduced in Section 6.1. We will explain the idea, and you will then fill in the details in the exercises. First, we need some definitions.

In a string s, a *subsequence* is defined as a subset of the characters of s arranged in their original, relative, order. More formally, a subsequence of a string s of length m is specified by a list of indices $i_1 < i_2 < i_3 < \ldots < i_k$, for some $k \leq m$. The subsequence specified by this list of indices is the string $s(i_1)s(i_2)s(i_3)\ldots s(i_k)$. For example, *ATTATG* is a subsequence in

$$s_1 = ATTCGATACAGTG,$$

but it is not a substring. It can be specified by the list of indices: 1,2,3,6,7,11. But, that is not the only list of indices that works to specify it (see if you can find others).

For emphasis: A *subsequence* of a string s need *not* consist of *contiguous* characters in s, while the characters of a *substring* must be contiguous.[4] Of course, a substring also satisfies the definition for a subsequence.

Common Subsequences Given two strings s_1 and s_2, a *common subsequence* is a subsequence that appears both in s_1 and in s_2. The *longest common-subsequence (LCS) problem* is to find a common subsequence of s_1 and s_2 that is as long or longer than any other common subsequence of s_1 and s_2. For example, the subsequence "*ATTATG*" occurs in both strings:

$$s_1 = ATTCGATACAGTG,$$

$$s_2 = TATCTGAGGTGA,$$

so, it is a *common* subsequence of (s_1, s_2). But, it is not a *longest* common subsequence of (s_1, s_2). The subsequence *ATTGATG* is a common subsequence that is longer. We often refer to the longest common subsequence of two string as the LCS of the strings.

Exercise 10.4.1 *What is a list of indices that specifies the subsequence "ATTGATG" in s_1, and in s_2, shown above. Is this the only list of indices that works? Is ATTGATG a* longest *common subsequence of (s_1, s_2)?*

10.4.1 Relating the LCS Problem to the Simple RNA-Folding Problem

An abstract ILP formulation for the LCS problem can be developed by relating the LCS problem to the *simple RNA-folding* problem, discussed in Chapter 6, in

[4] The distinction between subsequence and substring is often lost in the biological literature. But algorithms for substrings are usually quite different in spirit and efficiency than algorithms for subsequences, so that the distinction is an important one.

Section 6.1. The LCS problem is not exactly the same as the RNA-folding problem, but the ILP formulation we developed for simple RNA folding can be used for the LCS problem as well, with a minor modification.

To see the connection between the LCS problem and the RNA-folding problem, write the two strings s_1, s_2 on a line next to each other, with s_1 to the left of s_2. But with s_1 written *correctly*, left to right, and s_2 written right to left, i.e., *reversed*. We call this combined string $S_{1,2}$. For example, with the strings s_1 and s_2 defined above, $S_{1,2}$ is

$$ATTCGATACAGTG \quad AGTGGAGTCTAT.$$

Just as in RNA folding, we define a *pairing* in $S_{1,2}$ as set of *disjoint* pairs of characters in $S_{1,2}$. A pair that is in a pairing is said to be a *matched pair*. Now, in RNA folding, only certain pairs of characters were permitted to form a matched pair: they had to be complementary characters, and be separated by a given minimum distance. In the LCS problem, it will also be true that only only certain pairs of characters are permitted to form a matched pair, but the rule for which pairs are permitted is different from the case of RNA folding.

Permitted Pairs for the LCS Problem We say that two characters in $S_{1,2}$ form a *permitted pair*, if and only if they are *identical*, and first character is in s_1, and the second character is in s_2. A pairing that only uses permitted pairs is called a *permitted pairing*. Further, analogous to what we did for RNA folding, we define a *non-crossing* permitted pairing in $S_{1,2}$ as a permitted pairing where there are no matched pairs $(i,j), (i',j')$ with $i < i' < j < j'$. In a non-crossing permitted pairing, we can draw string $S_{1,2}$ on a horizontal line and connect the two characters in each matched pair with a curved line, so that none of the lines cross.

The Punch Line Although the rules for which pairs of characters are permitted to match in the RNA folding problem are different from the LCS problem, the ILP formulation for RNA folding only needs to know which pairs are permitted – the program that produces the concrete ILP formulation can accommodate any given set of permitted pairs. So, the program also works for the permitted pairs specified for the LCS problem. This leads to the following claim:

> The *longest common subsequence* of *two* strings, s_1 and s_2, can be found by finding the largest non-crossing, permitted pairing in the *single* string $S_{1,2}$ (see Figure 10.4). Further, the program that creates concrete ILP formulations for the simple RNA-folding

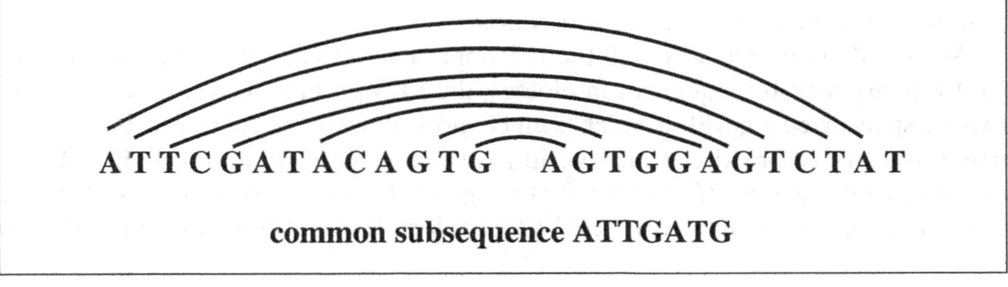

Figure 10.4 A Non-Crossing, Permitted Pairing in the String $S_{1,2}$ Created from String s_1 and the Reversal of String s_2.

problem can be used to create concrete ILP formulations for the LCS problem, provided only that the program is written so that the set of permitted pairs are given as input.

Exercise 10.4.2 *Find a largest non-crossing, permitted pairing in the string $S_{1,2}$ shown in Figure 10.4. It might be the one shown there.*

Exercise 10.4.3 *Give a complete explanation for why the ILP approach described above for the LCS problem is correct. In particular, why do we* reverse *the string s_2?*

Exercise 10.4.4 *Given the above discussion relating the LCS problem to the RNA-folding problem, and using on the ILP formulation for RNA folding, fully develop an abstract ILP formulation for the LCS problem. Use the binary variable $P(i,j)$ to specify whether or not the i'th character in s_1 is paired with the j'th character in s_2.*

Exercise 10.4.5 *Recalling the connection (discussed in Section 6.1.4) of the simple RNA folding problem to the maximum-clique problem, formulate a solution to the LCS problem in terms of the maximum-clique problem.*

Exercise 10.4.6 *An individual weight can also be given to each site in each string, to reflect the importance or reliability of that site, or the probability that the site should be in a meaningful subsequence. Then the weight of a common subsequence, s, is the sum of the weights, in each string, of the lists of indices defining s. This leads to the* maximum-weighted LCS problem, *of finding a common subsequence with weight larger or equal to the weight of any other common subsequence. Show how to modify the ILP formulation for the LCS problem to solve the weighted LCS problem.*

Later, in Section 18.1, we will develop an abstract ILP formulation for the problem of finding an LCS of *more than* two sequences.

10.4.2 Software and Practicality

The Python program, *LCS.py*, creates the concrete ILP formulation described in this chapter, to solve the LCS problem for two sequences of equal length. It can be downloaded from the book website.[5] Call this program on a command line as:

```
python LCS.py DNA_file.txt LP_outputfile.lp
```

where DNA_file.txt is the name of a plain-text file containing a DNA sequence over the alphabet {A,T,C,G} with at most 100 nucleotides, and LP_outputfile.lp names an output file for the concrete ILP formulation.

We should note that although the ILP approach to LCS is correct, and is practical for many realistic sequences in biology, the LCS problem is a case where there exists a specialized algorithm which is much more efficient, both in a provable worst-case sense, and in practice. That algorithm uses a general solution technique called *dynamic programming (DP)*. The Perl program, *LCS.pl* implements the dynamic programming approach to solve the LCS problem. It can also be downloaded from

[5] It was written by Jessica Au and Jessie Vuong as part of a class project in a course on ILP and computational biology. I thank those two students for allowing its distribution with this book.

the book website. You do not need to understand anything about Perl, computer programming, or dynamic programming to use the program. Just call it on a command line as:

```
Perl LCS.pl file-name
```

where "file-name" is the name of the file holding the two strings on two lines. There are many good discussions on the dynamic programming approach to LCS. One is in [83].

Exercise 10.4.7 *Make up two equal-length sequences from the DNA alphabet {A,T,C,G}, and solve the LCS problem for those sequences using LCS.py and Gurobi; and then solve the LCS problem for the same two sequences, using LCS.pl. Make sure that the two approaches find the same LCS length. Try different inputs with different lengths to see how the times used by LCS.py, by Gurobi, and by LCS.pl, are affected by the sequence lengths.*

Spoiler Alert: The dynamic programming (DP) approach is vastly faster than the ILP approach for the LCS problem. This is a bit disappointing, but valuable to know. And if we had efficient DP algorithms for all of the problems considered in this book, we would never need ILP. So, why do we learn ILP? More specifically, why include LCS in this book?

10.5 OPTIMAL PATHOGEN AND SPECIES BARCODING

10.5.1 Epidemics and Barcodes

When a epidemic (or perhaps a biological terrorist attack) hits a population, it is critical to identify the causal pathogen (a virus, bacteria, or fungus) as quickly as possible. Culturing bacteria, viruses, or fungi (extracted from blood, tissues, mucus, etc.) in a laboratory can take days, and does not always lead to a clear identification of the pathogen. But, every living (sub)species has sufficient *unique* DNA, or RNA, to distinguish it from other subspecies, and this can be exploited for very rapid identification of a pathogen whose genome has *previously* been sequenced.[6]

Identifying Pathogens in the Field It may soon be possible to quickly identify a causal pathogen by sequencing its *entire* genome, and then comparing it to the genome sequences of known pathogens. Ideally, this approach would use an inexpensive device that can sequence DNA or RNA quickly under harsh conditions outside of a laboratory. But more realistically, several research efforts have focused on using only a *part* of a genome sequence. Those efforts have designed devices (e.g., hybridization chips) that can identify pathogens using a *barcoding system* for a *predefined* set of *known* pathogens whose genomes have previously been

[6] The ILP approach described here was published in 2002 [160]. An ILP approach to a related design problem, where *multiple* pathogens must be detected with chips, was developed in in 2004 [108]. Since then, the use of barcoding in life sciences has expanded well beyond its original application. For example, this week saw the publication [173] of work using barcodes to deduce that whales had been hunted by early New Zealand Maori, and to identify which species (some now extinct there) had been hunted. DNA was extracted from whale bones found at archaeological sites, and the species of whale were identified using an established barcoding system for marine life. "By revealing species from once meaningless bone fragments, barcodes are 'opening new windows into historical ecosystems' " [150]. Thus barcodes are having a major impact outside of pathogen detection – in ecology and other areas of biology, and outside of biology in archaeology, anthropology, and criminal forensics, etc. One ecologist is in awe of barcodings power. "Its like a miracle, she says. You send them a little bit of dust and they send you back results." [150]

sequenced. Currently, the genomes of tens of thousands of subspecies of harmful viruses, bacteria, and fungi have been sequenced.

Barcoding Pathogens Suppose $\mathcal{P}$ represents the set of pathogens whose genomes have previously been sequenced. Each pathogen in $\mathcal{P}$ will be associated with a *unique* binary string, called its *barcode*. Then, when faced with an unidentified pathogen $\tilde{p}$ that is causing a disease outbreak, we want to quickly find the barcode of $\tilde{p}$, and compare it to the known barcodes. This will identify $\tilde{p}$, *if* $\tilde{p}$ is in $\mathcal{P}$. If $\tilde{p}$ is not in $\mathcal{P}$, we will at least be able to exclude all of the pathogens in $\mathcal{P}$.

So, what exactly is a barcode and how do we find it? Roughly, a barcode for a pathogen p, is a binary string identifying which of several (predefined) *substrings*, from a set $\mathcal{S}$, are contained in the genome sequence of pathogen p. In more detail, each position i in the barcode for p specifies whether or not a specific substring $s_i \in \mathcal{S}$ is contained in the genome sequence for p. For a barcoding system to be useful, each pathogen in $\mathcal{P}$ must have a *distinct* barcode.

An Example For an example (that is much too small to be realistic), suppose $\mathcal{P}$ consists of three "genome" sequences: *"cgatgc," "cagttc,"* and *"cagtgac."* If we choose the set of substrings, $\mathcal{S}$, to be $\{tg, ag\}$ in that order, then the barcodes for the sequences in $\mathcal{P}$ are 10, 00, and 11, respectively. This barcode system works because *cgatgc* contains the substring *tg*, but does not contain the substring *ag*; *cagttc* does not contain *tg*, but it does contain *ag*; and *cagtgac* contains both *tg* and *ag*.

So, if we have a chip that can rapidly tell whether or not the genome sequence of an unidentified pathogen, $\tilde{p}$, contains the substring *ag*, and whether or not it contains the substring *tg*, we can rapidly determine the barcode of $\tilde{p}$, identifying the pathogen – if $\tilde{p}$ is one of the three known pathogens in $\mathcal{P}$. In reality, chips need to distinguish between *tens of thousands* of known pathogens, not just three.

A Notational Comment We use the word "barcode" to refer to a specific binary string assigned to a specific pathogen in $\mathcal{P}$, and use "barcode system" to refer to the entire set of barcodes assigned to the pathogens in $\mathcal{P}$.

10.5.2 How the Chip Works

We don't need or want to go into great detail on how a chip identifies which substrings in $\mathcal{S}$ are contained in the DNA of an unidentified pathogen, but it helps to have some understanding of the idea. Recall that a DNA molecule is double stranded, where the strings in the two strands are *complementary*, i.e., (generally) a nucleotide A in one strand is opposite a T in the other strand; and a C is opposite a G. Two complementary strings (that are long enough) will *hybridize*, meaning that they will stick together, with complementary pairs opposite each other, as shown in Figure 10.5.

A chip made for a set, $\mathcal{S}$, of substrings from $\mathcal{P}$, has a cell for each substring s in $\mathcal{S}$, and that cell will have a physical DNA molecule consisting of substring s, attached to the cell. And, for each cell, we know which substring is attached to it. Then, given a double-stranded DNA molecule (from some pathogen), many copies of the molecule are made and denatured, so that the two strands of each copy separate. This results in many copies of each strand of the original DNA molecule.

5' TCGACCGCTCGA 3'

5' TCGAGCGGTCGA 3'

Figure 10.5 Hybridizing DNA Strings. Each nucleotide hybridizes with its complementary nucleotide: A with T, C with G. The $5'$ to $3'$ orientations of the two strands are the opposite of each other. One strand is shown with its $5'$ end on the left and its $3'$ end on the right; and the other strand is shown with the opposite orientation. When drawn this way, the former is sometimes called the "Watson" strand, and the latter is called the "Crick" strand.

By some chemistry, each strand is extended with a fluorescent tag that will light up when a particular frequency of light is shined on the chip. All the strands are contained in a liquid, which is next poured onto the chip. If a strand contains the substring that is complementary to a substring attached to a specific cell on the chip, the complementary substrings will hybridize, and the poured strand will stick to that cell. Finally, we shine the light, and the cells on the chip that light up identify which substrings in $\mathcal{S}$ are contained in the original DNA molecule. The barcode for the pathogen contains 1s exactly in the positions in the barcode system that correspond to the substrings identified by the chip.

10.5.3 Barcode Design

Above, we introduced barcoding of pathogens, and explained how to *use* a barcode system to identify an unidentified pathogen in $\mathcal{P}$. But, we didn't explain how to *design* a good barcode system. How can we design one? This is

> **The Barcode Design Problem** Given the genome sequences for every pathogen in $\mathcal{P}$, what *set* of substrings, $\mathcal{S}$, should we use for the barcodes?

Before we can develop a solution method, we have to specify the features we want in a barcode system. Then, we can explain how integer programming is used to design an optimal one.

Properties of a Good Barcode System The most essential properties of a barcode system is that each pathogen in $\mathcal{P}$ is assigned a *distinct* barcode; and that the barcode for p must have a 1 in a position i, if and only if the genome sequence for p contains the substring in $\mathcal{S}$ associated with position i. Note that the positions in a barcode system given to the substrings in $\mathcal{S}$ is arbitrary, but once established, the individual barcodes must be created in agreement with that order.

The next most important properties for a barcode system are that lengths of the substrings in $\mathcal{S}$ should not be "too long," nor "too short," nor "too similar" to each other. Also, there should not be "too many" substrings in $\mathcal{S}$. In addition to these properties, there are several other properties that have been proposed and examined, but the ones listed here are sufficient for our discussion. An interested reader can consult [160] for more properties and how to implement those properties using ILP.

There is a limit to the number of cells that can be on a chip, and a range of substring lengths that allow proper hybridization. So, given the *full genome* sequences of a set of pathogens, $\mathcal{P}$, a precise statement of the *Barcode Design* problem is:

> Select the *smallest* set, $\mathcal{S}$, of substrings of appropriate lengths, so that the genome sequence for each pathogen p in $\mathcal{P}$ contains a *distinct* set of substrings in $\mathcal{S}$.

A Solution A solution to an instance of the barcode design problem for a given set of pathogens, $\mathcal{P}$, consists of a set of substrings, $\mathcal{S}$, that have the properties stated above. So the logic of the ILP formulation that solves the problem must consider the set of substrings contained in the pathogens in $\mathcal{P}$. Let β denote the *entire* set of substrings in the genomes of the pathogens. But, as noted, the substrings selected for $\mathcal{S}$ should not be too short or too long, so we will only consider the substrings from β whose length is in the "Goldilocks range."[7] Let $\overline{\beta}$ be a list of the substrings in β with length in the appropriate range. Then, our approach to solving the barcode design problem, and the basis for the ILP formulation is the following

> **Key Insight:** For each *pair* (p,q) of pathogens in $\mathcal{P}$, the substrings chosen for $\mathcal{S}$ must contain at least one substring s from $\overline{\beta}$ that *is* in the genome of one of the pathogens but is *not* in the genome of the other. This condition is not only *necessary* for $\mathcal{S}$ to be a feasible solution – it is *sufficient*.

Exercise 10.5.1 *A different condition that is sufficient for a feasible solution, but not necessary, is: For each pair (p,q) of pathogens in $\mathcal{P}$, $\mathcal{S}$ has at least one substring from $\overline{\beta}$ that is contained in p, but not in q; and has at least one substring from $\overline{\beta}$ that is in q but not p.*

How does this condition differ from the condition in the Key Insight? Explain why this condition is sufficient, i.e., if $\mathcal{S}$ satisfies that condition, then $\mathcal{S}$ is a feasible solution to the barcode design problem. Remember that the term "feasible solution" does not mean it is necessarily optimal.

Why is the condition not a necessary condition?

Exercise 10.5.2 *Explain why the Key Insight is both necessary and sufficient. That is, explain why the condition must hold for any feasible solution $\mathcal{S}$, and why any set of substrings $\mathcal{S}$ from $\overline{\beta}$ that satisfies the condition, will be a feasible solution.*

10.5.4 An ILP Formulation for the Barcode Design Problem

The ILP Variables The abstract ILP formulation will use one binary ILP variable, $B(s)$, for each substring s in $\overline{\beta}$. The value of $B(s)$ indicates whether or not substring s in $\overline{\beta}$ will be chosen to be in $\mathcal{S}$. Variable $B(s)$ will be given value 1 if and only if s will be chosen; and will be given the value 0 otherwise.

The ILP Inequalities To implement the key insight, let $\beta(p,q)$ be the set of all substrings in $\overline{\beta}$ that are contained in *exactly one* of the two genomes $\{p,q\}$. Then, for each pair of genomes, (p,q), include the single inequality:

$$\sum_{s \in \beta(p,q)} B(s) \geq 1, \tag{10.8}$$

which says that at least one substring in $\beta(p,q)$ must be chosen to be in $\mathcal{S}$.

[7] Of course, the desired range must be specified in order to create a concrete ILP formulation, but we don't need to be specific when discussing an abstract ILP formulation.

The Objective Function Finally, the objective function is:

$$\text{Minimize} \sum_{s \in \overline{\beta}} B(s). \tag{10.9}$$

10.5.5 Added Constraints

The basic ILP formulation for the barcode design problem is given by the inequalities in (10.8), and the objective function (10.9). However, the barcode system that would be generated from solving this concrete ILP would probably not be ideal. There are several additional properties that are desirable. An important one is that no pair of substrings in $\mathcal{S}$ should be "*too similar.*" An exact definition of what is too similar must be specified, but assuming that has been done, the program that creates the concrete ILP formulations for problem instances should examine all pairs of substrings in $\overline{\beta}$. When two substrings, s and s' say, are found to be too similar, the following inequality should be added to the ILP formulation:

$$B(s) + B(s') \leq 1. \tag{10.10}$$

Another important extension is to change the right-hand side of the inequalities in (10.8) from 1 to 2 or larger. This change makes the barcode system more robust, allowing proper identification of any pathogen in $\mathcal{P}$, even if there are a few "false negatives" (i.e., cells that should light up, but don't) when the liquid is washed over the chip.

Exercise 10.5.3 *Suppose the inequalities specified by* (10.8) *are changed to*

$$\sum_{s \in \beta(p,q)} B(s) \geq k,$$

for some integer $k > 1$. *Explain in detail how to use the resulting chip to identify specific pathogens in* $\mathcal{P}$.

10.5.6 Reducing the Number of Variables

The ILP formulation for the barcode design problem has one variable for each substring in $\overline{\beta}$. Note that even if the same substring s appears in more than one genome in $\mathcal{P}$, or appears multiple times in a single genome in $\mathcal{P}$, substring s is included *only once* in the set of substrings in $\overline{\beta}$. Still, the formulation contains a very large number of ILP variables, since $\overline{\beta}$ comes from thousands of genomes. However, the number of variables can be significantly reduced with the method we describe next.

Removing Redundant Substrings Suppose two substrings, s and s' in $\overline{\beta}$ are contained in *exactly* the same set of genomes in $\mathcal{P}$. Then, the set of genomes in $\mathcal{P}$ that they are *not* contained in, must also be the same. Therefore, a barcode system $\mathcal{S}$ for $\mathcal{P}$ would *never* need to contain both s and s', and in the ILP formulation, one of the variables $B(s), B(s')$ can be removed.

Clearly, we want the computer program, call it C, that creates concrete ILP formulations for the barcode design problem to efficiently recognize as many such pairs as possible. How can we do that? We will concentrate on one simple, but effective, type of pairs: pairs of substrings (s, s'), where s' is a substring of s. But how do we efficiently determine if s and s' are contained in exactly the same set of genomes in $\mathcal{P}$. This seems very time consuming, but, there is actually an easy way.

Even without trying to reduce the number of variables, the computer program $\mathcal{C}$ must find, for each substring $s \in \overline{\mathcal{B}}$, the set of the genomes in $\mathcal{P}$ that contain s. That is required in order to create the inequalities in 10.8. Now, let s' denote the *prefix* of s, which is one character shorter than s (so s' is just s after removing the last character in s). And, suppose we organize program $\mathcal{C}$ so that it next finds the genomes in $\mathcal{P}$ that contain s'. We want to know if s and s' are contained in *exactly* the same *set* of genomes in $\mathcal{P}$. But, every genome in $\mathcal{P}$ that contains s also contains s', so if the *number* of genomes in $\mathcal{P}$ that contain s is *equal* to the number of genomes in $\mathcal{P}$ that contain s', then the sets will be identical. In fact, if s' is a prefix of s, then the number of genomes in $\mathcal{P}$ that contain s' will be equal to the number that contain s, if and only if s and s' are contained in exactly the same set of genomes in $\mathcal{P}$. In this way, we can identify many pairs of substrings, (s, s'), where one of the variables $B(s)$, or $B(s')$ can be removed from the ILP formulation. The determination of which one to remove can depend on whether we want to use smaller strings, which may reduce the cost of making a chip; or if we want to use longer strings, which may reduce the frequency of errors made when using the chip.

A more efficient and effective way to reduce the number of substrings that might be chosen for the barcode system is through the use of a data structure called a *suffix tree*. Discussion of suffix trees is out of the scope of this book. The interested reader can find details of suffix trees in [83], and details of their uses in the barcode design problem in [160].

11

Metabolic Networks and Metabolic Engineering

Recall the introduction to metabolism and the description of metabolic networks in Section 2.1.1. In this chapter we go beyond the *description* of a metabolic network, to address questions of *modifying* the network to improve its performance, or to reduce negative effects. This general topic is called *Metabolic Engineering*.

> Metabolic engineering is the practice of optimizing genetic and regulatory processes within cells to increase the cells' production of certain substances. These processes are (described by) chemical networks ... Metabolic engineering specifically seeks to mathematically model these networks, calculate a yield of useful products, and pin-point parts of the network that constrain the production of these products. Genetic engineering techniques can then be used to modify the network. ... [220]

So, given the metabolic network that details the genetic and chemical activities leading to the production of certain desired end-products, we create an *optimization problem* whose solution suggests ways to increase the production of the desired end product(s), or reduce the cost of producing them.

Metabolic engineering can also be used in a *reverse* fashion. If the network represents a biochemical process that results in *undesired* products (with or without desired products), we may want to intervene to alter the behavior of the network so that it reduces the production of undesired products, while maintaining or increasing the production of any desired products.

ILP and Metabolic Engineering There are many examples in metabolic engineering where the optimization problem is framed and solved as an *integer linear program* (for example, see [129, 159, 190, 196]), and even more where it is framed as a pure linear programming problem.

11.1 BOOLEAN NETWORKS

In many applications, a fully detailed metabolic network is *not needed* – a simpler network may suffice to allow the optimization problem to be framed and solved more easily. In this section, we introduce one type of network that simplifies a fully detailed metabolic network, and show how ILP is used to optimize the behavior of those networks.

A *Boolean metabolic network* is a directed graph where each node represents either a *chemical reaction*, a *chemical compound*, or an *enzyme*. A chemical reaction occurs when certain chemicals are combined; a *compound* is a chemical that is produced by a chemical reaction in the network, or is input to the network; and an *enzyme* is a biochemical that is necessary to facilitate or accelerate a chemical reaction. Nodes that have no edges coming into them are called *sources*, and nodes that have no edges out of them are called *end product* nodes. An edge directed out of a reaction node must go to a compound node, and an edge directed out of compound node must go to a reaction node. Edges out of an enzyme node must go to reaction nodes (see Figure 11.1).

Enzymes are always supplied externally, i.e., are represented by source nodes, and so the inputs to the network are compounds and enzymes; the end products of the network, represented by end product nodes, are compounds; and the internal nodes in the network represent both compounds and reactions. But, these elements are not what makes Boolean networks distinctive, since reactions, enzymes, and compounds are common elements in all metabolic networks.

What is distinctive about *Boolean networks* is that the *input and output behaviors* of compound and reaction nodes implement logical OR and AND operations, respectively. To explain this, it helps to think of assigning a logical value – either *true* or *false* – to each node. Then an assignment of true/false values to nodes is a *valid assignment* if it obeys the following:

(a) A source node can be set either true or false. (Setting a source node true means that the compound or enzyme represented there is available for use. Setting it false means it is not available.)

(b) A compound node that has incoming edges is set true if and only if *one or more* of the nodes pointing into it is set true. (A compound node implements an OR operation. The compound can be produced in any reaction represented by the nodes pointing into the compound node.)

(c) A reaction node is set true if and only if *all* of the nodes pointing into it are set true. (A reaction node implements an AND operation. The reaction needs all of the compounds and enzymes represented by the nodes pointing into the reaction node.)

For example, Figure 11.1 shows a hypothetical Boolean metabolic network. Nodes with names starting with "C," "R," and "E," are compound, reaction, and enzyme nodes, respectively. (Also, the shape of the node – a circle, a triangle, or a square – specifies its type. Figure it out for yourself!). The nodes enclosed by *solid* shapes are set *true*, while that nodes enclosed by *dashed* shapes are set *false*. That setting is a valid assignment.

Exercise 11.1.1 *Verify that the setting described above for Figure 11.1 is a valid assignment.*

Exercise 11.1.2 *In Figure 11.1, if we change the truth settings of E_1 to* true, *and E_2 to* false, *what must the setting for C_7 be in a valid assignment? Is there a valid assignment where node C_5 is set true, and one of the source nodes, C_1, C_2, C_3, C_6, E_2, is set* false*?*

Exercise 11.1.3 *In Figure 11.1, is it true that by setting the value of any single source node to false, the value of node C_5 must also be false in any valid assignment? Explain.*

Exercise 11.1.4 *Suppose that in Figure 11.1, the end product represented at node C_5* is undesired, *but the other end products* are desired. *Then, the source nodes represent potential* drug targets, *whose inhibition will inhibit the production of C_5. However, depending on which source node(s)*

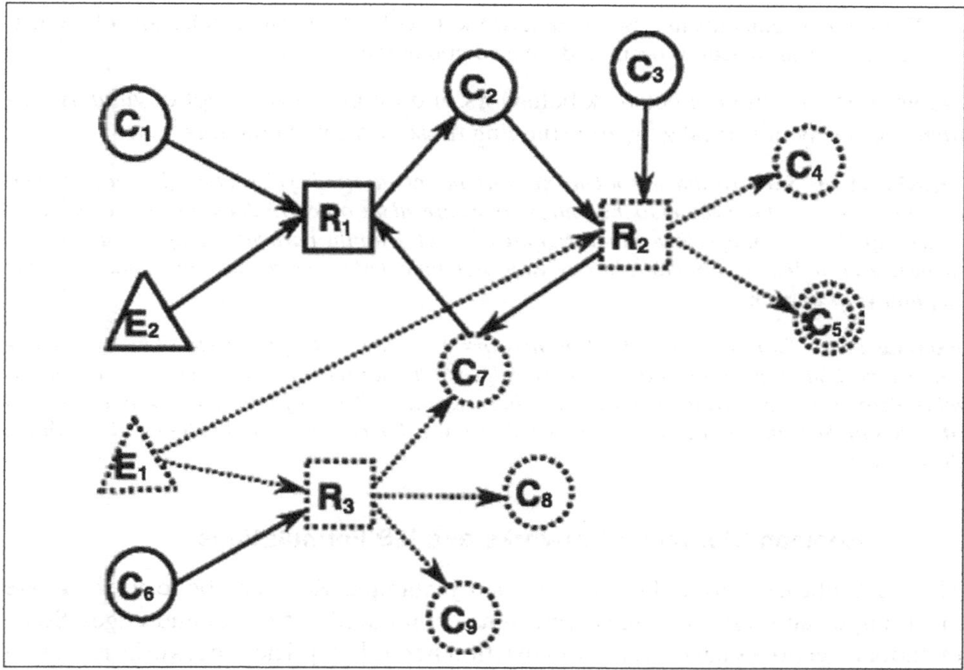

Figure 11.1 A Small, Hypothetical Boolean Metabolic Network. The figure displays the consequences of removing enzyme E_1 from the network input, i.e., setting the value of node E_1 to *false*, and all the other source nodes to true. The dashed lines lead to other nodes that are forced to be set false in any valid assignment. This figure is from figure 1 in [196].

are inhibited, different desired end products will also be inhibited. What source node(s) should be inhibited so that C_5 is inhibited, and the minimum number *of desired end products are also inhibited?*

Exercise 11.1.5 *Is it true that in a Boolean metabolic network, setting the value of each node to* true *is always a valid assignment? Similarly, is it true that setting the value of each node to* false *is always a valid assignment? Explain both answers.*

A More Physical Picture To make the workings of a Boolean network a bit more physical, we can think of each edge as transporting one of two possible *messages*: *True* or *False*. Each compound node and each reaction node receives the messages sent to it, and applies the OR or AND operations, respectively, to determine which message (True or False) to send along the edges out of it. Note that this is purely *qualitative* model. A node might receive one message and send out ten, for example. There is no conservation of messages, or of bits used to encode the message. We will come back to the issue of conservation later.

11.1.1 Boolean Networks As Models of Metabolic Processes

A Boolean metabolic network is a very *simplified model* of a full metabolic network, abstracting away many of its details. But, for several applications, Boolean networks capture enough of the reality of metabolic processes to be useful. In particular:

The valid assignments in a Boolean network describe the possible behaviors of the network, i.e., what reactions occur and what compounds are made.

We are interested in the network behaviors, and we use the concept of *valid assignments* as a way of formalizing and studying those possible behaviors.[1]

Exercise 11.1.6 *Suppose that a Boolean network has* no *directed cycles. Then given a particular assignment of true/false values to the source nodes, are all the other true/false assignments* forced *(i.e., uniquely determined) in a valid assignment? That is, for the particular assignment of values to the source nodes, can there only be* one *valid assignment of values to the nodes in the network? Explain your answer.*

Exercise 11.1.7 *Suppose that a Boolean network* does *have a directed cycle. Give an example where a particular assignment of true/false values to the source nodes does* not *uniquely determine all of the other true/false values in a valid assignment? That is, for the particular assignment of values to the source nodes, there are* two different *valid assignments of values to the nodes in the network.*

11.1.2 Boolean Metabolic Networks and ILP Formulations

The metabolic network in Figure 11.1 is a toy example. Real metabolic networks are much larger and more complex, with tens or hundreds of nodes and edges. Some metabolic networks have over 2,000 reaction nodes [189]. These networks represent metabolic processes with many pathways, redundancies, and undesired, as well as desired, end products. So, it is not a trivial task to analyze and understand such networks' behaviors, and determine what would happen if some of the nodes were removed, or some new reactions or enzymes were added. This is where it is valuable to have a simpler *Boolean* network representing the metabolic process, and to create an *ILP formulation* to model and interrogate its behavior. We develop that ILP formulation next, building on the approach in [159, 190].

ILP Representation We can capture the possible behaviors of any Boolean metabolic network using integer linear inequalities, with one binary variable for each node. We name the ILP variable corresponding to a node with the name of the node, but move the subscript up, into parentheses. For example, "$R(2)$" will be the name of the ILP variable corresponding to node R_2. Also, in ILP formulations we encode "true" with the integer 1, and "false" with the integer 0 in the inequalities.

Then, as an example, for node R_2 in the network in Figure 11.1, the following inequality implements the logic that: If the value of *each* of the nodes C_2, C_3 and E_1 is true, then the value of node R_2 *must* also be true.

$$R(2) + (1 - C(2)) + (1 - C(3)) + (1 - E(1)) \geq 1. \tag{11.1}$$

Conversely, the inequalities:

$$\begin{aligned} R(2) &\leq C(2), \\ R(2) &\leq C(3), \\ R(2) &\leq E(1). \end{aligned} \tag{11.2}$$

implement the logic that: If the value of *any* of the nodes C_2, C_3, and E_1 is *false*, then the value of node R_2 *must* also be *false*. Together, the inequalities in (11.1) and (11.2)

[1] This is a great example of *mathematical modeling* of biological phenomena, something that I said in the Introduction could only be taught through seeing and critiquing models. There is no well-developed theory or set of techniques for effective modeling.

implement the logic that node R_2 must be set to true if and only if *each* of the nodes pointing into node R_2 is set true. Hence, these inequalities express what is necessary for a valid assignment of true/false values to reaction node R_2, and the nodes that have edges pointing into R_2.

Exercise 11.1.8 *State and explain the analogous inequalities that implement the OR relation at any compound node, say, for node C_7 in Figure 11.1.*

Exercise 11.1.9 *State the inequalities that capture the behavior of* all *of the non-source nodes in Figure 11.1.*

Generalizing and Abstracting Generalizing from (11.1) and (11.2), we can state abstract inequalities for the behavior of a reaction node, R_i, with incoming edges from a set of compound nodes and a set of enzyme nodes, denoted $CR(i)$ and $ER(i)$ respectively:

$$R(i) + \sum_{j \in CR(i)} (1 - C(j)) + \sum_{r \in ER(i)} (1 - E(r)) \geq 1, \tag{11.3}$$

and for each node j in $CR(i)$:

$$R(i) \leq C(j), \tag{11.4}$$

and for each node r in $ER(i)$:

$$R(i) \leq E(r). \tag{11.5}$$

Exercise 11.1.10 *State and explain the analogous abstract inequalities that implement the OR relation for any compound node C_i.*

Exploring and Optimizing the Network with ILP We use $\mathcal{I}$ to denote all of the inequalities in (11.3)–(11.5), *together* with the inequalities in your answer to Exercise 11.1.10. Then, when 0/1 values are set for the variables corresponding to source nodes, the set of feasible solutions to the inequalities in $\mathcal{I}$ capture all of the possible behaviors of the Boolean network. (By Exercise 11.1.7, there may be more than one feasible solution if the network has a cycle.)

So, we can construct ILP formulations containing $\mathcal{I}$, possibly augmented with additional inequalities that express more biochemical constraints. Then, the feasible solutions to the ILP formulations *answer questions* about the Boolean network and the metabolic processes that it represents. We can also construct ILP formulations that allow us to *optimize* or *change* the behavior of those processes. This is best explained through some examples.

Example 1. Suppose $C_1, C_2, \ldots, C_d$ denote all of the nodes representing *desired* end products in a Boolean network, and $C_{d+1}, \ldots, C_f$ denote all of the nodes representing *non-desired* end products. Let $S_1, \ldots, S_k$ denote all of the source nodes – some representing compounds and others representing enzymes. Then, consider the ILP formulation that contains all the inequalities in $\mathcal{I}$, plus the following:

$$\sum_{i=1}^{d} C(i) = d,$$

$$\sum_{i=d+1}^{f} C(i) = 0,$$

with the objective function:

$$\text{Minimize} \sum_{j=1}^{k} S(j).$$

A feasible solution to this ILP formulation will indicate that it *is* possible to produce all of the desired end products *without* producing any undesired end products. An *optimal* solution (assuming a feasible solution exists) specifies the *minimum* number of source compounds and enzymes that need to be made *available* to achieve the production of the desired end products, without producing any of the undesired end products. Or, we could change the objective function from a minimization to a *maximization*, in order to minimize the number of sources that need to be *inhibited* for the same end result.

One illustration of this kind of problem was recently mentioned in *The New York Times*. The molecule *Coenzyme Q10*, (whose name sounds too much like something made up by a marketing department to be real – but it is) "helps muscle cells generate energy." So, it is a desireable end product. On the other hand, too much *cholesteral* can contribute to heart disease, and so is undesirable. Taking a *statin* regularly can reduce the production of cholesteral. But "... cholesteral and coenzyme Q10 are synthesized by the same biochemical pathway. As a result, statins not only lower cholesteral levels, they also deplete the body's stores of coenzyme Q10" [200]. So, we would like to alter that pathway to inhibit the production of cholesteral, while not reducing the production of coenzyme Q10.

Example 2. Suppose that all of the source compounds and enzymes are available to the network, so that all of the variables corresponding to source nodes are set to 1. Suppose also that all of the end products of the network are *undesired*, and we have the ability to inhibit any chosen set of *reactions* in the network (through withholding required enzymes, or through genetic engineering, or through the use of some additional chemicals or biological agents). Inhibition of reaction R_i corresponds to forcing the value of node R_i to false. But, inhibiting a reaction has some cost, and perhaps some unknown side effects, so we want to pick the *smallest* set of reactions to inhibit so that *none* of the end products are produced. How can we figure out which reactions to inhibit?

The answer is a bit sneaky. For each reaction node R_i, we add a *new* source node called F_i ("F" is for Fake) with an edge directed into node R_i. Since, in a valid assignment, R_i can be set to true if and only if all of the nodes directed into it are set to true, setting the fake source F_i to false forces R_i to also be set false. Reflecting those fake source nodes, consider the ILP formulation containing the inequalities in $\mathcal{I}$, together with the following.

For each real source node S_j:

$$S(j) = 1,$$

and

$$\sum_{\text{end product } C_i} C(i) = 0,$$

with the objective function:

$$\text{Maximize} \sum_{\text{fake source } F_i} F(i),$$

where $F(i)$ is the binary variable associated with node F_i.

An integer optimal solution to this ILP identifies a way to inhibit reactions in the network that disallow the production of any of the end products, while inhibiting the *fewest* reactions.

Exercise 11.1.11 *Explain explicitly how the above ILP* minimizes *the number of reactions that will be inhibited, since the actual objective function is a* maximization.

11.2 EXTENDING BOOLEAN NETWORKS, WITH ILP

There are two major ways that Boolean networks can and have been extended. One way is to incorporate more *biological phenomena* and detail into the network; and the other way is to express and solve more involved *analytical* questions about the network. Of course, these two extensions are often related. Many such extensions have been examined in the field of biological network analysis.[2]

One significant extension is to include *gene expression* and *gene regulation* into the networks. This is a natural extension because the production level of a protein in a cell is strongly affected by the expression level of the genes that code for the protein, or the peptides that make up the protein. Conversely, gene regulation, affecting whether or not a gene is expressed and/or at what rate, is strongly affected by proteins that act to regulate the gene. The biological reality is of a complex network of *interacting genes and proteins*, along with other possible regulatory elements, such as microRNAs.

This extension of biological phenomena is significant, but, typically does not require a large change in the mathematical and algorithmic techniques used to analyze the networks. So, ILP is still valuable with this extension. For an example of integrating both genetic regulatory networks with metabolic networks, and then using ILP to solve analysis problems on those networks, see [182].

Incorporating More Complex Logic Another significant extension is the introduction of more complex *logical* functions. In a pure Boolean network, each reaction node implements an AND relation, and each compound node implements an OR relation. But the behavior of biological elements in a metabolic or gene regulatory network might be better modeled by other logical relations. For example, a reaction node might have 10 activation edges incoming to it. But instead of setting the node's value to true if and only if *all* 10 of the nodes pointing to it are set true, a more accurate biological model might be to set the node's value to true if *at least* five, say, of the nodes pointing into it are set true. Many extensions of this type of logic are biologically appealing, and their implementation in an ILP formulation is generally straightforward.

[2] In fact, if we consider the *chronology* of such work, many of these "extensions" predate the use of Boolean networks in biology, or are independent of Boolean networks. But for simplicity the exposition, we will consider these alternatives models and formalizations to be extensions of the Boolean model.

Exercise 11.2.1 *Show how to implement the "at least five" relation discussed above, in an ILP formulation. You should be able to figure this out just by thinking and fiddling, but an idiom that solves this problem is discussed in Section 12.1 in Chapter 12.*

Extending Inhibition Even more significant to proper biological modeling is the way that *inhibition* is formalized. Recall from Section 2.1.1 where metabolic networks were first introduced, that an edge can be either an *activating* or an *inhibiting* edge (see Figures 2.7, 11.4, and 11.5).

In pure Boolean networks, there are only activation edges, so inhibition of a compound node, C_i, occurs when *all* of the nodes pointing into C_i are set to false; and inhibition of a reaction node, R_j, occurs when *at least one* node that points into R_j is set to false. Biochemically, this kind of inhibition occurs when an element is, or elements are, *not* delivered to the inhibited node.[3] But, there is a more biologically accurate and common kind of inhibition, called *negative feedback*.[4] Negative feedback is represented by an *inhibiting* edge.

In the extreme case, negative feedback means that when there is an inhibiting edge from a compound node, C_i, to a reaction node R_j, and node C_i is set *true*, then R_j must be set *false*. Biochemically this models the situation that compound C_i *actively interferes* with the reaction modeled at node R_j.

In a more *semiquantitative* case, such as the "at least five" example from above, negative feedback might mean that R_j will be set true if and only if the number of nodes set true, that point into R_j on activating edges, *minus* the number of nodes set true that point into R_j on inhibiting edges, is *at least five*, for example. Or perhaps, R_j will be set true if and only if there are more activating inputs to R_j (from nodes set true, along activating edges) than there are inhibiting inputs to R_j (from nodes set true, along inhibiting edges). Biological models of this type are realistic, and straightforward to implement with ILP formulations.

Exercise 11.2.2 *Show how to implement this new "at least five" relation for a node R_j, in an ILP formulation. This assumes that the number of activating edges into R_j is at least five.*

Show how to implement in an ILP formulation the relation that there are more activating inputs to R_j than inhibiting inputs to R_j.

Another simple extension is to introduce *costs* for the sources, and costs for adding or inhibiting reactions. Then, the objective function incorporates the costs, so that an optimal solution to the ILP formulation describes the minimum cost way to achieve a desired behavior.

11.2.1 Quantitative Extensions

Finally, a natural extension is to move from qualitative models, such as Boolean networks (even with some semiquantitative logic as above), and move to more *quantitative* network models. Quantitative models encode more biological or chemical reality than do Boolean networks. More realistic quantitative models use experimentally established *numerical* parameters, together with physical or biological constraints, and mathematical optimization, to *infer* unknown values, and to *optimize* the behavior of the network by altering some of the values and parameters.

[3] This might be called "inhibition by omission."
[4] Which we might call "inhibition by commission."

Metabolic Flow and Flux Many quantitative models incorporate numerical modeling of physical *flow* of metabolites (molecules involved in metabolism) along the edges of the network, subject to different kinds of constraints. The formal word used in this context is "flux," which can be interpreted as the "rate of flow." There are several specific ways that flux in a metabolic network is modeled. One of the major ones (which we only mention without a deeper discussion) is called *Flux Balance Analysis*. It relies on *linear programming* and *integer* linear programming, for the analysis and optimization of the network.

> Flux balance analysis (FBA) is a widely used approach for studying biochemical networks, in particular the genome-scale metabolic network reconstructions ... These network reconstructions contain all of the known metabolic reactions in an organism and the genes that encode each enzyme. FBA calculates the flow of metabolites through this metabolic network, thereby making it possible to predict the growth rate of an organism or the rate of production of a biotechnologically important metabolite. [142]

Flux balance analysis is typically implemented using *linear programming*, but there are analysis questions where *integer* linear programming has been needed and used (see, for example, [90, 124]). Once the behavior of a metabolic network is encoded by linear inequalities, and an objective function is given, say to maximize the production of some desired product, we can change parts of the network to answer "what if" questions. In this way, the model becomes a powerful predictive tool for metabolic engineering.

11.2.2 A Mass-Flow Model

Here we describe in detail an approach to quantitative metabolic modeling that is similar in spirit to FBA, but differs in detail. The approach is called *mass-flow analysis*, and is detailed in [127].

The directed graphs used to represent metabolic networks for mass-flow analysis, are identical to the directed graphs used for Boolean metabolic networks. As before, each node in the network corresponds to a metabolite (i.e., a compound or an enzyme), or to a reaction. And again, each metabolite node has edges out to reaction nodes, and each reaction node has edges out to metabolite nodes. However, what flows along an edge is not a True/False message (value) – it is a specified collection of molecules. And, instead of implementing AND/OR logic at reaction and compound nodes, we implement physical, chemical *conservation and balance laws* at each node.

We will assume that there is only one type of molecule that flows on any given edge. For example, an edge might have a flow of water molecules, but it can't also have a flow of dioxygen molecules – those might flow on some other edges. Also, we can assume that there is at most one edge into any metabolite node. After seeing the entire network model, you should verify for yourself that these assumptions are not limiting in any way.

Conservation at Nodes To be more precise about conservation, the number of a particular atom that flows (in one time unit) into a node, must be equal to the number of that atom that flows out of the node, in the same time unit. This is a conservation law, when there is no accumulation of mass at a node. For example, consider the classic chemical reaction that converts six *carbon dioxide* molecules and six *water* molecules into one *glucose* molecule and six *dioxygen* molecules. The chemical formula for this reaction is:

$$6CO_2 + 6H_2O \Rightarrow C_6H_{12}O_6 + 6O_2,$$

where some energy must also be added to make the reaction work. (But we are only modeling the flow of atoms and molecules, not the balance of energy, so that is ignored.)

We see that the input to this reaction contains $6 \times 2 + 6 = 18$ oxygen atoms, and 6 carbon atoms, and $6 \times 2 = 12$ hydrogen atoms. The output from the reaction is 6 carbon atoms, and 12 hydrogen atoms, and $6 + 6 \times 2 = 18$ oxygen atoms. So the chemical reaction reorganizes the atoms into different molecules, but the number of each type of atom is completely conserved. How do we encode this in linear constraints?

Consider the reaction node R_g for the chemical reaction specified above. Per some time unit, some number of *molecules* of carbon dioxide flow into R_g on one edge, e_1, and some number of molecules of water flow into R_g on another edge, e_2. We use variables $X(e_1)$ and $X(e_2)$, respectively, to denote the number of molecules that flow into R_g on edges e_1 and e_2.

Similarly, in that same time unit, some number of glucose *molecules* flow out of R_g on one edge, e_3, and some number of dioxygen molecules flow out of R_g, on another edge, e_4 (see Figure 11.2). We use variables $X(e_3)$ and $X(e_4)$ to denote those numbers. Then, conservation of *atoms* at node R_g is encoded by:

$$\begin{aligned} 12X(e_1) + 6X(e_2) &= 6X(e_3) + 12X(e_4), \\ 6X(e_1) &= 6X(e_3), \\ 12X(e_2) &= 12X(e_3). \end{aligned} \tag{11.6}$$

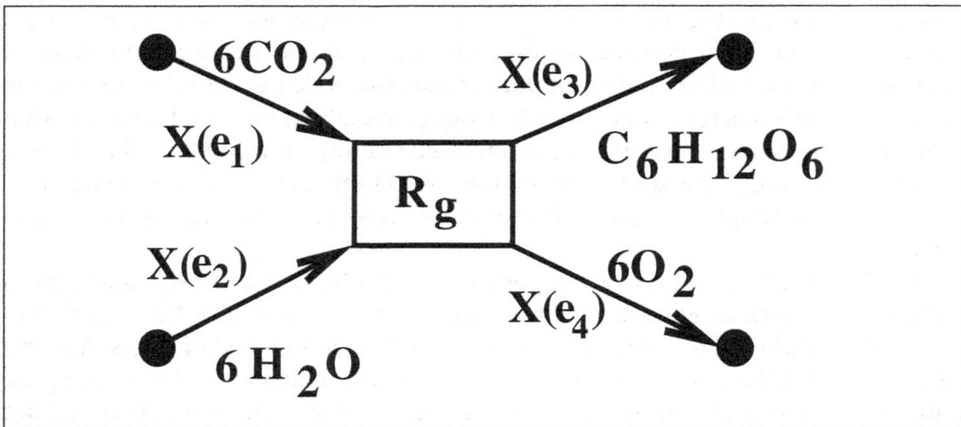

Figure 11.2 A Reaction Node, R_g, in a Mass-Flow Network That Takes in Carbon Dioxide, Water, and a Little Pinch of Energy – and Creates Glucose and Dioxogen. The values of variables $X(e_1)$, $X(e_1)$, $X(e_3)$, and $X(e_4)$ are, respectively, the number of carbon dioxide *molecules* that flow on edge e_1 (in a fixed time unit); the number of water molecules that flow on edge e_2; the number of glucose molecules that flow on edge e_3; and the number of dioxygen molecules that flow on edge e_4.

The first equation in (11.6) encodes the conservation of oxygen *atoms* through R_g; the second encodes the conservation of carbon atoms through R_g; and the third encodes the conservation of hydrogen through R_g. These conservation equations for *atoms* are *necessary* for the correct modeling of the reaction at R_g. But, they aren't sufficient.

Balance of Molecules In addition to conserving atoms at nodes of the networks, we also need to encode the correct *balance* between the arrival of carbon-dioxide *molecules* and the arrival of water *molecules* (per time unit), so that all of the arriving atoms are used in the reaction. Similarly, we need to encode the correct balance between the exit of glucose molecules and the exit of dioxygen molecules. These balance requirements are encoded by:

$$\begin{aligned} 6X(e_1) &= 6X(e_2), \\ 6X(e_3) &= X(e_4). \end{aligned} \tag{11.7}$$

Exercise 11.2.3 *Explain completely why these balance requirements are needed, in order to model the behavior of the network, and the mass flow through it. That is, why is it not enough to conserve the atoms that flow into and out of each node?*

Rate Limits To complete a realistic model of network behavior, we need to encode a *maximum rate* for each chemical reaction in the network. That is, at each reaction node, there is an *upper bound* on the number of reactions (per fixed time unit) that are possible. For example, suppose that per time unit, at reaction node R_g, only seven reactions creating glucose and dioxygen are possible. How do we encode this? Since one reaction at R_g requires six water molecules, which come in on edge e_2, the inequality

$$X(e_2) \leq 42,$$

will limit the number of reactions (per time unit) at R_g to at most seven.

Exercise 11.2.4 *Explain 42.*

Exercise 11.2.5 *If we didn't include any rate limits in the network model, would there be any limit on the amount of end product that is produced per time unit?*

Exercise 11.2.6 *Why is it sufficient to limit just $X(e_2)$? That is, why don't we have to have an analogous inequality for each edge in to and out of R_g? Conversely, could we arbitrarily choose any edge in to or out of R_g to limit? Is water special? (Yes, water is extremely special, but in the context of this model, would any other single edge limit work as well?)*

There may also be limits on the number of each type of molecule that comes into the network at the source nodes. These are easy to incorporate.

Exercise 11.2.7 *So far, we have not modeled flow through a metabolite node. Each metabolite node takes in a number of molecules, all of the same type, from reaction nodes, and directs those molecules to other reaction nodes, or to end product nodes. Since the molecules remain intact at a metabolite node (there are no reactions there), and there is only one type of molecule that flows into it, the conservation/balance laws at a metabolite node are simpler* *than for a reaction node. How are they simpler? Using similar edge notation to what we used in modeling conservation at reaction nodes, state the inequalities needed for a metabolite node.*

Putting It All Together

To do mass-flow analysis for an entire network, we write the conservation and balance equations for each node in the network; write the inequalities based on reaction-rate limits; include any bounds on the numbers of molecules at the source nodes; and restrict the values of all the variables to be integral. These inequalities capture the possible behaviors of the network. Then, if we add an objective function, we will have a complete ILP formulation that can be solved to optimize some behavior of the metabolic network.

What objective functions are meaningful? If the end products of the network have quantitative values (biomass, money, or utility), the objective function could be a linear function of the end product variables, each multiplied by a constant that reflects the value of that end product. Then, the objective function will be to *maximize* the total value. Similarly, the molecules input at the source nodes may each have a cost, in which case, the objective function could be to maximize the total value of the end products, minus the total cost of the source molecules. Or, we could have some fixed targets for the amounts of the end products, and minimize the costs of the source molecules input to the network. Or, we might want to limit the production of some intermediate chemical, or undesired end product ... or ... or ...,

Back to LP We have described mass-flow network analysis in terms of discrete objects, i.e., atoms and molecules, so the numbers of these objects are integers. Hence, the optimization of these objects uses *integer* linear programming. I believe that this is the conceptually clearest approach. However, in [127], the (integer) number of an object was converted into a fractional *molecular weight* of that object. So, what were conserved and maximized were weights of molecules. This converted an *integer* linear programming problem to a pure *linear* programming problem. Even though in general, linear programs solve faster than integer programs, I conjecture that in this case, the ILP formulation will solve in about the same time as the LP formulation. For a student who can write a computer program that creates concrete ILP formulations for mass-flow analysis, testing this conjecture would make a good project.

11.3 FANTASY NETWORK ANALYSIS

In this section we examine two totally *fictional, artificial* networks in order to discuss a type of network model and network analysis that might plausibly occur, and how to do the appropriate analysis with linear programming and ILP.

Consider the network in Figure 11.3. Nodes X and Y represent sources, and W and Z represent end products. Suppose there are only 2,000 units of X and 180 units of Y available. To produce one unit of W requires two units of X and three units of Y. Similarly, to produce one unit of Z requires four units of X and two units of Y. The cost of a unit of X is 2 (somethings), and the cost of a unit of Y is 3. The value of a unit of W is 120 (somethings), and the value of a unit of Z is 90. Further, the source material at X gets split between W and Z, so that 70% goes to W and

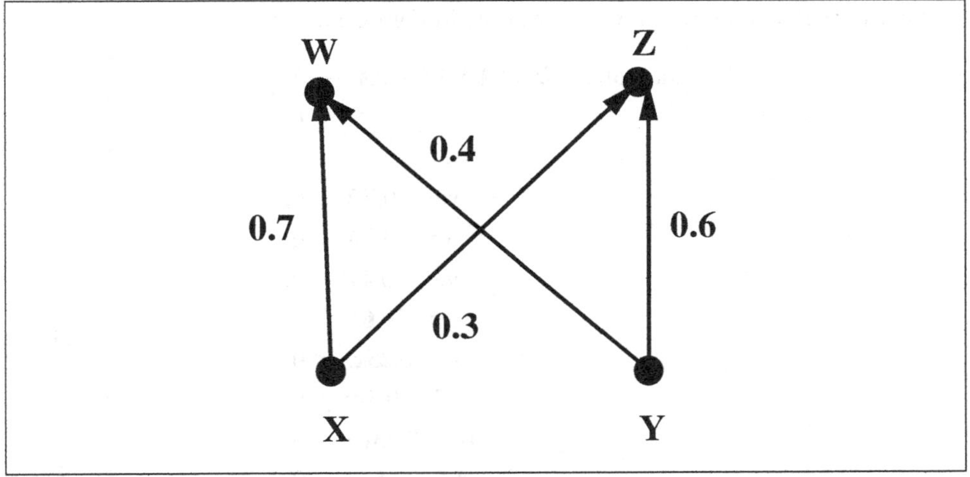

Figure 11.3 A Network with Only Activation Edges.

30% goes to Z. Similarly, 40% of the source material at Y goes to W, and 60% goes to Z. Suppose also, that the values for X, Y, W, Z must be *integral*. With all of this information, how much of X and Y should be used to *maximize* the value of what is produced?

Of course, this problem can be framed as an integer linear programming problem, and most of the ILP formulation is straightforward. The only interesting part is how to implement the requirements that one unit of Z requires four units of X *and* two units of Y; and to implement the analogous constraints for W. We use the variables xw and xz for the number of units of X that are made available for W and Z to use, respectively. The analogous variables, yw and yz are used for the number of units of Y that are made available for W and Z. Then, it might be reasonable to try to express the requirements for Z as:

$$Z = 4xz + 2yz,$$

but that will not work.

Exercise 11.3.1 *Explain why the inequality $Z = 4xz + 2yz$ will not work to express the requirement that each unit of Z requires four units of X,* and *two units of Y.*

Exercise 11.3.2 *Explain why the inequalities*

$$Z \leq \frac{xz}{4}, \qquad (11.8)$$
$$Z \leq \frac{yz}{2},$$

will work.

So, the ILP formulation for the example in Figure 11.3 is:

$$
\begin{aligned}
&\text{maximize } 120W + 90Z - 2X - 3Y \\
&\text{s.t.,} \\
&xw - 0.7X = 0, \\
&xz - 0.3X = 0, \\
&yw - 0.4Y = 0, \\
&yz - 0.6Y = 0, \\
&Z - 0.25xz \leq 0, \\
&Z - 0.5yz \leq 0, \\
&W - 0.33yw \leq 0, \\
&W - 0.5xw \leq 0, \\
&X \leq 2{,}000, \\
&Y \leq 180, \\
&\text{All variables are integral.}
\end{aligned}
\tag{11.9}
$$

This ILP has an optimal solution of value 5,640, which is achieved by setting $W = 23, Z = 54, X = 720$, and $Y = 180$. Then, $xw = 504, xz = 216, yw = 72$, and $yz = 108$.

Exercise 11.3.3 *Verify that the values stated above form a feasible solution to the ILP formulation in (11.9).*

Inhibition Now we add inhibition into the network model. See Figure 11.4, where two dashed edges with empty arrow heads have been added to the network from Figure 11.3. These edges represent *inhibition*. Of course, we need more detail on the

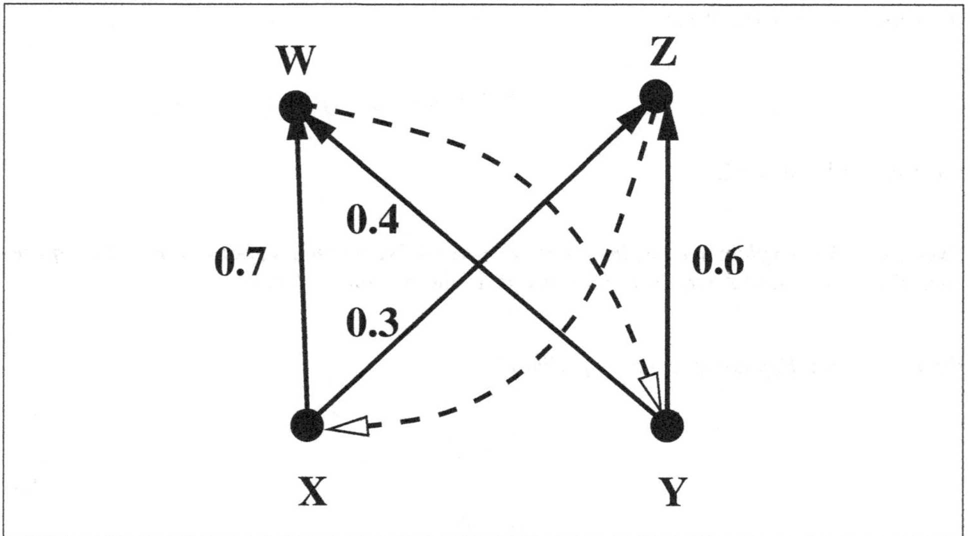

Figure 11.4 Network with Both Activation and Inhibition Edges. Inhibition edges are shown dashed, with white arrowheads.

level of that inhibition. For example, suppose that the effect of the W to Y edge is to *reduce* the availability of Y by $2 \times W$ units. Similarly, the effect Z to X edge is to reduce the availability of X by $15 \times Z$ units. How does the network behave now, and how can we figure this out?

It's Head Spinning It almost makes my head spin to try to think this through. It seems that when Z increases, the inhibition effect of Z on X increases, so X decreases, so the activating effect of X on W decreases, so W decreases, so its inhibiting effect on Y decreases, so Y increases, and so its activating effect on Z increases. So that cycle has the overall effect of continually driving the value of Z up. But, there is another cycle that affects Z. When Z increases, X decreases, so the activating effect of X on Z decreases, so this cycle has the effect that: *When the value of Z increases, the value of Z decreases* – Ouch!! How can we make sense of all this?

Steady-State Equilibrium The answer is to look for *equilibrium* or *steady-state* behavior of the network. That is, some values for X, Y, W, and Z, which can remain constant, exactly *balancing out* the activating and inhibiting effects on each node. In particular, we want to find values for those four variables to *simultaneously satisfy* the following inequalities:

$$
\begin{aligned}
xw - 0.7X &= 0,\\
xz - 0.3X &= 0,\\
yw - 0.4Y &= 0,\\
yz - 0.6Y &= 0,\\
Z - 0.25xz &\leq 0,\\
Z - 0.5yz &\leq 0,\\
W - 0.33yw &\leq 0,\\
W - 0.5xw &\leq 0,\\
X + 15Z &\leq 2{,}000,\\
Y + 2W &\leq 180,\\
\text{All variables are integral.}
\end{aligned}
\tag{11.10}
$$

Exercise 11.3.4 *The inequalities in (11.10) are the same as those in (11.9), except for the last two. So, those must reflect the action of inhibition. Explain in detail how those two inequalities model the inhibition in the network.*

It is not obvious that the inequalities in (11.10) have a *steady-state* solution, or how the system would get to that equilibrium (although that is a different issue). But, in fact there is a feasible solution to those inequalities, so there is a steady-state, equilibrium solution. Then, in order to find an *optimal* solution, we add the objective function from (11.9), and solve the resulting concrete ILP formulation.

Exercise 11.3.5 *Use Gurobi to find an integer optimal solution to the ILP formulation specified by (11.10) and the objective function in (11.9).*

11.4 TIME: THE FINAL FRONTIER

The final extension we will discuss is that of explicitly adding in the *time* needed for reactions and for transport of metabolites along the edges of the network, and the

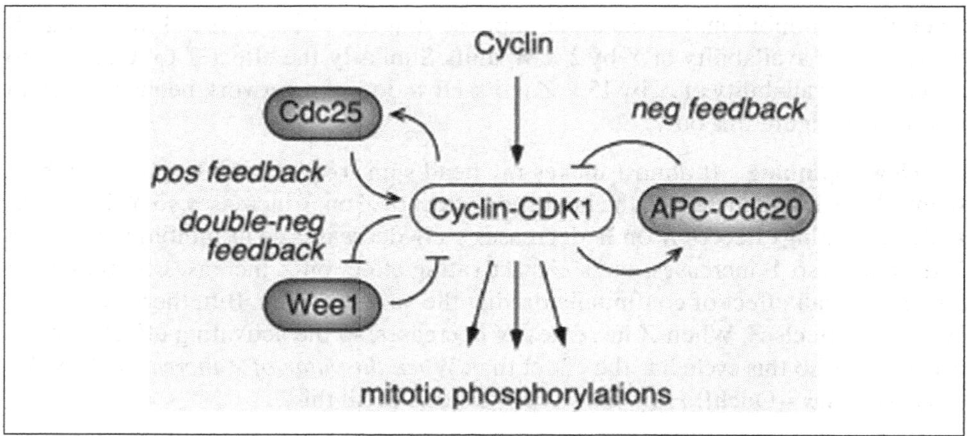

Figure 11.5 Mitotic Oscillator in Xenopus (African Frog) Embryos and Xenopus Egg Extracts. The edges with arrow ends represent *activation*, and edges with perpendicular ends represent *inhibition*. Depending on the parameters of positive and negative feedback (i.e., activation and inhibition), and how they are synchronized, the production of the end product (phosphorylations) can oscillate with time. Are there parameters that make the production of the end product stable, i.e., reaching some steady state (other than in the totally dead state)? This figure is figure 6 in [67].

possibility that different reactions can occur at different times. In this case, we are not looking for a steady-state behavior, since there may not be one, but for *dynamic* behavior over time (for example, see Figure 11.5).

Cyclin-CDK1 is produced, at some rate, in the light oval, given *cyclin* and *Cdc25* as input. *Cdc25*, *Wee1*, and *APC-Cdc20* are produced, at some rates, in the dark ovals. The inhibition in this network is more complex and confusing than in the network in Figure 11.4, because *Cyclin-CDK1* inhibits *Wee1*, which, in turn, inhibits *Cyclin-CDK1*. It is harder now to imagine a steady-state behavior. However, if *time* is added into the model, and movement along an edge takes one time unit, then the action of *Cycline-CDK1* on *Wee1* takes place in one time unit, followed by the action of *Wee1* on *Cycline-CDK1* in the next time unit, repeating in this way for many cycles. The effect is that the behavior of this network *oscillates* with time.

Exercise 11.4.1 *We can again use ILP to analyze the behavior of the Xenopus network. The key is to use variables for the values of the nodes at* odd *and* even *time steps, where the values at an odd time step influences the values at the next even time step, and vice versa. Try to figure this out, and write the inequalities that capture this time-based behavior of the network.*

12

ILP Idioms

In Chapter 4 we discussed particular *If-Then* and *Only-If* idioms for *binary* variables; and in Chapter 8 we discussed XOR idioms for binary variables. In this chapter we discuss more general versions of those idioms; discuss several other ILP idioms; and discuss a general approach to creating and exploiting ILP idioms.

12.1 GENERAL *IF-THEN* IDIOMS FOR LINEAR FUNCTIONS WITH BINARY VARIABLES

Suppose L is a linear function of *binary* variables, where all the coefficients in L are *integers*,[1] and the value of b is an integer. The following is a general *If-Then* idiom involving L.

12.1.1 The Less-Than Case

> *If* the binary ILP variables in the linear function L are set so that the value of L is *less than or equal* to an integer b, *then* the binary variable, z, *must* be set to value 1.

For example, consider the statement:

$$\begin{aligned} &\texttt{If } 5x_1 + 7x_2 - 8x_3 + 2x_4 \leq 2 \\ &\texttt{Then } z \texttt{ must be set to } 1, \\ &\texttt{where all of the variables } x_1, ..., x_4, z \texttt{ are binary.} \end{aligned} \tag{12.1}$$

In this example, L is the linear function: $5x_1 + 7x_2 - 8x_3 + 2x_4$; and b is 2.

Variable z in the idiom is called an "indicator" variable. Note that each occurrence of such an *If-Then* idiom must have its own z variable. For example, if an ILP formulation has two occurrences of this type of idiom, with two different linear functions, L_1 and L_2 say, then the respective indicator variables might be called z_1 and z_2, to distinguish them.

[1] See Section 1.1.2 for the needed definitions.

Implementation How do we create linear inequalities to implement the *If-Then* idiom stated above? To start, note that the variables in L are assumed to be binary, so there is a *limit* on the *smallest* value that L can attain, and it is easily determined, as follows: for every term in L with a *positive* coefficient, set the associated variable to 0; and for each term in L with a *negative* coefficient, set the associated variable to 1. In most of the topics considered in this book, all the coefficients in L will be positive, in which case, the smallest value that L can attain is zero.

Knowing the smallest value that L can attain, we define m as the smallest value that $L - b$ can attain. Equivalently, the smallest value that L can attain is given by $m + b$. (In the example above, m is $-8 - 2 = -10$.) Then, the following abstract inequality implements the general *If-Then* idiom for binary variables in the less-than case:

$$L - (m - 1) \times z \geq b + 1. \tag{12.2}$$

For example, with L and b as defined above, the inequality is:

$$5x_1 + 7x_2 - 8x_3 + 2x_4 - (-11 \times z) \geq 3.$$

Correctness To see that inequality (12.2) correctly implements the idiom, consider the situation where the value of L *is* less than or equal to b. Then, something strictly positive must be added to L, in order for inequality (12.2) to be satisfied. Since z is a binary variable, it follows that z *must* be set to 1 in order to even have a *hope* of satisfying inequality (12.2). But, will the inequality (12.2) be satisfied when z is set to 1?

When z is set to 1, then in order to satisfy (12.2), it must be that $L-(m-1) \geq b+1$. The smallest value that L can attain is $m + b$, so $L - (m - 1) \geq (m + b) - (m - 1)$, which simplifies to $L - (m - 1) \geq b + 1$, as desired. So, when $L \leq b$, z is forced to be set to 1, and inequality (12.2) is satisfied. Notice, the analysis showing that (12.2) is satisfied when z is set to 1, did not use the assumption that $L \leq b$.

To finish showing that inequality (12.2) correctly implements the idiom, we must also show that it has *no* bad *side effects*. So, we need to consider what happens when the value of L is *strictly greater* than b. We have seen above that when z is set to 1, (12.2) is satisfied even if $L > b$. So, to show there are no bad side effects, we only have to examine what happens when $L > b$ and z is set to 0. When z is set to 0, inequality (12.2) becomes the requirement that $L \geq b + 1$. But, since the variables in L are binary, the coefficients of L are integers, and the value of b is an integer, when $L > b$, the value of L must actually be greater or equal to $b + 1$. So, inequality (12.2) will again be satisfied when z is set to 0. So, when $L > b$, inequality (12.2) will be satisfied whether z is set to 0 or to 1. Therefore, inequality (12.2) does correctly implement the general *If-Then* idiom for L in the less-than case.

Exercise 12.1.1 *Where, exactly, did we need the assumption that all of the coefficients in L are* integer, *and that the value of b is an integer?*

12.1.1.1 An δ-Relaxation

Above, we assumed that all of the coefficients are integer, and that b is an integer. That was actually to simplify the initial presentation of the implementation. When some of the coefficients in L are *not* integer, and/or b is not an integer, but the variables are still binary, the *If-Then* idiom can be implemented as:

$$L - (m - \delta) \times z \geq b + \delta, \tag{12.3}$$

where δ is a "sufficiently" small positive number.

Exercise 12.1.2 *Verify the that inequality 12.3 correctly implements the If-Then idiom for the less-then case, when the coefficients in L are not necessarily integer, and/or b is not an integer.*

Does the value of δ matter when z is set to 1? Does it matter when z is set to 0? Are there any constraints on the size of δ? Hint: Since the variables in L are binary, there is a definite, and knowable, smallest value greater than b, that L can attain.

12.1.2 The Greater-Than Case

We again assume that all the coefficients in L are integers, and also assume that the value of b is a *nonnegative* integer. The general *If-Then* idiom for a linear function with binary variables, in the *great-than* case is:

> *If* the binary ILP variables in a linear function L are set so that the value of L is *greater than or equal* to a *nonnegative integer, b, then* the binary variable, z, *must* be set to value 1.

For example, consider the statement:

$$\texttt{If } 5x_1 + 7x_2 - 8x_3 + 2x_4 \geq 2 \texttt{ Then } z \texttt{ must be set to } 1, \tag{12.4}$$

where all of the variables $x_1, ..., x_4, z$ are binary.

Implementation As before, note that the variables in L are assumed to be binary, so there is a limit on the *largest* value that L can attain, and it is easily determined. Let M denote the largest value that the linear function L can attain. (In the above example, M is 14.) Then, the following abstract inequality implements this general *If-Then* idiom for linear functions of binary variables, in the greater-than case:

$$L - (M + 1) \times z \leq b - 1. \tag{12.5}$$

For example, with L and b as defined above, the desired inequality is:

$$5x_1 + 7x_2 - 8x_3 + 2x_4 - (15 \times z) \leq 1.$$

Exercise 12.1.3 *Explain explicitly how M is determined given a specific linear function L.*

Correctness To see that inequality (12.5) correctly implements the *If-Then* idiom for linear functions in the greater-than case, consider the case that the value of L is greater or equal to b. Then something strictly positive must be *subtracted* from the value of L, in order to satisfy inequality (12.5). It follows that z *must* be set to 1. But, will the inequality (12.5) be satisfied?

The largest value that L can attain is M, and when $z = 1$ the amount subtracted from the value of L will be exactly $M + 1$, so the left-hand side of (12.5) will have value *less than or equal* to -1. Then, since $b \geq 0$, it follows that $b - 1 \geq -1$, and so inequality (12.5) will hold. In summary, if the value of L is greater than or equal to b, then variable z *must* be set to 1, and this will satisfy inequality (12.2).

Exercise 12.1.4 *So far, where did we use the assumptions that the coefficients of L are integers, and that b is a nonnegative integer? Hint: Nowhere, that's where. Explain.*

Exercise 12.1.5 *Explain why inequality (12.5) does not have any bad side effects. That is, if it happens that the value of L is strictly less than b, then inequality (12.5) will not affect the optimal solution to the ILP. In more detail, show that when the value of L is strictly less than b, then inequality (12.2) will be satisfied, whether z is set to 0 or 1.*

Exercise 12.1.6 *If we allow the coefficients of L, and the value of b, to be fractional (non-integer), then the If-Then idiom for linear functions in the greater-than case can be implemented by a modification of inequality (12.5), similar to the way it was done in inequality (12.3) Find this modification and explain why it is correct.*

12.2 GENERAL *ONLY-IF* IDIOMS FOR LINEAR FUNCTIONS AND BINARY VARIABLES

The inequalities (12.2) and (12.5) implement *If-Then* idioms for linear functions with binary variables, but they do not implement *Only-If* idioms. In this section we develop implementations for those idioms.

12.2.1 The Greater-Than Case

We start with the following idiom:

$z = 1$ *only* if the value of L is *greater than or equal to* b.

Notice that this is equivalent to:

If $z = 1$, *then* the value of L *must* be greater or equal to b.

To implement this idiom, we might first try:

$$L + z \geq b + 1.$$

This looks promising, since when $z = 1$, the inequality becomes $L \geq b$, which can be satisfied *only if* the value of L is greater or equal to b. However, the inequality has a bad *side effect*, because when $z = 0$, the inequality becomes $L \geq b + 1$, which may incorrectly constrain the set of feasible solutions. Instead, we need an inequality that imposes the desired constraint, $(L \geq b)$ when $z = 1$, and is *automatically satisfied* (i,e., becomes *redundant*) when $z = 0$.

To construct such an inequality, let s be the *smallest* value that function L can achieve,[2] and set $m = s - b$ (i.e., $s = m + b$). Then, we can correctly implement the *Only-If* idiom with the inequality:

$$L + m \times z \geq m + b. \tag{12.6}$$

[2] Remember that the variables in L are binary, so s is easily determined by the computer program creating a concrete ILP formulation.

Exercise 12.2.1 *Write out the concrete inequality given by (12.6) for the example where L is* $5x_1 + 7x_2 - 8x_3 + 2x_4$*; and b is 2.*

Verify that the concrete inequality does correctly implement the Only-If *idiom, and that it does not have any bad side effects.*

Exercise 12.2.2 *Fully explain why the abstract inequality in (12.6) correctly implements the Only-If idiom, and that it has no bad side effects. To do that, check and explain what happens when when* $z = 1$*, and when* $z = 0$*.*

12.2.2 The Less-Than Case

We next want an implementation for the following idiom:

$z = 1$ *only* if the value of L is *less than or equal to b*.

Recall that M denotes the largest value that L can attain. Then the following inequality implements the desired idiom:

$$L + M \times z \leq M + b. \tag{12.7}$$

Exercise 12.2.3 *Explain why inequality (12.7) correctly implements the Only-If idiom for linear functions in the less-than case. Be sure to explain why it has no bad side effects. Also, could we replace M by a smaller value in the implementation? If so, how small can it be? Does the implementation in (12.7) require any constraints on L or b, e.g., that b must be integer, or that the coefficients in L must be integers?*

12.2.3 An Extension from Binary Variables to Bounded Variables

So far in this chapter we assumed that all the variables in L were binary, and that assumption is appropriate for most of the biological topics we discuss in this book. That assumption was used to establish upper and lower bounds, M and s, on the value of L, and the related number m. These three numbers were the key elements in establishing the correctness of the idioms.

However, if the variables in L are *not* necessarily binary, but are *bounded* (i.e., for each variable in L, there are *known* upper and lower bounds on the values that the variable can attain), then the values that L can attain are also bounded, both above and below.[3] Hence, the inequalities (12.2) and (12.5) also correctly implement the *If-Then* idioms; and the inequalities (12.6) and (12.7) also correctly implement the *Only-If* idioms, for any *bounded variables*; not just for binary variables.

12.3 EXPLOITING THE IDIOMS

The *If-Then* and *Only-If* idioms are more useful than it might at first appear, because we can build additional idioms from them. We will discuss several examples of such extensions.

The NOT-AND (NAND) Idiom for Inequalities Let L_1 and L_2 be linear functions whose variables are bounded, and consider the linear inequalities: $L_1 \geq b_1$, and

[3] Using the known bounds on the values of the variables in L, M, s, and m are determined, when needed, by the computer program that creates the concrete ILP formulations.

$L_2 \geq b_2$. Suppose we require that *at most* one of the two linear inequalities is satisfied. This is the NOT-AND (NAND) idiom for *inequalities*.[4]

The NAND idiom can be implemented by using the *If-Then* idiom (12.5) *twice*: once, so that *if* $L_1 \geq b_1$, *then* variable z_1 is set to value 1; and once so that *if* $L_2 \geq b_2$, *then* variable z_2 is set to value 1. Next, we add the inequality:

$$z_1 + z_2 \leq 1, \tag{12.8}$$

which ensures that *at most* one of the two variables, z_1 and z_2, can be set to value 1, and hence at most one of the two linear function can be satisfied.

The OR Idiom for Inequalities Similarly, suppose we require that *at least one* of the inequalities is satisfied. This is the OR idiom for *inequalities*.[5] To implement this, we use the *Only-If* idiom (12.6) *twice*: once, so that z_1 is set to 1 *only if* $L_1 \geq b_1$; and once so that z_2 is set to 1 *only if* $L_2 \geq b_2$. Then, we replace inequality (12.8) with:

$$z_1 + z_2 \geq 1. \tag{12.9}$$

Exercise 12.3.1 *Fully explain why the NAND idiom for inequalities uses the If-Then idiom, and the OR idiom for inequalities uses the Only-If idiom.*

Exercise 12.3.2 *With the above implementation of the OR idiom, it is easy to extend the idiom to more than two inequalities: Given any number of linear inequalities, we require that at least one of the inequalities is satisfied. Fully detail how to implement this extended OR idiom.*

Returning to the case of only two linear inequalities, if neither of the linear functions L_1 or L_2 can ever have a negative value, then the *OR* idiom for two inequalities can be implemented more simply as:

$$\begin{aligned} L_1 &\geq (1 - z)b_1, \\ L_2 &\geq z \times b_2, \end{aligned} \tag{12.10}$$

where z is a binary variable.

Exercise 12.3.3 *Explain why the inequalities in (12.10) correctly implement the OR idiom when neither L_1 nor L_2 can take on negative values. Can this implementation be easily extended to more than two linear inequalities?*

An XOR Idiom for Inequalities Suppose we have two inequalities, one involving the linear function L_1, and the other involving the linear function L_2.[6] To require that *exactly* one of the two inequalities is satisfied, we use the appropriate *If-Then* idioms for both L_1 and L_2; and the appropriate *Only-If* idioms for both L_1 and L_2. Then, we replace inequality (12.8) with:

$$z_1 + z_2 = 1. \tag{12.11}$$

This idiom is called the *Exclusive-Or (XOR)* idiom for *inequalities*.[7]

[4] The NAND idiom for *binary variables* is simpler. That idiom is: for two binary variables X and Y, *at most one* of the two variables can be set to 1. It is implemented by: $X + Y \leq 1$.

[5] The OR idiom for *binary variables* is simpler. That idiom is: for two binary variables X and Y, *at least one* of the two variables must be set to 1. It is implemented by: $X + Y \geq 1$.

[6] For example, the two inequalities might be $L_1 \geq b_1$, and $L_2 \leq b_2$. So, there are four cases covered by this discussion.

[7] The XOR idiom for two *binary variables*, X and Y, is simpler. There we require that *exactly* one of the two variables be set to 1, and the other set to 0. It is implemented by: $X + Y = 1$.

An IMPLIED-SATISFACTION Idiom for Inequalities We can also express the construct:

$$\texttt{if } L_1 \geq b_1 \texttt{ then } L_2 \geq b_2,$$

by using the appropriate *If-Then* idiom for the first inequality, and the appropriate *Only-If* idiom for the second inequality, and then replacing inequality 12.8 with

$$z_1 \leq z_2, \tag{12.12}$$

which forces z_2 to have value 1 if z_1 has value 1.

Note that we have stated the IMPLIED-SATISFACTION idiom for only one of four possible cases, based on the relation of L_1 to b_1, and L_2 to b_2. The other three cases differ only in which appropriate *If-Then* and *Only-If* idioms are used.

A NOT-EQUAL Idiom for Integer Variables In many ILP formulations, it is natural to *require* that two *integer* variables, X and Y (whose values can be larger than 1), must take on *different* values. This is easily implemented using the *OR* idiom. The inequality:

$$(X - Y \geq 1) \; OR \; (Y - X \geq 1), \tag{12.13}$$

is satisfied *if and only if* the values of X and Y are different.[8]

Forcing vs. Testing If X and Y are binary variables, then the inequality $X + Y = 1$ *forces* X and Y to have different values. However, sometimes we only want to *test* if two binary variables have different values. That is, we want the logical construct that implements:

> *If* binary variable X is NOT-EQUAL to binary variable Y, *then* binary variable z must be set to 1.

This is implemented by:

$$\begin{aligned} z &\geq X - Y, \\ z &\geq Y - X. \end{aligned} \tag{12.14}$$

Note that these inequalities do not *prevent* z from being set to 1 even though X and Y have the same value. If the objective function *minimizes* the number of such z variables set to 1, then this is not a problem, since the ILP solver will only set a z variable to 1 when forced to. But, if we need to explicitly prohibit z from being set to 1 when X and Y have the same value, we need to implement an *Only-If* idiom; that is, that z is set to 1, only if X and Y have different values.

Exercise 12.3.4 *Develop inequalities to implement the idiom that z is set to 1, only if X and Y have different values.*

The observant reader may have found the *NOT-EQUAL* idiom for *binary* variables looks familiar. In fact, we *have* seen this idiom already. In the context of the tanglegram problem, this idiom was called the *If-XOR* idiom, and discussed in Section 8.6. We discus it again here because of its relation to more general *NOT-EQUAL* and *XOR* idioms for integer variables and for linear functions.

8 The situation is simpler when X and Y are *binary* variables; then the NOT-EQUAL idiom is trivially implemented as $X + Y = 1$.

NOT-EQUAL Idioms for Linear Functions More generally, suppose L_1 and L_2 are *linear functions* of variables whose values are *bounded* from both above and below. Then, the values of L_1 and L_2 are also bounded above and below. If further, both L_1 and L_2 can only attain integer values, then the functions $L_1 - L_2$ and $L_2 - L_1$ have bounded values that are always integers. In that case, we can express the NOT-EQUAL idiom, $L_1 \neq L_2$ as:

$$(L_1 - L_2 \geq 1) \ OR \ (L_2 - L_1 \geq 1). \tag{12.15}$$

Of course, when creating a concrete ILP formulation, the OR idiom must be implemented with linear inequalities, as explained earlier.

A Special Case A particularly simple case is that each linear function, L_1 and L_2, is just a *single integer variable* whose value is bounded between 1 and n. Then, starting with the statement in (12.15), and expanding the OR idiom, we can derive the following inequalities that implement the requirement that $L_1 \neq L_2$:

$$\begin{aligned} L_1 - L_2 - n \times (z - 1) &\geq 1, \\ L_2 - L_1 + n \times z &\geq 1, \end{aligned} \tag{12.16}$$

where z is a *binary* variable. To see that the inequalities in 12.16 correctly implement the NOT-EQUAL idiom, we must check the consequence of setting z to 1, and of setting z to 0. When z is set to 1, the first inequality in (12.16) becomes

$$L_1 - L_2 \geq 1,$$

and the second inequality becomes

$$L_1 - L_2 \leq n - 1,$$

which is trivially satisfied. So, there are no bad side effects when $z = 1$. When z is set to 0, the second inequality in (12.16) becomes

$$L_2 - L_1 \geq 1,$$

and the first inequality in (12.16) is trivially satisfied (check this), so again there are no bad side effects. Since z is binary, it must either have value 1 or value 0, so the inequalities in (12.16) implement $(L_1 - L_2 \geq 1) \ OR \ (L_2 - L_1 \geq 1)$, with no bad side effects. Hence, they implement the NOT-EQUAL idiom for integer variables with bounded values.

Exercise 12.3.5 *Check that the inequalities in (12.16) have no bad side effects, no matter what value is assigned to binary variable z.*

Exercise 12.3.6 *Explain why inequality (12.15) correctly implements the NOT-EQUAL idiom when L_1 and L_2 only take on* integer *values. Why is it necessary to assume that L_1 and L_2 can only take on integer values?*

12.3.1 A Small Interruption: Back to Second Largest or Second-Largest Clique

In Section 2.2.7 we saw how to create a single ILP formulation to find two cliques K and K' in an undirected graph G, were K is a largest clique in G, and K' is a largest clique in G that contains *no* nodes in K. But, we did not develop an ILP formulation for the less restricted problem of finding two cliques K and K' in G, where K is a largest clique in G, and K' is a largest clique containing *at least* one node *not* in K. That is, K' can contain some nodes of K, but can't be contained in K. That problem can now be solved, using the extended OR idiom.

As in Section 2.2.7, we use a binary variable $C(i)$ to record whether node i is chosen to be in K, and use a binary variable $D(i)$ to record whether node i is chosen to be in K'. Then, what we want is $D(i) - C(i) = 1$, for *at least* one node i. But, since both variables $D(i)$ and $C(i)$ are binary, this is equivalent to $D(i) - C(i) \geq 1$, for *at least* one node i. This requirement is easily implemented by using the extended OR idiom, as detailed in Exercise 12.3.2, to ensure that at least one of the inequalities, $D(i) - C(i) \geq 1$, is satisfied.

Exercise 12.3.7 *Fully detail the abstract ILP formulation for the above variant of the second-largest clique problem, and explain why it is correct.*

12.4 THE KEY TO THE IDIOMS

There is a common form to all of the idioms for inequalities that we have seen. They first implement the logic that when a specific inequality is satisfied, an *indicator variable*, such as z, z_1, z_2, is set to value 1; or conversely, they implement the logic that when an indicator variable is set to 1, a specific inequality must be satisfied; or both. Then, using an individual indicator variable for each inequality, we add constraints on the indicator variables to implement relations between the inequalities. Those relations are implemented by idioms for variables.

The idioms described here a just a few of the ways that *If-Then* and *Only-If* idioms can be exploited. We will see other ways further along in the book. Also, it often happens that one can exploit properties of a problem to derive alternate (perhaps simpler or more understandable) implementations for an idiom already described here in a general form. Hence, it is important to try to understand the *logic* of the implementations given here (rather than just memorizing the templates), so that one can think through and understand other implementations that may differ somewhat from the ones given here. So, we end this chapter with a series of exercises that both expand your inventory of idioms, and help develop your skill at thinking through their logic.

Exercise 12.4.1 *Given $k > 2$ inequalities, explain how to implement the requirement that at least two of the k inequalities must be satisfied.*

Exercise 12.4.2 *Given two sets of inequalities, K_1 and K_2, explain how to implement the requirement that the number of inequalities in K_1 that are satisfied is equal to the number of inequalities in K_2 that are satisfied.*

Exercise 12.4.3 *Explain how to implement the idiom:*

$$L_2 \geq b_2 \texttt{ only if } L_1 \geq b_1.$$

Exercise 12.4.4 *Earlier we saw the NAND idiom for binary variables. We also will need an AND idiom for binary variables, where we want binary variable z to to be set to 1 if and only if both binary variables X and Y are set to 1. Do the following inequalities correctly implement this AND idiom, without bad side effects? Explain.*

$$\begin{aligned} z &\geq X + Y - 1, \\ z &\leq X, \\ z &\leq Y. \end{aligned} \tag{12.17}$$

Now, suppose we only want to implement the idiom: If both X and Y are set to 1, then z must be set to 1. What inequalities implement this idiom? What inequalities implement the only-if variant: z is set to 1 only if both X and Y are set to 1?

Exercise 12.4.5 *Later in the book, we will need to implement the idiom:*

If $z = 1$ *Then X* must *be equal to Y, where z is a* binary *variable, and X and Y are both* integer *variables whose values are from 1 to a given maximum, n.*

We call this the conditional forced equality idiom for binary variables. To implement this idiom, we first rewrite the logic as:

$z = 1$ *only if* $((X - Y \leq 0)$ *AND* $(Y - X \leq 0))$*, where X, Y and z are as above*

We claim this logic can be implemented with the following two inequalities:

$$\begin{aligned} (X - Y) + (n-1)z &<= n - 1, \\ (Y - X) + (n-1)z &<= n - 1. \end{aligned} \tag{12.18}$$

Derive the inequalities in (12.18), using the implementation of the Only-If idiom shown in inequality (12.7), and using the observation made in Section 12.2.3 that the inequalities in (12.7) work for bounded *variables, as well as for binary variables.*

Can you also derive the inequalities in (12.18) from other idioms we have already discussed?

Exercise 12.4.6 *Another idiom that will be needed later in the book concerns two ILP variables, X and Y. The idiom is:*

If $X = 0$ *then* $Y = 0$.

For example, in a metabolic network, X might correspond to an enzyme necessary to catalyze a particular chemical reaction, and Y might correspond to the end product of that reaction. If the quantity of the enzyme, X, is zero, then the quantity of the product, Y, will necessarily be zero. How do we implement this idiom with integer, linear inequalities? It depends on the kind of variables that X and Y are. We call this idiom the conditional zero idiom

(a) *Suppose that X and Y are both* binary *variables. Show an inequality that implements the idiom in that case.*

(b) *Now suppose that X is binary but Y can take on any nonnegative value, i.e., greater or equal to zero. For example, suppose again that X corresponds to a necessary enzyme, which is either input to the chemical reaction, or it is not. Y now represents the* amount *of the end product of the reaction. If the enzyme is missing, the amount of the end product will be zero, but if the enzyme is there, the amount of the end product will be determined by other factors and could be an amount other than 1. Does your answer to question* (**a**) *still work? If not, why not?*

(c) *Again, suppose that X is a binary variable, but Y can take on any nonnegative value* up to a specified *value called* Y_{max}. *Consider the following inequality:*

$$Y + (1 - X) \times Y_{max} \leq Y_{max}.$$

Does this inequality correctly implement the idiom that if X is not present (i.e., has value 0), then the value of Y must be 0? Does it have any bad side effects? Be sure to check what happens when $X = 0$, and *when* $X = 1$.

(d) *Now, suppose that both X and Y can take on nonnegative values up to some limit, denoted M.*[9] *This situation occurs when X represents the quantity of enzyme X. The idiom in this case just expresses the extreme case that if the quantity of X is zero, then the quantity of Y will also be zero. But if the quantity of the enzyme is greater than zero, the amount of the end product might depend on the quantity of X, and other factors.*

Does your answer to question **(c)** *still work? If not, why not?*

Consider the following inequalities, where Z is a binary variable:

$$X - (Z \times M) \leq 0,$$

$$Y \leq (1 - Z) \times M.$$

Do these inequalities implement the idiom in the case when both X and Y can take on nonnegative values? Do those values have to be integer? Do the inequalities have any bad side effects?

Exercise 12.4.7 Max of Two Values *Let* X_1, X_2, *and f be variables that can take on any values. The function* $f = \max(X_1, X_2)$ *sets f to the* largest *of the values of* X_1 *and* X_2. *How do we implement this* max *idiom with linear inequalities? We call this the max of two values idiom.*

One approach is to use the inequalities:

$$\begin{aligned} f &\geq X_1, \\ f &\geq X_2, \end{aligned} \tag{12.19}$$

along with an ILP formulation for:

$$(f \leq X_1) \; OR \; (f \leq X_2). \tag{12.20}$$

(a) *Explain why the above logic will correctly set f to the maximum value in* X_1 *and* X_2.

(b) *Give an ILP implementation (i.e., the linear inequalities) for the statement in (12.20). There are many ways to do this. Use what you have learned from the idioms already discussed.*

(c) *Show how to generalize the ILP implementation to set f to the maximum of any fixed number of variables, i.e., more than two.*

Exercise 12.4.8 Min of Two Values *Find and explain an ILP implementation that sets a variable f to the* minimum *value of two variables. Then generalize to the minimum value of any fixed number of variables.*

It's in Your Hands Now There are many additional idioms that are used in integer linear programming. A few will be detailed later in the book. However, with the idioms and ideas already presented here, many readers will be able to develop additional idioms on their own. Just do it, and enjoy!

[9] For simplicity, for now we assume that the limit is the same for X and Y.

PART II

13

Communities, Cuts, and High-Density Subgraphs

13.1 COMMUNITY DETECTION IN A NETWORK

Recall the goal of the network analysis in the *mountain lion* study, discussed in Section 2.1.1:

> "to quantify the extent to which the network was composed of distinct communities ... *within* which many edges occurred, and *between* which few edges occurred." (italics added) [64]

The key point in this analysis is not only to find large *high-density* subgraphs, as we did earlier, but to find ones with great *contrast* between the *interior* density of the subgraph, and the density of its *connections* to the rest of the graph.

> Merely the finding that a network contains tightly knit groups at all can convey useful information: if a metabolic network were divided into such groups, for instance, it could provide evidence for a modular view of the networks dynamics, with different groups of nodes performing different functions with some degree of independence. [140]

One might ask if it is necessary or helpful to make such an explicit distinction between the interior and exterior densities, because a *largest* high-density subgraph is likely to have few connections to nodes outside of it, or else it could have been enlarged. But, that contrast is not guaranteed. The *community detection* problem makes that contrast and the goal of finding it, explicit.

13.1.1 Modularity

> Modularity is considered to be one of the main organizing principles of biological networks. A biological network module consists of a set of elements (e.g., proteins/ reactions) that form a coherent structural subsystem and have a distinct function. [113]

But, how exactly should we define a single community or the collection of communities or subgraphs in a graph? A widely used and studied formalization of the concept of a community comes from the definition of the *modularity* of a graph, first proposed in [140].

13.1.1.1 Definition of Modularity

Here, we will develop a formal definition of *modularity*,[1] which was proposed in [140], and is very widely used. This definition is very technical, and it may be hard to see what it means. So, we will develop it slowly.

For an undirected graph $G = (V, E)$ of n nodes, let $d(i)$ denote the *degree* of node i, i.e., the number of edges that touch node i, and let A denote the adjacency matrix for graph G. That is, $A(i,j) = A(j,i) = 1$ if and only if there is an edge in G between nodes i and j; and $A(i,j) = A(j,i) = 0$ otherwise.

Next, let the set $\mathcal{C}$ denote a *partition*[2] of the nodes of G into an unknown number of disjoint subsets. Although we don't know how many subsets there are, we denote that number by k, so the partition $\mathcal{C} = (C_1, C_2, \ldots, C_k)$. Each of the sets C_p is formally called a class of partition $\mathcal{C}$, but in the context of computing modularity, these are the *modules* or *communities*.

For each pair of nodes, $i < j$, in G, let $\mathcal{C}(i,j)$ be an *indicator variable*, which has value 1 if and only if nodes i and j are *together* in the *same* class of $\mathcal{C}$. We don't know which class that might be, but i and j are in it together.

The Key Definition For a pair of nodes (i,j) in G, we define

$$M(i,j) \equiv A(i,j) - \frac{d(i) \times d(j)}{2|E|}, \tag{13.1}$$

and we will motivate this definition below.

Next, we define the *modularity of a partition* $\mathcal{C}$ as:

$$Q(\mathcal{C}) \equiv \frac{1}{2|E|} \sum_{i<j:\ i,j \texttt{ are nodes in } G} M(i,j) \times \mathcal{C}(i,j). \tag{13.2}$$

Finally, we define the *modularity of graph* G as the *maximum* $Q(\mathcal{C})$ over all ways of partitioning the nodes of G. More formally,

$$Q(G) \equiv \max_{\texttt{partition } \mathcal{C} \text{ of } V} Q(\mathcal{C}). \tag{13.3}$$

Notice that the *number* of classes in the partition that determines $Q(\mathcal{C})$ is *not specified* in the problem input. Rather, the number of classes is determined in order to *maximize* the value of $Q(\mathcal{C})$. In contrast, many alternative clustering and partitioning methods require specifying the number of classes in a partition as part of the input of a problem instance. This flexibility is one of the strengths of modularity and part of the explanation for its popularity. Another reason is an efficient *greedy* algorithm [20] that, in practice, is reported to compute a good *approximation* of the modularity of a graph, and a good partition.

[1] When we refer to the word "modularity," we will be talking about the technical definition stated in this section. Elsewhere, as in the above quotes, we use similar words, such as "modular" or "module," which are used in a broader, more colloquial way.

[2] Formally, a *partition* $\mathcal{C}$ of a set S is a division of the elements of S into some number of subsets, called *classes*, so that each element of S is in *exactly* one class of $\mathcal{C}$. It follows trivially that the classes are *disjoint*.

13.1.1.2 What Do These Definitions Mean?

The definition of modularity is not easy to understand! I can only give a partial explanation for it, and I have never seen what I consider to be a compelling explanation.[3] But, some explanation is better than no explanation. Still, this will be a hard section, so if you are mostly interested in how to *compute* modularity using ILP, you can skip to Section 13.1.2.

First, notice that the multiplicative term $\frac{1}{2|E|}$ in (13.2) is a *constant* number, the same for every partition $\mathcal{C}$, so we can ignore it when trying to find a partition that maximizes $Q(\mathcal{C})$, i.e., one that defines $Q(G)$.

Next, consider a specific partition $\mathcal{C}$ of the nodes of G. Notice that if nodes i and j are *not* in the same class of $\mathcal{C}$, then $\mathcal{C}(i,j)$ has value 0, so the term for node pair (i,j) contributes *nothing* to $Q(\mathcal{C})$. The value of $Q(\mathcal{C})$ is therefore only affected by node pairs (i,j) where i and j are in the *same* class of $\mathcal{C}$. So, $Q(\mathcal{C})$ reflects only what is happening *inside* each of the specified classes. That means that the heart of the definition of modularity is the quantity $M(i,j)$. Where does it come from? What does it mean?

A Hand-Waving Answer Focus on a particular class in $\mathcal{C}$, say C_p, and the nodes in C_p. In G there are some edges between pairs of nodes in C_p, and if C_p was chosen well, there will be a *high density* of edges between the nodes in C_p. But generally, not all possible edges between nodes in C_p will be in G. For a pair of nodes (i,j) in C_p, what $M(i,j)$ tries to measure is how "surprising" or "abnormal" or "unexpected" it is that edge (i,j) is in C_p, or is not in C_p. If there are many edges between nodes in C_p that are "unexpected," then we want the numerical contribution of C_p to $Q(\mathcal{C})$ to be large. Conversely, if most of the edges between nodes in C_p are "expected," then we want that numerical contribution to be small. So, the partition $\mathcal{C}$ that maximizes $Q(\mathcal{C})$ is one that groups nodes into classes (modules) with many "unexpected" edges.

OK, But *how* do we actually determine if an edge is *expected*? What is *surprising* or *abnormal* depends on which graphs G is *compared* to. The classic answer is to compare G to some set of *randomly generated* graphs. Then, $M(i,j)$ compares the *reality* (i.e., whether edge (i,j) is actually in G) to what we would expect in these *randomly* generated graphs.

OK, But *which* randomly generated graphs? This is where *good modeling* and *art* come in. One general idea is to specify a set of random graphs that are each *similar* to G in many ways, but are not generated in a way that *biases* the density of its induced subgraphs.[4]

OK, But it's too vague. So, let's look at a particular multi-set of random graphs, call it $M(G)$, generated in some well-defined way. Since we will be comparing G to the graphs in $M(G)$, and we want those graphs to be similar to G, we will randomly

[3] I have seen several that make incorrect mathematical statements, or don't address the key modeling questions.

[4] In general when we want to measure how unexpected it is that a graph G has some particular property, we want to compare G to a set of graphs that are randomly generated to be as similar to G as possible, but making sure that the graph generation process does not bias the frequency that the generated graphs have, or don't have, the property of interest.

generate graphs with the *same* number of nodes, n, and the *same* number of edges, $|E|$, as G has. Recall that $d(k)$ denotes the *degree* of node k.

Then, for each node k in G, we define

$$p(k) \equiv d(k)/2|E|.$$

So, $p(k)$ is *proportional* to the degree of node k, divided by the total number of edges in G. In fact, $p(k)$ is exactly *half* that ratio.

Exercise 13.1.1 *Another way to interpret $p(k)$ is based on the famous fact called*

The Handshake Lemma: *In any graph G, $\sum_{node\ k} d(k) = 2|E|$.*

Using this lemma, explicitly state the alternative interpretation of $p(k)$.

With those definitions, given G, we generate a random graph with the following

Procedure M(G):

Repeat the following $|E|$ times:

Pick a node randomly from the nodes in G. In particular, each node k is picked with probability $p(k)$. Then, pick a second node randomly from the nodes in G. Again, each node k is picked with probability $p(k)$. Call the first picked node i, and the second node j.

Then, put the edge (i, j) into the random graph being constructed. Note that j might be the same node as i, in which case the edge generated is a self-loop at node i.

Note that for a given pair of nodes (i, j), the number of (i, j) edges that could be generated in an execution of this procedure is anywhere from 0 to $|E|$ (unlikely, but possible). Similarly, although the graph generated will have n nodes and $|E|$ edges, the degree of any node i in the new graph might be different from $d(i)$.

With this procedure, we can precisely define $M(G)$ as the *multi-set* of graphs generated if we run the $M(G)$ procedure an *infinite* number of times.

Exercise 13.1.2 *Above, we simply used $p(k)$, for each node k, as a* probability. *But, for the set of $p(k)$ values to be properly considered as probabilities, each must be between 0 and 1 (inclusive), and the sum $\sum_k p(k)$* must *be equal to 1. Verify that the $p(k)$ values are proper probabilities.*

Averages Of course we can't actually run the procedure an infinite number of times.[5] But if we run it a thousand-gazillion or more times, the multi-set of graphs created would converge to $M(G)$, and we could use that multi-set of graphs to compare to the actual graph G, determining how surprising or unexpected it is that a particular edge (i, j) is in G, compared to graphs in $M(G)$. For that comparison, we count the number of (i, j) edges in each of the randomly generated graphs, and then compute the *average* number of (i, j) edges over all the created graphs. That average will converge to what is called the "expected number" of (i, j) edges.[6] We use $E(i, j)$ to denote that expected number. Since in each execution of procedure $M(G)$, the number of (i, j) edges created could be more than one, it is possible that $E(i, j)$ will be greater than one.

[5] And, anyway, I don't know what infinity is, which is why I gravitate to *finite* math.

[6] The term "expected number" here has a precise, technical meaning in probability theory, and does not have exactly the same meaning as it does in colloquial speech. We won't attempt a formal definition of the term, but the concept of the converging average is sufficiently correct for our purposes.

Recapping, we have defined a multi-set of random graphs $M(G)$, and a specific process for generating many random graphs in $M(G)$, and the concept of the expected number of times, $E(i,j)$, that the edge (i,j) appears in a generated graph. So, $E(i,j)$ could be used to measure how surprising it is that the pair (i,j) is an edge in the given graph G, compared to being in a graph in $M(G)$. Following that viewpoint, we would redefine $M(i,j)$ as

$$A(i,j) - E(i,j),$$

and use those $M(i,j)$ values in the definitions of $Q(\mathcal{C})$ and $Q(G)$. But, actually generating a bazillion graphs in $M(G)$ in order to determine $E(i,j)$ values is wildly impractical.

Fortunately, we can calculate $E(i,j)$ *analytically*, i.e., by a formula, *without* actually having to run procedure $M(G)$ even once. Here is the derivation of the formula for $E(i,j)$.

Deriving $E(i,j)$ Looking at procedure $M(G)$, we see that in a *single* pick of an edge, the probability that the particular edge (i,j) is picked is equal to the probability that node i is chosen as the first node, times the probability that node j is chosen as the second node; plus the probability node j is chosen as the first node, times the probability that node i is chosen as the second node. That probability is exactly:

$$2 \times p(i) \times p(j) =$$

$$\frac{2 \times d(i) \times d(j)}{(2 \times |E|)^2}.$$

Now in procedure $M(G)$, there are $|E|$ edges that are picked. The edge picks are independent of one another, so the probability that edge (i,j) is picked is the same on each edge pick. A theorem in probability theory, called the *linearity of expectation*, says that in order to determine $E(i,j)$, we can just multiply the number of edge picks by the probability that edge (i,j) is picked in a single edge pick. This gives,

$$E(i,j) = |E| \times \frac{2 \times d(i) \times d(j)}{(2 \times |E|)^2} =$$

$$\frac{d(i) \times d(j)}{2|E|}.$$

We Conclude What we have derived is that if we define $M(i,j)$ as $(A(i,j) - E(i,j))$, we can *calculate* $M(i,j)$ as

$$A(i,j) - \frac{d(i) \times d(j)}{2|E|},$$

which is exactly the term for the node pair (i,j) that appears in (13.1), the definition of $M(i,j)$. And that is the best I can do to explain the *derivation* of $M(i,j)$.

All of this this should make good intuitive sense. For example, if nodes i and j both have high degrees, it shouldn't be terribly surprising if there is an edge between them, and conversely, it would be somewhat surprising if there was no (i,j) edge. This corresponds well to the definition of $M(i,j)$, because when both nodes have high degree, $\frac{d(i) \times d(j)}{2|E|}$ will be large. So if (i,j) is an edge, $M(i,j)$ will be a small number, and when (i,j) is not an edge, $M(i,j)$ will be even smaller.

Exercise 13.1.3 *Thinking through the case when both nodes i and j have high degree, as above, to see that the formal definitions and derivations agree with intuition, is something that I call a "sanity check." Do a sanity check for the case that both nodes have low degrees.*

Is $M(G)$ the Best Class to Compare with? We said that we want to compare G to graphs in a class of random graphs that are similar to G, but are generated without biasing the density of subgraphs. The procedure that defines $M(G)$ does produce graphs with the same number of nodes and edges as G, and without any explicit consideration for the distribution of subgraph densities. Moreover, if we define $Ed(i)$ as the *expected degree* of node i, over the graphs generated in $M(G)$, $Ed(i)$ is $2 \times |E|$ times the probability of picking node i in any node pick, which is

$$2 \times |E| \times \frac{d(i)}{2 \times |E|} = d(i).$$

So, the multi-set of graphs generated are similar to G in this way also. But is that good enough so that the given definition of modularity is useful? That requires an empirical answer.

Actually, we might be able to do better. It is easy to create a procedure[7] that generates random graphs where in each graph, the degree of any node is exactly its degree in G. That is, the degree of a node, i, will be $d(i)$ in *every* graph generated, although the number of copies of an edge (i,j) can vary. So, those graphs are even *more* similar to G than are the graphs in $M(G)$, and yet they are still randomly generated. Why don't we compare G to that set of graphs instead of to $M(G)$?

I think the answer is that we (or at least I) don't know an analytical way to compute $E(i,j)$ over that multi-set of graphs. If we wanted to know those $E(i,j)$ values, we would have to actually run the generation procedure a gazillion times. So, the use of the definition of $M(i,j)$ given in (13.1) may be the result of a compromise between the best modeling, and what is computationally more efficient. And that is the best I can do to explain the meaning of $M(i,j)$.

Whew!! OK, this is not easy going. If you don't yet have a feel for $M(i,j)$, $Q(\mathcal{C})$ or $Q(G)$, just consider $Q(\mathcal{C})$ to be a value that can be computed, given $\mathcal{C}$; and $Q(G)$ to be a value that can also be computed, in principle, given G; and realize that many people who work on the community detection problem think these values have meaning for that problem.

13.1.2 Computing Modularity By ILP

We next show that $Q(G)$ can be determined using integer linear programming. Note that when we compute $Q(G)$, we are mostly interested in determining the *classes* in the partition that determines $Q(G)$, although the *value* of $Q(G)$ can also be informative.

[7] The basic idea is as follows: Start with G and cut each edge in the middle, so that each edge becomes two *stubs*. The number of stubs at this point is exactly $2|E|$, since there were $|E|$ edges, and each one becomes two stubs. Then if we *randomly* pair up the unattached ends of the stubs to form $|E|$ edges, the result will be a graph G' where each node has the same degree in G' as in G, since the attached end of each stub didn't change. It is possible that G' would be graph G again, but there are many other possibilities as well. This class of graphs allows parallel edges and self-loops.

The abstract ILP formulation we develop must define a partition C of the nodes of G when any concrete ILP formulation is created and solved. For that, it would be natural to create variables and inequalities so that each node is *explicitly* assigned to a *named* class when a concrete ILP formulation is solved. However, there is a simpler, but a bit more subtle, approach to creating a partition. This was developed in [1], although the details we use here are a bit different from that original ILP.

The Variables For each pair of nodes (i,j), we create the binary ILP variable $X(i,j)$, which will record *whether or not* nodes i and j are *together* in some (unnamed) class. Variable $X(i,j)$ will be given the value 1 to indicate that nodes i and j *are* together in some (unnamed) class; and $X(i,j)$ will be given the value 0 to indicate that they are not. These are the only variables used in the ILP formulation.

For any given input graph, G, once the concrete ILP formulation for the modularity of G is solved, the values of the $X(i,j)$ variables will define the classes of a partition C, and the assignment of nodes to the classes. In this way, partition C is established, even without any explicit assignment of nodes to specific (named) classes.

The Inequalities We need inequalities to enforce the requirement that the nodes are *partitioned* into a set of classes, but the inequalities we will use do this in a somewhat subtle, *indirect* way. The key is to ensure that the variables obey *transitivity*. That is, if $X(i,j)$ is set to 1; and $X(i,k)$ is set to 1; then $X(j,k)$ *must* also be set to 1. If the variable values obey transitivity, then the classes of the partition will be clearly defined from the values of the X variables. Transitivity is easily implemented with the following inequalities.

For each triple of nodes, i,j,k:

$$X(i,j) + X(i,k) - X(j,k) \leq 1. \tag{13.4}$$

Exercise 13.1.4 *Suppose C is a partition of the nodes in G, and that the values of the X variables are consistent with the classes in C. That is, if nodes i and j are together in the same class of C, then $X(i,j)$ is set to 1. Explain why the values of the variables satisfy the transitivity requirement.*

Conversely, suppose the values of the X variables are set based on partition C. Explain how to use those X values alone to reconstruct partition C.

The Objective Function

$$\texttt{Maximize} \sum_{i<j:\, i,j \texttt{ are nodes in } G} [M(i,j) \times X(i,j)]. \tag{13.5}$$

Once that maximum has been computed, $Q(G)$ is obtained by multiplying it by $\frac{1}{2|E|}$.

The Abstract ILP The abstract formulation to compute the *modularity* of a graph is shown in Figure 13.1.

13.1.3 Tests

The computation of modularity has been used to find a partition of the members in the (now iconic) 34-node *Zachary Karate Club* graph [212], where the nodes

$$\text{Maximize} \quad \sum_{i<j:\ i,j \text{ are nodes in } G} [M(i,j) \times X(i,j)]$$

Such that for each triple of nodes, i, j, k:

$$X(i,j) + X(i,k) - X(j,k) \leq 1$$

Every variable $X(i,j)$ is binary.

Figure 13.1 An Abstract ILP Formulation to Compute the Modularity of a Graph.

represent the members of the club, and edges represent friendships.[8] Zachary's Karate Club graph has been an obligatory test case for every community detection method ever suggested, and the received wisdom is that the graph partitions into the *four* modules shown in Figure 13.2.

In my test, the ILP formulation for modularity solved in 0.24 seconds, and identified exactly the same four modules. The partition obtained from the computation of modularity on Zachary's graph is considered a success for the formal definition of modularity given in (13.3). But, it is only one small test, and a fairly easy one. How well do modularity computations find meaningful partitions in other applications? For example, how well do they identify known protein complexes in PPI networks? Modularity, as formally defined by $Q(G)$ in (13.3), also can have some bizarre behavior.

Bizarre Results It is known that when the modules of interest are small compared to the size of the graph, the partition found by computing modularity does not correctly identify the modules of interest. Other oddities have also been observed. For example, consider the graph with one large clique of size d, and one extra edge (u, v) connecting a node u in the clique to a node v of degree one, outside the clique. In this case, the partition associated with modularity partitions the nodes into *two* classes, one containing all nodes other than u and v (forming a clique of size $d - 1$), and the other class containing nodes u and v. But, for any natural application I can think of, the partition containing the clique of size d in one class, and node v in another, seems more meaningful.

As another example of the peculiarity of the definition of modularity, consider again Zachary's Karate Club depicted in Figure 13.2, and the four modules associated with $Q(G)$. With that node partition, there are 57 edges inside the modules, and 21 edges crossing between modules. But, if we move nodes 24 and 28 from the module they are in, to the module whose nodes are represented by circles, the number of crossing edges falls to 19, and hence the number of edges inside the modules increases to 59. Further, before the move, nodes 24 and 28 had only two edges to other nodes

[8] OK, I know this doesn't seem like a graph that arises in biology, but it passes for one when you remember that each of the nodes in the graph represents a member of the species H. sapiens, in the genus Homo, in the family Hominidae, in the suborder Haplorhini, in the order Primates, in the phylum Chordata, in the kingdom Animalia, in the domain Eukarya, which sounds like (high school) biology to me.

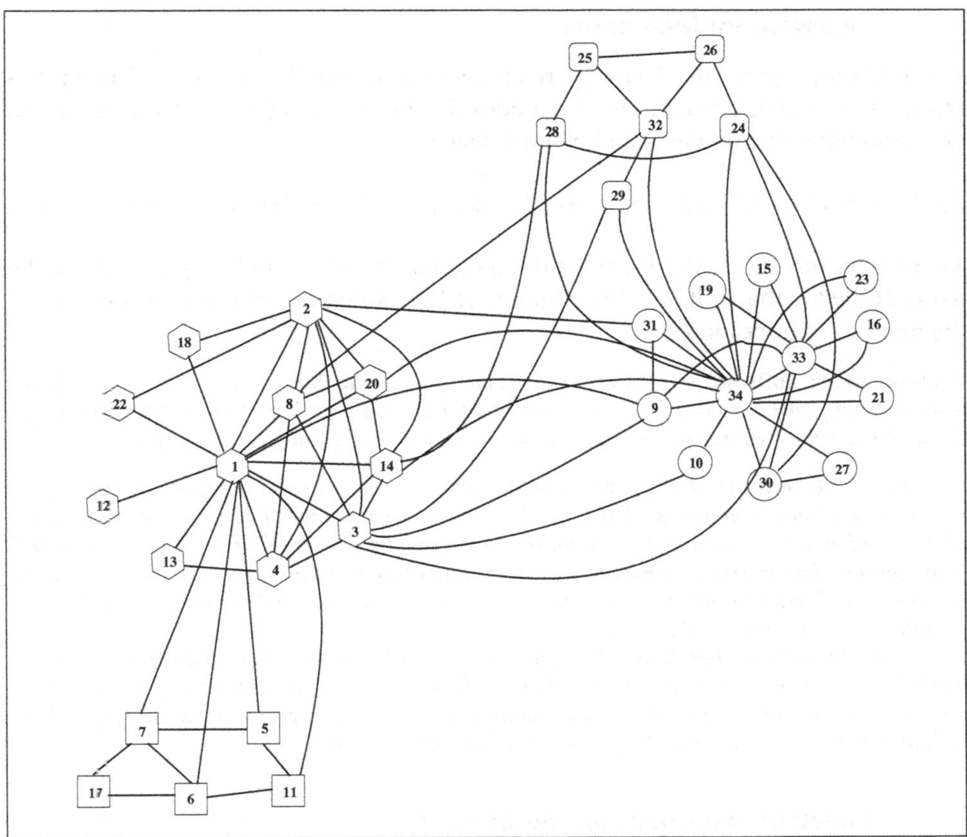

Figure 13.2 Zachary's Karate Club. The partition that determines the modularity of this graph divides the nodes into four modules. The nodes in each module are identified by the geometric shape used for the node. The shapes are circles, squares, hexagons, and squares with rounded corners.

in their module, while after the move, the have four. So, it is difficult to argue that the node partition associated with modularity is actually the most meaningful or natural one.

More Trouble In [105], there is a more precise and mathematical treatment of some problems with the node partition given by modularity. There, an infinite set of graphs is defined, where the difference (based on a precise mathematical definition) between the modularity partition and the obvious "informative" partition[9] can be made *arbitrarily* large. So, the modularity partition can be made to look arbitrarily bad. Still, the modularity definition from [140] is overwhelmingly *the* most commonly used definition in the community-finding literature.

[9] This is a subjective matter, but I completely agree with the authors that the partition they propose is the more natural and informative one.

13.1.4 Software for Modularity

The Python program *modular2.py* reads in an undirected graph G and computes $Q(G)$, along with the partition of the nodes that determines $Q(G)$. Call the program on a command line in a terminal window using:

```
python modular2.py input-file output-file.lp number-of-nodes
```

where *input-file* holds the binary adjacency matrix for G, and *output-file.lp* is the name of the file that will hold the concrete ILP formulation, and *number-of-nodes* is the number of nodes in G.

Exercise 13.1.5 Software *Use program* modular2.py *and the adjacency matrix for Zachary's Karate Club (zacharymatrix.txt), to verify that the partition into four classes, shown in Figure 13.2, is the partition associated with the modularity of the adjacency matrix.*

Exercise 13.1.6 Software *Use program* randomgraph.py *to generate random graphs, varying the number of nodes and edge densities. Then use program* modular2.py *to create concrete ILP formulations to compute the modularity of the graphs, and use Gurobi to solve those ILP formulations. Summarize the time needed for* modular2.py *and for Gurobi, as a function of the graph size and edge density. Compare those times to the time needed to find dense subgraphs with the programs discussed in Chapter 2.*

Random graphs are not the ideal graphs to use to explore issues of modularity, because they likely have much less modular structure than do the graphs for which modularity computations are intended. Search the internet to find adjacency matrices for graphs with strong modular structure, and test out modular2.py *and Gurobi on those graphs.*

13.1.5 Labeling and Counting the Classes

Integer linear programming allows easy extensions of what we want to extract from modularity computations, and then allows additional constraints to be added into the formulations. The following extensions to the modularity model are new, illustrating the versatility of ILP models, and of ILP solvers.

The computation of modularity, $Q(G)$, has the side effect of defining a partition of the nodes of G, but only *indirectly*, as explained earlier. So, for example, there is no variable in the ILP formulation that is set to the *number* of classes in the partition found by solving the ILP. Further, the ILP computation of modularity does not return a *naming* of the classes in the partition, or an *explicit* assignment of the nodes to (distinctly) named classes of the partition. Determining such values when a concrete formulation is solved, allows other modifications to the ILP formulation that might help finding more meaningful partitions.

For example, if the ILP formulation has a variable for the *number* of classes in the partition, we can add inequalities to the formulation that put required *bounds* on the number of classes, both from below and above. Or, we can add inequalities that set the *minimum* or *maximum sizes* of any class in a partition. We can also add inequalities to bound, from above or below, the number of edges *between* classes. These kinds of additions seem likely to improve the biological meaning of a partition. In this section, we detail how these modifications can be achieved in an ILP formulation.

The Key Idea We assume that the nodes in the input graph G are labeled by unique integers.[10] Then, the key idea is to name each class in the partition found by the ILP,

[10] This is a trivial assumption, since the natural labeling of the nodes in a graph with n nodes is $1, \dots n$.

by the *smallest* label of any node in the class. For example, a class consisting of nodes $\{4,7,22,88\}$ will be named *class 4*.

To achieve this naming, we introduce a new binary variable, $m(j)$ for each node j in the input graph $G = (V,E)$. We want variable $m(j)$ to be assigned value 1 if and only if j is the smallest node label of all the nodes in the class that node j is in. To implement this, we use two types of inequalities.

First, for each node j in V:

$$\sum_{i<j} X(i,j) + m(j) \geq 1, \tag{13.6}$$

which forces variable $m(j)$ to have value 1 if j is the minimum node number in the class containing node j.

Second, for each node j and each node $i < j$:

$$X(i,j) + m(j) \leq 1, \tag{13.7}$$

which forces variable $m(j)$ to have value 0 if j is together in a class with a node, i, that has node label less than j. Then,

$$NC = \sum_{j} m(j), \tag{13.8}$$

sets the integer variable NC to the *number of classes* in the partition determined by the computation of modularity. With it, we can bound the number of classes from above, with

$$NC \leq a, \tag{13.9}$$

or from below, with

$$NC \geq b. \tag{13.10}$$

Explicit Assignment So now we have named the classes. How do we implement an explicit assignment of nodes to those classes?

For each node j and each $k \geq j$, we will use the binary variable $c(k,j)$ to indicate whether or not node k is in the class named j. Variable $c(k,j)$ will have value 1 to indicate that k is in class j, and will have value 0 to indicate that it is not. The correct assignment to the $c(k,j)$ variable is accomplished using the following two inequalities.

$$\sum_{j=1}^{k} c(k,j) = 1, \tag{13.11}$$

which says that node k must be in exactly one class, whose label is *less than or equal* to k. And, for each $j < k$,

$$X(k,j) + m(j) - c(k,j) \leq 1, \tag{13.12}$$

which assures that *if* k is in the same class as j, *and* j is the name of the class (because j is the smallest node label in that class), *then* variable $c(k,j)$ must be set to 1. This establishes that if k is in class j, then the variable $c(k,j)$ must be set to 1.

Conversely, we claim that if $c(k,j)$ is set to 1, then node k must be in the same class as node j, and that class must be named j. To see this, suppose that k is in some other class, say $j' \neq j$. By what we established above, $c(k,j')$ must be set to 1. But, $c(k,j)$ is assumed to be set to 1, and inequality 13.11 only allows one c variable to be set to 1, leading to a contradiction, thus establishing the claim.

So, $c(k,j)$ will have value 1 if and only if node k is in the class named j. Therefore, the c variables correctly act to label each node with the name of the class that the node is contained in.

Counting Size To count the size of the class named j, we can use the integer ILP variable $N(j)$ and the following:

$$N(j) = \sum_{k \geq j} c(k,j). \tag{13.13}$$

Note that $N(j)$ will be zero if there is no class named j in the partition.

Bounds Using these variables, it is easy to *bound* the size of a class from above. Suppose we want a class to be no larger than than the number d. We can use the inequality, for each j:

$$N(j) \leq d. \tag{13.14}$$

Bounding the size of any class from *below*, so each class has size *at least* d, is a bit more subtle. For each j, the inequality

$$d \times m(j) - N(j) \leq 0, \tag{13.15}$$

will do it.

Exercise 13.1.7 *Explain why inequality (13.15) will force class j to be at least size d. What happens when there is no class named j?*

More Explicit Naming Now, if we want an even more explicit reporting of the (integer) name of the class that each node is in, we can use the following.

For each node k :

$$c(k) = \sum_{j=1}^{k} c(k,j) \times j. \tag{13.16}$$

For example,

$$c(5) = c(5,1) \times 1 + c(5,2) \times 2 + c(5,3) \times 3 + c(5,4) \times 4 + c(5,5) \times 5.$$

Exercise 13.1.8 *Explain why the inequalities in (13.16) correctly assign each variable $c(k)$ to the name of the class that node k is in.*

Exercise 13.1.9 *If you are able to program in Python, extend the program* modular2.py *to implement the extensions discussed in this section.*

13.2 CUTS: MAX, MIN, AND MULTI

Problems involving *cuts* in graphs, particularly *minimum* or *maximum weight*, or *multi-cuts*, are ubiquitous in mathematical modeling and optimization. This is also true for computational problems in biology: many biological problems can be cast as cut problems on graphs. In most cases, the computed cuts give a biologically interesting *partition* of the biological elements represented by the nodes of the graph. Hence these problems and solutions are related to, and sometimes a competitor for, the kinds of computations discussed in the previous section on *community detection*.

Here, we discuss several cut problems that arise in biology, and their solution using integer linear programming. First, we need some definitions.

13.2.1 Definitions

We define an *edge cut* (usually just referred to as a *cut*) in an undirected graph G, as a *partition* of the *nodes* of the graph into *two* non-empty sets. This is also called a *bipartition*. For example, in Figure 13.3, the two sets $\{1,2,3\}, \{4,5,6\}$ form a bipartition of the nodes, and hence define an *edge cut*.

The definition of an *edge* cut may seem strange since it does not explicitly mention *edges*. But any bipartition of the nodes *uniquely* defines the set of edges that *cross* between the two sets of nodes in the bipartition. An edge *crosses* the cut if and only if its two end points are in different sets of the bipartition. For example, the edges crossing the cut defined by the bipartition $(\{2,3,4\}, \{1,5,6\})$, are $\{(1,2), (1,3), (1,6), (3,4)\}$.

Another way to think of an *edge cut* in a *connected* graph G, is that the cut is a set of edges whose removal from G results in a *disconnected* graph.[11] If the removal of the edges in an edge cut results in *exactly* two disconnected components, then the edge cut is also uniquely defined by the *edges* that cross it.

If the graph has edge *weights*, then the *weight* or *capacity* of a cut is the *sum* of the weights on the *edges* that *cross* the cut. For example, in Figure 13.3 the weight of the cut defined by the node partition $(\{2,3,4\}, \{1,5,6\})$, is 12. If the edges are unweighted, then the weight of a cut is just the *number* of edges that cross it. If the weight of a cut is *smaller or equal* to the weight of all other cuts in G, then it is called a *minimum weight cut*, or just a *minimum cut*. Similarly, if the weight of a cut is *larger or equal* to the weight of all other cuts in G, then it is called a *maximum weight* cut, or just a *maximum* cut. For example, in Figure 13.3, the cut $(\{1,2,3,4,6\}, \{5\})$ is a minimum cut, of weight 9; and the cut $(\{1,2,3\}, \{4,5,6\})$ is a maximum cut, of weight 19.

In a connected graph G, if two nodes, s and t, are specified, an edge cut that separates s from t is called an *(s,t) cut*. Equivalently, an (s,t) cut is a *bipartition* of the nodes of G such that one class in the bipartition contains s, and the other class contains t. The weight of an (s,t) cut is the sum of the weights of the edges that cross the cut. A maximum (s,t) cut is an (s,t) cut whose total weight is larger than, or equal to, all other (s,t) cuts. Similarly, a minimum (s,t) cut is an (s,t) cut whose weight is smaller than, or equal to, all other (s,t) cuts. For example, in Figure 13.3, when s is node 3, and t is node 6, the minimum (s,t) cut has weight 12, and the maximum (s,t) cut has weight 19.

[11] A graph G is *connected* if, for any pair of nodes, there is a path connecting those two nodes, and is *disconnected* otherwise. If a graph G is disconnected, it consists of two or more *connected components*, which are each a connected subgraph (possibly a single node) of G.

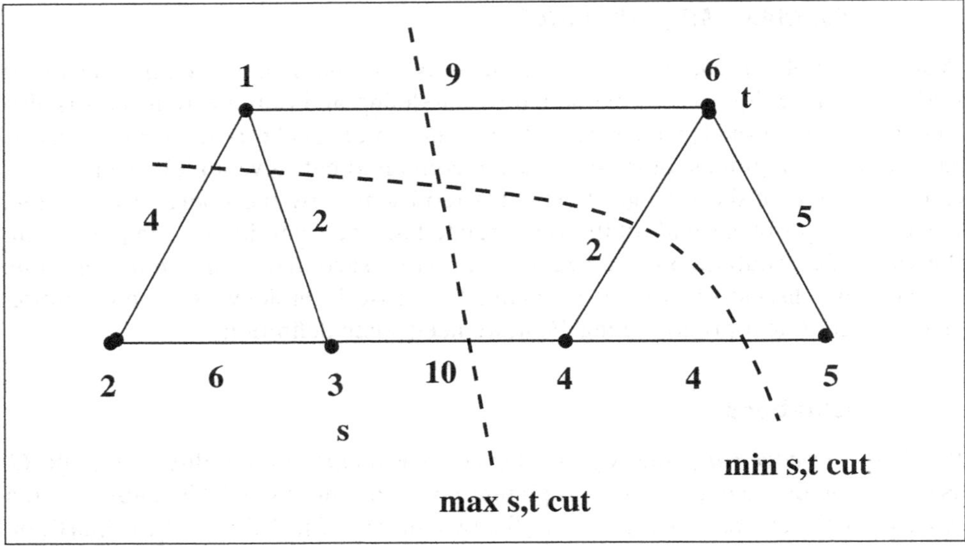

Figure 13.3 Graph G, Illustrating Edge Cuts, Both Maximum and Minimum. Those cuts are also respectively maximum and minimum (s,t) cuts, where $s = 3$ and $t = 6$.

A set of edges, E_k, in a *connected* graph G, is called a *k cut* if the removal of E_k from G breaks G into exactly k *connected* components. If set K of k specific nodes are specified in G, then a set of edges, E_K, is called a K *cut* if the removal of E_K creates a k cut that separates *every* pair of nodes in K. That is, each node in K ends up in a different connected component after the removal of the edges in E_K. The concept of a K cut generalizes an (s,t) cut, which is a two cut that separates node s from t.

13.2.2 Examples of (s,t) Cuts in Protein Graphs

In the next sections we will discuss in depth two applications of cuts, and the solution of the associated cut problems with integer programming. But first we briefly discuss three applications of minimum (s,t) cuts that arise in computational and systems biology.

In [192], graphs and cuts are used in the detection of protein *hotspots*, which are regions in a protein with a high concentration of residues that can *bind* to residues in other proteins, forming protein complexes. A *residue contact graph* is constructed, where each node represents a residue in the protein, and an edge between two nodes represents that the fact that the residues associated with the nodes can be in physical contact. Each edge has a weight that reflects the strength of that contact, or bond. The role of an (s,t) cut is explained in the following:

> In the residue contact graph, the minimum weight cut between two residues illustrates the minimum total contact potential to separate these two residues into two disconnected subgraphs.

That is, a minimum weight (s,t) cut in the residue contact graph identifies a set of contacts needing the smallest amount of energy to separate the residues represented by nodes s and t.

A similar approach is used in [114], where *successive* (s,t) minimum cuts in a *disconnectivity graph* are computed[12] to study *protein folding*, based on *free energy surfaces* of peptides. Each node in the graph represents an amino acid residue in a given peptide,[13] and an edge between two nodes represents the fact that the associated amino acids are in contact (bind) when the peptide is folded. The weight on the edge is essentially the amount of energy necessary to allow the two amino acids to move relative to each other, to disconnect, or separate. Then, given two nodes, s and t, that represent two amino acid residues, say A and B, that are *not* in contact, a minimum-weight (s,t) cut in G identifies a minimum weight set of contacts whose disconnection will separate residue A from B. Biologically, this identifies the minimum amount of energy, and where to apply it, to separate those two residues.

A third paper [207] uses a similar graph, and minimum (s,t) cut computations to identify protein *domains*, which are intervals in the protein that can fold and function independently of the rest of the protein.

13.2.3 Guilt By Association: *K* Cuts in PPI Networks

Here we look in more depth at a biological problem whose solution uses multi-cuts and integer programming.

In most organisms, there are a large number of proteins whose functions are *unknown*. For example, around 10 years ago, no function was known for about *one third* of all the proteins in *yeast*, one of the most thoroughly studied organisms in all of life. So, several methods have been developed to *deduce* the function of a protein from the *known* functions of other proteins.

In [139], protein-protein interaction (PPI) networks are used to assign (conjectured) functions to proteins whose functions are not yet known. Recall that the nodes in a PPI network represent proteins, and an edge between two nodes indicates a relationship between the proteins they represent, e.g., that the proteins physically interact, or are expressed at the same time in a cell, or are expressed in the same part of a cell, or are together in some functional pathway, etc. The method in [139] builds on the understanding that proteins in a high-density subgraph of a PPI network likely *work together* in a *protein complex* to achieve some biological function. So, if many of the proteins in a high-density subgraph have known functions, and those functions are either the same or highly related, then it is natural to conjecture that *most or all* of the proteins in that subgraph have similar functions. This reasoning is called *guilt by association.*

Further, if a high-density subgraph is "large enough" and a large percentage of its proteins have known functions, then it is natural to conjecture that each protein in the subgraph whose function is *unknown*, actually has the same function as one of the proteins with a known function. There may also be edges in the PPI network between two nodes with known, but *different*, functions, although the number and density of such edges will generally be small in comparison to the number of edges between nodes with the same or related function. This reflects the view that proteins are organized into high-density complexes, which then have some, but limited, interactions with other proteins in different complexes.

12 And organized into a Gomory-Hu cut tree, which is outside the scope of this book.
13 For our purposes, we can define a "peptide" as a short section of a protein.

What Is the Problem? Now, how exactly should we use the logic of *guilt-by-association* to form a *precisely stated* computational problem? And, how can we solve that computational problem when we have real data? We call the computational problem the *Function Assignment* problem.

13.2.3.1 Formalizing the Function Assignment Problem

Given a PPI network with n nodes, we start by giving a *function label* to the nodes whose function is already known.[14] For convenience, function labels are integers from 1 to k, where k is the number of distinct protein functions that are known. Note that many nodes may be given the same label, because they have the same known function.

Next, we want to assign one of the k function labels to each node whose function is *unknown*. That assignment should express a *plausible* conjecture for the functions of the proteins whose functions are unknown. But how? Following the model of protein complexes discussed above, the approach in [139] is to solve the following

> **Function Assignment Problem:** Given a PPI network where some of the nodes have function labels, assign a function label to each of the nodes without one, *minimizing* the number of edges whose end nodes have *different* function labels.

Another way to express the function assignment problem is to *partition* the nodes into exactly k subsets so that in each of the k subsets, all of the nodes with known function have the same function, minimizing the number of edges whose end points are in *different* subsets. The partition of the nodes into k subsets defines a *K cut*.

Exercise 13.2.1 *We defined the general K cut problem as one of partitioning the nodes of a graph into $k = |K|$ subsets, so that each of the k nodes in K is in a different subset. The function assignment problem is similar to the K cut problem, and in both, a solution is a k cut. In what way does the Function Assignment Problem differ from the K cut problem?*

13.2.3.2 An ILP Formulation for the Function Assignment Problem

The ILP formulation will have one binary variable, $X(i,g)$, for each node/function pair (i,g). If a node i has a known function, and hence already has a function label, say g, then we add the inequality

$$X(i,g) = 1,$$

to the ILP formulation.

For each node i that does *not* have a known function, and hence has no function label, variable $X(i,g)$ will be set to 1 to indicate that node i will be assigned function g; and otherwise $X(i,g)$ will be set to 0. Then, for each node i, we have the inequality:

$$\sum_{g=1}^{g=k} X(i,g) = 1, \tag{13.17}$$

which says that each node must be assigned exactly one of the k known functions.

[14] More correctly, the function of the protein represented by the node, is known. But, for simplicity, we will just refer to the "function of the node."

The ILP formulation will also have one binary variable, $Z(i,j)$, for each edge (i,j). Variable $Z(i,j)$ must be set to value 1 if nodes i and j are assigned *different* functions. This is implemented by using the NOT-EQUAL or XOR idioms, discussed in Section 8.6. For each function g, the formulation will have:

$$\begin{aligned} Z(i,j) &\geq X(i,g) - X(j,g), \\ Z(i,j) &\geq X(j,g) - X(i,g). \end{aligned} \tag{13.18}$$

To understand the inequalities in (13.18), note that when *neither* node i nor j is assigned function g, $X(i,g) = X(j,g) = 0$, so each inequality becomes

$$Z(i,j) \geq 0,$$

which is trivially true. This is also the case when *both* nodes i and j *are* assigned function g. But, when *exactly* one of the nodes is assigned function g, then one of the inequalities becomes

$$Z(i,j) \geq -1,$$

and the other becomes

$$Z(i,j) \geq 1,$$

which forces $Z(i,j)$ to be set to value 1.

Finally, the objective function is:

$$\mathit{Minimize} \sum_{i<j} Z(i,j). \tag{13.19}$$

Note that the inequalities in (13.18) force $Z(i,j)$ to have value 1 *if* nodes i and j are assigned different functions, but there are no inequalities that enforce the *only-if* direction. This causes no problem, because the objective function *minimizes* the number of Z variables set to value 1.

General K Cuts The function assignment problem can be specialized to the general K cut problem. Simply give each of the k nodes to be separated a different function, and solve the resulting function assignment problem. So, the ILP formulation given above can be used to solve instances of the K cut problem.

Exercise 13.2.2 *Suppose we want an ILP formulation for the* maximum-*weight K-cut problem, rather than* minimum-*weight K-cut problem. Will it work if we just change the objective function in the ILP formulation we have presented, from "minimize" to "maximize"? If not, what changes to the inequalities are necessary? Write out the full abstract ILP formulation for the maximization problem.*

Now, write out an abstract ILP formulation for the minimization problem, so that the only change needed to convert it to a maximization problem is to change the word "minimize" to "maximize."

13.2.3.3 Software for Cuts

The Python program *kcutf.py* can be downloaded from the book website. It reads a file describing a graph, and another file specifying the functions of some of the nodes, i.e., the set of nodes K, and their functions. Then, the program creates the concrete ILP for the minimum *K-cut* problem.

Call the program on a command line in a terminal window as:

```
python kcutf.py graph-file function-file ilp-file.lp number-of-nodes
```

An example graph-file is *zacharymatrix.txt* and an example function file is *zachfuncs*. Each line of *zachfuncs* specifies a node and a function number.

An idiosyncrasy of the program is that the functions must be numbered 1 through k, and the last node in the listing must be a node with function k.

Exercise 13.2.3 *Type in the adjacency matrix for the graph shown in Figure 13.3. Then use program* kcutf.py *to find the minimum-weight three cut that separates nodes 1, 3, and 6.*

13.2.3.4 Generalizations and Specializations

First, suppose that each edge has a given *weight*, and we want to assign function labels (equivalently, partition the nodes into classes) to minimize the total weight of the edges whose end points are in *two* different classes. This only requires a trivial change to the objective function, which we leave to the reader.

Exercise 13.2.4 *Explain how to modify the ILP formulation to implement edge weights, as described above.*

Second, with realistic protein functions, and for any particular function g, there will generally be several proteins that have function g. So, we will often require that each function label be assigned to *at least* some specified percentage of the nodes. For example, if function g must be assigned to at least 10% of the nodes (including those whose function is initially known), then the ILP formulation should include the inequality:

$$\sum_i X(i,g) \geq n/10.$$

Third, suppose there are only *two* functions, and exactly *one* node, s, is initially assigned the first function, and exactly *one* node, t, is initially assigned the second function. Then a solution to the function assignment problem on this input produces a *bipartition* of the nodes into two groups, one containing node s, and the other containing node t. So, the ILP solution defines a *minimum-weight* (s,t) *cut*, i.e., a minimum weight *edge cut* whose removal separates node s from node t. And, as noted in Exercise 13.2.2 we write the ILP formulation so that it can solve either the minimization or the maximization versions of the problem, simply by the change of a word.

Exercise 13.2.5 *What happens in the function assignment problem, if the no node is initially assigned a function? Does the concrete ILP formulation still define a meaningful problem?*

Exercise 13.2.6 *k* **Partition** *The following problem is called the k-*partition *problem. Divide the nodes of an edge-weighted graph G into k groups, where each group has some minimum number of nodes, d (e.g., at least three), minimizing the total weight of edges whose end points are in two different classes. When $k = 2$, and $d = 1$, this problem is called the* global min-cut problem.

We use the term "k partition" instead of "k cut," because a k cut need not have the additional requirement that each class contains at least d nodes.

Modify the abstract ILP formulation for the function assignment problem, to obtain an abstract ILP formulation for the k-partition problem.

Next, modify that formulation to obtain an abstract ILP formulation for the maximum total weight *k-partition problem.*

Software for four Cuts and four Partitions The Python program *4part-random.py* can be downloaded from the book website. It reads a file describing a graph, and then randomly chooses nodes, with probability 0.5 each, which will be assigned initial functions. The function assigned to one of the chosen nodes, is chosen uniformly at random from four functions. Then, the program creates the concrete ILP for the resulting maximum four-cut problem. This program allows the user to experiment with different graphs and study the utility of four-cuts in graphs.

Exercise 13.2.7 *Use program* graphgen.py *to randomly generate graphs with different numbers of nodes and edge densities, and then use program* 4part-random.py *as described above. Explore how efficiently Gurobi solves the four-cut ILP formulation as a function of graph size and edge density.*

13.2.4 Maximum Cuts in Population Genomics: More Guilt By Association

Here we discuss a problem in genomics, which is cast and solved as a maximum cut problem in a graph.

Association mapping is a widely used, population-based approach to try to efficiently locate genes and mutations that influence genetic traits of interest (diseases or important commercial traits). In this section, we introduce the general idea of association mapping; *genome-wide association studies (GWAS)*; the problems caused by population (sub)structure; and an approach to handling those problems through the use of *maximum cuts* in graphs, and the use of integer linear programming.

Candidate Regions Association mapping was first developed for applications where there was already a *candidate region* conjectured to contain a gene or mutation that contributes to a genetic trait of interest. A locus or site with such a contributing gene or mutation is called a *causal locus* or *causal site*. In those applications, association mapping is used to verify or refute the conjecture, or to more precisely locate a causal site. For example, see [62].

Genome-Wide Association Studies (GWAS) When *no* candidate region is known, a more ambitious form of association mapping, called a *genome-wide association study* (GWAS) begins *without* a candidate region, scanning the *entire* genome for a site or locus that may contribute to the trait of interest. Today, GWAS is the dominant way that association mapping is done.

We next explain a bit more about association mapping, leading to an important problem that has been addressed using integer linear programming.

13.2.4.1 Cases, Controls, and Their Use

Suppose we want to locate sites in the genome, which influence a specific genetic trait. Individuals who possess that trait are called the *cases*, and individuals who don't possess the trait are called the *controls*.

The key idea in association mapping is to study (partial) genome sequences in a large number of cases and controls, to find places in the genome where those two groups differ in some systematic way. In particular, to find sites (typically SNP sites)[15] where the *allele frequencies* differ between the cases and controls. For example, it may be that at a SNP site, nucleotide A occurs in 80% of the cases, and nucleotide C occurs in the other 20% of the cases; while the A/C frequencies are reversed in the controls.[16] Such a situation would strongly suggest that the causal mutation is at, or *near*, that SNP site.

Note that we deduce that a site is causal, or near one, when we see large differences in allele frequencies between cases and controls, *even if* we don't have any mechanistic understanding or conjecture for how such mutations would cause the trait of interest. Indeed, that is precisely what makes association mapping so valuable.

13.2.4.2 Difficulties

There are several difficulties, both technical and conceptual, with the association mapping approach to gene finding. First, in most real situations, the data is not as clean, nor is the association as strong, as in the above example. Sophisticated statistical methods are then necessary to determine how strongly the data suggests that a site is a causal site, or near a causal site. Second, there is a large, systematic problem called *population (sub)structure* that has to be addressed to make association mapping successful.

Population (Sub)structure The logic of association mapping relies on the assumption that a genome site where allele frequencies in the cases are *very different* from the allele frequencies in the controls, likely identifies a causal site, or is near a causal site, for the trait of interest. But, differences in allele frequencies can occur for other reasons. The most serious is the fact that different (sub)populations have many genomic variations that are only *historical* mutations, unrelated to the trait of interest.[17] These mutations are often *neutral*, i.e., having no influence on *any* observed trait. Then, differences seen between the cases and controls may only be due to differences in the populations, rather than to mutations that cause the trait. This issue is particularly significant if the difference of allele frequencies *between* populations are much larger than the differences in allele frequencies between the cases and controls *inside* the same population. To further emphasize these points, we quote from [187]:

> In association studies, the discrepancies in the SNP-allele frequencies between the cases and the controls are believed to imply an association of the SNP with the disease, but if the cases and controls were collected from two very different populations, this discrepancy may be explained by the difference between the two populations, and hence the SNP is not necessarily associated with the disease.

15 "SNP" stands for *single nucleotide polymorphism*. A SNP site in a genome is a nucleotide site with high allele variability across the population. But it is generally assumed that in the population, only *two* different nucleotides appear at that site in high frequency. Recall that "allele" is the word used in genetics for *state* or *variant*. It was introduced in Section 3.1.

16 There are actually some diseases, where *all* of the cases have one particular nucleotide at the SNP site, and *none* of the controls have that nucleotide there.

17 In fact, such historical mutations are exploited to identify the ancestry of individuals, by companies such as 23 and Me or Ancestry.com.

Even subtle differences in the population structures of the cases and the controls may result in spurious associations.

Another way to state the issue is that in association studies we would *ideally* want the cases to be genetically identical to the controls, *except* at the site(s) that are causal for the trait of interest. More realistically, we would like the *distribution* of allele frequencies at any *noncausal* SNP site to be the same in the cases and the controls. Those best-case conditions are hard to achieve when the sample comes from two or more (sub)populations.

A Continuing Problem Even though the *issue* of population (sub)structure is well understood in the abstract, it remains a serious *practical* problem because we don't always know how the allele frequencies differ between the (sub)populations, and we don't always know the (sub)population ancestries of the individuals in the association study. Without knowing those two things, and because a true causal SNP site might only have a small genetic influence on the trait, the problem of population (sub)structure can be significant – leading to spurious conclusions.

What to Do? What is typically done to reduce the impact of the (sub)structure problem, is to look at *all* of the SNP data of the individuals in the study (before looking at differences between cases and controls), to *subdivide* the individuals into groups whose *total* collections of SNPs are very similar. The expectation is that this will divide the individuals into groups from the same (sub)populations, even if we don't know what those (sub)populations are, or don't know much about the allele frequencies in any particular (sub)population.

The key point is that two individuals in the same (sub)population have the same alleles at *many* SNP sites, even if one of the individuals is a case and the other is a control, since there are only a *few* sites that contribute to the trait of interest. And, at many of the SNP sites where the two individuals are identical, the alleles there differ from the alleles possessed by individuals *outside* of that (sub)population.

After the individuals have been divided into groups, differences between cases and controls can be examined *inside* each group. This reduces the problems caused by population (sub)structure. The first, and most widely used method for dividing the individuals into groups is a computer program called STRUCTURE [154], which does not use integer programming. However, a later method [187] frames the problem of dividing the individuals into groups as a *maximum-cut* problem on a graph.

What Did They Actually Do? After this long discussion of GWAS and the problem of (sub)population structure, how does all this relate to *cuts* and *ILP*?

The method in [187] first calculates a *genomic distance*, $D(i,j)$, between each pair of individuals (i,j), based on *all* of the known SNP alleles of the individuals i and j. The choice of the distance measure is important, and discussed in detail in [187]. But, for our purposes, we don't need to know the details of the distance.[18]

Second, the method constructs a graph $G = (V,E)$ where each individual in the study is represented by a single node of G; and with an edge (i,j) of weight $D(i,j)$ between each pair of nodes (i,j).

[18] The only thing we will note here is that they found that the simple *Hamming distance* did not perform well.

Then, assuming that there are k distinct (sub)populations in the study, they compute a *maximum-weighted k partition* in G. Removing any edges that cross between classes of the partition, the resulting connected components divide the individuals into k groups, which are taken as the k (sub)populations in the study. To implement this approach, we can use the ILP formulation for the *k-partition* problem (discussed in Exercise 13.2.6).

It is also straightforward to add constraints to make the division more biologically plausible. For example, to require that each group in the division have a size *between* some anticipated minimum and maximum.

If k is not known in advance, different values of k can be tried, and the partitions examined to see which one seems the most informative. In most cases, k will be under *five*. In fact, in [187], the focus is on $k = 2$.

Why Maximum and Not Minimum? Clearly, instead of computing a measure of *difference* between the genomes of individuals in the study, we can compute a measure of *similarity* between individuals. Then, the problem of dividing the individuals in groups becomes a *minimum* cut problem rather than a *maximum*-cut problem. For $k = 2$, which is the focus in [187], there are famous worst-case efficient algorithms (provably and practically) for finding a minimum cut, in contrast to the case of *maximum* cut where *no* such algorithms are known. So, why not take this approach, casting the division problem as a *minimum*-cut problem, allowing the use of a provably efficient algorithm?

I am not sure of the answer, but my guess is that it is harder to get a similarity measure that leads to a biologically meaningful cut. Taking the case of $k = 2$, we generally want both cuts to be of significant size, and sometimes even of roughly the same size. The cuts with an equal number of nodes in both groups (the most *balanced* cut) will have $n^2/4$ edges crossing the cut, while a cut that has one group with *only one* node (the most *unbalanced* cuts) will only have $n - 1$ edges that cross the cut. So, assuming that the edge weights are uniformly distributed (just to see the point), a minimum weight cut will likely have *many fewer* edges, and will likely be much *more unbalanced*, than a maximum weight cut. Hence, a maximum weight cut may be more biologically meaningful. Whatever the reason, the approach in [187] uses maximum-weight cuts, where integer linear programming is an attractive solution technique.

13.3 HIGH-DENSITY SUBGRAPHS: A REFINEMENT FOR LARGE, SPARSE GRAPHS

In Section 17.1 we discussed the use of high-density subgraphs to find disease-related proteins, and showed how that problem is framed and solved via integer programming in [133]. The specific computational problem was to find the largest subset of nodes, V', in a graph $G = (V, E)$, such that each node in V' is adjacent to at least half of the nodes in V'.

Earlier, however, we discussed different definitions of high-density subgraphs, and in Section 4.5 we developed an ILP formulation for

The Largest High-Density Subgraph Problem: Given an undirected graph $G = (V, E)$, and a density threshold d between 0 and 1, find an induced subgraph,

$G' = (V', E')$ in G, with the maximum number of nodes, such that the density of G' is greater or equal to d.

Here, we discuss a technique from [133] that allows practical solution of the largest high-density subgraph problem in large, *sparse* graphs. The technique greatly reduces the number of variables used, compared to the ILP formulation from Section 4.5.

Removing Variables The ILP formulation for the maximum-density subgraph problem in $G = (V, E)$, detailed in Section 4.5, has a variable $P(i,j)$ for *each pair* of nodes (i,j) in G, whether or not (i,j) is an edge in G. For a graph with n nodes, the number of pairs of nodes is $\binom{n}{2} = n(n-1)/2$. However, the ILP formulation for the maximum-density subgraph problem only has a variable $E(i,j)$, *if* the node pair (i,j) is an *actual* edge in G. And, many graphs that reflect biological phenomena are very *sparse*, with a number of edges only a *few percent* of $\binom{n}{2}$. In that case, the number of $E(i,j)$ variables is much smaller than the number of $P(i,j)$ variables. In very large, sparse graphs, the number of $P(i,j)$ variables, in contrast to the number of $E(i,j)$ variables, is *too large* for a practical implementation, so we want an ILP formulation that avoids $P(i,j)$ variables.

The $P(i,j)$ variables appear in the denominator of inequality (4.20), and as a consequence, they appear on the right-hand side of inequality (4.21). The denominator in (4.20) *counts* the number of *pairs* of nodes in a selected subgraph; so if we can compute that count some other way, we can avoid the use of $P(i,j)$ variables altogether. The approach developed in [133] uses the integer variable q, and sets its value with the inequality:

$$q = \sum_{i=1}^{n} C(i). \tag{13.20}$$

So, the value of q is exactly the number of nodes selected for the subgraph. Hence the value of the denominator in (4.20) equals $\binom{q}{2} = q(q-1)/2$. Then, if we add inequality (13.20) to the formulation in Figure 4.4, it would be mathematically correct to change (4.20) to:

$$\frac{\sum_{i,j \in V, i<j}[E(i,j)]}{q(q-1)/2} \geq d, \tag{13.21}$$

and change (4.24) to:

$$\sum_{i \in V, j \in V, i<j} E(i,j) - (d \times q(q-1)/2) \geq 0. \tag{13.22}$$

These changes remove all of the $P(i,j)$ variables, which is what we wanted to do. However, there is a problem. The problem is that $q(q-1)/2 = (q^2 - q)/2$ is not a *linear* function of variable q – it is a *quadratic* function of q – and integer *linear* programming does not allow quadratic functions. Only linear functions are allowed. So, how do we fix this problem? We fix it with one added *assumption* that is justified in most biological applications.

13.3.1 The Assumption and Its Use

The assumption we will need is that there is a known *upper bound* on the number of possible nodes in a biologically realistic near clique. That is, there is a known upper bound, u on the value of the ILP variable q. This assumption is realistic in many biological applications. For example, in [133], high-density near cliques are conjectured to identify proteins involved in certain diseases such as epilepsy. In those applications, graph G may be large, but the largest near clique of proteins with genetic contributions to epilepsy is believed to have at most 20 proteins.[19]

How does the upper bound, u, allow the removal of the $P(i,j)$ variables? First, for each positive integer value of c, up to u, we add inequalities that set a *binary* variable $a(c)$ to value 1, if and only if the value of q *equals* c. This is achieved through the addition of two binary variables, $g(c)$ and $s(c)$, and the use of the *If-Then* idiom for binary variables. In detail, for each integer c from 2 to u, create binary *variables* $s(c)$ and $g(c)$, and use them in the following inequalities:

$$s(c) \leq \frac{q}{c}, \tag{13.23}$$

$$(c - q) + s(c) \times u \geq 1, \tag{13.24}$$

$$g(c) \leq \frac{c}{q}, \tag{13.25}$$

$$(q - c) + g(c) \times u \geq 1, \tag{13.26}$$

and

$$\begin{aligned} a(c) &\leq g(c), \\ a(c) &\leq s(c), \\ g(c) + s(c) - 1 &\leq a(c). \end{aligned} \tag{13.27}$$

Explaining the Inequalities For each permitted integer c, if $q < c$, then inequality (13.23) ensures that binary variable $s(c)$ is set to 0, since q/c would be less than 1, and $s(c)$ can only take on values of 0 or 1. Conversely, if $q \geq c$, then inequality (13.24) ensures that $s(c)$ is set to 1. Taken together, these two inequalities set $s(c)$ to 1 *if and only if* $q \geq c$. Similarly, inequality (13.25) sets $g(c)$ to 0 if $q > c$; and inequality (13.26) sets $g(c)$ to 1 if $q \leq c$. Taken together, these two inequalities set $g(c)$ to 1 if and only if $q \leq c$. Then, the set of inequalities in (13.27) guarantees that $a(c)$ is set to 1, *if and only if* the value of q is *exactly* c.

Clearly, the value of $a(c)$ is 1 for *exactly* one permitted integer c. The value of $a(c')$ is 0 for every other integer c'. Hence, the desired value, q^2, is equal to the value of the following *linear* function of the $a(c)$ variables:

$$\sum_{c=2}^{u} [c^2 \times a(c)]. \tag{13.28}$$

[19] However, there may be several disjoint near cliques with genetic contributions to epilepsy; each of size up to 20 proteins.

Note that each c^2 term is a *constant*, not a variable. With this idea, the denominator in (4.20) can be replaced by the *linear* function of q and the $a(c)$ variables:

$$\left(\sum_{c=2}^{u}[c^2 \times a(c)] - q\right)/2. \tag{13.29}$$

Exercise 13.3.1 *Give a complete explanation for why the inequalities in (13.27) guarantee that $a(c)$ is set to 1, if and only if the value of q is exactly c. It is helpful to explain the* if *direction separately from the* only if *direction. Show also that these inequalities do not have any bad side effects.*

14

Character Compatibility with Corrupted Data and Generalized Phylogenetic Models

Biological data almost always has some level of corruption – noise, errors (both systematic and random), and omissions, etc. Those issues were *partly* dealt with in Chapter 3, where the approach was to remove columns (characters) from the input matrix, M, in order to leave the largest set of pairwise-compatible[1] columns in M. That was formalized as the *Minimum Column Removal (MCR) problem.*

In this chapter we return to the MCR problem and questions about character compatibility and how to handle data error and missing data. Then, we return to the discussion of *pancreatic cancer* and its relation to the MCR problem, first introduced in Section 3.5.1. Finally, we also discuss a *relaxation* of the perfect phylogeny model, and show that problems defined on that model are solved by relating them to compatibility problems with missing data. This chapter relies heavily on the material on the *minimum column removal (MCR)* problem, and the reader may need to review Section 3.4.

14.1 HANDLING MISSING AND CORRUPTED DATA IN CHARACTER COMPATIBILITY PROBLEMS

In Chapter 3, Section 3.4, we defined a *perfect phylogeny* and developed an ILP formulation for the minimum column removal (MCR) problem. In that approach, we assumed that *no* values in the input matrix M were missing. That is, each cell in M had a value, either 0 or 1. However, real biological data often has many *missing* values as well as *errors*, and a proper treatment of biological data must address those issues. Many current examples of (assumed) perfect-phylogeny evolution, where the data has a large amount of "noise," come from *single-cell* sequencing done in order to determine the development and fate of individual cells. This is becoming a huge area in molecular genetics and development. It was first used in the study of cancer, as introduced earlier in the book (see, for example, [115]), but has expanded to cover many biological phenomena concerning normal cell growth and development. A characteristic of this single-cell sequence data is that it has a large amount of missing and incorrect values.

[1] The definition of compatibility is given in Section 3.2.2.

Here, we consider perfect-phylogeny problems when some entries of M are missing and/or in error. These, and related ILP formulations were first developed in [87].

14.1.1 Imputing Missing Values

We start by stating two problems that we will solve using ILP.

> **Problem IM:** Given a binary input matrix M where some cells are *missing* data, can those cells be given binary values so that in the resulting matrix M', every pair of columns is compatible, i.e., so that the sequences in M' can be generated on a perfect phylogeny?

Problem IM is of interest if the full (but unknown) data underlying M is from a perfect phylogeny, and the remaining data has no errors. Otherwise, the answer to most instances of Problem IM will be "no," and we will learn very little from that result. However, Problem IM can be generalized to the following more informative

> **Problem IMM**: Given a binary matrix M with some entries missing, fill in the missing entries so as to *minimize* the number of incompatible pairs of sites in the resulting matrix M'.

Clearly, for any input, Problem IM can be solved by solving Problem IMM and checking if the solution value is zero, so we concentrate on Problem IMM.

If any pairs of columns in M are already incompatible, based on the *known* values in those two columns, then the answer to the instance of problem IMM will be the number of incompatible column pairs in M, *plus* the minimum number of *additional* pairs of columns that are made incompatible when missing values are set.

In the discussion of the IM and IMM (and other) problems, M will always refer to the *input* matrix, and M' will refer to a resulting matrix after all of the missing entries in M have been given values. We introduce the ILP formulation for Problem IMM through the example shown in Figure 14.1.

14.1.2 A Concrete ILP Formulation for Problem IMM

The ILP formulation for problem IMM will have one binary variable $Y(i,j)$ for each cell (i,j) in M that is missing a value. For example, for matrix M in Figure 14.1, the ILP formulation will have variables $Y(2,1)$, $Y(4,1)$, $Y(4,2)$, $Y(5,1)$, and $Y(6,2)$. The value, 0 or 1, given to $Y(i,j)$ in a solution to the concrete ILP formulation will be the value assigned to cell $M(i,j)$. More formally, we say that the value of $Y(i,j)$ is the *imputed* value for cell $M(i,j)$.

M	p	q
1	0	0
2	?	1
3	1	0
4	?	?
5	?	0
6	0	?

Figure 14.1 Six Rows and Two Columns in the Input M. A missing value is denoted by a question mark.

Avoiding Incompatibilities The two columns in Figure 14.1 already have the binary pairs $(0,0)$ and $(1,0)$, but the columns are not incompatible[2] in M since the *known* values in columns (p,q) do *not* have the binary pairs $(1,1)$ and $(0,1)$. However, depending on how the Y variables are set, columns (p,q) can become incompatible. For example, if variable $Y(2,1)$ and $Y(6,2)$ are each assigned value 1, then row 2 will have the pair $(1,1)$ and row 6 will have the pair $(0,1)$, making columns p and q incompatible in the resulting matrix M'. And, there are other assignments of values to the Y variables that make those columns incompatible. So, we will use a binary variable, $B(p,q,1,1)$, to record whether the Y variables have been set in a way that creates a row in M' with the binary pair $(1,1)$. Similarly, we use a binary variable, $B(p,q,0,1)$, to record whether columns (p,q) have the binary pair $(0,1)$ in M'.

Variable $B(p,q,1,1)$ is set with the inequalities:

$$\begin{aligned} Y(2,p) &\leq B(p,q,1,1), \\ Y(4,p) + Y(4,q) - B(p,q,1,1) &\leq 1, \end{aligned} \tag{14.1}$$

which forces $B(p,q,1,1)$ to have value 1 when the missing value in the *second* row is set to 1, or the two missing values in the *fourth* row are set to 1. The number of these inequalities is equal to the number of rows in column pair (p,q) that can *possibly* have values $(1,1)$ in M'.

Similarly, the inequalities to set variable $B(p,q,0,1)$ are:

$$\begin{aligned} Y(2,p) + B(p,q,0,1) &\geq 1, \\ Y(4,p) - Y(4,q) + B(p,q,0,1) &\geq 0. \end{aligned} \tag{14.2}$$

Exercise 14.1.1 *Explain why the inequalities (14.1) set $B(p,q,1,1)$ to 1, if the Y variables are set so that binary pair $(1,1)$ appears in columns (p,q) in M'. Do these inequalities allow $B(p,q,1,1)$ to be set to 1 even if the binary pair (1,1) does not appear in columns (p,q) in M'? Similarly, explain the conditions for $B(p,q,0,1)$ to be set to 1.*

Exercise 14.1.2 *Suppose column pair (p,q) in M is missing the binary combination $(0,0)$ and there is a row r where both columns p and q are missing values. What is the correct inequality used to set $B(p,q,0,0)$ corresponding to row r?*

More of the ILP Formulation The ILP formulation will also have a binary variable, $I(p,q)$, for each pair of columns in M that can be made incompatible in M'. We want variable $I(p,q)$ to have a value of 1 in the ILP solution, *if* the Y variables are set in a way that makes columns (p,q) incompatible. This is done for the example in Figure 14.1 as follows:

$$B(p,q,1,1) + B(p,q,0,1) - I(p,q) \leq 1. \tag{14.3}$$

Exercise 14.1.3 *Does inequality (14.3) allow $I(p,q)$ to be set to 1 even if columns $\{p,q\}$ have not become incompatible in M'?*

The Objective Function The following objective function, added to the inequalities already explained, creates a correct ILP formulation for the IMM problem:

$$\text{minimize} \sum_{(p,q)} I(p,q).$$

[2] Note that it would be wrong to say these columns are "compatible." Hence the hated double negative.

Exercise 14.1.4 *Given the objective function for Problem IMM, explain why the inequalities (14.1), (14.2), and (14.3) are sufficient, despite the answers to Exercises 14.1.1 and 14.1.3.*

14.1.3 The Abstract ILP Formulation for IMM

Generalizing from the example in Figure 14.1, given the input matrix M, let P be the set of column pairs, which are not incompatible in M, but have some missing value(s) in either column p or q. For each pair of columns $(p,q) \in P$, we let $d(p,q)$ be the set of *ordered* binary pairs from $\{(0,0),(0,1),(1,0),(1,1)\}$ that *do not* appear in column pair (p,q) in M. The term "$d(p,q)$" stands for (deficient) binary combinations in column pair (p,q).

Then the ILP formulation will have a binary variable $B(p,q,a,b)$ for each column pair $(p,q) \in P$ and each ordered binary combination a,b in $d(p,q)$. We want inequalities to force $B(p,q,a,b)$ to have value 1 if the ordered combination a,b has been created (through the setting of the Y variables) in *some* row in columns (p,q).

The ILP formulation must also have a binary variable $I(p,q)$ for each $(p,q) \in P$, and inequalities to force $I(p,q)$ to have value 1, if $I(p,q,a,b)$ has been set to 1 for *every* combination a,b in $d(p,q)$. Therefore, $I(p,q)$ will be set to 1 if (but not only if) the imputations of the missing values in columns (p,q) cause those sites to be incompatible.

Exercise 14.1.5 *In the example above, $d(p,q)$ contained two binary combinations. Suppose that $d(p,q)$ contains three binary combinations. How does that change inequality (14.3)? How does the constant in the right-hand side of the inequality depend on the size of $d(p,q)$?*

Exercise 14.1.6 *Each of the four inequalities in (14.1) and (14.2) (setting values for B variables), is determined by a binary pair that is missing from column pair (p,q), and by the pattern of values in M of a row in (p,q). For example, the binary pair (1,1) is missing, and row 2 of M has the pattern ?,1. This leads to the inequality $Y(2,p) \leq B(p,q,1,1)$ in (14.1).*

For each of the four ordered binary pairs, denoted (a,b), there are exactly three possible patterns of values in a row of M, where (a,b) can appear in columns (p,q) in M'. Write out these 12 cases, and for each case, determine the correct inequality to set $B(p,q,a,b)$.

Exercise 14.1.7 *Write out in full the concrete ILP formulation to solve Problem IMM for the matrix M shown in Figure 14.1.*

Exercise 14.1.8 *Write out in full the abstract ILP formulation to solve Problem IMM.*

Exercise 14.1.9 *What is the solution to the IMM problem in:*

```
1?0
111
001
010
```

Exercise 14.1.10 *Write out in full the abstract ILP formulation to solve Problem IM.*

Creating Concrete Formulations In creating a concrete ILP formulation for an input matrix M, a computer program (or the person, if creating the concrete formulation manually) identifies the pairs of columns in P, and the set $d(p,q)$ for each column pair (p,q) in P.

If M is an n by m matrix, the above ILP formulation for problem IMM creates at most $n \times m$ Y variables, $2m^2$ B variables, $\frac{m^2}{2}$ C variables, and $O(nm^2)$ inequalities, although all of these estimates are *worst* case and the numbers are typically much smaller. The ILP formulation as described might have redundant or useless inequalities, but we have left them in our description for conceptual clarity. In practice, the Gurobi preprocessor will often determine that they are redundant, and remove them, and sometimes having redundant inequalities can reduce the time needed for solving the ILP formulation.

Exercise 14.1.11 *In the ILP formulation developed in Section (14.1.3) for Problem IMM, P is the set of all column pairs that are not incompatible in M, and have some missing values. The ILP formulation has inequalities for each column pair (p,q) in P. But if it is not possible to make (p,q) incompatible in any imputation of the missing values, those variables and inequalities for the column pair (p,q) are useless. For example, the column pair*

```
01
10
?1
1?
```

is in P, but no matter how the missing values are set in M', the binary pair $(0,0)$ will not appear, and the column pair will be compatible in any M'. So, variables $B(p,q,0,0), B(p,q,0,0)$ and $I(p,q)$ are useless, and inequalities involving them are useless. A computer program that creates concrete ILP formulations for the IMM problem could omit those variables and inequalities, but I claim that no harm is caused by including them.

Explain in full detail why this is.

14.1.4 Software for Missing Values

The Perl program *bbmisstest.pl* can be downloaded from the book website. It takes in a matrix M with entries from $\{0,1,2\}$, where "2" indicates a missing value, and creates the concrete ILP formulation to solve the IMM problem for M. Call the program on a command line in a terminal window as:

```
perl bbmisstest.pl data-file ILP-file.lp
```

The example data file, *missing-bb*, can be downloaded from the book website. The example in Exercise 14.1.9 is also informative.

14.2 HANDLING BOTH MISSING AND CORRUPTED DATA

Here we discuss a natural imputation problem to handle data with *both missing and corrupted* values. Recall that the MCR problem is motivated by the assumption that the *true* evolutionary history underlying the given sequences is representable by a perfect phylogeny, but because of data errors, the sequences *cannot* be derived on a perfect phylogeny. The perfect-phylogeny *signal* has been corrupted. The solution to the MCR problem extracts the largest number of columns in the data where the signal of a perfect phylogeny still remains. But the MCR problem assumes that the input, M, contains no missing data. When the input has *both* errors and missing data, the following problem is a natural way to recover as much signal of the underlying perfect phylogeny as possible.

Problem IMCR: Over all matrices created by imputing missing values in M, find a matrix M' to *minimize* the number of columns of M' that must be removed, so that the remaining data can be derived on a perfect phylogeny. That is, find an M' to minimize the solution value of Problem MCR on M'.

An ILP formulation for Problem IMCR is easily obtained from the ILP formulation for Problem IMM as follows:

Given M, start with the ILP formulation for the IMM problem. Let $D(i)$ be a new binary variable used to indicate whether or not column i should be removed. A value of 1 for $D(i)$ means that the column i should be removed. Then, for each pair $(p,q) \in P$, add the inequality

$$D(p) + D(q) - I(p,q) \geq 0, \tag{14.4}$$

which says that if the missing values are imputed so that column pair (p,q) becomes incompatible, then either site p or site q (or both) must be removed.

Exercise 14.2.1 *Explain in detail why inequality (14.4) is correct. Does the inequality allow $D(p)$ or $D(q)$ to be set to 1 even if $I(p,q)$ is set to 0?*

Exercise 14.2.2 *How should the objective function in the ILP formulation for Problem IMM be changed in the ILP formulation for Problem IMCR?*

Exercise 14.2.3 *Some pairs of columns might be incompatible in M (based on the known values). Let F be the set of such pairs. Explain how to modify the ILP formulation to handle these in Problem IMCR.*

Flipping Values Another approach to handling data M where some column pairs are incompatible, is to *change* individual entries in M, so that all the column pairs become compatible. The minimum number of entries in M that must be changed can also be used as a measure of how close the data is to fitting the perfect-phylogeny model.

Exercise 14.2.4 Minimum Flipping Problem *Describe the abstract ILP formulation to find the minimum number of individual entries in M that must be flipped (0 to 1, or 1 to 0), so that the resulting data contains no incompatible pairs of columns. If you need to, use the ideas given in the following hint.*

Hint: One approach is to use the abstract ILP for the IM problem. The idea is to consider every *cell in M to have a missing value. Explain why this ILP formulation will necessarily have a feasible solution.*

Of course, if we stop here, the ILP formulation will have a feasible solution, but it may have no *relation to M, or to the minimum flipping problem. That can be addressed in the objective function, which is the sum of terms, one term for each cell in M. Suppose that in the input M, $M(i,j)$ is 0. Then, the term for (i,j) in the objective function will be $Y(i,j)$. But if $M(i,j)$ is 1 in the input M, then the term for (i,j) in the objective function will be $(1 - Y(i,j))$. So, the objective function will be*

$$\textit{Minimize} \sum_{i,j:M(i,j)=0} Y(i,j) + \sum_{i,j:M(i,j)=1} (1 - Y(i,j)).$$

Explain in detail how this abstract ILP formulation now correctly solves the minimum flipping problem.

A deeper and broader discussion of perfect-phylogeny flipping problems is found in [43]. The authors present theoretical and empirical results using different ILP formulations and several advanced ILP-solution methods.

14.3 BACK TO THE ARTIFACT PROBLEM IN PANCREATIC CANCER

In Chapter 3, Section 3.5.1, we described how the artifact problem in cancer could be approached using the *weighted MCR* problem on the input matrix M. This is exactly the first part of the approach to the artifact problem in [193]. It is also related to the approach in [163]. In Section 3.5.1, we said that both papers use additional ideas that will be explored later in the book. Here, we return to those two papers, discussing in more detail how they go beyond solving the weighted MCR problem.

14.3.1 Completing the Subclone Identification Solution

Recall that the first part of the approach in [193] is to find a largest subset of pairwise-compatible columns, $\mathcal{K}$ in the input matrix M. That was done by solving an instance of the *MCR* problem. In the second part of the approach, i.e., once $\mathcal{K}$ has been found, each of the other columns will be successively *modified* to become compatible with the current $\mathcal{K}$, and then added to $\mathcal{K}$. At the end of this process, all columns will be in $\mathcal{K}$, but the columns not initially in $\mathcal{K}$ will have been modified. In more detail, the following problem is solved for each successive column, c, not initially in $\mathcal{K}$:

Flipping-in-One-Column Problem Given column c and a set of columns $\mathcal{K}$, *flip* (0 to 1, or 1 to 0) the *fewest* entries in column c of M, so that the resulting column c is compatible with each column in $\mathcal{K}$. If there is a specific weight $w(i, c)$ for changing the entry in row i of column c, then find the *minimum total weight* way to flip entries so that the resulting column c is compatible with each column in $\mathcal{K}$.

Exercise 14.3.1 (a) *Why is it always possible to flip values in column c, so that c becomes compatible with all the columns in $\mathcal{K}$? That is, why does the* flipping-in-one-column problem *always have a feasible solution?*

(b) *Describe and explain an abstract ILP to solve the* flipping-in-one-column problem.

(c) *Actually, in [193], it is assumed that the ancestral sequence, at the root of the perfect phylogeny, must be the* all-zero *sequence. In that case, the condition for two columns (p, q) to be incompatible, is that the binary pairs 0,1; 1,0; and 1,1 all appear in columns p, q. That is, the 0,0 pair need not appear in columns (p, q).*

Explain how to modify the ILP formulation for the flipping-in-one-column *problem in this case.*

(d) *Now suppose that for each cell (i, j) in M, there is a cost $w(i, j)$ for flipping the value in cell $M(i, j)$. Modify the ILP formulation so that it solves the problem of finding the minimum cost set of flips which make c compatible with every column in $\mathcal{K}$.*

14.3.2 Completing the Metastatic Seeding Solution

To explain the full approach in [163], I will start with another procedure that is closer to what the authors suggested, but is *not yet* the full method they developed.

Almost the Correct Method In [163], instead of directly solving the weighted MCR problem on M or $G(M)$ (i.e., on the given data alone), the authors start with the following procedure:

First, note that each column in M contains one of 2^n possible binary patterns (binary strings). For *each* column c in M, and each binary pattern α, compute a *reliability score* reflecting how likely it is that the "correct" data for column c is pattern α (which might not be the actual observed pattern for c). We won't discuss how these scores are obtained, but, if the reliability score is sensible, then α should get a *high* score if it *is* a pattern observed in some column of M, particularly if it is in column c. Conversely, α should get a *low* score if it is very different from any of the observed binary patterns in M, particularly in column c.

Second, create a node-weighted graph $G'(M)$ with $m \times 2^n$ nodes, one node for each *column-pattern* combination, (c, α); and give the node a weight equal to the reliability score for that combination, i.e., how likely it is that the *correct* pattern for column c is α. Then, put an undirected edge between two nodes represented, for example, by (c, α) and (c', α'), if and only if the binary patterns α and α' are *compatible*, and $c \neq c'$.

Next, using ILP, find a *maximum-weighted clique*, $\mathcal{K}$, in $G'(M)$ with the added condition that $\mathcal{K}$ *has exactly* m nodes, the number of columns in M. For each column, c, clique $\mathcal{K}$ will have exactly one node from the 2^n nodes associated with c. Suppose c is associated with the column-pattern combination (c, α). Then, assign the pattern α to column c. After doing this for every column c in M, the result is a new matrix, called M'. If a column, c, in M' differs from the column c in M, the pattern in M' is considered to be a "correction" to the original one. Mutations (from state 0 to state 1) in column c that are removed from M indicate "false positives" in M, and mutations that are added indicate "false negatives" in M.

Since the binary patterns in M' come from clique $\mathcal{K}$ in $G'(M)$, the columns of M' will be pairwise compatible, so there is a unique perfect phylogeny, T', for M'. This procedure is expected to not only find a biologically meaningful evolutionary history, T', of the tumors and mutations, but to also identify and "correct" errors and artifacts in the input data, M.

Exercise 14.3.2 **(a)** *You might worry that $G'(M)$ doesn't have a clique of size m. Explain why $G'(M)$* must always *have a clique of size m.*

(b) *Explain why a clique with m nodes contains* exactly *one node from each set of 2^n nodes associated with the same column.*

(c) *What must be changed in the ILP formulation of the weighted MCR problem to implement this new procedure?*

(d) *Give the best argument you can, for and then against, the expectation that M' is "more correct" than M.*

(e) *Do you have any biology based arguments against this approach?*

14.3.3 The Correct Method

OK, finally, the actual procedure used in [163].

First, for each of the 2^n binary patterns, use the data in M to create a *single* reliability score for that pattern. This is in contrast to the procedure above, where a reliability score is obtained for each column-pattern combination. We won't detail how those scores are obtained, but if the reliability scores are sensible, a high score will be given to a pattern that occurs *frequently* in M, and a low score will be be given to a pattern that is very different from any observed data.

After the reliability scores have been computed, create a node-weighted graph $G''(M)$ with exactly 2^n nodes, where each node is associated with one of the 2^n binary patterns, and is weighted by the reliability score for that pattern. For every pair of nodes (p, q) in $G''(M)$, add an edge between nodes p and q if and only if the patterns associated with nodes p and q are *compatible*.

Then, using ILP, find a maximum weight clique, $\mathcal{K}$, in $G''(M)$. The patterns associated with the nodes in $\mathcal{K}$ will be pairwise compatible, as will any *subset* of those patterns. Let $\mathcal{P}(\mathcal{K})$ denote the set of patterns associated with the nodes in $\mathcal{K}$.

Next, for each column, c, choose the pattern, α in $\mathcal{P}(\mathcal{K})$, with the *highest* reliability score, discussed above, for the *column-pattern* combination, (c, α); and assign the binary pattern α to column c in M. Let M'' denote the resulting matrix, after all of the m pattern assignments have been made. If the pattern α is different from the original pattern in column c of M, then α is considered the *corrected* data for column c.

Finally, since all of the patterns in M'' come from $\mathcal{P}(\mathcal{K})$, they will be pairwise compatible, and so a perfect phylogeny can be constructed for M''. That perfect phylogeny is the reported solution to the artifact problem, and the proposed history of tumor seeding in pancreatic cancer.

Exercise 14.3.3 **(a)** *Give the best argument you can, for and then against, the expectation that M'' is "more correct" than M and/or M'.*

(b) *Do you have any biology based arguments against this approach?*

14.4 AN EXTENSION OF PERFECT PHYLOGENY TO LESS RESTRICTED MODELS

The key biological assumption that leads to the perfect-phylogeny model is that in the evolutionary history of the taxa, each character mutates from the zero state to the one state *exactly* once, and *never* from the one state *back* to the zero state. Hence, every character c labels *exactly one* edge in a perfect phylogeny for M, indicating the point in the evolutionary history of the taxa when character c mutated. This assumption has proven valuable in many biological contexts, but it is not appropriate in every context. Further, there are biological contexts where the perfect-phylogeny assumption had been accepted, but is now being questioned. One such context is in the study the development and progression of cancer [27, 116].

14.4.1 Dollo Parsimony and Persistent Phylogeny

Several extensions of the perfect-phylogeny model have been proposed in order to address a wider range of evolutionary phenomena. In the *Dollo* (Parsimony) model [57, 66, 166], each character can mutate from the 0 state to the 1 state at most *once* in the history (as in the perfect-phylogeny model), but the character can mutate from the 1 state *back* to the 0 state at *any* point where the character has state 1. This models evolutionary characters that are gained with low probability, but that are *lost* with much higher probability. The Dollo model is appropriate

> ...for reconstructing evolution of the gene repertoire of eukaryotic organisms because although multiple, independent losses of a gene in different lineages are common, multiple gains of the same gene are improbable. [166]

Recently, a more limited version of the Dollo model was proposed, where any character can mutate from state 0 to state 1 *at most once* in the history; and *symmetrically*, it can mutate from state 1 to state 0, but *at most once* in the history. This model is called the *persistent-phylogeny* model. It models evolutionary characters that are gained with low probability, and then are lost with low (but not-zero) probability. The persistent-phylogeny model was first proposed and further examined in the papers [23, 24, 25, 26, 155, 156, 214]. The persistent-phylogeny model leads to the

Persistent-Phylogeny Problem: Given an n by m binary matrix M, determine whether M can be represented by a persistent phylogeny for M, and if so, build one.

14.4.2 Solving the Persistent-Phylogeny (PP) Problem By Integer Programming

As we will see, the persistent-phylogeny problem can be solved by a *constrained variant* of the IM problem, and hence the ILP formulation for the PP problem is a small modification of the formulation for the IM problem. To explain this, we start with some definitions and facts.

Given a binary matrix M, the *extended matrix* M_e for M contains *two* columns, j_1 and j_2, for *each* column j in M. Column j_1 of M_e is derived from column j in M by replacing every occurrence of "0" in column j of M with "?" in column j_1 of M_e. Column j_2 of M_e is derived from column j_1 by replacing every occurrence of "1" in j_1 with "0." For example, in Figure 14.2, the first column of M_e is the same as the first column of M, but every "0" in the first column of M has been replaced by a "?" in the first column of M_e. Then, the second column of M_e is the same as the first column of M_e, but every "1" in the first column M_e has been replaced with a "0" in the second column of M_e.

Note that for any pair of columns (j_1, j_2) in M_e that are derived from the same column j in M, column j_1 contains a "?" in a row r if and only if column j_2 also contains a "?" in row r. We call such cells *twin cells*.

Completing an Extended Matrix A *completion*, M'_e, of an extended matrix M_e derived from binary matrix M, is obtained by replacing each "?" in M_e with "0" or "1," subject to the *added*

> **Completion Constraint** For any column j in M, if $M_e(r, j_1) = M_e(r, j_2) =$ "?," then $M'_e(r, j_1)$ and $M'_e(r, j_2)$ must be equal.

That is, the values given to twin cells must be the *same*, either both 0 or both 1 (see Figure 14.2).

M	M_e	$M'_e =$ Completion of M_e
1110	101010??	10101000
0111	??101010	11101010
0000	????????	00000000
1010	10??10??	10001000
1100	1010????	10101100
1111	10101010	10101010

Figure 14.2 Matrices M, M_e, and a Completion M'_e. Matrix M cannot be represented by a perfect phylogeny since it violates the perfect-phylogeny theorem (for example, columns 1 and 2 are incompatible), but matrix M'_e satisfies the perfect-phylogeny theorem. So M'_e can be represented by a perfect phylogeny, and therefore, M can be represented as a persistent phylogeny. The trees are shown in Figure 14.3.

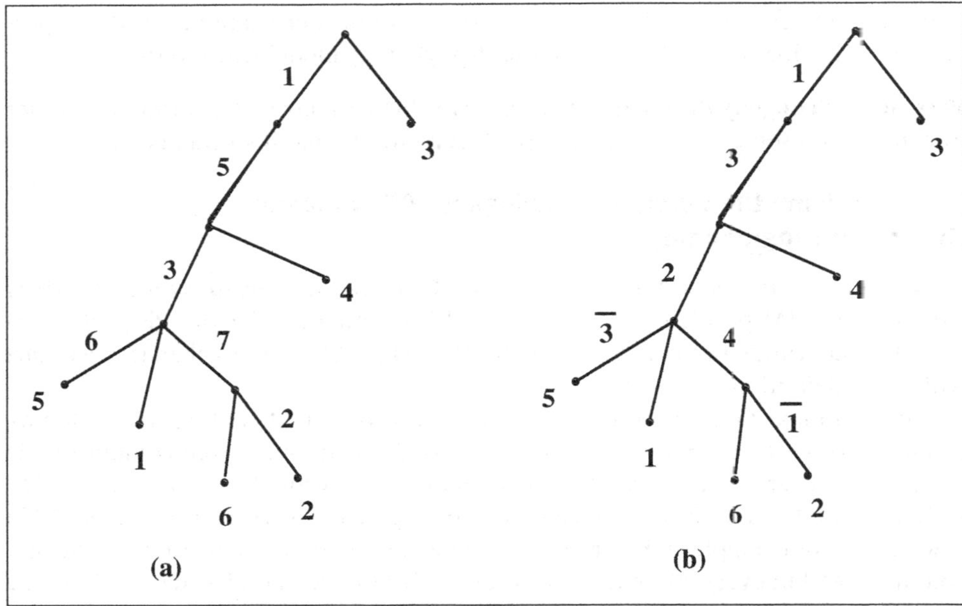

Figure 14.3 Part (a) Shows the Perfect Phylogeny for Matrix M'_e, and Part (b) Shows the Persistent Phylogeny for the Input Matrix M from Figure 14.2. A character c written on an edge without an over-bar represents a mutation of character c from state 0 to state 1; a character c with an over-bar represents a back mutation of character c from state 1 to state 0. In the persistent phylogeny shown here, characters 1 and 3 back mutate. The character labeling an edge of the persistent phylogeny is derived from the character labeling that edge on the perfect phylogeny. To convert a character c on the perfect phylogeny to the character on the persistent phylogeny, divide c by two and call the result d. If d is integral, place a bar over d. If d is fractional, round up to the next integer.

For a column j in M, we can think of column j_1 in M_e as answering the question: "has character j *mutated* from state 0 to state 1"; and we can think of column j_2 in M_e as answering the question "has character j *back-mutated* from state 1 to state 0."

The following theorem, stated and proved in [23], is the key to the solution and to the ILP formulation of the persistent-phylogeny problem.

Theorem 14.4.1 (Completion Theorem) *Let M_e be the extended matrix obtained from binary matrix M. Then M can be represented by a persistent phylogeny if and only if there is a completion M'_e of M_e such that M'_e can be represented by a perfect phylogeny. Further, any perfect-phylogeny T for M'_e can be converted to a persistent phylogeny for M (see Figures 14.3a,b).*

So, the persistent-phylogeny problem on M can be solved as an IM problem on M_e, constrained by the completion constraint. We will not try to justify Theorem 14.4.1, but you should be able to get a feel for why it is correct from looking at a few examples, keeping in mind the comments above about the meaning of columns j_1 and j_2 in M_e. With the above discussion, and our prior discussion of problem IM, it should now be straightforward to describe an abstract ILP formulation for the persistent-phylogeny problem.

Exercise 14.4.1 *Modify the abstract ILP formulation for problem IM to obtain an abstract ILP formulation for the persistent-phylogeny problem. Write out the concrete ILP formulation for the problem instance M shown in Figure 14.3.*

14.4.2.1 Software for the Persistent-Phylogeny Problem

The Perl program *spersistent.pl* is a Perl script that takes in the name of a file with a binary matrix, and calls the programs *sperfreduce.pl, sprepare-persistent.pl*, and *persistent.pl*, finally producing, in file *"persistent.lp,"* the concrete ILP formulation to solve the persistent phylogeny problem for input matrix in the data-file. All of these Perl programs can be downloaded from the book website.

Call program *spersistent.pl* as:

```
perl spersistent.pl data-file
```

An example data-file that has a persistent-phylogeny is *per-data*. An example data-file that does not have a persistent-phylogeny is *char-data*.

In More Detail Script *spersistent.pl* first calls program *sperfreduce.pl*, which removes duplicate rows and every column that is compatible with all other columns. Removing those sites usually reduces the size of the problem significantly, and the resulting ILP solves faster.

If used outside of *spersistent.pl*, call *sperfreduce.pl* on a command line in a terminal window as:

```
perl sperfreduce.pl data-matrix
```

Script *spersistent.pl* next calls the Perl program *sprepare-persistent.pl*, which processes a binary matrix, read from a file, to create the file containing the *extended matrix*, with 2s (representing ?s).

If used outside of *spersistent.pl*, call on a command line in a terminal window as:

```
perl sprepare-persistent.pl data-file
```

The data-file can hold any binary matrix. However, if the binary matrix has first been passed through *sperfreduce.pl*, then it will be smaller, and the ultimate concrete ILP formulation will be smaller and will solve faster. So, it is recommended that the data-file be one that is the output of sperfreduce.pl.

Finally, script *spersistent.pl* calls the Perl program *persistent.pl*, which creates a concrete ILP to solve the persistent phylogeny problem for the binary matrix that was first input to *sperfreduce.pl*. The ILP formulation is output to file *persistent.lp*.

If there is a persistent phylogeny, then the optimal solution for *persistent.lp* will have value 0. If there is no persistent phylogeny, then the ILP will be infeasible.

If used outside of *spersistent.pl*, call *persistent.pl* as

```
perl persistent.pl matrix-file
```

where "matrix-file" holds an extended matrix, with values of 0, 1 and 2, usually created by the program *sprepare-persistent.pl*.

Exercise 14.4.2 *Use* spersistent.pl *on input* per-data *and then* char-data *to produce the concrete ILP formulations. Then use Gurobi to verify that one has a persistent phylogeny and the other does not.*

Now, instead of using spersistent.pl, *sequentially use the programs sperfreduce.pl, sprepare-persistent.pl, and persistent.pl with input data* per-data. *Use the output of one program as input to the next. Be sure you get the same end result you got when using* spersistent.pl.

15

More Tanglegrams, More Trees, More ILPs

15.1 MINIMIZING SUBTREE EXCHANGES WITHIN AN OPTIMAL SOLUTION

In Chapter 8, we developed an abstract ILP formulation to find a tanglegram minimizing the number of line *crosses* between the two trees in the tanglegram. With that formulation, Gurobi found a tanglegram for the trees in Figure 8.1, using only *nine* crosses. In the reported optimal solution, Gurobi set the value of 16 of the X variables to 1, meaning that it specified 16 subtree exchanges to transform the tanglegram in Figure 8.1 to the tanglegram in Figure 8.4, which we denoted as $\mathcal{T}^*$.

When I tried to verify that those 16 exchanges did create $\mathcal{T}^*$, I readily got lost in the details. But, there is another optimal solution (i.e., using only nine crosses), which needs only nine (coincidentally) subtree exchanges. These are detailed in Exercise 8.4.2 on page 150. Since it is desirable to have an optimal solution to the tanglegram problem that also uses the *fewest* subtree exchanges, we have the following problem:

> **The Bi-Objective Tanglegram Problem** Given an input tanglegram $\mathcal{T}$, find an optimal solution to the tanglegram problem on $\mathcal{T}$, and if there is more than one optimal solution, find an optimal solution $\mathcal{T}^*$, to *minimize* the number of subtree exchanges required to convert $\mathcal{T}$ to $\mathcal{T}^*$.

This is an example of a *multi-objective* problem, with two objectives. In optimization, multi-objective problems are common, and can have more than two objectives. In the *bi*-objective tanglegram problem, the *primary* objective is to solve the tanglegram problem; and the *secondary* objective is to find, among all tanglegrams solving the primary objective, one that minimizes the number of subtree exchanges.

An ILP Solution to the Bi-Objective Tanglegram Problem The approach we take will illustrate an approach to solving muli-objective problems in general. The starting point in explaining this approach is the abstract ILP formulation for the tanglegram problem developed in Section 8.4. In that ILP formulation, we let $\aleph$ denote the set of all the $X1$ and $X2$ variables in the formulation. Then, add to that formulation two new variables, C and A, and the following two inequalities:

$$C = \sum_{i<j} C(i,j), \tag{15.1}$$

$$A = \sum_{X \in \aleph} X. \tag{15.2}$$

Variable C simply counts the number of crosses in an ILP solution, and variable A counts the number of subtree exchanges specified by the solution. Next, we change the objective function to:

$$\texttt{minimize } C.$$

The resulting formulation is just a (different looking) formulation for the tanglegram problem, still *without* any incentive to find a solution with few subtree exchanges. The natural next step is to change the objective function to

$$\texttt{minimize } C + A. \tag{15.3}$$

This adds the term, A, for the number of subtree exchanges, so the optimal ILP solution is encouraged to reduce that number as well as the number of crossovers. But it is not clear how the *overall* optimal solution will balance the two objective. To minimize the sum, an optimal solution might increase the number of crossovers above the minimum possible, in order to reduce the number of exchanges by a larger amount. So the objective function in (15.3) is not yet the right one. A correct objective function is:

$$\texttt{minimize } (m+1) \times C + A, \tag{15.4}$$

where m is the number of non-leaf nodes in two input trees.

Why Is This Correct? The key to seeing that the objective function in (15.4) does what we want, is that the multiplier $m + 1$ is *larger* than the maximum number of subtree exchanges that are possible. So, an increase of even a single crossover will result in an increase in the sum by $m+1$, an amount that cannot be offset by a decrease in the number of subtree exchanges. Hence an optimal solution *must minimize* the number of crossovers. But since the term A is part of the objective function, any optimal solution must minimize the number of subtree exchanges over all solutions that minimize the number of crossovers. This is exactly what is required of a solution to the bi-objective tanglegram problem.

Exercise 15.1.1 *Another approach to the bi-objective tanglegram problem is to* first *solve the tanglegram problem using only the primary objective, determining the minimum number, call it C^*, of crosses possible in any tanglegram for the input trees.*

Explain how to next use C^* in an ILP to solve the bi-objective tanglegram problem.

15.2 A DISTANCE-BASED OBJECTIVE FUNCTION FOR TANGLEGRAMS

Recall that the tanglegram problem resulted from the biological objective of measuring the similarity of two phylogenetic trees. A different way to evaluate similarity is by the *sum* of the *distances* between the corresponding leaves in T_1 and T_2. For this, we assume that the leaves of a tree are positioned on their leaf-line with *uniform* spacing between consecutive leaf positions. In a tree, T, in a given tanglegram, the order of the leaves (starting from the top) specifies a *position number* for each leaf in T.[1]

Recall that $\overline{T_1}$ and $\overline{T_2}$ denote the drawing of the trees T_1 and T_2 after all of the node exchanges have been made. We use $P1(i)$ to denote the position in $\overline{T_1}$ of the leaf labeled i; similarly, we use $P2(i)$ to denote the position in $\overline{T_2}$ of the leaf labeled i. Then the *Interleaf Distance* of $\overline{T_1}$ and $\overline{T_2}$ in a tanglegram is defined as:

$$\sum_{i=1}^{i=n} |P1(i) - P2(i)|,$$

where the vertical bars around $P1(i) - P2(i)$ indicate the *absolute value*.[2] Then, we have

The Interleaf-Distance Tanglegram Problem: Given two input trees T_1 and T_2, choose subtree exchanges, creating drawings $\overline{T_1}$ and $\overline{T_2}$ in a tanglegram, to *minimize* the Interleaf Distance between $\overline{T_1}$ and $\overline{T_2}$.

The interleaf-distance tanglegram problem is certainly related to the original tanglegram problem, and a tanglegram that is good under one criterion is likely to be good in both criteria. But, the two criteria are not identical, and it useful to have computational methods for each.

Exercise 15.2.1 *Is the following true: If the interleaf distance of trees $\overline{T_1}$ and $\overline{T_2}$ in a tanglegram $\mathcal{T}$ is zero, then the number of crosses in $\mathcal{T}$ must be zero. Is the converse true, i.e., that if there are no crosses in $\mathcal{T}$, then the interleaf between the trees in $\mathcal{T}$ must also be zero? Explain both answers fully.*

We discuss the interleaf-distance tanglegram problem in detail out of an interest in interleaf distance as a measure of similarity, and also because the ILP implementation introduces a new ILP idiom.

15.3 AN ILP FORMULATION

We will explain in detail the variables and the inequalities that are defined for tree T_1, with the understanding that analogous variables and inequalities must be defined for tree T_2. Variables from the two trees will only be together in the objective function, and in the implementation of the absolute value function, needed for the objective function.

The Core Variables As in the discussion in Chapter 8, we use the variable $X1(v)$ for each non-leaf node v in T_1, and set its value to 1 to specify a subtree exchange at node v, and to 0 to specify no exchange at v. These choices determine the order of the leaves in the final drawing $\overline{T_1}$ of T_1. In the solution to the original tanglegram problem,

[1] Recall that a position number in a leaf ordering is not the same as the label of the leaf. For example, in Figure 8.1, the leaf labeled 4 in T_2 is in position 3, and the leaf labeled 12 in T_2 is in position 8.

[2] The "absolute value" of a number x is just x, if x is *positive*, and is $-x$, if x is negative.

the ILP only needed to determine the *relative* order of every pair of leaves in the final tanglegram, However, in the interleaf-distance tanglegram problem, the ILP we develop will determine the *actual position* of each leaf in the two resulting trees. Knowing the positions of the leaves allows the ILP to then determine the distance between the leaf i in T_1 and the leaf i in T_2.

For each i ranging from 1 to n, we create an *integer* variable $P1(i)$ whose value will be the position of leaf i in $\overline{T_1}$.

The First Inequalities We need to express the requirement that each leaf in T_1 appears at a *distinct* position from 1 to n, after the subtree exchanges are made in T_1. For that, it is sufficient to require that the value of each variable $P1(i)$ be between 1 and n, and that for every pair of leaves (i,j) in $T_1, P1(i) \neq P1(j)$. Clearly, for each leaf i in T_1, the first requirement is implemented by:

$$\begin{aligned} P1(i) &\geq 1, \\ P1(i) &\leq n. \end{aligned} \tag{15.5}$$

To implement the second requirement, that $P1(i) \neq P2(i)$, we rephrase it as: For each pair of leaves (i,j) in T_1, either $P1(i) > P1(j)$ or $P1(j) > P1(i)$. And since $P1(i)$ and $P1(j)$ must have positive integral values, we can rephrase this as:

For each pair of leaves (i,j) in T_1,
either $P1(i) - P1(j) \geq 1$ or $P1(j) - P1(i) \geq 1$.

Because the differences have bounded value, this can be implemented using the *OR* idiom for inequalities (12.9), discussed in Section 12.3. Applying that idiom, using two indicator variables, $z1(i,j)$ and $z'1(i,j)$, gives the inequalities:

$$\begin{aligned} P1(i) - P1(j) + (-n+1) \times z1(i,j) &\leq 0, \\ P1(j) - P1(i) + (-n+1) \times z'1(i,j) &\leq 0, \\ z1(i,j) + z'1(i,j) &= 1. \end{aligned} \tag{15.6}$$

The first inequality is for the case that $P1(i) > P1(j)$; the second inequality is for $P1(i) > P1(j)$; and the third inequality enforces that exactly one of those inequalities holds.

Exercise 15.3.1 *Verify that the first two inequalities in (15.6) correctly implement the OR idiom for inequalities.*

Additional Requirements The inequalities given in (15.5) and (15.6) ensure that each leaf i in T_1 is assigned to a *distinct* position between 1 and n. However, it is not true that *every* leaf order can be achieved by subtree exchanges in T_1. For example, in Figure 8.2, the leaf order 1, 3, 2, 4, 5 is impossible to achieve by subtree exchanges. Hence, we need to add inequalities to the ILP formulation to restrict the assignment of values so that the leaf order specified by the $P1$ values can be achieved by subtree exchanges in T_1. The needed inequalities are simple to state, as we show next, but it is a bit subtle to see that they are correct.

Inequalities to Restrict *P1* As before, for two leaves labeled i and j in T_1, let u denote the least common ancestor, $lca(i,j)$ in T_1. For each leaf pair (i,j) where i

appears *before* j in the leaf order of the input T_1 (whether or not i and j are numerically in order or out of order), create the inequalities to implement the relation: $P1(i) < P1(j)$ if and only if $X1(u) = 0$, which is equivalent to $P1(i) \leq P1(j) - 1$ if and only if $X1(u) = 0$.

We will examine the two directions individually. We first examine the relation: *If* $P1(i) \leq P1(j) - 1$ *then* $X1(u) = 0$. Even though $P1(i)$ and $P2(j)$ are not binary variables, their values are *bounded* between 1 and n, so that their difference is bounded by $1 - n$ and $n - 1$. So, following the implementation ideas for the *If-Then* idiom developed in Section 4.3.1, and simplifying a bit, we obtain:

$$P1(i) - P1(j) + n \times (1 - X1(u)) \geq 0. \tag{15.7}$$

Inequality (15.7) ensures that *if* $P1(i) \leq P1(j) - 1$ *then* $(1 - X1(u))$ *must* be 1. To see this, note that when $P1(i) \leq P1(j) - 1$, $P1(i) - P1(j) \leq -1$, so some positive amount must be added so that the sum is greater or equal to zero, hence $(1 - X1(u))$ must be set to 1. But then, $X1(u)$ must be set to zero, as claimed.

We also need to check that inequality (15.7) is actually satisfied when $X1(u) = 0$. To see that, note that the smallest that $P1(i) - P1(j)$ can be is $1 - n$, so adding n to $P1(i) - P1(j)$ makes the sum positive. Further, we need to check that inequality (15.7) cannot cause any bad side effects. So, consider the case when the condition $P1(i) \leq P1(j) - 1$ does *not* hold. In that case, $P1(i) > P1(j) - 1$. This is equivalent to $P1(i) - P1(j) \geq 0$, in which case inequality (15.7) will be satisfied no matter how $X1$ is set. Hence, inequality (15.7) does not have any bad side effects.

Now we consider the opposite direction, i.e., *if* $X1(u) = 0$ *then* $P1(j) < P1(i)$. This is implemented by:

$$P1(i) - P1(j) + n \times (1 - X1(u)) + 1 \leq n, \tag{15.8}$$

which ensures that if the value of $X1(u)$ is zero, then $n \times (1 - X1(u)) + 1$ will have value $n + 1$, which is greater than n. So in order to satisfy the inequality in that case, $P1(i) - P1(j)$ must contribute something that is strictly *negative*, and hence it must be that $P1(i) < P1(j)$. Therefore, inequalities (15.7) and (15.8) together implement the relation

$$P1(i) < P1(j) \texttt{ if and only if } X1(u) = 0. \tag{15.9}$$

Note that in inequalities (15.5), (15.6), (15.7), and (15.8), "n" represents the number of leaves in the input trees, and is a *constant* number that is known for any specific problem instance. The program, or the person, creating a concrete formulation for a problem instance, must replace "n" with that constant, everywhere "n" appears in the abstract formulation.

Exercise 15.3.2 *Show that inequality (15.8) does not have any bad side effects.*

Is It Correct? We now need to convince ourselves that the inequalities (15.7) and (15.8) ensure that a feasible solution to the ILP formulation specifies an order of the leaves that can be achieved by subtree exchanges in T_1 and T_2. And conversely, we need to be convinced that any order of the leaves that can be achieved by subtree exchanges in T_1 and T_2, specify values of the $P1, P2$ and $X1, X2$ variables that form a feasible solution to the ILP formulation.

The converse direction is the easiest to verify. Suppose $\mathcal{T} = (\overline{T_1}, \overline{T_2})$ is a tanglegram for the input trees T_1, T_2. We use the order of the leaves in $\mathcal{T}$ to set the $P1$ and $P2$ variables. Then, given the input representations of trees T_1 and T_2, $\mathcal{T}$ is defined by a set S of subtree exchanges made on T_1 and T_2. For each node v in T_1, we set $X1(v)$ to 1 if and only if S contains a subtree exchange at v. Clearly, inequality (15.9) holds for any actual tanglegram, and so holds for $\mathcal{T}$. Hence, inequalities (15.7) and (15.8) will hold for the values of the $P1$ and $X1$ variables that come from $\mathcal{T}$. The argument for $P2$ and $X2$ variables is the same.

Exercise 15.3.3 *Explain the other direction. That is, explain that the inequalities (15.7) and (15.8) specify a leaf ordering that can be achieved by node exchanges starting with the input trees T_1 and T_2.*

Finally, the objective function for the ILP formulation is:

$$\texttt{minimize} \quad \sum_{i=1}^{i=n} |P1(i) - P2(i)|. \tag{15.10}$$

15.3.1 The Absolute-Value Idiom

The objective function (15.10) uses the *absolute-value* function. But absolute value is not a linear function, so can we express absolute value using only integer *linear* inequalities? The answer is another ILP idiom, the *minimizing absolute values idiom*.

Since the objective function consists of the *sum* of absolute values of differences, we only need to explain how to implement the absolute-value function for a single difference. So, we look at the problem:

$$\texttt{minimize} \quad |P1(i) - P2(i)|,$$

for just a *single* value of i.

The optimal solution to the following ILP will solve the problem of minimizing $|P1(i) - P2(i)|$. Remember that $P1(i)$ and $P2(i)$ are already constrained to have positive values.

$$\texttt{Minimize} \quad s(i), \tag{15.11}$$

subject to:

$$\begin{aligned} P1(i) - P2(i) &\le s(i), \\ P2(i) - P1(i) &\le s(i), \\ s(i) &\ge 0. \end{aligned} \tag{15.12}$$

Explaining the Idiom Clearly, to minimize $s(i)$, we want to set $s(i)$ as small as possible. But, all of the inequalities must be satisfied, so, $s(i)$ should be set to the *maximum* of zero and $P1(i) - P2(i)$, and $P2(i) - P1(i)$. Because $P1(i)$ and $P2(i)$ are both positive, *exactly one* of the two differences in (15.12) will be strictly positive and the other will be strictly negative; or both differences will be zero. Suppose $P1(i) - P2(i)$ is negative and hence, $P2(i) - P1(i) > 0$. Then, $s(i)$ will be set to $P2(i) - P1(i)$, which will be equal to $|P1(i) - P2(i)|$. A similar analysis holds if $P2(i) - P1(i) < 0$. When both differences are zero, $s(i)$ will be set to 0, which is the absolute difference in that case. Hence, in all cases, $s(i) = |P1(i) - P2(i)|$, as claimed.

Using the Idiom in the Interleaf-Distance Tanglegram Problem The objective function for the interleaf-distance problem is

$$\texttt{minimize} \sum_{i=1}^{i=n} |P1(i) - P2(i)|,$$

so, for each i, the term $|P1(i) - P2(i)|$ is *independent* of any other term in the objective function. Hence, using the *absolute value* idiom for each i, the objective function can be written as

$$\texttt{minimize} \sum_{i=1}^{i=n} s(i).$$

Then, the full, abstract ILP formulation is given in Figure 15.1.

15.3.2 Absolute-Value Idiom in An Inequality

The Question In the ILP formulation for the interleaf distance tanglegram problem, the absolute value function is part of the objective function, and we saw in the previous section how to implement the idiom in that case. But, in that implementation it was critical that the absolute value function was part of a minimizing objective function. So, how can we implement $|X|$ for a variable X, when X is used in an *inequality*, rather than in the objective function?

An Answer When we know lower and upper *bounds* on the permitted value of X, implementing absolute value is fairly simple, and provides a nice application of the *OR* idiom for inequalities with bounded variables, discussed in Section 12.3. We use a new variable A, standing for *absolute value*. Below are inequalities that will set A to $|X|$, where the value of X is bounded.

$$\begin{aligned} A &\geq 0, \\ A &\geq X, \\ A &\geq -X, \\ (A \leq X) \; OR \; (A &\leq -X). \end{aligned} \tag{15.13}$$

To see how this works, try the case when X has value 5, and then the case that X has value -5. In either case, the only way to satisfy all of the inequalities is to set the value of A to 5, i.e., $|X|$. More generally, divide the analysis into the cases where X is strictly positive, strictly negative, or zero.

Exercise 15.3.4 *Explain why the inequalities in (15.13) correctly set the value of A to $|X|$. Why is is important that the value of X be bounded?*

To completely translate the statements in (15.13) to *linear inequalities*, we recall that when the value of X is bounded, we can implement the *OR* idiom for variables using linear inequalities, as discussed in Section 12.3.

Exercise 15.3.5 *Applying the details of the OR idiom from Section 12.3, write out in full detail the linear inequalities to implement the absolute-value idiom for bounded variables.*

$$\text{Minimize} \sum_{i=1}^{i=n} s(i)$$

Subject to
For each i from 1 to n:

$$P1(i) - P2(i) \leq s(i)$$
$$P2(i) - P1(i) \leq s(i)$$
$$s(i) \geq 0$$

$$1 \leq P1(i) \leq n$$
$$1 \leq P2(i) \leq n$$

For each leaf pair (i,j) in T_1, where i appears before j in the leaf order of the input T_1, and $u = lca(i,j)$:

$$P1(i) - P1(j) + n \times (1 - X1(u)) \geq 0$$

$$P1(i) - P1(j) + n \times (1 - X1(u)) + 1 \leq n$$

For each leaf pair (i,j) in T_2, where i appears before j in the leaf order of the input T_2, and $u = lca(i,j)$:

$$P2(i) - P2(j) + n \times (1 - X2(u)) \geq 0$$

$$P2(i) - P2(j) + n \times (1 - X2(u)) + 1 \leq n$$

For each pair i and $j > i$:

$$P1(i) - P1(j) + (-n + 1) \times z1(i,j) \leq 0$$

$$P1(j) - P1(i) + (-n + 1) \times z'1(i,j) \leq 0$$
$$z1(i,j) + z'1(i,j) = 1$$
$$P2(i) - P2(j) + (-n + 1) \times z2(i,j) \leq 0$$
$$P2(j) - P2(i) + (-n + 1) \times z'2(i,j) \leq 0$$
$$z2(i,j) + z'2(i,j) = 1$$

The $X1, X2$, and the z variables are binary. All other variables are integral.

Figure 15.1 **An Abstract ILP Formulation for the Interleaf-Distance Tanglegram Problem. The indicator variables are *z1(i,j)*, *z'1(i,j)*, *z2(i,j)*, and *z'2(i,j)*.**

15.3.3 Another Variant: The Min-Max Distance Problem

Suppose that the objective function for the interleaf-distance tanglegram problem is changed to:

`Minimize the Maximum value of` $|P1(i) - P2(i)|$, `for i from 1 to n.`

That is, find a tanglegram for trees T_1 and T_2 that make the *largest* leaf-pair distance as *small* as possible. Thus, a *sum* in the objective function is changed to a *Max*. How do we implement this with linear inequalities? The answer is to add:

For each i from 1 to n,

$$|P1(i) - P2(i)| \leq Q, \tag{15.14}$$

where Q is an integer variable. And, of course we have to use the correct absolute-value idiom to express each $|P1(i) - P2(i)|$ with linear inequalities. Then, we simply change the objective function to:

$$\texttt{minimize } Q. \tag{15.15}$$

15.4 ROOTED SUBTREE-PRUNE-AND-REGRAFT (rSPR) DISTANCE

The tanglegram shown in Figure 8.1 was first published in the paper [77] by M. Goddard and A. Burt, and their main interest in that tanglegram was discussed in Chapter 8. But, for an additional analysis, they wanted to determine the frequency of a genomic phenomenon called *horizontal transmission*.[3] The approach the authors took was to compute the minimum number of *subtree-pruning-reconnection (SPR)* operations needed to convert one of the input trees, T_1, into the second input tree, T_2. Today, this operation is called a *rooted subtree prune and regraft (rSPR)* operation; and the problem of converting one specified tree to another specified tree, using the minimum number of *rSPR* operations is called the *rSPR distance problem.*

15.4.1 Definitions

Let T be a rooted, leaf-labeled tree. Informally, a *rooted subtree-prune-and-regraft (rSPR)* operation modifies tree T by *cutting* or breaking away one edge e and the subtree below e, and then *reattaching* e and its subtree to some point (a node or a point in the interior of an edge) in the remaining part of T.

If edge e is reattached to the interior of an edge, then a new node is created and it is given a node label distinct from the other node labels in T. All the other node labels remain the same in the two trees. Essentially, an rSPR operation changes an edge (u, v) to an edge (u', v), where u' might be a newly created node (see Figure 15.2).

The computational approach that Goddard and Burt used to solve the *rSPR* distance problem was essentially *brute force.* Starting from tree T_1 in Figure 8.1, they applied all possible *rSPR* operations to generate 462 trees that are one operation away from T_1. Then they applied all possible *rSPR* operations to each of those 462 trees, etc., until tree T_2 was generated. They found that five *rSPR* operations were

[3] For our purposes here, you need not understand what horizontal transmission is.

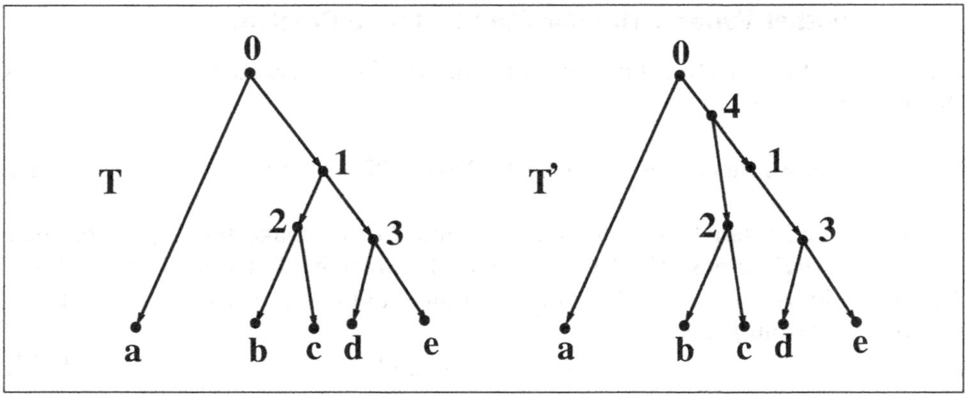

Figure 15.2 An rSPR Operation Removes the Edge *(1,2)* from the Rooted Tree *T* and Reattaches It in the Interior of Edge *(0,1)*, Creating the New Node 4 and the New Tree *T′*. Essentially, the *(1,2)* edge has been converted to the *(4,2)* edge. As drawn, we can picture the rSPR operation as a "pivot" of edge *(1,2)*.

sufficient. On this problem instance, their approach generated $(462)^4$ trees. While this kind of brute-force approach can work for small trees, it will quickly become impractical as the size of the tree increases.

But ILP Works Well What is of interest here, is that compact ILP formulations that solve the *rSPR* distance problem have been developed [86, 204]; and instances of the *rSPR* distance problem, which would be impossible to solve *exactly*, by brute force, can be solved by these ILP methods.

We will describe the ILP for the *rSPR* distance problem, but we will only sketch the ideas behind it. Unlike most computational biology problems that are solved with ILP, the Task A part of the solution to the *rSPR* distance problem is *not* straightforward – the underlying logic behind the solution method relies on nontrivial mathematical reasoning and structural theorems that are outside of the scope of this book. The interested reader can read [204] for the original ILP solution, which exploits the structural theorems, and can read [86], chapter 14, for an exposition of the needed structural theorems and an explanation of the ILP formulation given here, which refines the formulation in [204]. Although the logic of the method is not simple, *translating* that logic into an ILP formulation (Task B) *is* simple, and the ILP formulations are small.

15.4.2 The Key Fact

The optimal solution to the *rSPR* distance problem has a value that is *equal* to the *minimum* (equal) *number* of edges to *remove* from both T_1 and T_2, so that the two resulting sets of *subtrees* are the same, after removing any node with only one out edge.

The key fact actually needs to be stated more precisely to be completely correct, but the statement is true enough for our purposes here. We will not prove the fact in this book, but will exploit it to give an abstract ILP formulation for the *rSPR* distance problem. Use of that fact is at the heart of Task A in the solution to the *rSPR* distance problem (see Figure 15.3 for an example).

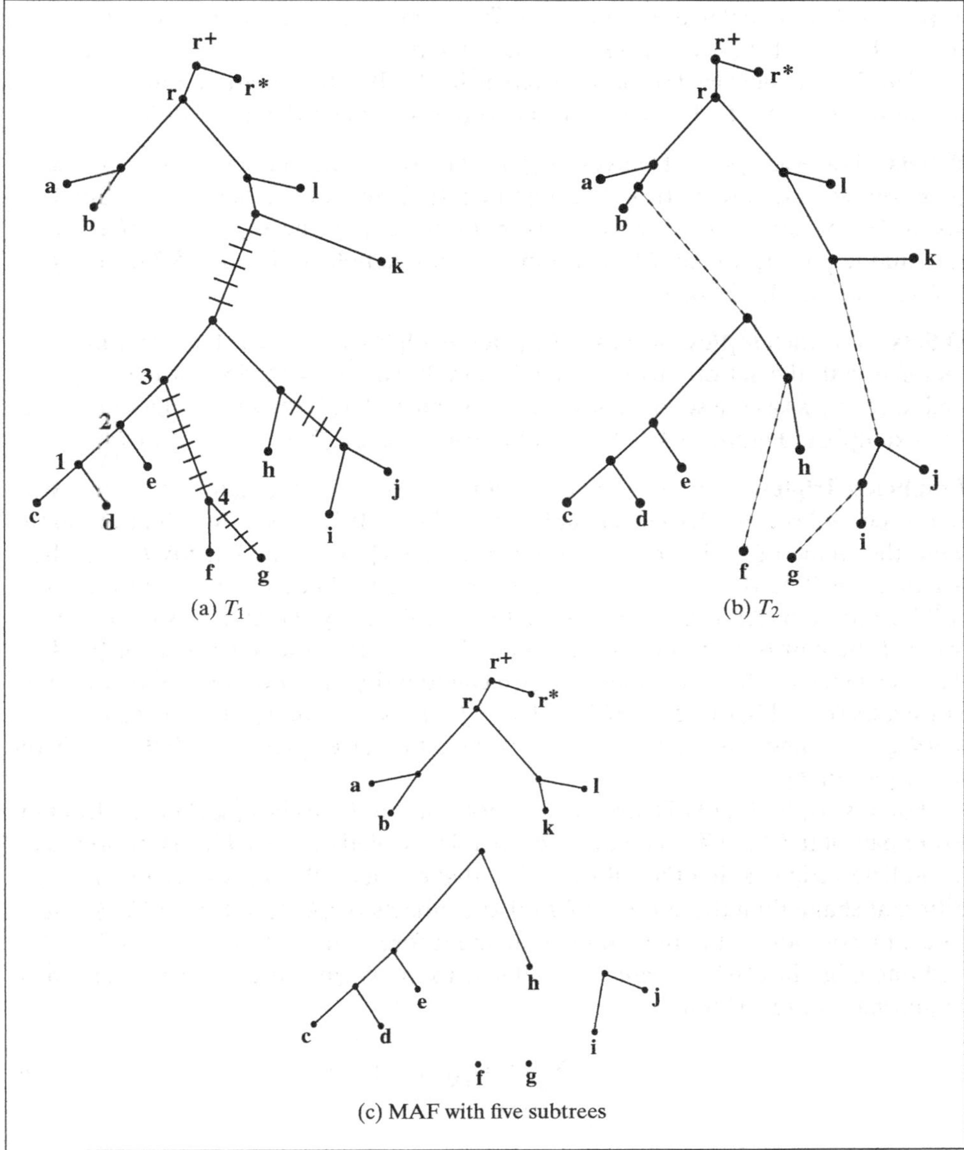

Figure 15.3 Panels (a) and (b) Show Two Trees T_1 and T_2 with the Same Leaf Set. Although it is not shown, the edges are directed away from the root, labeled r^+ in both T_1 and T_2. Deleting the four edges in T_1 marked with perpendicular lines, and then removing any node with only one out edge, results in the five subtrees shown in panel (c). The same five subtrees result after deleting the dashed edges in T_2, and then removing any node with only one out edge. The subtrees are called a *Maximum Agreement Forest (MAF)* of (T_1, T_2).

15.4.3 An Abstract ILP Formulation for the rSPR Distance Problem

The Variables For each edge e_1 in T_1, we create a binary ILP variable $D_1(e_1)$; and for each edge e_2 in T_2, we create a binary ILP variable $D_2(e_2)$. We want an ILP formulation where variable $D_1(e_1)$ (or $D_2(e_2)$) will be set to 1 to indicate that edge

e_1 (or e_2) *should be* deleted from T_1 (or T_2) to create a MAF for (T_1, T_2); and set to 0 to indicate that the edge *should not* be deleted.

The D variables are the only variables in the ILP formulation, but we need to define several sets of edges and nodes in order to explain the inequalities.

P Sets For each *pair* of leaves (i,j) in T_1, we define the set, $P_1(i,j)$, of edges (considered as undirected) on the unique path from leaf i to leaf j in T_1. Similarly, given T_2, for each pair of leaves (i,j) in T_2, we define the set $P_2(i,j)$ of edges on the unique path from leaf i to leaf j in T_2. For example, in Figure 15.3a, $P_1(d,f) = \{(d,1),(1,2),(2,3),(3,4),(4,f)\}$.

Q Sets For each *triple* of leaves $\{i,j,k\}$, define $Q_1(i,j,k)$ as the set of edges in $P_1(i,j)$ *unioned* with the set of edges in $P_1(i,k)$. Similarly, for each triple of leaves $\{i,j,k\}$, define $Q_2(i,j,k)$ as the set of edges in $P_2(i,j)$ unioned with the set of edges in $P_2(i,k)$. For example in Figure 15.3a, $Q_1(d,f,g) = \{(d,1),(1,2),(2,3),(3,4),(4,f),(4,g)\}$.

Conflicted Triples Consider the set of *three* leaves, say $\{i,j,k\}$, in T_1; and consider the three nodes $lca(i,j), lca(i,k)$, and $lca(j,k)$ in T_1. It is easy to see, that no matter what the shape of T_1 is, or where the leaves $\{i,j,k\}$ are in T_1, exactly *two* of those *lca* values will be the same (call it node u), and the other *lca* value (call it node v) will be different. And, it is easy to see that u will always be an *ancestor* of v in T_1. We call the pair of leaves in $\{i,j,k\}$ whose *lca* is node v, the *lower* pair of $\{i,j,k\}$ in T_1. Using exactly the same reasoning, we can identify the lower pair of $\{i,j,k\}$ in T_2. For example, in Figure 15.3a, if $i = d, j = f$, and $k = g$ in T_1, then u is $lca(d,f) = lca(d,g) = 3$, and v is $lca(f,g) = 4$, so (f,g) is the lower pair in T_1, but (d,f) is the lower pair in T_2.

For any triple $\{i,j,k\}$ of leaves, if the lower pair of $\{i,j,k\}$ in T_1 is *different* from the lower pair of $\{i,j,k\}$ in T_2, then the triple $\{i,j,k\}$ is called *conflicted*. The importance of a conflicted triple is that the subtree of T_1 that connects the leaves $\{i,j,k\}$ has a very different shape than the subtree of T_2 that connects $\{i,j,k\}$ (see Figure 15.4). Therefore, in a solution to the maximum agreement forest, at least one edge in $Q_1(i,j,k)$ and one edge in $Q_2(i,j,k)$ must be deleted to break up those two subtrees. Those requirements are written as:

$$\sum_{e_1 \in Q_1(i,j,k)} D_1(e_1) \geq 1. \tag{15.16}$$

$$\sum_{e_2 \in Q_2(i,j,k)} D_2(e_2) \geq 1. \tag{15.17}$$

It will turn out, that for any triple $\{i,j,k\}$, only one of those inequalities is actually needed, because the other one will be implied by the other inequalities in the ILP formulation. So in the final formulation, we only include inequality 15.16.

Also, for any MAF for (T_1, T_2) if node pair (i,j) are *not* together in the same subtree of T_1 after the deletion of the specified edges in T_1, it must be that (i,j) are also *not* together in the same subtree of T_2 after the deletion of the specified edges in T_2. That requirement can be written as:

$$\sum_{\text{edge } e_1 \in P_1(i,j)} D_1(e_1) \;\leq\; |P_1(i,j)| \;\times \sum_{\text{edge } e_2 \in P_2(i,j)} D_2(e_2), \tag{15.18}$$

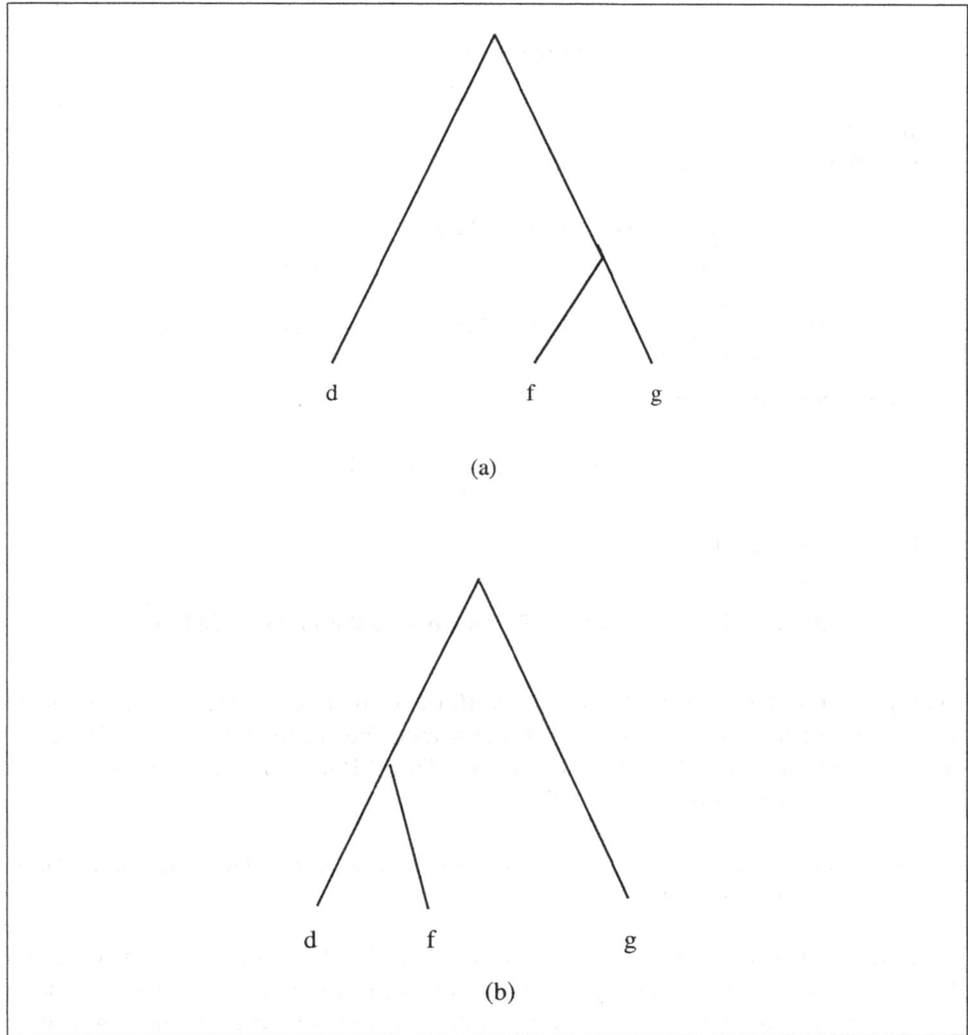

Figure 15.4 Panel (a) Shows the Leaf Triple {*d,f,g*} in T_1, with Lower Pair *(f,g)*. Panel (b) Shows the Same Triple in T_2, with Lower Pair *(d,f)*. Thus, the triple {*d,f,g*} is conflicted.

$$\sum_{\text{edge } e_2 \in P_2(i,j)} D_2(e_2) \leq |P_2(i,j)| \times \sum_{\text{edge } e_1 \in P_1(i,j)} D_1(e_1). \tag{15.19}$$

Inequalities (15.16), (15.17), (15.18), and (15.19), express conditions that *must* hold in a MAF, but in fact they are also *sufficient* to define a MAF. That is, if the D variables are set so that the four conditions are satisfied, then the set of edges whose associated D value is 0, will be a MAF for (T_1, T_2). We will not prove that in this book. Figure 15.5 shows the abstract ILP formulation for the *rSPR* distance problem, using the logic developed above for finding a MAF.

This abstract ILP formulation has been implemented in a computer program program (written in C++ by Yufeng Wu) that creates concrete ILP formulations

$$\text{Minimize} \sum_{e_1 \in T_1} D_1(e_1)$$

subject to:
For each pair of leaves (i,j),

$$(i) \quad \sum_{\text{edge } e_1 \in P_1(i,j)} D_1(e_1) \leq |P(i,j)| \times \sum_{\text{edge } e_2 \in P_2(i,j)} D_2(e_2)$$

$$(ii) \quad \sum_{\text{edge } e_2 \in P_2(i,j)} D_2(e_2) \leq |P_2(i,j)| \times \sum_{\text{edge } e_1 \in P_1(i,j)} D_1(e_1)$$

For each conflicted triple $\{i,j,k\}$,

$$(iii) \quad \sum_{e_1 \in Q_1(\{i,j,k\})} D_e \geq 1$$

All variables are binary.

Figure 15.5 An Abstract ILP Formulation to Compute *rSPR*(T_1,T_2).

when given the description of two trees with the same leaf set. We ran the program with the input trees from Figure 8.1. It confirmed the result stated in [77], that the minimum number of rSPR operations is five. The ILP has 409 inequalities, and was solved in 0.05 seconds by Gurobi 6.5.

Exercise 15.4.1 *Explain why inequality (15.17) is not needed in the abstract ILP formulation for the rSPR distance problem.*

In our discussion of the approach to solving the *rSPR* distance problem, we stated that we could solve it by finding the minimum (*equal*) number of edges to remove from both T_1 and T_2, so that the two resulting sets of subtrees are the same, after removing any node with only one out edge. However, the abstract ILP in Figure 15.5 does not have any inequalities that explicitly enforce the requirement that the number of edges removed from T_1 equals the number removed from T_2. That equality would be implemented by the addition of

$$\sum_{e \in T_1} D_1(e) = \sum_{e \in T_2} D_2(e).$$

Exercise 15.4.2 *It turns out that this explicit* equality *is not needed in the ILP formulation for the* rSPR, *because the other inequalities in the formulation force it to hold. Explain why this is true.*

16

Return to Steiner Trees and Maximum Parsimony

In Chapter 5, Section 5.1, we discussed the *maximum parsimony* problem in phylogenetics, and gave two ILP formulations to solve it. We said there that those ILP formulations were essentially formulations for the more general *Steiner-tree* problem on graphs, restricted to graphs that are hypercubes. In this chapter, we make a more explicit connection between the general Steiner-tree problem on graphs and several problems that arise in computational biology. This connection will be pursued in the next chapter as well. We start with definitions and exercises, before discussing biological applications.

16.1 THE STEINER-TREE PROBLEM AND EXTENSIONS

In the general Steiner-tree problem, the input is an *undirected* graph $G = (V, E)$, and a subset of nodes $V' \subseteq V$. A Steiner tree for V' is a tree, T, contained in G that connects *all* of the nodes in V', possibly using some nodes not in V', i.e., in $\{V - V'\}$. Nodes in T that are not in V' are called *Steiner nodes*. Note that in this definition, there is *no* specified root node. If the application requires a specific root node r, as in the maximum parsimony problem, then r can be added to V if it is not already in it.

An *optimal* Steiner tree T for V' is a Steiner tree for V' the minimizes the *number* of edges in the tree over all Steiner trees for V'. Equivalently, an optimal Steiner tree for V' minimizes the number of Steiner nodes over all Steiner trees for V'. The *Steiner-tree problem* on graphs is the problem of finding an optimal Steiner tree, given a graph $G = (V, E)$ and a subset $V' \subseteq V$.

Exercise 16.1.1 *We claim that the* maximum parsimony problem (MPP) *discussed in Section 5.1.4 is a specialization of the Steiner-tree problem, where given an n-by-m matrix M of binary sequences, the graph G is the hypercube H_m, and the nodes in V' correspond to the sequences in the rows of M, along with the node for the all-zero sequence. Explain and justify this claim in detail.*

Conversely, we can see that the ILP formulation given for the MPP, can be used to solve the general Steiner-tree problem.

Exercise 16.1.2 *Although the MPP is defined on a graph G that is a hypercube, it should be clear that this assumption is not essential in its ILP solution, so the abstract ILP formulations in Figure 5.3 and 5.5 can be used to solve the general Steiner-tree problem on any graph G.*

Explain this claim in detail. The ILP formulations in Figures 5.3 and 5.5 refers to a root and to a set of leaves, so you will need to specify which nodes in the general Steiner-tree problem are considered leaves, and what the root node should be.

Exercise 16.1.3 *Generalizing the Steiner-tree problem further, sometimes the edges in the given graph* $G = (V, E)$ *are* directed, *rather than undirected, and it is possible that for a pair of node* (u, v)*, only one of the directed edges* $< u, v >$ *or* $< v, u >$ *is in G. Then, the problem comes in two variations: either a specific root node r is given, or a root r is determined during the solution of the Steiner-tree problem. In either case, the input contains a set of nodes* V'*. It is required to find a smallest set of edges* $E' \subseteq E$ *so that every node in* V' *can be reached via some* directed *path from r. Such a directed tree is also called an* arborescence.

Create ILP formulations for these two variants of the Steiner-tree problem, by modifying both of the ILP formulations in Figure 5.3 and 5.5.

Exercise 16.1.4 *In our next variation of the Steiner-tree problem, the edges in the input graph* $G = (V, E)$ *are* directed, *and each edge has a given* weight. *Then, given a root r and a subset of nodes* $V' \subseteq V$*, it is required to find an arborescence, T, that contains a directed path from r to each node in* V', minimizing *the* sum *of the weights of the edges in T. This is called a* minimum-weight arborescence for V'. *Now what must be changed in the answer to the previous exercise?*

Sometimes $V' = V$*, in which case an arborescence connecting the root to the nodes in* V' *is called a* spanning arborescence. *Is the abstract ILP formulation to find the minimum-weight spanning arborescence different in any significant way from the ILP formation for the general minimum-weight arborescence problem?*

Exercise 16.1.5 *In some applications in biology, the Steiner-tree variant in the previous exercise is extended as follows. In the objective function, each directed edge* $< u, v >$ *contributes its weight, if* $< u, v >$ *is in the selected tree T; plus a cost* $c(u, v)$ *times the number of paths in the solution that traverse edge* $< u, v >$*. This objective function is called a* local-global hybrid.

Detail an abstract ILP formulation this local-global variant of the Steiner-tree problem.

On to Biological Problems We next discuss two additional problems in computational and systems biology which are essentially variants of the Steiner-tree problem and are solved using integer linear programming.

16.2 IPOINT: DEDUCING PROTEIN PATHWAYS

Knockouts A classic technique in molecular genetics, widely used to deduce molecular pathways and subnetworks, is to *disable* a particular gene in some organism, and then see what, if any, consequences there are, under different experimental conditions (temperature, abundance of food, etc.). The disabled gene is called a *knockout*, and one of the possible consequences is the *reduced* or *increased* expression of some other gene(s). When this happens, the conclusion is that the knockout gene is involved (upstream) in some molecular pathway(s) which *ultimately* lead to (downstream) gene(s) whose expression level is changed. But, there likely will be many unseen steps (involving many genes, proteins, or regulatory elements) in the pathway(s) leading from the knockout gene to the gene(s) with changed expression. Data from many knockout experiments are then required in order to fully deduce the pathways, and in such projects, many thousands of individual knockout experiments are conducted. Additional information is sometimes obtained by knocking out a *pair* of genes together at one time.

Knockout experiments are often difficult, tedious, slow, and expensive. Moreover, without a set of candidate genes or proteins that *might be* in the pathways of interest, many of the knockout experiments will contribute no information about those pathways. This knockout approach is a kind of brute-force, blind search. However, an easier (computational) approach has been proposed and explored. That approach is possible when good protein-protein interaction (PPI) networks, or gene-transcription networks, are available for the proteins or genes in the organism. One such approach that uses integer programming is discussed in [11], and implemented in a program called *iPoint*.

The Goal We denote the knockout gene as g, and the set of genes whose expression levels change after the knockout of g, as $\mathcal{F}(g)$. The goal is to computationally deduce plausible pathways that extend from gene g to the set of genes $\mathcal{F}(g)$.

The Key Idea The idea, first articulated in [209], is to cast the problem as a Steiner-tree problem where the graph G is the *known* interaction network (e.g., a PPI network or a gene-transcription network). The root node, r, for the problem instance, is the knockout gene, g, (or its protein product in the case of a PPI network); and the set of nodes V' are the nodes corresponding to the genes in $\mathcal{F}(g)$. An optimal Steiner tree for that data is then a conjectured, possible subnetwork, or set of overlapping pathways, that connect the knockout gene to the affected genes. This is plausible because the edges in the interaction network represent known biological interactions between genes and/or proteins in the organism. It is therefore plausible that a *true* pathway from the knockout to an affected gene is a path in the interaction network.

An optimal Steiner tree from root g to the nodes corresponding to the affected genes, represents the *smallest, most reduced* plausible set of paths that can explain the observed data. And even if the Steiner tree is not correct in all details, it identifies genes and proteins that are plausibly on the true pathways, and hence identifies good candidates for further knockout experiments.

But There Are Always Complications The key ideas and the basic framework of the method are stated above. But, as in almost all data analysis problems in biology, there are many complicating issues. First, *weights* for edges in the interaction network should be added, and are added in [11], to reflect levels of confidence in the data, or how strongly or frequently an interaction has been observed under the experimental conditions of interest, etc.

Second, some data in interaction networks only indicate that two proteins and/or genes interact, without more causal detail; but some data indicate the temporal or causal order of the interaction. The former case is represented by an undirected edge in the network, and the latter case corresponds to a *directed* edge. So, the computational problem to solve is the (partially directed), edge-weighted Steiner-tree problem.

Third, computational experiments in [11] done with real data where the causal pathways are known and can be used to evaluate the fidelity of the Steiner-tree approach, suggest that a *local-global* objective function (introduced in Exercise 16.1.5) is more effective at identifying plausible pathways. Finally, the optimal Steiner tree produced by the ILP computation will generally not be unique. In [11], up to 35 *co-optimal or near-optimal* Steiner trees were found, and their edge sets merged, creating a subnetwork rather than a single tree. And, of course, there is always the issue of false negative data (i.e., missing edges), and/or false positive data (i.e., edges in the network that should not be there).

Empirical Evaluation Two biological systems were examined in [11], including *Huntington's* disease, a neurodegenerative disease in humans caused by progressive mutations in the Huntington (HTT) gene. The PPI network that was used had over 10,000 proteins (nodes), close to 45,000 protein-protein interactions (edges), and over 4,000 protein-DNA interactions (additional edges). While the number of edges seems large, it is helpful to realize that this network is extremely sparse.[1]

The variant of the Steiner-tree problem discussed in Exercise 16.1.5 was solved using an ILP formulation that is similar to, but a bit different from, the formulation in Figure 5.5. The node for the HTT gene was the root of the Steiner tree and V' consisted of nodes corresponding to proteins whose expression level in humans with Huntington's disease showed at least a two-fold difference from expression levels in humans without the disease. The computation time to find 35 different solutions took under 2 hours, showing the practicality of this approach. The way that the solution was validated, using known biological information, is discussed in [11], and is outside the scope of this book.[2]

16.3 MAXIMUM PARSIMONY AND DUCTAL CARCINOMA PROGRESSION

The maximum parsimony criterion, together with integer linear programming, was used in [36] to deduce plausible evolutionary histories of *invasive ductal carcinoma (IDC)*, a form of breast cancer, along with *ductal carcinoma in situ (DCIS)*, a precursor of IDC. There were several notable biological features of this problem and data, including one particular feature that makes integer linear programming uniquely suited and practical for this problem. These biological features were exploited in [36] to convert what at first looks like a standard Steiner-tree problem on an impractically huge graph, to a simpler problem on a much smaller graph. Thus, this example illustrates the productive interplay between biological modeling and ILP engineering.

16.3.1 Biology and Data

The Biology The first notable feature of this study is that it used data on *copy number variations (CNVs)*. As the name suggests, genomes can have multiple copies of particular genes, and that number can change through duplications or deletions of genome intervals. Moreover, certain stresses on the organism, including cancer, induce changes in copy numbers. So, the number of copies of a particular gene or region can be used as a *nonbinary* phylogenetic character to deduce an evolutionary tree representing a putative history of the genes.

In particular, as DCIS and IDC progress, certain genes (called *oncogenes*) that encourage uncontrolled cell replication, increase in number; and certain genes (called *tumor suppressor genes*) that normally control cell replication, decrease in number. Of note, the tumor suppressor gene, *TP53*, which is central in the progression of many types of cancer, is very important in the development of IDC. Hence, several

[1] The number of node pairs, and hence potential edges, is close to 50 million.
[2] But, the authors conclude that the pathways found were biologically meaningful. That is not surprising. If the results weren't good, we would not be talking about this paper.

properties of the change in the copy number of TP53 are given special attention in the model of IDC development in [36].

> Numerous studies have shown ... losses of TP53 mechanistically promote instability in the genome, and hence variations of the copy number of TP53 deserve special attention.

The second notable feature of this study is that the data is obtained by taking *single-cell* samples from individuals with IDC. Single-cell sampling is in contrast to simpler, prior, *bulk-sampling* methods that sample many cells (up to 1 million) at a time. When many cells are sampled together in a mixture, the data shows the mutations that are present in some of the cells in the mixture, but can't identify which mutations occur together in any *single* cell. This makes it difficult to interpret the history of the mutations if cells in the sampled mixture are not identical. While single-cell sampling is currently more difficult and expensive than bulk sampling, single-cell cancer data is increasingly being collected, and will be more common as the sampling technology improves and the clinical value of single-cell sampling is established.

The third biological feature, and most important in its relation to integer programming, is that in great contrast to the evolutionary history of most species, cancer cells usually contain *both ancestral* and *descendant* taxa. In most other evolutionary systems, we can only see the *extant* taxa – the ancestral taxa is extinct and must be *deduced* from the extant taxa and an evolutionary model. For example, we see animals that exist today, and we know a bit about ancestral animals from bones and fossils, but generally any animal that is an ancestor of an animal species we see today, is extinct. However,

> ... in tumorigenesis, ancestor and descendant clones may co-exist throughout tumor progression. [36]

So, the problem of deducing the evolutionary history of DCIS and IDC progression, is exclusively a problem of deducing a tree structure and temporal order of the *known* CNV data. We do not, in addition, need to deduce unseen, putative ancestral taxa.

The Data Each data item (taxon) in this study is a list of *eight* numbers, which are the number of copies of five oncogenes and three tumor-suppressor genes in a sampled cell.[3] Such a list of eight numbers (a taxon) is called a *CNV list*. A dash in a CNV list separates the counts for oncogenes from the counts for tumor suppressor genes. A healthy cell should have the *healthy CNV* list $(2,2,2,2,2--2,2,2)$, since a healthy cell should have exactly two copies of each gene (one on each of the two copies of each chromosome). Then, for example, CNV mutations in the cell might change a healthy CNV list to $(2,2,4,7,2--0,2,2)$. Note that the number of copies of an oncogene can *decrease* for a while, and the number of copies of a tumor suppressor gene can *increase* for a while. In fact, there is a phenomenon (called a *doubling-loss event*), where the number of TP53 genes increases in a cell; and then the number of copies of most genes in a descendant cell doubles; followed by a loss of copies of one or more genes. Life is not always simple.

[3] The oncogenes are COX2, MYC, HER2, CCND1, and ZNF217; and the tumor suppressor genes are TP53, DBC2, and CDH1.

Distances between Taxa The focus of the study in [36] is on using fine-grained single-cell CNV data to deduce the *temporal progression* of the CNV mutations in cancerous cells. A *distance* function is proposed, which gives a distance from one CNV list to another CNV list. Suppose one CNV list is denoted $(s_1, s_2, .. , s_8)$ and the other is denoted $(t_1, t_2, \ldots, t_8)$. Then the *naive distance* from the first to the second CNV list is:

$$\sum_{i=1}^{i=8} |s_i - t_1|.$$

For example, the naive distance from the healthy CNV list to and list $(2, 2, 4, 7, 2 - -0, 2, 2)$ is $0 + 0 + |-2| + |-5| + 0 + |2| + 0 + 0 = 9$.

The naive distance is *symmetric*, but a more biologically informed distance need not be. That is, the distance from one list A to another list B might be different from the distance from list B to list A, since the biological events that increase a copy number may be different from the biological events that decrease it.

The actual distance between CNV lists used in [36] is an extension of the naive distance, but is more biologically informed and more mathematically complex. We will not discuss further details of that distance, but will assume that it is not symmetric, and note that the distance is designed to model what is known about changes in copy numbers in cancer cells.

16.3.2 More on the Biological Model

In addition to the feature already noted, that "invasive tumors still contain cancer cells from earlier progression steps," the model of CDIS and IDC progression in [36] contains several useful features. One is that the copy number of a gene can increase or decrease, but if ever it goes to *zero*, it does not change. A second is that taxa contained in cells with DCIS should occur *before* (in time) taxa contained in cells with IDC, although real cell samples do not always catch the time point where the cell has DCIS but not IDC. A vital third feature is that the density of the sample is high enough in each cell so that

> ... the cells sampled in each dataset are a reasonable representation of the whole tumor. [36]

The model also makes several very precise assumptions about the effect of changes in the copy number of TP53, which we will not discuss here. The authors include these, and other, assumptions in order to make the biological model more accurate, but also

> ... to illustrate the kind of complex constraints for which the ILP approach is especially well suited. [36]

The Maximum Parsimony Approach to CNV Reconstruction Given the input CNV lists of the observed copy numbers in a cell, the goal is to construct a *directed* tree, rooted at the healthy CNV list, where each node in the tree is labeled by a distinct CNV list, and each CNV list labels a distinct node. The distance of a directed edge in the tree is the distance (discussed above) given by the CNV lists that label the two end points of the directed edge. The tree must produce each of the input CNV lists, using changes (mutations) in copy numbers that obey the biological model, and *minimize* the total distance on the edges in the tree. This is the *maximum parsimony problem for CNV reconstruction.*

Recall that maximum parsimony is the criterion used earlier in Section 5.1, where the maximum parsimony problem was cast as a *Steiner-tree* problem on a *hypercube*. The hypercube represented the space of all possible taxa, both observed and unobserved.

Exercise 16.3.1 *Now that we have a more formal definition of the Steiner-tree problem, express the maximum parsimony problem for CNV reconstruction as an instance of the Steiner-tree problem. That is, what is the underlying graph, and what is the set V', and what is the root?*

Spoiler Alert: We effectively give the answer next.

In principle, the approach used for the maximum parsimony problem could be used to solve the maximum parsimony problem for CNV reconstruction. That approach would use a graph that again represents the space of all possible CNV taxa. Such a graph would be a generalization of a hypercube, allowing more than binary values at each site. Note however, that the graph would be *impractically* large. Even though there are only eight sites, the range of possible copy numbers at any site is large – up to 15 in the examined data in [36]. Denoting the range of possible values as z, the number of nodes in the generalized hypercube would be z^8, which is more than 2.5 billion for $z = 15$. So, at first, it appears that the maximum CNV parsimony criterion will not allow practical computation. But here is where we see the productive interaction of the underlying biology, and practical issues in solving ILP formulations.

The Key to Practicality The two biological assumptions are that "ancestor and descendant clones may coexist throughout tumor progression" and "the cells sampled in each dataset are a reasonable representation of the whole tumor." These two assumptions imply that a generalized hypercube containing nodes for *all* possible taxa is *not* needed. Instead, we can find the maximum CNV parsimony tree on a graph $G = (V, E)$ that *only* contains nodes for the healthy CNV list, and the *observed* CNV lists, i.e., CNV lists that are in the problem input. So, we can think of this as a Steiner-tree problem on a weighted, directed graph where *every* node in the graph is in the input set V' of nodes that must be in the tree. Hence, the problem can be viewed as a variant of the *minimum-weight spanning arborescence* problem.

The ILP Formulation Given the above comments, the ILP formulation developed in Exercise 16.1.4 could be the basis for a solution to the maximum parsimony problem for CNV reconstruction. This would be practical, depending on the size of the graph G. In the case of the data examined in [36], there were at most 207 taxa, so the graph would only have 208 nodes and under 100,000 directed edges (one for each ordered pair of nodes). That is a manageable number of nodes and edges for current ILP solvers, using the ILP formulations for minimum-weight spanning arborescence problem.

The ILP formulation used in [36] essentially follows this approach, but is a bit different from the formulation suggested by Exercise 16.1.4. As discussed earlier, the model of CNV evolution contains several very specific constraints that are not part of the general Steiner-tree problem, and the ILP formulation needs to reflect those constraints. If all of the biology of the CNV evolution problem could be encoded in the edge weights of the underlying graph, then the maximum parsimony problem for CNV reconstruction could be solved as an instance of the minimum-weight spanning arborescence problem.[4] Unfortunately, not all of the biological constraints can be encoded in edge weights, and it is these complex constraints that require the use of integer programming: "ILP tools provide a way to describe and efficiently solve for optimal tree models in the presence of such complex constraints" [36].

16.3.3 Biological Conclusions

The work reported in [36] is mainly methodological, developing the approach to study CNV evolution via maximum parsimony and integer programming. The methods should reveal more biology when more data becomes available. And, the methods are also expected to find application beyond the study of IDC and DCIS. Still, what are the current biological conclusions?

Unfortunately, the biological conclusions are very preliminary, and somewhat speculative, due to limited data. But the authors state that empirical results suggest that "... progressions estimated from ... affected individuals are non-random and classifiable into several categories seemingly distinguished by distinct selective pressures and distinct mechanisms ..." And, "The complex and heterogeneous evolutionary landscape they reveal may have important implications for strategies for cancer treatment." The reader who is more interested in the biology than in the methodology, is invited to consult the original paper.

[4] This would be great, because the minimum-weight spanning arborescence problem on an arbitrary, directed graph can actually be solved by a (worst-case) efficient, and practical algorithm [123].

17

Exploiting and Leveraging Protein Networks

A huge amount of data about the biology and chemistry of proteins has been collected and organized in the form of *protein-protein interaction (PPI)* networks. But the importance of those networks is far greater than just repositories of raw data. Many creative ways have been devised to *exploit* PPI networks to answer questions of great interest in biology, and to *leverage* what has been learned from observations and experiments, to make deductions and conjectures where observations and experiments have *not* been conducted. In this chapter, we examine three examples, and detail the role of integer linear programming in those efforts.[1]

17.1 EXAMPLE 1: EXPLOITING PPI NETWORKS TO FIND DISEASE-RELATED PROTEINS

> Proteins and other biological molecules rarely work alone; they brush up against one another in fleeting interactions or band together to form complex cellular machines. Only through such partnerships can proteins perform their many functions. Breakdowns in those interactions can affect human health. [68]

In the work described in [133], the goal is to find proteins that are involved in certain human diseases, where that involvement was *not* previously known. Two things are assumed to be known:

(a) A large PPI network $\mathcal{N}$, where each node represents a specific protein, and each edge represents a pair of proteins that are known to interact in some biologically significant way.

(b) A subset of proteins, S, in network $\mathcal{N}$ that are *known* to be involved with a specific disease, denoted D.

Network $\mathcal{N}$ and subset S are used to try to identify *additional* proteins that are involved with disease D. The high-level *idea* is to find proteins *not* in S that *co-occur* with proteins in S, in cliques, or in other high-density subgraphs of $\mathcal{N}$. The following quote from [133] explains the biological basis for this idea:

[1] Leveraging of protein networks was also illustrated in the discussion of program *iPoint* discussed in Section 16.2 of the previous chapter; and in Section 13.2.3, in the discussion of the *function assignment problem*.

> Many studies, ... link diseases to dysfunctions of *assemblies of proteins* working in concert. ... Therefore, a more systematic understanding of certain disorders could be achieved by looking directly for related protein *complexes*, rather than focusing on single proteins. (italics added)

17.1.1 In More Detail

The PPI networks were obtained by integrating information from several existing protein and disease databases containing thousands of proteins and over 100 diseases. For a given disease D, and a given PPI network $\mathcal{N}$, the set S is the subset of nodes (proteins) in $\mathcal{N}$ that are known to be associated with disease D. Then, two types of weights are computed: Each node i in $\mathcal{N}$ is weighted by a measure of its *distance* in $\mathcal{N}$ from the nodes in S; and each edge (i,j) is weighted to reflect the amount of *evidence*, or our *confidence*, that proteins i and j interact in the development of disease D. Thus, edges with large weights are likely to identify protein interactions related to D, while edges with small weights are *not known* to be related to D, but might be. Both the node weights and the edge weights contribute to the weight of a subgraph of $\mathcal{N}$. See [133] for the biological and mathematical details of the weighting.

For our purposes, we just need to know that $\mathcal{N}$ is a node and edge-weighted graph, where the weight of a node reflects the distance of the node from nodes in S, and the weight on an edge (i,j) reflects an estimate of how much proteins i and j are *known* to interact in the development of disease D. Finally, given weighted graph $\mathcal{N}$, the method in [133] finds *high-weight, high-density subgraphs*. Any node in such a subgraph that is *not* in S, and hence is not already known to be associated with disease D, identifies a protein p that is *conjectured* to be associated with disease D. Additional computational or laboratory investigations can then verify or refute that conjecture.

Epilepsy The work in [133] examined several diseases in detail. Figure 17.1 shows 10 high-density subgraphs found in the study of *epilepsy*. Nodes represented by diamonds indicate proteins already known to have a role in epilepsy; nodes represented by ovals indicate additional proteins conjectured to play a role in the disease. The weight of an edge is represented by its width.

17.1.2 Which High-Density Subgraphs?

We stated that high-weight, high-density subgraphs in $\mathcal{N}$ are used to identify proteins of interest, but, specifically, which definition of high-density subgraph is productive for this purpose? In the end, that is a question of good biological modeling, and needs to be evaluated empirically. In fact, if we find that certain kinds of subgraphs are better or worse at (correctly) identifying new proteins associated with a disease, that tells us something about the mechanics or nature of the disease, as well as the proteins involved with it. So, deviating from the specific details used in [133], one could try many specific definitions, from cliques to near cliques to subgraphs with high ratios of the number (or total weight) of edges to the number of nodes, etc.

But, getting back to integer programming, and to the spirit of the approach in [133], we define *high-density* in a way that is close to the one used there.

> Find the largest subset of nodes, V', in a network $\mathcal{N} = (V, E)$, such that each node in V' is adjacent to *at least half* of the nodes in V'.

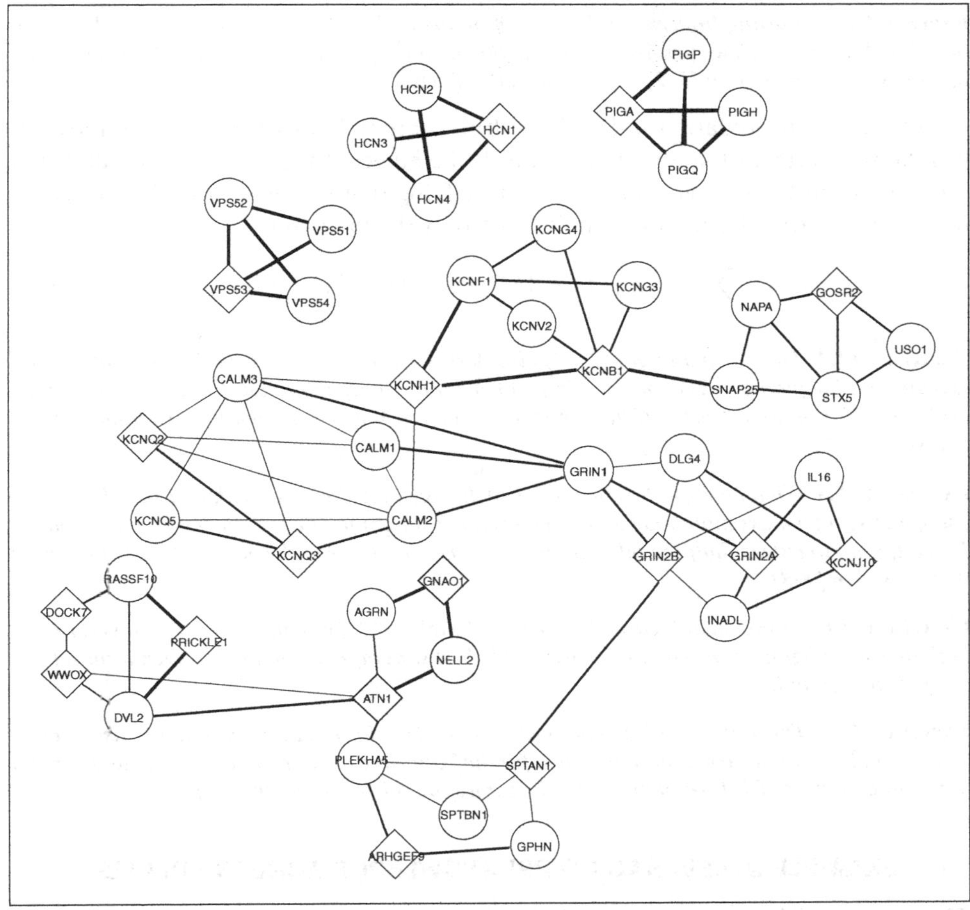

Figure 17.1 **Ten High-Density Protein Clusters Associated with Epilepsy. Diamonds represent proteins known to be associated with epilepsy; ovals represent additional proteins conjectured to be associated with epilepsy. (The figure is similar to one in [133]).**

17.1.3 ILP Implementation

How do we implement this as an ILP? As before, for each node $i \in V$, the ILP variable $C(i)$ will be set to one if and only if node i is chosen to be in V', and variable $E(i,j)$ will be set to one if and only if both nodes i and j are in V'. We introduce the integer ILP variable q and set its value in the equality:

$$q = \sum_{k=1}^{n} C(k). \tag{17.1}$$

So, the value of q will be $|V'|$, the size of V'. Then, the requirement that each node in V' be adjacent to at least half of the nodes in V' can be expressed as:

For each node i, variable $C(i)$ is set to value one, *only if* $\sum_j E(i,j) \geq q/2$.

To implement this requirement, we use the *Only-If* idiom for linear functions and binary variables, implemented in inequality (12.6). Then, recalling that $|V| = n$, and simplifying a bit, we get the inequality:

$$\sum_j E(i,j) - q/2 + n \times (1 - C(i))/2 \geq 0. \tag{17.2}$$

Exercise 17.1.1 *Starting from the Only-If idiom in inequality (12.6), derive the inequality given in (17.2). Also, although we verified in general that inequality (12.6) has no bad side effects, verify that the specific inequality (17.2) has no bad side effects.*

The actual formulation in [133] is a bit different from what is stated above. It has the requirement that every node i in V' be adjacent in $\mathcal{N}$ to at least half of the *other* nodes in V'. That is equivalent to specifying that $C(i)$ is set to value one, *only if* $\sum_j E(i,j) \geq (q-1)/2$. This is implemented by the inequality:

$$\sum_j E(i,j) - (q-1)/2 + n \times (1 - C(i)) \geq 0. \tag{17.3}$$

Exercise 17.1.2 *The inequality in (17.3) is also derived from (12.6). Verify that derivation, and explain that inequality (17.3) correctly implements the requirement that if node i is in V', then node i is adjacent to at least half the* other *nodes in V'. Also, verify that the inequality has no bad side effects.*

Exercise 17.1.3 *What needs to be changed if each edge in $\mathcal{N}$ has a weight, and the objective is changed to find a maximum node and edge-weighted subgraph H of $\mathcal{N}$, such that every node in H is adjacent to at least half the other nodes in H. This is closer to the actual objective function that is used in [133].*

Exercise 17.1.4 *As shown in Figure 17.1, the actual goal in [133] is to find a set of nonoverlapping high-density subgraphs, instead of just the single best subgraph. Explain how such a subset can be efficiently found.*

Exercise 17.1.5 *The authors of [11] comment that the ILP formulation used there (discussed in Section 16.2) is similar to the one developed for the Topo-free query problem. Detail and explain the ways that these ILP formulations are the same, and the ways that they differ.*

17.2 EXAMPLE 2: LEVERAGING PPI KNOWLEDGE ACROSS SPECIES

The Success The great success of bioinformatics and computational biology was initially due to the fact (discussed in Chapter 10) that many molecular *sequences* are very similar, or even the same, throughout a huge range of life. A *hemoglobin* molecule in humans is almost identical in sequence, function, and structure, to the *hemoglobin* molecule in chickens, for example.[2]

As computational biology matured, we learned that more than molecular sequence, structure, and function are conserved. Higher-order *patterns of interactions*, *pathways*, and *relationships* between molecules are also conserved. Sometimes such a pattern is called a *motif*, sometimes a *protein complex*, and sometimes a *system*. Often they occur in highly integrated, functional *pathways* consisting of a series of interacting proteins and/or genes, promoters, and transcription factors. Hence, complex relationships among proteins are likely to be the same, or very similar, in many different species. What we learn, perhaps at great expense in money, time, and cleverness, about molecular patterns in one species can likely be used to deduce patterns in other species. This *leveraging of knowledge* is the key to the success of modern computational biology.

The Disparity Much is known about certain species that are cheap and easy to work with (e.g., *yeast*), or are species of great interest to us (e.g., *H. sapiens*). But, there are

[2] Actually, I just wrote that without checking it, but I bet it is true.

a huge number of species that have gotten much less attention. Extensive protein-protein interaction data has been produced for certain species, but not for others. Given this reality, and the reality of pattern conservation (i.e., similarity), a major current goal in computational biology is to

> [u]se PPI networks that were developed *for one species*, and the patterns found in those PPI networks, to yield biological knowledge about protein interactions, complexes, pathways, and patterns in *other species*.

Here, we examine one such effort, from [31], which relies on integer linear programming (along with other optimization methods we will not discuss). The technical details of the ILP formulation are more involved than most of the formulations discussed in this book, but that adds to the pedagogical value of discussing them, in addition to the biological value of the work.

17.2.1 The Biological Motivation and Experimental Setup

The research in [31] concerns the identification and understanding of *protein complexes*. Recall that a protein complex is an *assembly* of proteins that work in *concert*. We will also need to remember the following biological observation:

> In a given species, a set of proteins, $\mathcal{P}$, that work together in a protein complex tend to be *highly connected* to each other in the PPI network for that species. At very least, with no errors in the PPI data, the proteins in $\mathcal{P}$ should form a *connected subgraph* in the PPI network.[3]

This observation, along with the similarity of protein complexes across species, are the two central biological assumptions in [31].

Finding Unknown Complexes As stated earlier, we have extensive PPI networks for proteins in *humans*, and *yeast* and *drosophila*, and certain other species. But, there are millions of species in the world, and PPI networks are lacking for most. So, suppose we have a set of proteins, $\mathcal{P}$, from a species for which we *do not* have a PPI network. For concreteness, suppose the species is *Mus Musculus* (a name that reminds me of *Mighty Mouse*). Suppose further, that we conjecture that the proteins in $\mathcal{P}$, or many of them, are actually part of a protein complex in *Mus* (that conjecture is related to how the proteins in $\mathcal{P}$ were collected). So, we *expect* that *if* we did have a PPI network for *Mus*, the proteins in $\mathcal{P}$ would be represented by a highly connected subgraph of that network. But, lacking a PPI network for *Mus*, how can we use a known PPI network, $\mathcal{N} = (V, E)$, from a related species (say *humans*) to help support or help refute the conjecture that the proteins in $\mathcal{P}$ are part of a protein complex in *Mus*? That is the question addressed in [31].

Assigning Proteins in $\mathcal{P}$ to Nodes in $\mathcal{N}$ The approach in [31] begins by finding high *sequence similarities* between the proteins in $\mathcal{P}$ and proteins represented by nodes in $\mathcal{N}$.[4] So, for each protein p in $\mathcal{P}$, there is a set of nodes, denoted $V(p)$, in $\mathcal{N}$, which represent proteins that are *highly similar* to p. And, for each node v in $\mathcal{N}$, there is a

[3] This is also true for molecules that work together in pathways, motifs, and systems, but the paper discussed here concerns protein complexes.

[4] The authors point out that similarities other than sequence similarity, might also be used, but are not in this study.

set of proteins, denoted $\mathcal{P}(v)$, in $\mathcal{P}$ that are highly similar to the protein represented by v. Note that $V(p)$ could be empty, and that $\mathcal{P}(v)$ will be empty for many nodes in $\mathcal{N}$. This leads to the following computational problem.

The Topo-Free Query Problem

> Find a *connected* subgraph, K, of $\mathcal{N}$ with the *largest* set of nodes, such that for each node v in K, we can assign to v a *distinct* protein p in $\mathcal{P}(v)$.

That is, we want to find an *assignment* from proteins in $\mathcal{P}$ to nodes in K, where each node v in K is assigned a *single* protein in $\mathcal{P}(v)$, and *no* two nodes in K are assigned the same protein. Moreover, the nodes in K must induce a connected subgraph of $\mathcal{N}$. The reason is the following: Based on the biological model, if the proteins in $\mathcal{P}$ that map to K are actually part of a protein complex, and the proteins represented in $\mathcal{N}$ have biological properties that are similar to the proteins that map to K, then K should *ideally* induce a *high-density subgraph* of $\mathcal{N}$. But, the data and model are not always ideal, so a simpler, but weaker condition is used, that K at least induces a *connected* subgraph in $\mathcal{N}$.

As an example, in Figure 17.2, $\mathcal{P}$ contains four proteins, and for each $p \in \mathcal{P}$, the set of nodes, $V(p)$, is listed to its right. There is no solution to this Topo-free query problem of size *four*, i.e., that assigns all four of the proteins in $\mathcal{P}$ to a *connected* set of nodes in $\mathcal{N}$. However, there is a solution of size three, where $K = \{b, d, e\}$, assigning protein 1 to node d; protein 2 to node b; and protein 3 to node e. There is a different solution where $K = \{a, b, j\}$, assigning protein 1 to node a; protein 2 to node b; and protein 3 to node j.

What Do We Learn from a Solution? Before we discuss *how* we solve the Topo-free query problem, we should be clear on its value. If we find that the optimal K is of size close to $|\mathcal{P}|$ (maybe even equal to $|\mathcal{P}|$), that is strong evidence that $\mathcal{P}$ does indeed come from a protein complex. Conversely, if the largest K is very small compared to the size of $\mathcal{P}$, we cannot conclude that $\mathcal{P}$ comes from a complex.[5] Moreover, if the optimal ILP solution supports the conjecture that the proteins in $\mathcal{P}$ are part of a protein complex, the specific edges and shape of the subgraph of $\mathcal{N}$ induced[6] by K might indicate the interaction pattern of the proteins assigned to the nodes of K. In fact,

> Our method also suggests new complexes for which no prior experimental evidence is available … [31]

Of course, the example in Figure 17.2 is a toy. In real applications, the PPI networks are much larger, although the sizes of $\mathcal{P}$ used in [31] were as small as 4, and no larger than 25. Those sets of proteins (i.e., choices for $\mathcal{P}$) came from *mice*, *rats*, and *cows*, which (at the time the paper was written) lacked extensive PPI networks. The paper examined 59 subsets of proteins from mice, 55 from rats, and 10 from cows, in addition to subsets from known protein complexes that were used for calibration and control. In total, about 600 protein subsets were tested. The established PPI networks

[5] We can more exactly calibrate the significance of the size of the optimal K, by solving the Topo-free query problem using *randomly selected* sets of proteins, comparing the results obtained to the size of K obtained for the original $\mathcal{P}$.

[6] See Section 2.1.3 for the definition of an induced subgraph.

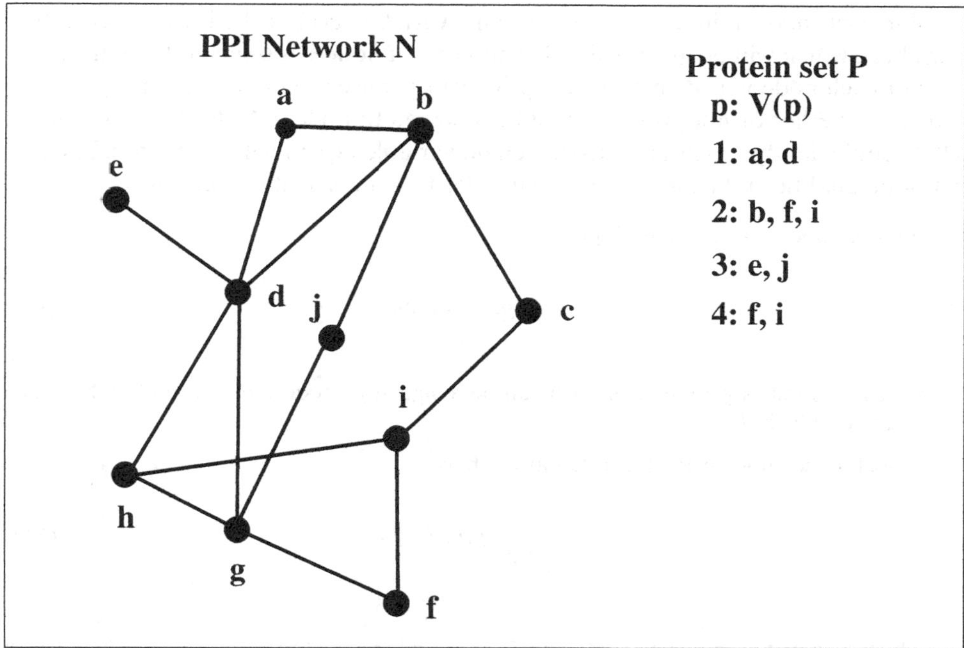

Figure 17.2 Example of the Topo-Free Query Problem. The name "Topo-free query" comes from the fact that there is no PPI network for the proteins in $\mathcal{P}$, and so their *topology* is unknown. They are free of a topology. The network is adapted from figure 1 in [31].

were from *yeast*, with 5,430 proteins and 39,936 interactions; *drosophila*, with 6,650 proteins and 21,275 interactions; and *human*, with 7,915 proteins and 28,972 interactions. Around five percent of nodes in $\mathcal{N}$ represented proteins that were similar to proteins in $\mathcal{P}$.

17.2.2 The ILP Formulation

First, we simplify somewhat the statement of the problem, and develop an ILP formulation for this simplified problem. We will later extend that formulation to solve the full Topo-free query problem. The simplification is that in addition to $\mathcal{P}, \mathcal{N}$ and the protein similarity data, the input will contain a *constant* number t, called the *target*. The problem then is to find, if there is one, a solution to the Topo-free query problem where the set K contains exactly t nodes.[7]

All of the inequalities used in the ILP formulation for this variant (but with a different objective function) are shown in Figure 17.3. Some readers might prefer to look at it now to see the overall structure of the formulation, but I will start by discussing the variables and inequalities a little bit at a time.

17.2.2.1 Node-Selection and Assignment Variables

There will be *five* types of variables in the ILP formulation. We first discuss two of them: the *node-selection* variables, and the *assignment* variables.

[7] This variant is closer to the actual version solved in [31].

For each node v in V, we use a binary variable, $c(v)$ (called a node-selection variable), which will be given value 1 if and only if node v is selected to be in set K.

For each node v in V, and protein p in $\mathcal{P}(v)$, we use the binary variable $g(p, v)$ to indicate whether or not protein p will be assigned to node v. Note that if p is not in $\mathcal{P}(v)$ (equivalently, v is not in $V(p)$), then no variable $g(p, v)$ will be created. The first set of inequalities relate assignment variables to node-selection variables.

For each node $v \in \mathcal{N}$ and each protein $p \in \mathcal{P}(v)$:

$$g(p, v) \leq c(v), \tag{17.4}$$

which says that a protein p in $\mathcal{P}(v)$ can be assigned to node v *only if* node v has been selected to be in K.

Also, for each node v, the formulation will have:

$$\sum_{p \in \mathcal{P}(v)} g(p, v) \leq 1, \tag{17.5}$$

which says that at most one protein in $\mathcal{P}(v)$ can be assigned to node v. Similarly, for each protein p:

$$\sum_{v \in V(p)} g(p, v) \leq 1, \tag{17.6}$$

which says that a protein, p, can be assigned to at most one node in $\mathcal{N}$; and that node must be in $V(p)$.

A Pause. Where Are We Now? The c and g variables defined above, together with inequalities (17.4), (17.5), and (17.6) will define a mapping from *some* of the proteins in $\mathcal{P}$ to *some* of the nodes in $\mathcal{N}$. The nodes in $\mathcal{N}$ that are mapped to, define the set K. That mapping assigns at most one protein in $\mathcal{P}$ to a node in $\mathcal{N}$, and each protein in $\mathcal{P}$ maps to at most one node in $\mathcal{N}$.

But, the critical condition needed for a solution to the Topo-free query problem, is that the set of nodes, K, that are mapped to, must induce a *connected* subgraph of $\mathcal{N}$. How can that requirement be implemented in the ILP formulation? The high-level idea is to create ILP variables and inequalities to ensure that the subgraph of $\mathcal{N}$ induced by K contains a *rooted, directed tree* that reaches every node in K, when the edges of the subgraph are directed appropriately. We next discuss that implementation.

17.2.2.2 The Root Variables and Inequalities

A solution to the ILP formulation will describe a rooted, directed tree consisting of directed paths running from a *root node*, denoted r, to each of the nodes in the selected set K. So, the ILP solution must identify the node r. To do that, the ILP formulation will have, for each node v in $\mathcal{N}$, a binary variable, $r(v)$, which indicates whether or not (and, it is "not" for all but one node) node v is the chosen root node, r. That is implemented with the following inequalities.

For each node v in $\mathcal{N}$:

$$\sum_{v \in \mathcal{N}} r(v) = 1. \tag{17.7}$$

The root must be in K, so for each node v in $\mathcal{N}$, we have

$$r(v) \leq c(v). \tag{17.8}$$

These inequalities say that exactly one node must be chosen to be the root, and a node v can be chosen as root *only if* v is also selected to be in K.

17.2.2.3 Edge and Path Variables and Inequalities

Edge Variables Next, the ILP formulation must specify which edges are *allowed* to be in the rooted, directed tree. For each (undirected) edge (u, v) in $\mathcal{N}$, we use a binary variable $e(u, v)$, which will be given value 1 if and only if *both* of its end points have been selected for K. That is, if and only if edge (u, v) is in the subgraph of $\mathcal{N}$ *induced* by the nodes in K. So, the ILP formulation will have for each edge (u, v) in $\mathcal{N}$:

$$\begin{aligned} c(u) + c(v) - e(u, v) &\leq 1, \\ 2 \times e(u, v) - c(u) - c(v) &\leq 0. \end{aligned} \tag{17.9}$$

These inequalities implement the familiar *If-Then* and *Only-If* idioms for binary variables implemented in inequalities (4.8) and (4.10), introduced in Sections 4.3.1 and 4.3.2. They ensure that $e(u, v)$ will be set to value 1 if and only if both $c(u)$ and $c(v)$ are set to value 1.

Path Variables Next, we define variables that (indirectly) specify the rooted, *directed* paths that we claimed will be specified by a feasible solution to the ILP formulation. The purpose of these paths is to ensure that the subgraph of $\mathcal{N}$ induced by the nodes in K form a *connected* subgraph of $\mathcal{N}$. Note that the paths can (and usually will) overlap and share nodes and edges.

For each edge (u, v) in $\mathcal{N}$, we use *two* integer-valued variables, $f(u, v)$ and $f(v, u)$, whose values are in the range 0 to $t - 1$ (remember that t is the fixed target). These variables are called *path* variables.

We will see that the path variables will determine a set of $t - 1$ rooted, directed paths in $\mathcal{N}$, starting at node r, with exactly one path *ending* at each node in the set $\{K - r\}$. But, we want the edges in the paths to only come from the subgraph induced by K. So, we need inequalities to ensure that a path variable $f(u, v)$ has value greater than zero, only if the corresponding edge variable $e(u, v)$ has value 1. This is implemented as: For each edge (u, v) in $\mathcal{N}$:

$$\begin{aligned} f(u, v) &\leq |\mathcal{P}| \times e(u, v), \\ f(v, u) &\leq |\mathcal{P}| \times e(u, v). \end{aligned} \tag{17.10}$$

Note that these inequalities, together with the inequalities in (17.9), ensure that a path only contains nodes in K.

Exercise 17.2.1 *At this point in the development of the ILP formulation, there is no harm in replacing the term $|\mathcal{P}|$ in (17.10) with the smaller term $t - 1$. Explain why? Later we will see why we use $|\mathcal{P}|$ instead of $t - 1$.*

17.2.2.4 The Cleverness

We have defined path variables and inequalities that ensure that the two end nodes for any path variable with strictly positive value, must be in K. But we want more than that. We want the edges associated with strictly positive path variables to form *directed paths* from r, ending at each node in $\{K - r\}$. How can we do that? It will follow from three requirements, presented next, that more fully specify properties that path (f) variables must obey.

Requirements That Force Directed Paths:

(Ra) For each node $u \in \mathcal{N}$ that is *not* selected to be in K, and for each node $v \neq u$:

$$f(v,u) = f(u,v) = 0, \tag{17.11}$$

(Rb) *If* node u is selected for the root node, r, then

$$\begin{aligned} \sum_{v \neq u} f(u,v) &= t - 1, \\ \sum_{v \neq u} f(v,u) &= 0, \end{aligned} \tag{17.12}$$

which says that there are $t-1$ paths directed *out* of node r; and no paths directed *into* node r.

(Rc) *If* node v is selected to be in K, but *not* selected to be the root, then:

$$\sum_{v \neq u} f(v,u) - \sum_{v \neq u} f(u,v) = 1, \tag{17.13}$$

which says that the number of paths from r *into* a node u in K (where $u \neq r$), must be *one larger* than the number of paths *out* of node u. So, for every node $u \neq r$ in K, *exactly* one path from r *ends* at node u. Then, since the paths start at the same node, r, and run to every node in $\{K - r\}$, the paths will define a *connected* set of edges.

Exercise 17.2.2 *Write up a more complete explanation for why the requirements* **(Ra), (Rb), and (Rc)** *imply that the subgraph induced by K must be connected. As part of that answer, explain why the edges whose corresponding path variables have strictly positive values, specify a set of $t - 1$ paths from r, where exactly one path ends at each distinct node selected to be in K.*

Exercise 17.2.3 *It may seem that requirement* **(Ra)** *is redundant, given the inequalities we have already developed for the formulation. Is it? Explain.*

ILP Implementation of the Requirements We have seen that requirements **(Ra)**, **(Rb)**, and **(Rc)** force the creation of $t - 1$ paths from r, and that each path ends at a distinct node in $\{K - r\}$. But, how are these three requirements implemented with linear inequalities in an ILP formulation? The answer given in [31] (in slightly different detail here) is the following.

For each node $u \in \mathcal{N}$:

$$\begin{aligned} \sum_{v \neq u} f(v,u) - \sum_{v \neq u} f(u,v) &= c(u) - t \times r(u), \\ \sum_{v \neq u} f(v,u) &\leq |P| \times (1 - r(u)). \end{aligned} \tag{17.14}$$

Justifications We claim that the inequalities in (17.14), along with the inequalities described earlier, implement the requirements in **(Ra), (Rb), (Rc)**, and have no bad side effects. We consider the requirements separately.

(a) When node u is *not* selected to be in K, $c(u)$ is set to 0, and by (17.8), $r(u)$ is also set to 0. Further, when $c(u) = 0, e(u, v) = 0$ by (17.9), and so $f(u, v) = f(v, u) = 0$, by (17.10), for any $v \neq u$. This is exactly the requirement stated in **(Ra)**.

Also, when $c(u) = r(u) = 0$, the inequalities in (17.14) reduce to

$$\sum_{v \neq u} f(v, u) - \sum_{v \neq u} f(u, v) = 0,$$

and

$$\sum_{v \neq u} f(v, u) \leq |\mathcal{P}|.$$

The first one is satisfied when $f(u, v) = f(v, u) = 0$, for any $v \neq u$. The second one is always true, so (17.14) has no bad side effects when $c(u) = r(u) = 0$.

(b) When node u *is* chosen for the root node, r, $r(u) = 1$, so by (17.8), $c(u) = 1$ also. So, the inequalities in (17.14) reduce to

$$\sum_{v \neq u} f(v, u) - \sum_{v \neq u} f(u, v) = 1 - t,$$

and

$$\sum_{v \neq u} f(v, u) \leq 0,$$

so

$$\sum_{v \neq u} f(u, v) = t - 1,$$

and

$$\sum_{v \neq u} f(v, u) = 0,$$

which are exactly the requirements in **(Rb)**.

(c) When node u is selected to be in K, but is *not* chosen to be the root, $c(u) = 1$, and $r(u) = 0$, so the first inequality in (17.14) reduces to

$$\sum_{v \neq u} f(v, u) - \sum_{v \neq u} f(u, v) = 1,$$

which is the requirement in **(Rc)**. The second inequality reduces to $\sum_{v \neq u} f(v, u) \leq |\mathcal{P}|$, which is always true.

Exercise 17.2.4 *In the justification for the inequalities in (17.14), we explicitly discussed why there are no bad side effects in case* **(a)**, *but we did not discuss that for the other two cases. Complete the discussion to show that (17.14) does not have any bad side effects.*

No Objective Function In the variant of the Topo-free query problem we have considered, there is *no* objective function. The problem only asks for *feasibility*, i.e., whether there is a set K of t nodes, which satisfies all of the requirements. But Gurobi (and other solvers) require something that looks like an objective function. Otherwise, it will give an error message. So, we can use: *"Minimize 0"* in place of a real objective function.

Generalizing with Weights As discussed earlier in the book, often an edge of a PPI network is weighted to indicate the reliability, or probability, of the relation represented by that edge. Further, there may be more than one feasible solution, K, to an instance of the Topo-free query problem, and we may want a way to choose between them. In that more realistic case, we will use an actual objective function, while not changing the rest of the abstract ILP formulation. The objective function will be:

$$\text{maximize} \sum_{(u,v)\in E} w(u,v) \times e(u,v), \tag{17.15}$$

where $w(u, v)$ is the weight given to an undirected edge (u, v).

Summarizing The complete abstract ILP formulation for this variant of the Topo-free query problem is shown in Figure 17.3.

Exercise 17.2.5 *The inequalities in (17.10) allow $e(u, v)$ to be set to 1, even if $f(u, v) = 0$. We claim that not only should this be allowed, but it is* necessary *for the ILP formulation to correctly solve the edge-weighted version of the Topo-free query problem. Explain.*

Exercise 17.2.6 *So far, we have assumed that the edges in $\mathcal{N}$ are undirected. Modify the ILP formulation for the Topo-free query problem to handle the case that some of the edges are* directed.

Exercise 17.2.7 *In the ILP formulation for the Topo-free query problem, any edge (u, v) has a weight, $w(u, v)$, which is unrelated to the direction that edge (u, v) might be traversed. This is appropriate for PPI networks where all of the edges are undirected. But in some PPI networks, some of the edges are directed, reflecting asymmetric biological information (e.g., that one protein activates a second). Modify the ILP formulation to implement this variant. More than the objective function needs to be changed.*

17.2.3 Back to the Original Problem

In the version of the Topo-free query problem that we started with, there was *no* target t given as input to a problem instance. Rather, we wanted to choose the *largest* set of nodes K in $\mathcal{N}$ that satisfied all of the requirements. To solve this version of the problem, we change the objective function to:

$$\text{maximize} \sum_{v\in\mathcal{N}} c(v),$$

and replace t with $\sum_{v\in V} c(v)$ in (17.21).

Exercise 17.2.8 *Verify that the modifications discussed above do correctly create an ILP formulation for the version of the Topo-free query problem originally stated. Explain why $|\mathcal{P}|$ was used instead of $t - 1$ in the inequalities in (17.10).*

$$\text{Maximize} \sum_{(u,v)\in E} w(u,v) \times e(u,v). \tag{17.16}$$

Subject to
For each node $v \in \mathcal{N}$ and each protein $p \in \mathcal{P}(v)$:

$$g(p,v) \leq c(v). \tag{17.17}$$

For each node v in $\mathcal{N}$:

$$\begin{aligned} \sum_{p\in\mathcal{P}(v)} g(p,v) &\leq 1, \\ \sum_{v\in\mathcal{N}} r(v) &= 1, \\ r(v) &\leq c(v). \end{aligned} \tag{17.18}$$

For each protein p:

$$\sum_{v\in V(p)} g(p,v) \leq 1, \tag{17.19}$$

For each edge (u,v) in E:

$$\begin{aligned} c(u) + c(v) - e(u,v) &\leq 1, \\ 2 \times e(u,v) - c(u) - c(v) &\leq 0, \\ f(u,v) &\leq |\mathcal{P}| \times e(u,v), \\ f(v,u) &\leq |\mathcal{P}| \times e(u,v). \end{aligned} \tag{17.20}$$

For each node $u \in V$:

$$\begin{aligned} \sum_{v\neq u} f(v,u) - \sum_{v\neq u} f(u,v) &= c(u) - t \times r(u), \\ \sum_{v\neq u} f(v,u) &\leq |\mathcal{P}| \times (1 - r(u)). \end{aligned} \tag{17.21}$$

Figure 17.3 The Abstract ILP Formulation for the *Topo-Free Query* Problem Where the Edges Are Weighted, and a Target *t* Is Given at Input. This is the variant of the Topo-free query problem discussed in [31]. The c, *g* and e variables are binary. The *f* variables are integer with values between 0 and *t-1*.

Exercise 17.2.9 *Suppose, instead of maximizing the* number of nodes *in set K, we want to choose a set K satisfying all of the ILP requirements, and maximize the* total weight of the edges *in the subgraph induced by K. Show what needs to change to create that ILP formulation.*

Inserting Nodes So far, we have omitted an important element of the assignment model used in [31], namely *node insertions*. In the variants of the Topo-free query problem that we have discussed, the subgraph induced by K *must* be connected. But sometimes that is too demanding a requirement, and by using it, we may miss important biological insights. (Remember that this problem originates in the context of limited, perhaps preliminary, or incomplete data, particularly about $\mathcal{P}$). A more relaxed and perhaps more biologically informative variant of the problem is to allow a *small* number of nodes, denoted b, to be included in the connected subgraph, even if they are not assigned a protein from $\mathcal{P}$. That is, the subgraph induced by the nodes in K, plus at most b additional nodes, must form a connected subgraph of $\mathcal{N}$.

For example, in Figure 17.2, if we set $b = 1$, then we can insert node g, and the set $\{b, d, f, g, j\}$ is a permitted choice for K that allows all of $\mathcal{P}$ to be mapped to nodes in K. Allowing a small number of insertions can have a dramatic effect. In [31], when the target t was set to $|\mathcal{P}|$, a solution K was found with no needed insertions (i.e., $b = 0$) in only 80 of the 600 test cases.

Exercise 17.2.10 *Show how to modify the ILP formulation for the Topo-free-query problem, to solve the Topo-free query problem with allowed insertions, when a value of b is given as input to a problem instance.*

Exercise 17.2.11 *In addition to insertions, explicit* deletions *of proteins from $\mathcal{P}$ are also allowed in [31]. Deletions are meaningful when $t = |\mathcal{P}|$. But, with the objective of maximizing the size of K, explicit deletions from $\mathcal{P}$ are not needed. Explain why not.*

17.2.4 Does the ILP Formulation for Topo-free Queries Seem Familiar? It Should!

The requirements **(Ra), (Rb)**, and **(Rc)** and their ILP implementations are similar to requirements and their ILP implementations in the *maximum parsimony* problem (the Steiner-tree problem on hypercubes) in Section 5.1.3, and in the Steiner-tree and *arborescence* problems discussed in Chapter 16. The important point is that the ILP formulation for the maximum parsimony problems required the construction of one path from the *prespecified* root node to *each* of the *prespecified* nodes in a set V'. The idea of directed paths used in the ILP for Topo-free query problem, is also similar and related to the paths used in explaining the GG formulation for the TSP problem in Section 9.6. There, the salesman had to deliver one unit of snake oil to each node in the graph, and the ILP formulation specified that the number of units taken into any node was one greater than the number of units taken out of the node, similar to requirement **(Rc)**. So, the GG ILP formulation can also be thought of as defining a set of directed paths from the root, where exactly one path ends at each node in the graph.

But, unlike the Steiner-tree problem where there is a *prespecified* set of nodes, V', that must be connected; and unlike the TSP where *every* node must be visited, and must be on a single path from the root, in the Topo-free query problem, the input does *not* specify *which* nodes of the PPI $\mathcal{N}$ must be connected. The set of nodes to be connected can be a strict subset of all the nodes in $\mathcal{N}$, and that subset is not specified in the problem input. Instead, the set of nodes, K, takes the place of V', and K is only determined by the *solution* to the ILP formulation. Similarly, no *root* is specified at input, and the root node r must also be determined in the ILP solution. Still, when the requirements **(Ra), (Rb)**, and **(Rc)** are satisfied, the edges in the subgraph of $\mathcal{N}$ induced by K will be *connected*, and hence the ILP formulation for the Topo-free query problem is similar to the formulations of the other problems we have seen, where the solution consists of a connected subgraph of a given graph. We will encounter another example in Chapter 19.

In the ILP literature, the kind of constraints used to *force* a connected subgraph are sometimes called *network flow inequalities*. This is particularly true for TSP formulations like the GG formulation. See [8] for a general discussion of ILP approaches to connected subgraph problems, and see [53] for the use of advanced ILP solution techniques to solve those problems. There are also several additional papers in the computational and systems biology literature which leverage PPI networks in ways that involve connected, Steiner-tree-like subgraphs, to identify important features or modules in sets of proteins. For one such paper, see [56], where lymphoma microarray data was used with a PPI network to identify functional modules in the network. A related ILP approach is used in [12] to identify deregulated subnetworks in gene regulatory networks, rather than in PPI networks.

Exercise 17.2.12 *The following problem concerns the design of conservation corridors. The biological background to this problem is that in wildlife conservation it is important that protected parcels of land be* connected, *if not* contiguous. *The connections can consist of narrow strips of land (corridors) that allow the wildlife to move in safety from one protected area to another. Of course, designs are constrained by costs and budgets, so decisions can be challenging. The following quote is from [53]:*

> *We look at the problem of designing so-called wildlife corridors to connect areas of biological significance (e.g. established reserves). Wildlife corridors are an important conservation method in that they increase the genetic diversity and allow for greater mobility (and hence better response to predation and stochastic events such as fire, as well as long term climate change). Specifically, in the wildlife corridor design problem, we are given a set of land parcels, a set of reserves (land parcels that correspond to biologically significant areas), and the cost (e.g. land value) and utility (e.g. habitat suitability) of each parcel. The goal is to select a subset of the parcels that forms a connected network including all reserves.*

Your Problem: *Give an explicit model for the wildlife corridors problem, and develop an abstract ILP formulation for it. In the model there should be a cost and a benefit for each potential* reserve; *and also a cost and benefit for each potential* corridor. *The input consists of an undirected graph, which shows the potential reserves and the potential corridors, with their costs and benefits, and with an overall budget target. One problem is to connect all of the potential reserves, with the minimum total cost. Another problem is to choose a connected subgraph to* maximize *the benefits while keeping the costs* under *the budget target. In some variants of these problems, some reserves have already been purchased. Then, the problem statements can either* require *that those reserves be included in the solution, or not.*

If all of the reserves are already purchased, and the remaining problem is to select corridors to connect the reserves, is that problem an instance of a pure Steiner-tree problem? Explain.

Exercise 17.2.13 *The problem of designing wildlife corridors and reserves is similar to the species-protection problem considered in Chapter 1. Explain how they differ and how that affects the ILP formulations to solve the problems.*

17.3 EXAMPLE 3: IDENTIFYING *DRIVER* GENES IN CANCER

> During the time course of cancer evolution, tumor cells accumulate numerous genomic aberrations, however only a few "driver aberrations" are expected to confer crucial growth advantage – and have potential to be used as therapeutic targets. The identification of these driver aberrations and the specific genes they alter poses a significant challenge as they are greatly outnumbered by functionally inconsequential "passenger" aberrations ... [183]

Thus, the mutation in a *driver* gene may, through the alteration of a pathway, affect the expression of a gene far *downstream* in the pathway, and that altered gene expression may initiate cancer or accelerate its growth. Of course, we don't know which genes are driver genes. So, we want analytical methods that can distinguish driver genes from all the other mutated genes, which may only be passenger genes, or genes whose function is unrelated to the cancer. A recent paper used PPI networks, and other data, along with an ILP formulation, to do that.

The study in [183, 184] looked at n patients with cancer to identify which of their genes had *mutated*, and which of their genes had aberrant *expression*.[8] The mutated genes are the *potential driver genes*, $\mathcal{D}$, and the genes with aberrant expression, $\mathcal{G}$, are called *outlier* genes, which may have been (later) affected by mutations in the driver genes. Then, a precise computational problem was stated, and an abstract ILP formulation was developed to solve the problem. The full approach is called *HIT'N'DRIVE*.

17.3.1 The Driver-Gene Problem

The *driver-gene* problem seeks to identify which of the potential driver genes are actual driver genes, and also, which of the outlier genes were likely affected by the driver genes. Input to an instance of the problem includes a network, $\mathcal{N}$, which is a PPI network, or a gene-protein-interaction network (or a combination of both). Input also includes a set of potential driver genes, $\mathcal{D}$, and a set of outlier genes, $\mathcal{G}$, with aberrant expression levels. It is assumed that we have specified $\mathcal{D}$ and $\mathcal{G}$ sufficiently so that the product (protein or regulatory RNA) expressed by any gene in $\mathcal{D}$ or $\mathcal{G}$, is represented by a node in the network $\mathcal{N}$.

Formally With a user-defined value, d, we say that a potential driver gene, dg, *can influence* an outlier gene og, if there is a path in $\mathcal{N}$ from the node for dg to the node for og, with d edges or less. Then, we have

> **The Simplified Driver-Gene Problem:** Find a *smallest set* of genes in $\mathcal{D}$ that can influence all of the genes in $\mathcal{G}$.

Of course, because of the parameter d, there might be a gene in $\mathcal{G}$ that cannot be influenced by any gene in $\mathcal{D}$. In that case, the simplified driver-gene problem has no solution.

Exercise 17.3.1 *Recall that the* set-cover *problem was defined in Section 7.3.1.2. Express the* simplified driver-gene *problem in terms of the set-cover problem, and write an abstract ILP formulation for the simplified driver-gene problem.*

Don't peek, but part of the answer is given below.

17.3.1.1 Improving the Biological Fidelity

The simplified driver-gene problem captures the *spirit* of the problem and the solution discussed in [184], but the actual method, in an attempt to better model biological reality, is more involved. The full problem expands on the simplified problem in three

[8] Recall that gene expression means the protein, or regulatory RNA, that the gene codes for, is produced.

significant ways. First, the way that *gene influence* is modeled and computed tries to capture more *global* properties of the network. Second, gene mutation data, and gene expression data is broken down into data from *individual patients*. Finally, instead of framing the driver-gene problem in terms of set cover, it is framed in terms of *weighted multi-set cover*, explained below. We discuss these three changes in turn.

Gene Influence We will not discuss in detail the way that influence is defined and computed in [184], but only quote the following passage from the paper.

> HITn'DRIVE uses a particular influence value of a potential driver gene on other (possibly distant) genes based on the (gene or protein) interaction network in use. In order to capture the uncertainty of interaction of genes with their neighbors, it considers a random walk process which propagates the effect of sequence alteration in one gene to the remainder of the genes through the network.

For our purposes, we only need to assume that for every pair (dg, og) of a potential driver gene and an outlier gene, we are given an *influence value*, denoted $I(dg, og)$, of dg on og. What is of more interest here is the way these values are used in modeling and solving the driver-gene problem.

Patient-Based Data and Bipartite Graph When the simplified gene-driver problem is viewed as a set-cover problem, as in Exercise 17.3.1, the data is represented by a bipartite graph, $\mathcal{B}$, with sides A and B. Each node in A represents a gene in $\mathcal{D}$, each node in B represents a gene in $\mathcal{G}$, and there is an edge between nodes for dg and og if and only if dg can influence og.

In the expanded gene-driver problem, the bipartite graph, $\mathcal{B}$, is expanded to a bipartite graph $\mathcal{B}^* = (A, B^*)$. The A side of $\mathcal{B}^*$ is the same as in graph $\mathcal{B}$, where each node represents one gene in $\mathcal{D}$. However, the opposite side of $\mathcal{B}^*$ is not equal to B. In $\mathcal{B}^*$, there is a node on side B^* for each pair (p_i, og), where og is a gene whose expression is altered in patient p_i. We refer to that node as (p_i, og). So, $\mathcal{B}^*$ represents the individual expression data for each patient, and each node in side B of $\mathcal{B}$ is (possibly) expanded into n nodes in $\mathcal{B}^*$. Finally, there will be a *weighted* edge in $\mathcal{B}^*$ between the node dg on the A side, and the node (p_i, og), on the B^* side, if (p_i, og) is in $\mathcal{B}^*$. The weight of that edge will be $I(dg, pg)$. However, in some variants of the problem, there is also a user-defined parameter Δ, and any edge with weight less than Δ is removed from $\mathcal{B}^*$.

The Weighted Multi-Set Problem The final expansion from the simplified driver-gene problem is that the full driver-gene problem is framed as a *weighted multi-set cover* problem, defined on the bipartite graph $\mathcal{B}^*$, as follows.

> **The Driver-Gene Problem:** Given the bipartite graph $\mathcal{B}^*$ defined above, and a user-defined value γ between 0 and 1, find a *smallest set*, X, of nodes in A, so that for every node (p_i, og) in B^*,
>
> $$\sum_{\text{node } (p_i, dg) \in X} I(dg, og) \geq \gamma \times \left[\sum_{\text{node } (p_i, og) \in \mathcal{B}^*} I(dg, og) \right].$$
>
> Note that each $I(dg, og)$ value is a constant.

That is, for each patient p_i, and each gene og whose expression level in patient p_i is aberrant, the nodes chosen for X that represent potential driver-genes must have a *total influence* that is *sufficient* to explain the altered expression level of og in patient

$$\texttt{Minimize} \sum_{dg \in A} x(dg)$$

such that:
For each variable $e(p_i, dg, og)$,

$$x(dg) = e(p_i, dg, og)$$

and for each node (p_i, og) in B^* :

$$\sum_{\text{node } dg \in A} I(dg, og) \times e(p_i, dg, og) \geq \gamma \times \left[\sum_{\text{node } (p_i, og) \in B^*} I(dg, og) \right].$$

Figure 17.4 The First Inequality Implements the Stated Relationship between *x(dg)* and *e*(p_i, *dg*, *og*). The second inequality implements the requirement that the total influence must be sufficient, for each (p_i, *og*). In the variant of the problem where the parameter Δ is used, it is possible that the ILP will be infeasible.

p_i. Sufficient in this context means that the total influence must be a fraction of at least γ of the total possible influence. Note that if the expression of gene og in patient p_i is normal, then there will be no edge from node dg to node (p_i, og).

Ouch! The problem statement sounds complex, and it certainly is more complex than the simplified driver-gene problem, but it should make sense if you parse the statement a few times.[9]

An ILP Formulation for the Driver-Gene Problem We use the binary variable $x(dg)$ to indicate whether or not node dg will be chosen to be in set X. We also use a binary variable $e(p_i, dg, og)$, which is set to 1 if and only if node (p_i, dg) is chosen for X, and *there is* an edge in B^* between node dg in A and node $(p_i, 0g)$ in B^*. The ILP formulation for the driver-gene problem is shown in Figure 17.4.

Exercise 17.3.2 *The full statement of the driver-gene problem in [183, 184] has an additional user-supplied parameter, α, which is between 0 and 1. In the full statement of the problem, it is not required to cover* every *potential passenger gene as we do in Figure 17.4, but only a fraction α of the potential passenger genes. The covered genes are then taken to be the actual passenger genes.*

How should the ILP formulation be changed to implement this change? What do you think is the biological meaning of α, when its value is less than one?

[9] Often, the biggest obstacle to understanding a mathematical statement is just parsing it.

18

More String and Sequence Problems Solved by ILP

In Chapter 10 we examined several string and sequence problems in biology that can be solved using integer linear programming. In this chapter we extend one of those string problems, develop an ILP solution to a more complex sequence problem, and discuss a related problem in phylogenetics.

18.1 THE LCS PROBLEM FOR MULTIPLE STRINGS

In Section 10.4, we developed an ILP solution for the problem of finding a *longest-common subsequence* (LCS) for *two* strings. Now we expand to the problem to find an LCS of *multiple* strings. First, some needed definitions.

Recall from Section 10.4 that a *subsequence* in a string s of length m is specified by a list of indices $i_1 < i_2 < i_3 < \ldots < i_k$, for some $k \leq m$. Given a set of $n > 2$ strings, $\mathcal{S} = \{s_1, s_2, \ldots s_n\}$, a string s is a *common subsequence* of $\mathcal{S}$ if s is a subsequence in *each* of the n strings in $\mathcal{S}$. A *longest common subsequence (LCS)* of $\mathcal{S}$ is a common subsequence s^* with length greater or equal to any other common subsequence of $\mathcal{S}$. As in the case of two strings, we can also define the *maximum-weighted common subsequence*, when each site in each string has a given weight.

The *length* of an LCS, or the maximum weight of a common subsequence, of $\mathcal{S}$, has been used as a measure of the overall *similarity* of the strings in $\mathcal{S}$, for example, to test if those strings have a significant family relationship, and if so, to highlight that relationship.

An Application: Map and Marker Integration The LCS (and the weighted version) have also been used to *integrate* marker information from multiple genomic maps that are somewhat inconsistent. Suppose we have several independent deductions (maps) of where a set of m distinct markers appear in a given chromosome. However, the maps are inconsistent, having even disagreements on the proper *linear order* of the markers. The inconsistent maps may have been created by different technologies with different types of errors, or different units of measurement, and/or by different labs with different levels of quality control. The LCS from the multiple strings identifies information that is *consistent* in the strings, at least with respect to linear order, and hence the LCS is considered more *reliable* than the parts of the strings outside of the LCS. This is a kind of "safety in numbers" or "crowd consensus" approach to extracting good data (maybe not all of it) from data whose quality is mixed. Representative examples of this kind of data extraction and marker integration appear in [2, 125].

> Combining data to construct an integrated map not only consolidates information from different sources onto a single map, but information contributed from each data set increases the accuracy of the map. [125]

18.1.1 An ILP Formulation for LCS of Multiple Strings

Consider three strings, $\mathcal{S} = \{s_1, s_2, s_3\}$. Suppose s is a subsequence that is common to the pair of strings (s_1, s_2); and also common to the pair (s_1, s_3). Then, s is in *all* three strings s_1, s_2, and s_3, and hence is a common subsequence of $\mathcal{S}$. More generally, we have

The Key Insight A string s is a common subsequence of a *set* of n strings $\mathcal{S} = \{s_1, s_2, \ldots s_n\}$ *if and only if* s is a common subsequence of each pair $(s_1, s_2), (s_1, s_3) \ldots (s_1, s_n)$.

So, the LCS problem for a set of n strings, $\mathcal{S}$, can be solved by finding the *longest* string s that is a common subsequence in the $n-1$ pairs $(s_1, s_2), (s_1, s_3) \ldots (s_1, s_n)$. Note the special role of string s_1 in the statement of the key insight: s_1 is part of *each* of the $n-1$ pairs. But the selection of s_1 for this role is arbitrary. Any other single string in $\mathcal{S}$ would work as well.

Exercise 18.1.1 *Explain, in your own words, why the above statements are correct.*

The key insight leads to the idea that an ILP formulation to find an LCS of $\mathcal{S}$ should contain $n-1$ sets of inequalities used to find $n-1$ common subsequences of the pairs $(s_1, s_2) \ldots (s_1, s_n)$. This is necessary but it is not enough.

Without additional constraining inequalities, the ILP solver might find a *different* common subsequence for each of the $n-1$ pairs. So, we need additional inequalities to ensure that in each of those $n-1$ common subsequence pairs, the same subsequence of s_1 is chosen. How do we do that? We will outline a method in the following exercises. You will fill in the details.

Exercise 18.1.2 *If we combine the inequalities in the $n-1$ ILP formulations that find a common subsequence in each pair (s_1, s_k), for k from 2 to n, and we add inequalities that say that the subsequence chosen from s_1 must be the same in each of those $n-1$ ILP solutions, would this combined ILP formulation specify that the solution* must *be a subsequence common to* all *of the n sequences?*

Expanding the Variable Set Recall from Exercise 10.4.4 in Section 10.4 that variable $P(i,j)$ is the binary variable used in the ILP formulation for the LCS of *two* strings s_1 and s_2. It is set to value 1 if and only if the character at site i in s_1 pairs with the character at site j in s_2. In that variable, the two indices, i and j, refer to positions in strings s_1 and s_2, respectively, and there is no need for indices to explicitly identify which strings are being referred to. But, with multiple strings, we need a more explicit variable. For the multiple string LCS problem, we let $P(i,j,k)$ be a binary variable set to value 1 if and only if the character at site i in string s_1, and the character at site j in string s_k, form a matching pair in a solution to the multiple-string LCS problem. And, each ILP formulation for finding a common subsequence in strings s_1 and s_k must replace any occurrence of a $P(i,j)$ variable with the variable $P(i,j,k)$.

Now, in order to fully implement the *key insight*, we will use those $P(i,j,k)$ variables to express the logic that $s_1(i)$ (the character at site i in string s_1) is part of the common subsequence chosen for s_1 and s_2, *if and only if* $s_1(i)$ is part of the common subsequence in *each* of the $n - 1$ common subsequences in pairs $(s_1, s_3), \ldots (s_1, s_k), \ldots (s_1, s_n)$.

As a first step, let $U(i,k)$, for $k \geq 2$, be a binary variable that is set to value 1 if and only if variable $P(i,j,k)$ has value 1, for *some* position j in s_k. That is, $U(i,k)$ indicates whether or not site i in s_1 is chosen to be in the common subsequence of s_1 and s_k. Note, however, that since $U(i,k)$ is a binary variable, when it has value 1, it doesn't record which site in s_k is paired with site i in s_1.

Exercise 18.1.3

(a) *In order to correctly implement the key insight, we need $U(i,k) = U(i,2)$ for every $k > 2$. Explain why.*

(b) *Now, using the variables $P(i,j,k)$, state the abstract inequalities that correctly set the values for each $U(i,k)$, for $k \geq 2$.*

Answer: For each $k \geq 2$,

$$\sum_j P(i,j,k) = U(i,k).$$

Explain.

(c) *Next, state the inequalities that implement that logic that $U(i,2)$ is set to 1 if and only if $U(i,k)$ is set to 1 for* every $k > 2$.

(d) *Suppose we add the inequalities specified in parts (a) through (c) of this exercise, to the inequalities that set the $P(i,j,k)$ variables describing $n - 1$ (possibly different) common subsequences in pairs $(s_1, s_2), (s_1, s_3), \ldots (s_1, s_n)$. Do the combined inequalities now assure that any solution to this ILP formulation is a subsequence in* all *of the n strings, i.e., a common subsequence of S? Explain.*

(e) *Assuming that the answer to last question (d) is "yes," what* objective function *should be used to find the longest common subsequence of S?*

Exercise 18.1.4 *Summarizing the above discussions, write out in full detail the abstract ILP formulation to find an LCS of $n > 2$ strings.*

Exercise 18.1.5 *Suppose that each of the n strings has length m. As a function of n and m, how many variables and inequalities are used in the ILP formulation detailed in the previous exercise?*

Exercise 18.1.6 *The LCS problem seems similar to the* string site-removal problem *discussed in Section 10.2, and both problems are intended to expose subtle similarities in multiple strings that might not be apparent when comparing just two strings. Notice that after removal of the sites determined by a solution to the string site-removal problem, the resulting strings are identical, and hence form a common subsequence of the original strings. Will that common subsequence necessarily be a longest common subsequence? Explain.*

Does knowing the LCS of S help in finding an optimal solution to the string site-removal problem on S?

Exercise 18.1.7 *The problem of finding a LCS in set of n strings, S, is discussed in [21], where a different ILP formulation is suggested. The approach there is an extension of the approach used for the LCS of two strings. Recall that $\mathcal{A}$ denotes the alphabet used for the n strings in S. Focus on one character in $\mathcal{A}$, say the character g, and select one site in* each *of the n strings, such that character g appears at each of those n sites. Let's call each such set a* thread for g, *so, for example, if $n = 10$ and there are five occurrences of character g in each string, there are 5^{10} threads for g. Similarly, for each character in $\mathcal{A}$, there will be a set of threads.*

Two threads, T_1 and T_2 are non-crossing *if in each of the n strings, the site in T_1 is strictly to the left of the site in T_2; or, in each of the n strings, the site in T_1 is strictly the right of the site in T_2.*

With those definitions, we claim that an LCS of $\mathcal{S}$ corresponds to the largest *subset of threads, $\mathcal{L}$, such that each pair of threads in $\mathcal{L}$ is non-crossing.*

Is this claim correct? Explain. How is this thread-based approach an extension of the approach used for the LCS of two strings?

Using one ILP variable for each thread, and no other variables, create an abstract ILP formulation for the LCS problem, and explain why it is correct.

Exercise 18.1.8 *We now have two different ILP formulations for the LCS problem. It is often the case that there are two or more formulations that are each correct, but very different. How do we choose between them? Is one preferable to the other?*

For a rough "back-of-the-envelope" analysis of the practicality of the two formulations for the LCS problem, assume that each of the n strings has length m and has $\frac{m}{|\mathcal{A}|}$ occurrences of each character in $\mathcal{A}$. How many ILP variables and inequalities will the thread-based formulation have? How many variables and inequalities will the earlier formulation have? What is your overall conclusion about the relative practicality of the two methods?

18.1.2 Software for the LCS of Three Strings

The Python program *LCS_Mult.py*, written by UC Davis students Jessica Au and Jessie Vuong, can be downloaded from the book website. This Python script generates the ILP formulation to solve the LCS problem for multiple strings. Currently, it only works for three strings.

Call the program on command line as:

```
python LCS_Mult.py DNA.txt LP_outputfile.lp minDist num-strings
```

The example files are: *DNA_file.txt* is a plain text file containing three DNA sequences, each with a max of 100 nucleotides. Set *minDist* to 0 for the LCS problem; and *num-strings* is the number of strings input. Currently, use 3.

The output will be in the file *LP_outputfile.lp*

Also on the class website are the example input files: *test3seqs.txt, multiple_string.txt*

18.2 TRANSFORMING GENE ORDER BY REVERSALS

So far, our discussions of sequence analysis have focused on analyzing fully detailed DNA sequences, over a four-letter alphabet. That focus is appropriate for comparing genes or for finding patterns in DNA, etc., when the biological phenomena of interest occur at the level of individual genes, gene regulators, or the proteins encoded by genes. However, there are important biological phenomena that occur at a *larger* scale, e.g., at *chromosomal* or *genomic* scales, and these require a higher-level viewpoint. One such informative biological phenomenon is *long chromosomal reversals (inversions)*,[1] where the DNA in a long interval on a chromosome *reverses* direction. For example, if we represent each gene by a distinct integer, the interval with 10 genes:

$$1\ 10\ 4\ 5\ 2\ 6\ 3\ 9\ 8\ 7$$

[1] Other high-level chromosomal phenomena are transpositions, translocations, inverted translocations, and prefix flipping.

becomes:

1 10 6 2 5 4 3 9 8 7

when the interval containing 4 5 2 6 is reversed. Since genes can be separated by long distances on a chromosome, what seems like a small reversal of just four integers actually represents a very long chromosomal reversal.

It is believed that long chromosomal reversals are much rarer, with longer periods of time between reversals, than are mutations of single nucleotides. Therefore, we can use long chromosomal reversals to look *farther back* into evolutionary history than we can by using only single nucleotide mutations in the DNA sequence. Then, the appropriate object of interest is not a DNA sequence itself, but a sequence of *distinct integers*, where each integer uniquely represents one distinct gene (or other distinct point of interest) on a chromosome.[2] So, what we are interested in is the *order* of the genes on the chromosome, and how that order changes over time – chromosomal reversals are relatively rare, but frequent enough to be evolutionary informative. Then, following the principle of *parsimony*, the computational problem of interest is:

> **The Sorting-By-Reversals Problem** Given two permutations, P_1 and P_2, of the integers 1 to n, find the *minimum* number of interval reversals that transforms P_1 to P_2.

For example, see Figure 18.1.

The focus on the use of reversals to deduce evolutionary history was started by Palmer and Herbon [145] in 1988, and first studied as an explicit computational

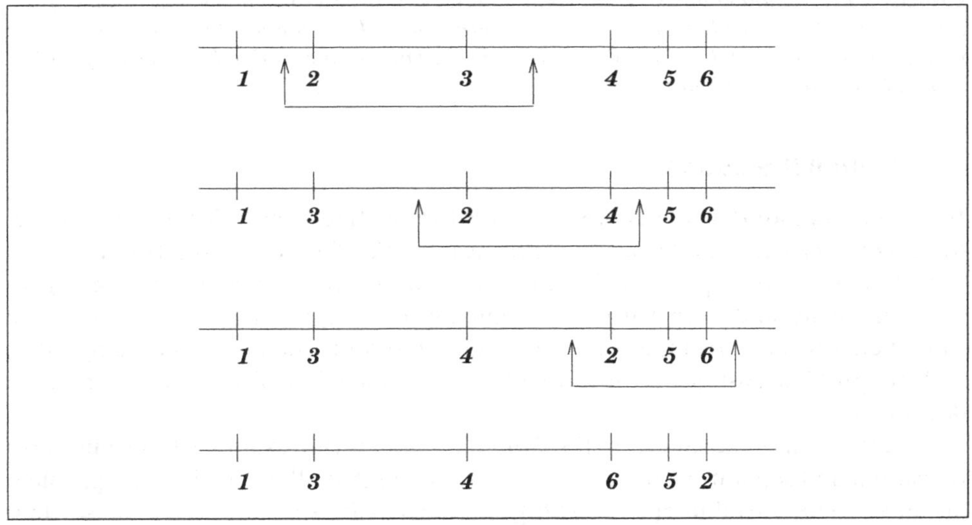

Figure 18.1 Each Line Represents a "Chromosome" with Six Genes. Three reversals transform the top chromosome to the bottom one. Two arrows enclose the interval to be reversed. The first chromosome is represented by the identity permutation 1,2,3,4,5,6; the second by 1,3,2,4,5,6; the third by 1,3,4,2,5,6; the last one is represented by 1,3,4,6.5,2.

2 There are genes that occur in multiple copies, but they are the exception, and we assume here that every gene is distinct, and so the genes can be given distinct integer labels.

problem by Kececioglu and Sankoff [103, 104]. That work was followed by a large body of research, most notably by Bafna, Hannenhalli, and Pevzner (for example, in [13, 88, 89]).

A Hard Problem There are no known algorithms to solve the sorting-by-reversals problem that are efficient in the worst-case sense,[3] and it certainly is not obvious how to solve sorting-by-reversal problems in practice. It is even less obvious how one could efficiently *prove* that an optimal solution had been found, even if you had one. So, it is natural to see if integer linear programming can be effective here, both to find optimal solutions and to establish the optimality of the solution.

Exercise 18.2.1 Turnips to Cabbage *One of the first real datasets that was studied, in [145], considered a section of the genomes of turnips and cabbage, two plants that are botanically closely related, even though they look quite different. Ten genes were identified in the two plants. In turnips, the genes were labeled so that their order is the natural order 1 through 10 (this is also called the identity permutation). Based on that numbering, the order of the analogous genes in cabbage is 1 10 4 5 2 6 3 9 8 7.*

Using the ILP formulation we will develop below, Gurobi 7.5 found this solution in about 28 seconds on my MacBook Pro; and that time was reduced to 12 seconds by a speed-up idea that we will also discuss below.

Verify that the following series of four reversals works: [4, 9], [2, 6], [2, 10], [5, 6]. *Each reversal is encoded by an ordered pair of numbers,* $[i, j]$, *where i indicates the* leftmost position *and j indicates the* rightmost position *of the interval to be reversed. For example,* [2, 6] *reverses the second through sixth integers in the permutation it is applied to, no matter what integers are in that interval.*

Exercise 18.2.2 *Another dataset from [145] describes the orders of eight analogous genes from tobacco and the flower lobelia. In this case, permutation P_1 is the identity permutation; and permutation P_2 is 7 1 2 4 5 3 6 8. Try to solve this instance of the sorting-by-reversals problem by hand. Gurobi solved it in 1.08 seconds.*

18.2.1 An ILP Solution

In this section, we develop an abstract ILP formulation from [120] for the *sorting-by-reversals problem*, and discuss its practicality. We will see that the ILP approach can effectively solve problem instances that would be extremely challenging to solve manually, and even harder to manually prove optimality. The instances we can efficiently solve include some real, older, biological datasets, but, many other realistic problem instances are too large for the *specific* ILP formulation that we develop here.

There are more advanced ILP *solution techniques* (for example, branching with relaxation and separation, discussed in Section 9.8) that allow much larger problem instances to be solved in practice [120], and other ILP formulations for the sorting-by-reversals problem [176]. That advanced material is beyond the scope of this book. So this section can be viewed as an introduction to the sorting-by-reversals problem; an introduction to modeling the problem by integer linear programming; and a discussion of a specific ILP formulation that is effective for *small to moderate-size* problem instances, including some realistic biological datasets. An interested reader

[3] Although there is an efficient algorithm for a closely related problem called the *signed* sorting-by-reversals problem [88, 89].

who wants to solve larger problem instances should consult [120]. Further exploitation of the integer programming approach for genomic reversals together with other genomic operations appears in [175,176,177,178]. Integer programming formulations for other types of chromosomal rearrangements, along with a framework to relate the ILPs, appears in [120].

18.2.1.1 The High-Level View of the Solution Method

Here, we address *Task A* discussed in the introduction of the book, i.e., developing the underlying *idea* that will be implemented into an ILP formulation. We start with two observations.

First Observation: Although the input to the sorting-by-reversals problem consists of two permutations, P_1 and P_2, of the integers from 1 to n, and there is no other restriction on what those permutations are, we can actually assume that P_1 is the *identity permutation*, where the integers are in their natural order $1, 2, 3, \ldots n$.

Exercise 18.2.3 *Explain why it is acceptable to assume that P_1 is the identity permutation. Stated differently, suppose we have a solution method for the sorting-by-reversal problem that only works when P_1 is the identity permutation, but P_2 can be any permutation. How can that solution method be used to solve the general sorting-by-reversals problem, i.e., when P_1 and P_2 can be any permutations of the integers from 1 to n?*

Second Observation: For any permutation P_2, at most $n - 1$ reversals are needed.

Exercise 18.2.4 *Describe a method to create any permutation P_2 starting with the* identity *permutation, using at most $n - 1$ reversals.*

We Imagine a Flow Based on those two observations, we describe the idea for the ILP formulation by imagining that the integers 1 through n are listed on a horizontal line called *Level 1*. Then the integers *migrate* or *flow* from Level 1 through $n - 1$ successive horizontal levels, 2 through n. At each successive level, the order of the integers is created from their order at the previous level by *one* reversal operation, or by an operation called a "NO-OP," which leaves the order of the integers unchanged.

In more detail, at each level there are n nodes, and each will be visited by a distinct integer. Also, for k from 1 to $n-1$, there is a directed edge from every node at a level k to every node at a level $k + 1$. This n-level graph is denoted $G(n)$. So, we view the integers as starting at Level 1 in their natural order, and flowing along edges of $G(n)$ (performing an interval reversal or NO-OP at each level) until the integers are in the order specified by permutation P_2. After that level, the integers flow to level n using only NO-OP operations at each level (see Figure 18.2).

Exercise 18.2.5 *We have assumed that permutation P_1 is the natural order of the integers 1 to n, and the sorting-by-reversals problem is to create a given permutation P_2 from P_1 by the fewest possible interval reversals. Suppose we have solved that problem instance for a given P_2. How would we solve the problem of transforming P_2 to P_1 with the fewest reversals. Explain fully.*

First ILP Variables and Inequalities

The ILP formulation will implement the imagined flow on graph $G(n)$, described above. We use a binary ILP variable $R(p, q, k)$ to indicate whether or not a reversal between positions p and $q > p$ (including positions p and q) will be performed at

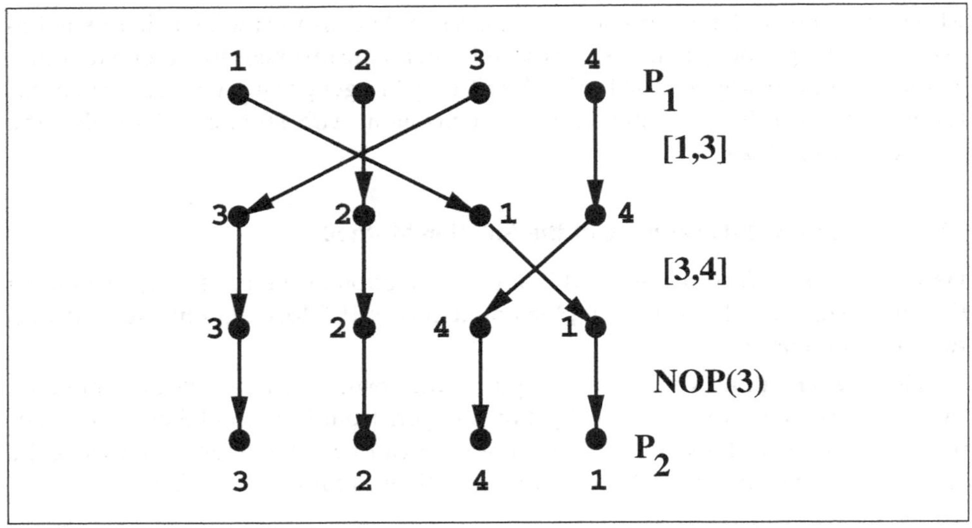

Figure 18.2 The Integers 1 Through 4, Starting in Their Natural Order, Flow from Level 1 to Level 4 in $G(4)$, Realizing the Permutation 3 2 4 1. In $G(4)$ there are 16 edges that go from Level k to Level $k + 1$, for k from 1 to 3. However, the figure only shows the edges actually used in the flow. The operations at levels 1, 2, and 3 are shown to the right of the figure, between two levels. The solution uses two reversals and one NO-OP.

level k. Variable $R(p,q,k)$ will be set to value 1 if that reversal will be performed, and will be set to value 0 otherwise. We also use a binary variable NOP(k) to indicate whether or not a NO-OP operation will be performed at level k. Variable NOP(k) will be set to value 1 to indicate that a NO-OP will be performed at level k; otherwise NOP(k) will be set to 0.

Exactly one reversal or NO-OP operation is performed at each level 1 through $n-1$, so the ILP formulation must contain the inequality:

$$NOP(k) + \sum_{(p,q):\, p<q} R(p,q,k) = 1, \tag{18.1}$$

for each k from 1 to $n-1$.

We also use a binary ILP variable, $X(k,i,p,q)$, for each edge (p,q) between levels k and $k+1$, in $G(n)$, and each integer, i, from 1 to n. In full detail, k specifies a level; i specifies an integer; and p and q represent a directed edge (p,q) which goes from node p in level k to node q in level $k+1$. Variable $X(k,i,p,q)$ is set to value 1 to indicate that integer i will flow *from* node p, *to* node q, *between* levels k and $k+1$.

Inequalities for Level 1 At Level 1, the integers are in their natural order, 1 through n. The ILP formulation needs inequalities whose effect is to flow each integer, i, from position i in Level 1 to some position (possibly i) in Level 2. For example, for integer 2 the formulation has the inequality:

$$\sum_{q=1}^{n} X(1,2,2,q) = 1, \tag{18.2}$$

which says that at Level 1, integer 2 must flow *from* node 2 in Level 1, *to* some node, q, at Level 2. More generally, for each integer i from 1 to n, we have:

$$\sum_{q=1}^{n} X(1, i, i, q) = 1. \tag{18.3}$$

Note that the inequalities in (18.3) will be the same no matter what the permutation P_2 is.

Inequalities for Level n Similar to the case of Level 1, Level n requires special inequalities. Recall that $P_2(z)$ specifies the integer that must be in position z at level n. For example, if $P_2 = 1, 3, 4, 6, 5, 2$, then integer 3 must be in position 2, and integer 6 must be in position 4, at level $n = 6$; so $P_2(2) = 3$ and $P_2(4) = 6$.

For the inequalities at Level n, we need the *inverse* information. We define $Q_2(i)$ as the position of integer i in permutation P_2. For example $Q_2(3) = 2$, and $Q_2(6) = 4$. More formally,

$$P_2(z) = i \text{ if and only if } Q_2(i) = z.$$

The flow viewpoint specifies that the integers must end up at Level n ordered by the permutation P_2. So, each integer i must end at node $Q_2(i)$ at Level n. To implement these requirements, for each i from 1 to n, the formulation will have the inequality:

$$\sum_{p=1}^{n} X(n-1, i, p, Q_2(i)) = 1, \tag{18.4}$$

which says that integer i must flow *into* node $Q_2(i)$ at level n, but it can flow *from* any node, p, at level $n - 1$.

Unlike the inequalities in (18.3), these inequalities *do* depend on the specific permutation P_2. Of course, the specific values of $Q_2(i)$ are known to the computer program that creates the concrete ILP formulation, when a concrete problem instance is specified.

Conservation in the Middle The ILP formulation must also specify that at each level, k, from 2 through $n-1$, *exactly* one integer will flow *into* each node (from level $k-1$); and at each level 2 through $n-1$, *the same* integer will flow *out of* that node. So, for each node q at a level k from 2 to $n-1$, we use the inequality:

$$\sum_{i=1}^{n}\Big[\sum_{p=1}^{n} X(k-1, i, p, q)\Big] = 1, \tag{18.5}$$

which implements the requirement that exactly one integer flows into node q, at level k. And, for each node p at a level k from 2 to $n-1$, we use the inequality:

$$\sum_{i=1}^{n}\Big[\sum_{q=1}^{n} X(k, i, p, q)\Big] = 1, \tag{18.6}$$

which implements the requirement that exactly one integer flows out of node p, at level k.

Exercise 18.2.6 *Explain why the inequalities in (18.5) are not needed for $k = n$. Similarly, explain why the inequalities in (18.6) are not needed for $k = 1$.*

We still need to implement the requirement that *if* integer i flows *into* a node p at a level $k = 2$ through $n - 1$, *then* i must be the integer that flows *out* of node p. To do that, we have the inequality:

$$\sum_{q=1}^{n} X(k-1,i,q,p) = \sum_{q=1}^{n} X(k,i,p,q), \tag{18.7}$$

for each integer i from 1 to n.

18.2.1.2 The Good Stuff: Inequalities Relating the *X* and *R* Variables

The most important, interesting and subtle inequalities connect the X variables with the R variables, so that the flow of the integers *between* successive levels of $G(n)$ is consistent with the chosen reversal or NO-OP operation.

To start, we examine a small example. Suppose $n = 4$, and that the integer in position 3 at Level 2 is also in position 3 at Level 3. The only way this can happen is if the operation performed at Level 2 was NOP(2), or $R(1,2,2)$, or $R(2,4,2)$ (see Figure 18.3).

So, we need an inequality that says that *if* the integer in position 3 at level 2 is the same as the integer at position 3 at level 3, *then* the operation at level 2 must one of those three operations. The following inequality implements that logic:

$$X(2,1,3,3) + X(2,2,3,3) + X(2,3,3,3) + X(2,4,3,3)$$

$$\leq NOP(2) + R(1,2,2) + R(2,4,2).$$

Why is this correct? Walking through the logic of this inequality, it follows from inequality (18.7) that

$$X(2,1,3,3) + X(2,2,3,3) + X(2,3,3,3) + X(2,4,3,3)$$

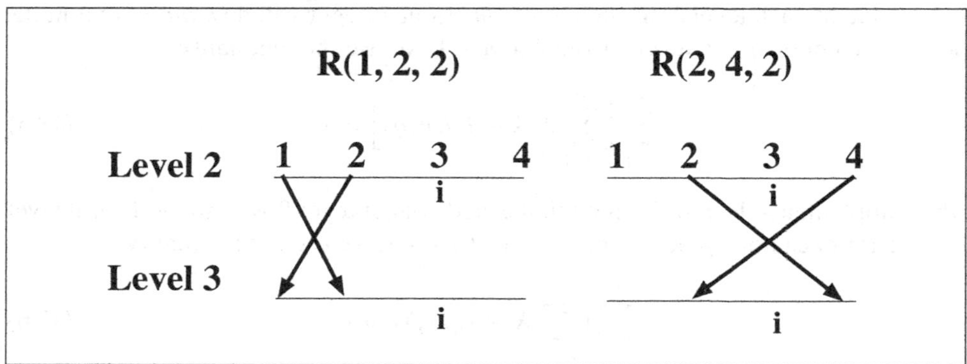

Figure 18.3 The Integer, *i*, in Position 3 at Level 2 Is Also in Position 3 at Level 3. This can be accomplished by a NOP(2) operation, or by the two reversal operations shown.

has value at most 1, and when that sum does have value 1, at least one of the variables $NOP(2), R(1,2,2)$, or $R(2,4,2)$ must have value 1 in order to satisfy the inequality. But, by inequality (18.1), $NOP(2) + R(1,2,2) + R(2,4,2) \leq 1$, so when

$$X(2,1,3,3) + X(2,2,3,3) + X(2,3,3,3) + X(2,4,3,3)$$

has value 1, *exactly* one of the variables $NOP(2), R(1,2,2)$, or $R(2,4,2)$ will have value 1.

The General Case That a Position Remains Unchanged Having looked at a small example, we can now develop the general case that a position remains unchanged. For any position w, if the integer at a position w is the same at both levels k and $k+1$, then, either a NO-OP was performed at Level k; or a reversal $R(p,q,k)$ was performed at Level k, and w is outside the interval $[p,q]$; or a reversal $R(p,q,k)$ was performed at Level k, and $q-p+1$ (the length of the interval) is odd, and w is exactly in the *middle* of the interval, i.e., $w = \frac{p+q}{2}$. Then, we need the following general inequality for every pair (k,w), where k is a level from 1 to $n-2$, and w is a position from 1 to n:

$$\sum_{i=1}^{n}[X(k,i,w,w)] \leq NOP(k) + \sum_{\{p,q\}:\ w\,=\,(p+q)/2} R(p,q,k), \qquad (18.8)$$
$$+ \sum_{p,q\,<\,w} R(p,q,w) + \sum_{p,q\,>\,w} R(p,q,w).$$

The Case That An Integer Moves So far, we have discussed the needed inequalities for the case that integer i does not change positions between two levels. Now we discuss the case when it does move.

If the integer at position w at level k moves to position $z > w$ at level $k+1$, then the only possible reversals at level k are: $[w,z], [w-1,z+1] \ldots [w-h,z+h]$, where either $w-h$ is 1, or $z+h$ is n. That is, w will move to z by any operation $[p,q]$ as long as $p \geq 1, q \leq n$, and the distance from p to w is the same as the distance from z to q (see Figure 18.4). A symmetric statement is true if $z < w$.

In the case that $w < z$, the following inequality implements the logic. For each level k from 1 to $n-2$, each position w from 1 to $n-1$, and each z, where $w < z \leq n$:

$$\sum_{i=1}^{n} X(k,i,w,z) \leq, \qquad (18.9)$$
$$\sum_{\substack{1\,\leq\,p\,\leq\,w\,<\,z\,\leq\,q\leq\,n \\ w-p\,=\,q-z}} R(p,q,k).$$

Exercise 18.2.7 *Write out the inequality that covers the case that $z < w$.*

The inequalities for the two cases can be combined into a *single* inequality. Let $a = \min(w,z)$ and $b = \max(w,z)$. Then, the ILP formulation can use the following inequality, for each level k from 1 to $n-2$, and each (unordered) pair of positions (w,z):

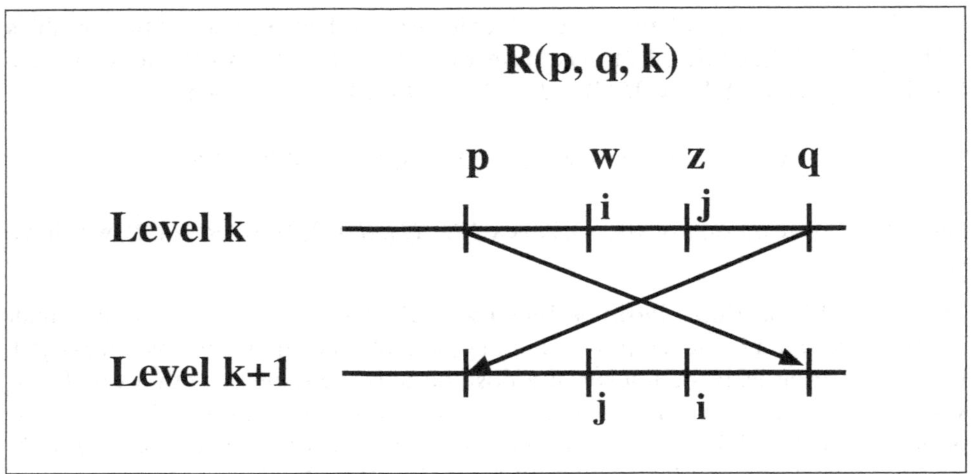

Figure 18.4 A Reversal That Exchanges the Integers, *i* and *j*, in Positions *w* and *z* > *w*, Must Be a Reversal of Interval *[p,q]*, Where the *p* < *w* and *q* > *z*, and *w* − *p* = *q* − *z*.

$$\sum_{i=1}^{n} X(k,i,w,z) \leq, \qquad \sum_{\substack{p=a-h\ <\ q=b+h \\ h \leq Min(a-1,n-b)}} R(p,q,k). \tag{18.10}$$

Exercise 18.2.8 *Explain how the inequality in (18.10) correctly combines the inequalities in the two cases, $w < z$ and $w > z$.*

The Objective Function Clearly, the objective function is to *minimize* the number of reversal operations performed at the first $n-1$ levels of $G(n)$. But, by inequality (18.1), exactly one reversal or NO-OP is performed at each of those levels, so the objective function can be written more compactly as:

$$\text{maximize} \sum_{k=1}^{n-1} NOP(k).$$

Also, we want the ILP solution to have the property that all of the reversal operations are done *first*, and then the remaining operations at successive levels are all NO-OPs. This is easily accomplished by including the inequalities:

$$NOP(k) \leq NOP(k+1), \tag{18.11}$$

for all k from 1 to $n-1$.

Exercise 18.2.9 *Explain exactly how the inequalities in (18.11) force an ILP solution to have the form described above.*

18.2.1.3 In Summary

Figure 18.5 shows the abstract ILP formulation for the sorting-by-reversals problem.

$$\text{Maximize} \sum_{k=1}^{n-1} NOP(k)$$

Such that:
(0) For each k from 1 to $n-1$:

$$NOP(k) + \sum_{(p,q):\, p<q} R(p,q,k) = 1$$

(i) For i from 1 to n:

$$\sum_{q=1}^{n} X(1,i,i,q) = 1$$

$$\sum_{p=1}^{n} X(n-1,i,p,Q_2(i)) = 1$$

(ii) For each node q at a level k from 2 to $n-1$:

$$\sum_{p=1}^{n} \Big[\sum_{i=1}^{n} X(k-1,i,p,q) \Big] = 1$$

For each node p at a level k from 2 to $n-1$:

$$\sum_{q=1}^{n} \Big[\sum_{i=1}^{n} X(k,i,p,q) \Big] = 1$$

(iii) For each integer i from 1 to n:

$$\sum_{q=1}^{n} X(k-1,i,q,p) = \sum_{q=1}^{n} X(k,i,p,q)$$

(iv) For each pair (k,w) with k from 1 to $n-2$, and w from 1 to n:

$$\sum_{i=1}^{n} [X(k,i,w,w)] \le NOP(k) + \sum_{\{p,q\}:\, w=(p+q)/2} R(p,q,k) + \sum_{p,q<w} R(p,q,w) + \sum_{p,q>w} R(p,q,w)$$

(v) For each level k from 1 to $n-2$, and each (unordered) pair of positions (w,z):

$$\sum_{i=1}^{n} X(k,i,w,z) \le \sum_{p=a-h<q=b+h:\, h\le Min(a-1,n-b)} R(p,q,k)$$

(vi) For each k from 1 to $n-1$:

$$NOP(k) \le NOP(k+1)$$

All variables are binary

Figure 18.5 An Abstract ILP Formulation for Sorting-By-Reversals.

18.2.2 A Simple, But Effective, Speedup

The length 10 problem shown in Exercise 18.2.1 was solved by Gurobi 7.5 in about 28 seconds on my MacBook Pro. However, there is a simple-to-compute *lower bound* on the minimum number of reversals needed, and when that was added to the ILP formulation, Gurobi solved the same problem in about 12 seconds. Other examples showed even more dramatic improvements, where the use of the lower bound reduced the computation time from minutes to seconds. So, here we discuss the lower bound method.

The Breakpoint Lower Bound Although we have defined the sorting-by-reversal problem as transforming a list of n integers in their natural order, P_1, into a different permutation P_2, the number of reversals that are needed is the same as to transform P_2 back to P_1. And, that viewpoint is much more convenient here, so we will adopt it from now on. Moreover, it is helpful to add a 0 at the left end of P_2 and $n+1$ at the right end of P_2, and consider P_1 to be the ordered list of integers 0 to $n-1$.

To explain the idea of the lower bound, consider the permutation $P_2 = 0\,5\,3\,2\,1\,6$, where $n = 5$. We want to transform it to $P_1 = 0\,1\,2\,3\,4\,5\,6$. Since 0 and 5 are *neighbors* in P_2 but *not* in P_1, we know that some operation will be required to break the neighborness of 0 and 5. Similarly, an operation is required to break the neighborness of 5 and 3, and of 1 and 6. A single reversal operation can break the neighborness of *at most two* neighboring integers, so transforming P_2 to P_1 must require at least $3/2 = 1.5$ reversals. But, the number of reversals is always an integer, and so we can conclude that *at least two* reversals are needed to transform P_2 to P_1. You might ask about the subinterval 3 2 1 in P_2. They are also out of order because they are in backward order. We don't add anything to the lower bound for them, because it *might* happen that they get reversed when some other undesired neighborness is broken.

More formally, two neighboring integers, i and j, in P_2 are called a *breakpoint* if $|i-j| > 1$. We use br to denote the *number* of breakpoints in P_2. Then, the number of reversals needed to transform P_2 to P_1 is *at least* $\lceil \frac{br}{2} \rceil$. This is called the *breakpoint lower bound*. In the above example, 0 5; 5 3; and 1 6 are breakpoints. Note that the *actual* minimum number of reversals needed might be *larger* than the breakpoint bound.

Clearly, given a specific permutation P_2, it is very simple to compute the breakpoint bound; and adding that information to a concrete ILP formulation for the sorting-by-reversals problem cannot hurt.

How to Incorporate the Breakpoint Bound In order to use the breakpoint bound, we need to make a few changes to the ILP formulation. We change the *objective function* to

$$\textit{minimize } S,$$

and add the equality

$$S = n - 1 - \sum_{k=1}^{n-1} NOP(k),$$

and the inequality

$$S \geq \left\lceil \frac{br}{2} \right\rceil.$$

A Final Example Another of the datasets from [145] shows the transformation of the *field mustard plant* genome to the *black mustard plant* genome. In that case, P_1 is again the natural order of the integers 1 through 12, and P_2 is 1 5 9 8 4 6 12 11 10 2 7 3. *Without* including the lower bound in the ILP formulation, Gurobi 7.5 found an optimal solution, with six reversals, in a little under 2 hours (6,973 seconds).

There are nine breakpoints in P_2, so the breakpoint lower bound implies that at least five reversals are needed to transform field mustard to black mustard. Incorporating that bound, making the simple changes discussed above, Gurobi 7.5 found the solution in about 43 minutes (2,572 seconds).[4] The difference in times is roughly the time used in the first computation before its *lb* value reached five. So, including the lower bound can have *significant* impact on the execution time.

Exercise 18.2.10 *The solution that Gurobi found for the* mustard *data is: R(4,7,1), R(3,4,2), R(2,9,3), R(2,4,4), R(2,5,5), R(7,12,6). Verify that this sequence of reversals does convert P_1 (which is the identity permutation) to P_2.*

18.2.3 Software for Sorting-By-Reversals

A Perl program, *breversals.pl*, creates the concrete ILP formulation when given an input permutation, P_2. It can be downloaded from the book website. It is called on a command line in a terminal window with the command:

```
perl breversals.pl number file-name
```

where "number" is the number of integers in the permutation, and "file-name" is the name of the file where permutation P_2 is written on one line. The permutation P_1 is assumed to be the identity permutation 1, 2, ... *number*. The ILP formulation is output to a file whose name begins with the letter "R," followed by the input file-name, followed by the extension ".lp."

Exercise 18.2.11 Software *Use program* breversals.pl *to create concrete ILP formulations for permutations you create, to determine (roughly) the maximum length permutation for which the sorting-by-reversals problem can be solved in "reasonable time" using Gurobi optimizer. Report.*

18.2.4 Signed Sorting-By-Reversals

There is a variant of the sorting-by-reversals problem where each integer in P_2 has a *sign*, either "+" or "-." In the initial, identity permutation, P_1, the sign of every integer is "+." A reversal operation changes the sign of every integer in the reversal interval. For example, +2 +5 -4 -1 +3 changes to +2 +1 +4 - 5 +3 when the interval with the middle three integers is reversed. The goal is to create the signed permutation P_2 from the (all positive) identity permutation P_1, using the minimum number of reversals.

This variant of the reversals problem is called the *signed sorting-by-reversal* problem, and is actually a *more* proper biological model than is the *sorting-by-reversals* problem. Further, after considerable deep investigation, an amazing result was obtained: a worst-case efficient algorithm for the problem was developed [88, 89].

[4] Gurobi 8 took about 33 minutes.

Despite the greater biological fidelity of the signed variant, and the existence of a worst-case efficient algorithm for it, the (unsigned) sorting-by-reversal problem remains important, because unsigned data is currently more available than signed data.

Exercise 18.2.12 *The* signed sorting-by-reversals *problem is an excellent example of a problem that has a worst-case efficient algorithm, but it is very complex, and finding it was an amazing accomplishment. In contrast, an ILP formulation for the problem is much easier to obtain.*

Modify the abstract ILP formulation shown in Figure 18.5 to obtain an abstract ILP formulation for the signed sorting-by-reversal variant.

18.2.5 Reversal Phylogenies

The sorting-by-reversal problem gives a *pairwise distance* between two permutations P_1 and P_2, which represent two gene orders. Knowing that distance for two genomes is of interest in itself, but pairwise distances are often used in more complex contexts, such as in building or labeling phylogenies.

One central problem in this context, is where we know a rooted phylogenetic tree T, where each non-leaf node has exactly two children. Each leaf of T is labeled by a given permutation of the integers from 1 to n, representing an ordering of n genes. We want to create a permutation of the integers from 1 to n, for each non-leaf node of T, which then defines a cost for each edge of T. In particular, if the two end points of an edge are labeled with permutations P_1 and P_2, then the cost for that edge is the minimum number of reversals needed to transform P_1 into P_2. The overall cost of the labeled phylogeny T is then the sum of the edge costs. Naturally, we can define

The Reversal-Phylogeny Problem: Given a phylogenetic tree T and a permutation for each of the leaves of T, create a permutation for each non-leaf node of T, to minimize the overall cost of the labeled phylogeny T.

A solution to the reversal phylogeny problem gives an estimate of the evolutionary history of the permutations at the leaves of T, assuming that the topology (shape) of T is already known.[5]

There is no known efficient algorithm for the reversal-phylogeny problem, and even the sorting-by-reversals problem is known to be NP hard [34]. So, some researchers have suggested a simpler approach that may still be biologically meaningful.

18.2.6 Breakpoints As Surrogates

Instead of using the minimum number of *reversals* needed to transform one permutation into the other, the simpler measure of the number of *breakpoints* defined by the permutations is used in [19,168]. Previously we assumed that one of the permutations is the identity permutation, but now we must relax that assumption, and that requires a small change in the definition of a breakpoint.

[5] The *reversal-phylogeny problem* might also be repeatedly solved when T is not known, and the problem is to create a best phylogeny T given only the leaf permutations.

Given two permutations P_1 and P_2 of n integers, a pair of integers, x and y, define a breakpoint in P_1 if they are neighbors (either x, y or y, x) in P_1, but not in P_2.[6] We say that the breakpoint is *defined* by (P_1, P_2), and use $B(P_1, P_2)$ to denote the *breakpoint distance* between P_1 and P_2, i.e., the number of breakpoints defined by (P_1, P_2). For example, if n is four, and P_1 is 4,2,3,1 and P_2 is 3,4,2,1, then $B(P_1, P_2)$ is two.

When transforming P_1 into P_2, if integers x and y define a breakpoint in P_1, there must be a reversal that breaks the adjacency of x and y in P_1. Hence, the number of reversals needed to transform P_1 to P_2 must at least be $B(P_1, P_2)/2$.

Exercise 18.2.13 *Find a simple argument to show that when given two permutations, P_1 and P_2, the number of pairs of integers that appear as neighbors in P_2 but not in P_1, is $B(P_1, P_2)$, as defined above. That is, we can interchange the roles of P_1 and P_2 in the definition of $B(P_1, P_2)$. Hint: The number of pairs of integers that are neighbors in P_1 is $n-1$, which is the same number of pairs of integers that are neighbors in P_2.*

Breakpoint Cost Given a fixed phylogeny T where each node is labeled by a permutation of the integers 1 through n, we define the *breakpoint cost* of an edge whose end points are labeled with P_1 and P_2, to be $B(P_1, P_2)$. Then, the *breakpoint cost* of the labeled phylogeny T is the *sum* of the breakpoint costs of the edges of T. With these definitions, the modified reversal-phylogeny problem becomes

> **The Breakpoint-Phylogeny Problem:** Given a phylogenetic tree T and a permutation for each of the leaves of T, create a permutation for each non-leaf node of T, to minimize the overall breakpoint cost of T.

The breakpoint-phylogeny problem has been addressed in several papers, including [19, 168, 137]. In those papers, a *heuristic* method is developed to try to solve the breakpoint phylogeny problem. We won't describe the heuristic method, but note that it must solve *many* concrete instances of the following problem:

> **The Breakpoint-Median Problem** Given *three* permutations P_1, P_2, and P_3 of the integers from 1 to n, construct a permutation, S, of the integers from 1 to n, which *minimizes* $B(S, P_1) + B(S, P_2) + B(S, P_3)$.

18.2.6.1 Finally, the TSP in All of This

OK, finally we can get to the punch line, which is that the breakpoint-median problem can be formulated and solved as a traveling salesman path problem, and hence it can be solved by an ILP formulation.

Solving the Breakpoint-Median Problem As a TS Path Problem Given three permutations P_1, P_2, P_3, we define the undirected graph G with nodes labeled 1 to n, and a weighted edge between each pair of nodes. An optimal TS path in this graph will define a permutation, S, of the integers from 1 to n. The cost given to the edge between nodes i and j will be the number of the permutations P_1, P_2, P_3, in which integers i and j are *not* adjacent. Call this edge cost $c(i,j)$.

To see the meaning of $c(i,j)$, note that *if* edge (i,j) is traversed in a TS path of G, then i and j will be adjacent in the resulting permutation, S', and will define a

[6] Note the asymmetry, although we could interchange P_1 and P_2 in the definition.

breakpoint in S', relative to each of the permutations P_1, P_2, and P_3 where i and j are *not* adjacent. So, the cost of that TS path will be exactly $B(S', P_1) + B(S', P_2) + B(S', P_3)$.

Conversely, any permutation, S', of the integers 1 to n defines a TS path with total cost of $B(S', P_1) + B(S', P_2) + B(S', P_3)$. Therefore, an optimal TS path in G will define a permutation of the integers from 1 to n, that optimally solves the breakpoint-median problem.

Exercise 18.2.14 *Consider the following three permutations.*

$$P_1 = 1234567$$

$$P_2 = 3546721$$

$$P_3 = 2431576$$

Create the seven-by-seven symmetric matrix M where the values of the cells $M(i,j)$ and $M(j,i)$ are $c(i,j)$. Then, using one of the previously discussed programs (which one?), create the concrete ILP formulation for the TS path problem with distance matrix M. Next, run Gurobi to solve that concrete ILP formulation, and extract the permutation, S, defined by the given optimal solution. Be sure you understand how to determine which integer is the first integer in S. What is the objective value for the optimal ILP solution? (If I did this exercise right, it should be seven. Is it?) Finally, explicitly check that the sum $B(S, P_1) + B(S, P_2) + B(S, P_3)$ equals the objective value Gurobi gave for the optimal ILP solution. Is it? Are they both seven (I hope)?

19

Maximum Likelihood Pedigree Reconstruction

The task of *family or pedigree reconstruction* is very active in many subfields of ecology, wildlife management, epidemiology, forensic science, and genetic medicine. The general problem is that we have genomic information about a group of individuals who are *thought* to be related, but their precise family structure is unknown. We want to determine from the genomic data, how the individuals are related, and build a multigeneration family pedigree for them.

As one example, family studies have been carried out in *fish* populations, where *multiple generations* can coexist, but unless the fish were observed and marked over time, their family structures are unknown. Knowing family structures in fish is useful in studying population dynamics and responses to environmental change, etc. So, genomic data from the fish is used to deduce family relationships. Similar studies could be carried out on a collection of mummies in order to deduce multigenerational family structures.

19.1 PEDIGREES

We will focus on *diploid* species, where each individual has one male and female parent. The complete specification of family relationships in a *set* of individuals is displayed on a *pedigree*, identifying the *founders* of the set (whose parents are not shown, and possibly unknown), and explicitly showing the two parents of each non-founding member of the set. In graph terminology, a pedigree is a *directed acyclic graph (DAG)* where each node represents an individual in the set, and each directed edge goes from a parent to one of their children. Hence, each node representing a non-founding member will have exactly two incoming edges, and each individual who has children, must be labeled as either male or female, so that each non-founding member has one male parent and one female parent. Further, if genetic information of each individual in the pedigree is known, then the information for any individual must be *inherited* from their parents' genetic information, obeying a specified *genetic-statistical model* of inheritance. This will be defined more completely in the next section. See Figure 19.1a for an example of a pedigree of five individuals – ignore for now the probabilities shown in the figure.

19.1.1 The Biological Inheritance Model: Haplotypes and Genotypes

Recall that in humans, and in other (*diploid*) species, each individual has two (nearly, but not completely identical) "copies" of each chromosome (humans have 23 pairs of chromosomes). More correctly, each of the two copies is called a "homolog."[1] The states (more properly "alleles") at any locus (e.g., nucleotide site, gene, or marker) might be different on the two copies, or they might be the same. Although an individual can only have two alleles at a given site, the set of possible alleles for a population of individuals might be larger. Here, we will assume the simplest case, where there are only two possible alleles, say 0 and 1 (or more commonly in basic genetics "*A*" and "*a*") in the *population*.[2] So at a given locus, an individual might have allele 0 on both chromosome copies; or allele 1 on both copies; or a 1 on one chromosome copy, and a 0 on the other copy.

The alleles at a *sequence* of loci on a *single* homolog (copy of a chromosome) is called a *haplotype* (see Tables 19.1 and 19.2, and also Figure 20.3 in the next chapter). So, an individual, I, has two haplotypes for each pair of homologs, one haplotype that was inherited from the mother of I, denoted M_I; and one haplotype, that was inherited from the father, F_I, of I. A parent of a child transmits one haplotype to the child, but it need not be an exact copy of either of the two of the parent's haplotypes. Rather, at each locus, the allele in the haplotype that a parent transmits to a child only needs to be an allele at that locus in *one* of the two haplotypes of that parent. This reflects a *very simple* model of *meiotic recombination*. Also, in the simplest model, there are *no mutations* of any of the alleles. The pair of haplotypes that an individual I has at a given chromosome is called the *genotype* of individual I for that chromosome. Note, however, that the genotype of an individual, I, does not specify *which* of the two haplotypes was transmitted from the father of I, and which from the mother of I.

For example, consider the pair of haplotypes of the father, F_I, shown in Table 19.1. Each consists of six sites, three of which have the same allele on both haplotypes. Therefore there are three sites with both a 0-allele and a 1-allele, and hence there are $2^3 = 8$ eight possible haplotypes the father can transmit to his child, I. These are called *F-permitted* haplotypes. Permitted haplotypes transmitted from the mother are called *M-permitted*.

Table 19.1 Two Haplotypes of One (Toy) Chromosome for F_I, the Father of Individual I. The pair of haplotypes is called F_I's genotype for that chromosome.

F_I :	0	0	1	0	1	0
	0	1	1	1	1	1

1 There is often an ambiguity about how the word "chromosome" is used. Often people use the word "chromosome" to refer to both of its homologs. For example "The human BRCA1 gene is located on the long arm of chromosome 17." But at other times, the word "chromosome" is used to refer to a single *physical molecule*. For example, "A telomere is a region at both ends of a chromosome." Both of these quotes are (somewhat modified) from Wikipedia. [221] Mostly, context will determine which meaning is intended. I will try to be more precise, but will sometimes rely on context as well.

2 While this is the simplest assumption, it is the most common assumption. Further, it is almost always true that there are at most two *common* observed alleles in a population.

For a given chromosome, an F-permitted haplotype together with an M-permitted haplotype is called a *permitted genotype* that can be transmitted to a child, I, of M_I and F_I.

19.1.2 The Genetic-Statistical Model of Inheritance

A genetic-statistical model for pedigree reconstruction must specify, in general, what genotypes a child can inherit from its parents, and how to calculate the *probability* of each possible genotype for the child, *given* the genotypes of the child's parents. Above, we assumed a simple model for the permitted genotypes, but there are other possible models, and the ILP approach to pedigree reconstruction can accommodate all of the models that I am aware of. Since the focus of this book is on the use of ILP, and not on pedigrees *per se*, we will only use the simple model. But, to fill out the genetic-statistical model, we still have to discuss the *probability* that an individual inherits a specified genotype.

19.1.2.1 The Probabilities of Inheritance

In the simplest model, if a specific parent has two *different* alleles at a locus, then the choice of which allele the parent transmits to a child is completely *random*. That is, each outcome has a probability of 1/2. But if both of a parent's alleles are the same at that locus, then that allele is transmitted with certainty (i.e., with probability one). Hence, if the two haplotypes of a parent differ at d loci, the *probability* is $\frac{1}{2^d}$ that the parent transmits a particular permitted haplotype to the child. The probability that the parent transmits a haplotype that is not permitted is 0.

Now, at any locus, the allele that one parent transmits to a child is completely *independent* of what the other parent transmits. That means that the haplotype, h_F, a child inherits from their father is independent of the haplotype, h_M, that the child inherits from their mother. So, assuming (for now) that we know which haplotype was transmitted from the father and which from the mother, the probability that the child inherits a particular *pair* of permitted haplotypes, h_F and h_M, is the probability of inheriting h_F from the child's father, *times* the probability of inheriting h_M from the child's mother.

Generally, we don't know which of a child's haplotypes is inherited from their father, and which from their mother. So, given the parents' haplotypes, we say that a pair of haplotypes, h_1 and h_2, is a *permitted pair* if one haplotype is F-permitted and the other is M-permitted. It is possible that both are F-permitted and/or both are M-permitted. For notation, we use $P(I|M_I, F_I)$ to denote the probability that individual I inherits the pair of haplotypes shown for I in the pedigree, given the specific haplotype pairs that his parents have. If individual I is a founder, then $P(I|M_I, F_I)$ is defined to have value 1. Otherwise, $P(I|M_I, F_I)$ equals:

(The probability that F_I transmits h_1) $\times$ (The probability M_I transmits h_2)
+ (The probability that M_I transmits h_1) $\times$ (The probability F_I transmits h_2)

Often, one of those two terms will be zero.

The reason that we *add* the two terms, is that they correspond to *disjoint* inheritance outcomes. The first term corresponds to the outcome that h_1 is inherited from the father, and h_2 from the mother; the second term corresponds to the outcome

Table 19.2 The Six-Loci Haplotypes of M_I, F_I, and I.

M_I :	0	0	1	0	1	0
	0	1	1	1	1	1
F_I :	0	1	1	0	0	0
	1	1	1	1	1	1
I :						
h_1 :	0	0	1	0	1	0
h_2 :	0	1	1	0	1	1

that h_1 is inherited from the mother, and h_2 from the father. We can add the terms because it is not possible for *both* outcomes to occur together. The formal probability rule used here is called the "additive rule for disjoint outcomes."

For example, consider the six-loci haplotypes of M_I, F_I, and their child I, given in Table 19.2. The probability that M_I transmits h_1 is:

$$1 \times 1/2 \times 1 \times 1/2 \times 1 \times 1/2 = 1/8.$$

The probability that F_I transmits h_1 is: 0.
The probability that M_I transmits h_2 is: 1/8.
The probability that F_I transmits h_2 is: 1/16.
So, $P(I|M_I, F_I)$, which is the probability that individual I receives the haplotype pair shown in the table, given the haplotypes of M_I and F_I, is: $(0 \times 1/8) + (1/8 \times 1/16) = \frac{1}{128}$. The positive term corresponds to the situation where M_I transmits h_1 and F_I transmits h_2.

19.1.3 Moving to a Pedigree

So now we have seen, in the simplest genetic-statistical model, how to calculate the probability that an individual receives a particular haplotype pair, given the haplotype pairs of the parents of I. But how do we calculate the probability of an entire, fixed *pedigree*, that specifies *both* the haplotype pair for each individual in the pedigree, *and* the *DAG* that shows how the individuals are related?

The Probability of a Pedigree For a pedigree $\mathcal{D}$ we use the notation $P(\mathcal{D})$ to denote the probability of pedigree $\mathcal{D}$, given a particular genetic-statistical model of inheritance. Then,

> $P(\mathcal{D})$ equals the *product* of the individual probabilities of the given haplotype pairs of each non-founding individual in $\mathcal{D}$.

More formally,

$$P(\mathcal{D}) = \prod_I [P(I|M_I, F_I)],$$

where symbol "$\prod$" means the *product* of the terms in the bracket, similar to the way that "$\sum$" means the *sum* of terms. Further, the subscript, I, of $\prod$, means that the terms being multiplied (i.e., inside the brackets) should vary over all possible individuals in the pedigree.

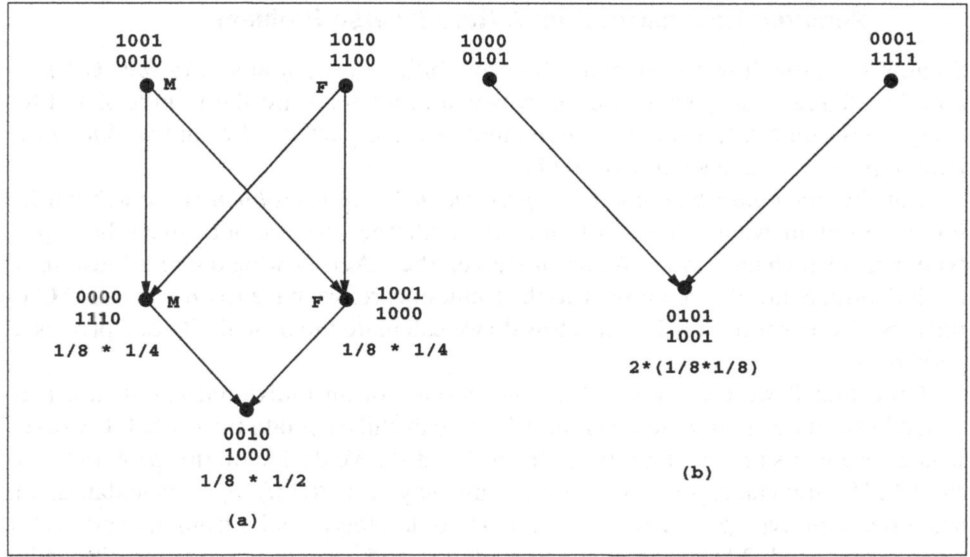

Figure 19.1 (a) A Pedigree of Five Individuals and the Probabilities of Each Haplotype Pair, Using the Simplest Genetic-Statistical Model of Inheritance. The characters "M" and "F" stand for *mother* and *father* respectively. The probability of the pedigree is the *product* of the probabilities at each node. The founders each have an assumed probability of 1. (b) A simple pedigree with three individuals. The probability of the pedigree is equal to the probability of the haplotypes of individual I. Note the multiplication by two in that probability calculation.

An Example Consider the pedigree in Figure 19.1a. In that pedigree there is only one founding couple, M_1 and F_1; their haplotype pairs are shown in the figure. Not knowing probabilities for the founder haplotype pairs, we simply take those to be 1 each. Given the haplotype pairs of the founders, the probabilities of the displayed haplotype pairs of their two children, M_2 and F_2, are calculated as explained above, and shown in the figure. We know that one of the two children is female, because that child is also a mother. Similarly, the other child is a male because that child is also a father. Together, those children, M_2 and F_2, are the parents of individual I (this is a highly incestuous family). The probability of the specified haplotype pair for individual I, given the haplotypes of its parents, is also shown in the figure.

Then, the probability of the pedigree shown in Figure 19.1a is

$$1 \times 1 \times \frac{1}{32} \times \frac{1}{32} \times \frac{1}{16} = \frac{1}{2^{14}},$$

the product of the individual probabilities shown in the figure.

Exercise 19.1.1 *In Figure 19.1, the probability of the haplotypes for individual I is given as* $2*(\frac{1}{8} \times \frac{1}{8})$. *(That is also the probability of this small pedigree, since I is the only non-founder.) It may seem that the stated probability is* twice *what it should be. Explain why the stated probability is correct.*

19.1.4 Pedigree Reconstruction: A Hard Inverse Problem

So now we know how to calculate the probability, $P(\mathcal{D})$, of any *fully specified proposed* pedigree $\mathcal{D}$, i.e., where we know which individuals are the founders and for every non-founder, I, which two individuals are the parents of I, and we know the haplotype pairs that each individual has.

But the *maximum-likelihood pedigree reconstruction problem* is a much harder *inverse* problem, where we do not know the pedigree – we are only given the haplotype pairs of each individual. We are not given the DAG showing the relationships of the individuals, nor do we know who the founders are. We have to *propose* a DAG to fully specify a proposed pedigree. How do we calculate the probability of a proposed pedigree?

Note that if we don't know the true parents of an individual I, but in a proposed DAG they are individuals I' and I'', we can still calculate the probability that I would receive its known haplotypes from I' and I''. We designate this probability as $P(I \mid I', I'')$, and calculate it *exactly* the same way that $P(I|M_I, F_I)$ is calculated, but using the haplotype pairs from I', I'' in place of haplotype pairs from M_I and F_I. So, given a proposed DAG $\mathcal{D}_p$, we can again define and compute the probability, call it $P(\mathcal{D}_p)$, that $\mathcal{D}_p$ generates the known haplotype pairs of the given individuals. That leads to

> **The Pedigree Reconstruction Problem**: Given a set of haplotype pairs, $\mathcal{H}$, for a set of individuals, and knowing the gender of each individual, and given the genetic-statistical model of inheritance, *construct* a proposed pedigree $\mathcal{D}_p$ for $\mathcal{H}$, that *maximizes* $P(\mathcal{D}_p)$ over all possible pedigrees that generate $\mathcal{H}$.

A pedigree $\mathcal{D}_p$ for $\mathcal{H}$ that maximizes $P(\mathcal{D}_p)$ is called a *maximum likelihood pedigree for* $\mathcal{H}$. Of course, we can never be sure we have found the historically correct pedigree for $\mathcal{H}$, but by solving the *maximum likelihood pedigree reconstruction problem* we *can* find a pedigree $\mathcal{D}_p$ that *best agrees* with the observed genetic information, $\mathcal{H}$.

How Can We Solve the Reconstruction Problem? The first thing to note is that a pedigree can be described by specifying the founders, and the gender of each individual in the pedigree, and for each non-founder, its two parents, one of whom is female and the other male. From this observation, we can see that the number of possible pedigrees is *finite*, and all the possible pedigrees can be generated in a mechanical way. So, one conceptually correct way to solve the pedigree reconstruction problem is to generate each possible pedigree for $\mathcal{H}$, and compute its probability, as discussed above.[3] Then, the pedigree with the highest probability is the solution. But, that approach would be hideously inefficient. Instead, we next develop an ILP formulation to solve the *maximum-likelihood pedigree reconstruction problem*. The formulation here is derived from the formulation developed in [51].

[3] It is not actually trivial to do this, since a pedigree must also be acyclic, but we won't be taking this brute-force approach, so we won't worry about how to make it work.

19.2 AN ILP FORMULATION FOR THE MAXIMUM LIKELIHOOD PEDIGREE RECONSTRUCTION PROBLEM

The input to an instance of the pedigree reconstruction problem consists of a set, $\mathcal{H}$, of n haplotype pairs for n individuals. The input also specifies the *gender* of each of the n individuals. We also assume that for each individual I and each *pair* of individuals (I', I''), where neither is I, and one is male and one is female, that the input specifies $P(I \mid I', I'')$. This is for convenience, since those probabilities are easily computed, as explained earlier. The optimal pedigree that is output will be described by specifying the founders, and the two parents for each non-founder, so that the "parent relation" is acyclic – that is, no individual is an ancestor of itself.

19.2.1 First Variables and Inequalities

The ILP formulation will need variables to specify the founders, and the parents of each non-founder. For that, it will have a *binary* ILP variable, $X(I, I', I'')$, for each individual I, and each pair of individuals (I', I''), where neither is individual I, and one is male and one is female. This variable will be set to 1 to indicate that pair (I', I'') are the parents of I; and set to 0 otherwise. So for each of the n individuals, I, the ILP will have the inequality:

$$\sum_{I \neq I', I \neq I''} X(I, I', I'') \leq 1, \tag{19.1}$$

which says that each individual can have *at most* one set of parents.

Founders The variables do not *explicitly* identify the founders. Rather, an individual I will be implicitly identified as a founder by the fact that

$$\sum_{I \neq I', I \neq I''} X(I, I', I'') = 0,$$

is in the solution.

A proposed pedigree where all the individuals are founders would not be meaningful, so the formulation will also have inequalities to limit the number of allowed founders. The inequality

$$\sum_I \left[\sum_{I', I''} X(I, I', I'') \right] = n - 2, \tag{19.2}$$

ensures that there are exactly $n-2$ *non-founders*, so there are exactly two *founders*. If we wish to allow more than two founders, but bound the number of founders between p and $q > p$, we can do so with the following inequalities:

$$\begin{aligned} \sum_I \left[\sum_{I', I''} X(I, I', I'') \right] &\leq n - p, \\ \sum_I \left[\sum_{I', I''} X(I, I', I'') \right] &\geq n - q. \end{aligned} \tag{19.3}$$

19.2.2 Avoiding Cycles

A pedigree is a DAG (directed acyclic graph) and so the parent relations (specified by the X variables that are set to 1) must not imply that an individual is their own ancestor. So, we must have variables and inequalities that enforce that restriction.

We say that an individual I' is an *ancestor* of individual I in a proposed pedigree $\mathcal{D}_p$ if there is a directed path in $\mathcal{D}_p$ starting at the node for I' and ending at the node for I. Then, we use a binary variable, $A(I',I)$ (called an *ancestry variable*) for each pair of individuals, where $I' \neq I$. Variable $A(I',I)$ will be set to 1 if individual I' is an ancestor of I in the proposed pedigree defined by the setting of the X variables. Then, the inequality:

$$A(I',I) + A(I,I') \leq 1, \tag{19.4}$$

forbids the situation where I' is an ancestor of I, and I is an ancestor of I'. By including such an inequality for each pair of distinct individuals, the proposed pedigree will be forced to be acyclic.

Ancestry Using ancestry variables avoids cycles, but how do we implement the ancestry relations? We use the following logic:

(a) *If* for three individuals $I, I', I'', X(I,I',I'')$ is set to 1 (so (I',I'') are the proposed parents of I), *then* both variables $A(I',I)$ and $A(I'',I)$ must be set to value 1.

(b) Further, using the logic of *transitivity*, *If* for three individuals $I, I', I^*, A(I',I)$ is set to 1, *and* $A(I^*,I')$ is set to 1, *then* $A(I^*,I)$ must be set to 1.

Exercise 19.2.1 *Does the logic stated in points* **(a)** *and* **(b)** *correctly express the desired meaning of the A variables? That is, does the logic correctly express the ancestry relation? Explain.*

Exercise 19.2.2 *State explicit linear inequalities to implement the logic in points* **(a)** *and* **(b)**. *If-Then idioms may be useful.*

Exercise 19.2.3 *Note that the logic above ensures that variable $A(I',I)$ will be set to 1* if I' *is an ancestor of I in the proposed pedigree defined by the setting of the X variables. But, the logic does not forbid $A(I',I)$ from being set to 1 even when I' is not an ancestor of I. Explain why that is true. Explain why that does not cause any problems for the full ILP formulation.*

19.2.3 The Objective Function

An optimal solution to the ILP formulation must create a pedigree $\mathcal{D}_p$ that maximizes $P(\mathcal{D}_p)$ over all pedigrees for $\mathcal{H}$. Recall that $P(\mathcal{D}_p)$ is defined as a *product* of probabilities. In more detail, if $\mathcal{D}_p$ is the pedigree specified by the values of the X variables in an optimal solution of the ILP formulation, then $P(\mathcal{D}_p)$ is:

$$\prod_{I,I',I''} [P(I \mid I',I'') \times X(I,I',I'')]. \tag{19.5}$$

Unfortunately, this objective function is *not* permitted in an ILP. The problem is that the objective function is *not* a linear function of the variables (X variables) that it contains. Each value $P(I \mid I',I'')$ is a *constant* number given as part of the input, so the multiplication of it by an X value is not the source of nonlinearity. Rather, it is the fact that all of the terms are multiplied together (the effect of $\prod$), so that the objective function multiplies an X variable by another X variable (actually many of them). In linear and integer linear programming, a variable is only allowed to be *multiplied* by a constant number, not another variable.[4]

19.2.3.1 The Fix

First, we think of the objective function a bit differently. We will think of it as:

$$\prod_{\{I,I',I'' \,:\, X(I,I',I'')=1\}} [P(I \mid I',I'')], \tag{19.6}$$

so, a $P(I|I',I'')$ term will only be part of the product in (19.6) if the value of the corresponding $X(I,I',I'')$ is set to 1. This way of thinking about the objective function makes intuitive sense and is mathematically correct. However, we still can't explicitly use (19.6) as the objective function in the ILP formulation, because the X variables that will have value 1 will only be determined during the *solution* of the ILP. So, we don't know which P terms to include at the time that the concrete formulation is being constructed.

A Further Fix Recall (probably from high school) that the *logarithm* of the *product* of two numbers, x and y say, is the *sum* of the logarithm of x and the logarithm of y, and this is true no matter what the base of the logarithm is. More formally,

$$\log(xy) = \log x + \log y, \tag{19.7}$$

so,

$$\log\left[\prod_{\{I,I',I'' \,:\, X(I,I',I'')=1\}} [P(I \mid I',I'')]\right] = \sum_{\{I,I',I'' \,:\, X(I,I',I'')=1\}} \log\,[P(I \mid I',I'')]. \tag{19.8}$$

Now, the logarithm (with any base) is an *increasing* function, so a proposed pedigree $\mathcal{D}_p$ for $\mathcal{H}$ that maximizes the function in (19.8) will be a pedigree that maximizes the functions in (19.6) and (19.5), and so will be a pedigree that defines an optimal solution to the *maximum likelihood pedigree reconstruction problem.*

Almost Done Unfortunately, the function in (19.8) *still* isn't in the correct form to be used as an objective function in an ILP formulation. But notice, that

$$\sum_{\{I,I',I'' \,:\, X(I,I',I'')=1\}} \log\,[P(I \mid I',I'')],$$

will always be equal to:

$$\sum_{I,I',I''} \log\,[P(I \mid I',I'')] \times X(I,I',I''), \tag{19.9}$$

which *finally is* a linear function of the X variables. So, we use the function in (19.8) as the objective function in the ILP formulation, and the optimal solution will be a maximum likelihood pedigree for $\mathcal{H}$.

[4] Although the *addition* of variables is allowed.

Exercise 19.2.4 *Following the logic and inequalities developed above, write a complete abstract ILP formulation for the maximum likelihood pedigree reconstruction problem.*

Exercise 19.2.5 *Certainly, the set of founders in a pedigree must include at least one female and at least one male. But in the ILP formulation given in this section, there are no explicit inequalities to require that. Why is this not a problem? Explain.*

Exercise 19.2.6 *The statement of the pedigree reconstruction problem used here assumes that the gender of each individual is known and given as part of the input to a problem instance. In contrast, in [51] it is assumed that gender information is not known. Then, in any proposed pedigree, a gender must be specified for each individual who is a parent. This constrains the possible solutions, since a consistent assignment is not always possible. For example, in a pedigree that proposes the parental pairs* $(1,2), (2,3), (1,3)$, *there is no way to assign gender to the individuals so that each pair has one male and one female. Hence, additional variables (to assign gender to individuals) and inequalities (to assure that each parental pair contains one male and one female) need to included in the ILP formulation. Describe the required variables and inequalities, and then show that these are correct.*

Exercise 19.2.7 *In the ILP formulation developed above for the pedigree reconstruction problem, there is no concept of age or generation, and no mechanism to disallow two individuals of very different ages (or generations) from parenting a child. In some datasets, this would allow the creation of a pedigree that is unrealistic, or even biologically infeasible. For example, it allows two individuals who could not be alive at the same time to "parent" a child. So, we want to develop logic and inequalities to avoid this problem.*

We define the minimum generation *and the* maximum generation *of an individual I as the* fewest *and* largest *number of ancestors, respectively, of I, along any path from a founder to I. We denote the first value as* minG(I) *and the second value as* maxG(I).

We want to implement the logic: If $X(I, I', I'')$ *has value 1,* Then maxG(I) =

$$\max\left[maxG(I'), maxG(I'')\right] + 1,$$

and minG(I) =

$$\min\left[minG(I'), minG(I'')\right] + 1.$$

For any individual I chosen to be a founder, the ILP formulation must have

$$maxG(I) = minG(I) = 0.$$

(a) *Using known idioms, show how to implement these relations with linear inequalities.*

Then, to implement the requirement that the parents of I are in roughly the same generation, we require that

$$maxG(I) - minG(I) \leq d,$$

for each individual I; where d can be chosen to be 1 or 2 for species such as humans, or it can be chosen to be larger for other species, such as certain fish.

(b) *Explain how this requirement discourages an ILP solution from deciding that two individuals of very different ages are the parents of a child. That is, why does the requirement likely do the right thing? But conversely, think of a scenario where the requirement is satisfied, and yet the parents are of unrealistically different ages to parent a child.*

(c) *Can you think of other ways to make the pedigree produced by the ILP solution be more biologically realistic, and how they can be implemented in the ILP formulation?*

19.2.4 The Backstory

I chose to include the topic of pedigree reconstruction topic in the book after giving a talk on the use of ILP in computational biology. A biologist in the audience said that most of her quantitative work and the work of other biologists who do quantitative biology of any sort, is based on *statistical inference*, either in a *maximum likelihood* or a *Bayesian* framework. She asked if ILP was useful there. The question of maximum likelihood is the easiest to address, because at its heart, it is an *optimization question*: under a specified statistical model, what structure best fits or generates the observed data? Further, what at first looks like a *multiplicative* objective function (finding a structure to maximize a *probability*) is easily turned into an *additive* objective function (through the use of logarithms) of the kind that ILP requires. So definitely, as demonstrated in this section, ILP can be used for maximum likelihood inference. The key issue is not that probability is involved, but rather, what the statistical model is, and what kind of structures are to be inferred. For many models and structures, ILP can be used to find the structure that, under the statistical model, maximizes the likelihood of producing the observed data.

In principle, ILP can also be used in *Bayesian analysis*, although there has been less use of ILP for Bayesian analysis than for maximum likelihood. The reason is that Bayesian analysis outputs a *probability distribution* over a set of possible structures, rather than one "best" structure, or even several co-optimal or close to optimal structures. But ILP can certainly be used in the inner loop of a method that seeks a range of structures and a calculation of the probability of each one, so the use of ILP in Bayesian analysis seems an interesting research topic to pursue.

19.3 INVERSE GENETICS PROBLEMS

Designing Pedigrees The goal of the *pedigree reconstruction* problem is to reconstruct history. But sometimes we want to *make* history. In particular, consider the problem where we have a set of founders, and also have a desired genomic genotype that none of the founders have. However, the genotype at each site of interest does occur in at least one of the founders. So, if we could *mix-and-match* (recombine) from the founders' genotypes, we could create an individual with the desired genomic genotype.[5]

> With the availability of genetic maps, containing the exact locations on the genome of genetic markers associated with desirable traits, selection at the genotypic level has become possible. This knowledge allows to design a schedule of crossings of individuals resulting ultimately in an individual with all alleles corresponding to desired favorable traits present. [33]

So, we want to create an individual with the desired genomic genotype, through a series of *genetic crosses* obeying classical Mendelian genetics. How should we design the schedule of crosses to minimize "the [expected] number of generations, the number of crossings, and the required populations size?" That problem is successfully addressed using ILP in [33].

Another Reverse Genetics Problem In the crossing problem discussed above, the founders are given, and the crosses are made by a plant breeder, according to a worked-out schedule. Another approach is to *select* the founders, and let them breed and cross as they will. Then, the optimization problem is which founders to select in order to create (with high probability) individuals with desired traits.

[5] This may sound creepy, but the application is not to human eugenics – its to plant breeding.

One application of this problem is to *captive breeding*, used to help save endangered species. This is best explained in the introduction of a paper that uses ILP for the *founder selection* problem:

> Methods from genetics and genomics can be employed to help save endangered species. One potential use is to provide a rational strategy for selecting a population of founders for a captive breeding program. The hope is to capture most of the available genetic diversity that remains in the wild population, ... In particular, we develop a mixed-integer linear programming technique that identifies a set of animals whose genetic profile is as close as possible to specified abundances of alleles (i.e., genetic variants), subject to constraints on the number of founders and their genders and ages. [135]

20

Two DNA Haplotyping Problems

20.1 INTRODUCTION

In Chapter 19 we introduced the concepts of *haplotypes*, and their role in pedigree construction. In this chapter we go deeper into issues of how haplotypes themselves are inferred, and the use of integer linear programming in solving those problems of inference.

Recall from the discussion in Section 19.1.1 that in *diploid* species, each individual has two *homologs* (nonidentical copies) of each chromosome, and that the DNA sequence on a single homolog is called a *haplotype*. Currently, it is difficult to separately learn the DNA sequences on only one of the two haplotypes. Instead, what can be easily obtained is information that *mixes* together parts of both haplotypes. However, knowing the separate sequences of the two haplotypes has great biological value. So *inferring* the two haplotype sequences from mixed information is a key computational problem in genomics and computational biology. The problem takes several forms, based on differing DNA sequencing technology and biological applications. The general reconstruction problem is called the *haplotype inference (HI), or phasing* problem.

In this chapter, we consider two variants of the haplotype inference problem: The *individual* variant, also called the *haplotype assembly* problem; and the *population* variant. The way that haplotype information is mixed is different in the two variants, but for both, we will show how integer linear programming has been used to address the HI problem.

20.2 HAPLOTYPE ASSEMBLY: THE INDIVIDUAL VARIANT OF THE HI PROBLEM

In this variant of the HI problem, we want to reconstruct (or estimate) the DNA sequences of the two haplotypes from a chromosome (or a fixed segment of it) of a *single, specific* individual. This variant frequently arises when doing a (relatively quick and cheap) sequencing of the genome (or parts of it) of an individual *human*.

In that technology, the sequencing reads are short, with a high error rate, but a mature "reference genome" for humans exists, and is used in the reconstruction.[1]

The input to this variant of the HI problem is a set of *reads* (short DNA sequences) from different intervals of the two (unknown) haplotypes. The (true) haplotype sequences are denoted H_0 and H_1. Each individual read is a sequence that comes from only one of the haplotypes, but, for any two reads, it is *not* known whether the reads originate from the *same* haplotype, or from the two *different* haplotypes. So, it is not immediate how to *assemble* the DNA sequences from the reads into two long sequences, H'_0 and H'_1, to best estimate the true haplotypes H_0 and H_1.

The key problem is how to *divide* the set of reads into two sets, with the interpretation that the reads in one set all come from one haplotype, and the reads in the other set all come from the other haplotype. This division is used to create H'_0 and H'_1. We assume that we know the *position* of each read, relative to the length of the chromosome (or segment of it) being sequenced. For example, we might know that the start of a particular read is located 300 bases from the 5' end of the chromosome of interest. This assumption is realistic for humans and other species where a finished reference genome is available. The individual reads may not agree completely with the reference (particularly because there is only a single reference sequence, but there are two nonidentical haplotypes), but the agreement between the reads and the reference sequence is sufficient to determine the correct location of the read. Then, we have the

> **Haplotype Assembly Problem:** Given a matrix R containing n (equal-length) reads that collectively span m positions in the genome, *create* DNA sequences H'_0 and H'_1 of length m each, and *place* each read into either H'_0 or H'_1, in the known location of that read, to *minimize* the total number of *differences* between the characters in the reads and the corresponding characters in H'_0 or H'_1, where the reads are assigned (see Figure 20.1 for an example).

The assumption that all the reads are of the same length is for convenience of exposition, and is easy to modify. If there are *no errors* in the reads, then the unknown sequences H_0 and H_1 would be an optimal solution to the haplotype assembly problem, i.e., $H'_0 = H_0$ and $H'_1 = H_1$. In that solution, each read would be assigned to the sequence it originated from. However, there are generally errors in the data, but if the errors are essentially random, and there are a "sufficient" number of reads, sequences H'_0 and H'_1, obtained from the optimal solution of the HI problem, should be good *estimates* for H_0 and H_1.

The haplotype assembly problem has been addressed with integer linear programming, using an ILP formulation that we will develop next.

20.2.1 An ILP Formulation for the Haplotype Assembly Problem

Here we develop an abstract ILP formulation for the haplotype assembly problem, following (essentially) the approach in [41].

[1] The reference genome is, of course, not identical to either of the haplotypes of the individual of interest. It is a reflection of humans as whole, accurate at sites where there is little variation over the population.

R	1	2	3	4	5	6	7	8	9	10
unknown sequence H_0	A	G	G	C	C	A	T	G	G	A
unknown sequence H_1	A	A	G	T	C	T	C	G	G	C
Read 1	A	G	G	T	-	-	-	-	-	-
Read 2	A	A	G	T	-	-	-	-	-	-
Read 3	A	G	G	C	-	-	-	-	-	-
Read 4	-	-	G	T	T	T	-	-	-	-
Read 5	-	-	G	C	C	A	-	-	-	-
Read 6	-	-	G	T	C	A	-	-	-	-
Read 7	-	-	G	C	C	A	-	-	-	-
Read 8	-	-	G	T	T	A	-	-	-	-
Read 9	-	-	-	-	C	A	T	C	-	-
Read 10	-	-	-	-	C	T	T	G	-	-
Read 11	-	-	-	-	-	A	T	C	G	-
Read 12	-	-	-	-	-	T	C	G	G	-
Read 13	-	-	-	-	-	A	C	G	G	-
Read 14	-	-	-	-	-	-	T	C	G	A
Read 15	-	-	-	-	-	-	C	C	G	C
Read 16	-	-	-	-	-	-	C	G	G	C
deduced sequence H'_0	A	G	G	C	C	A	T	C	G	A
deduced sequence H'_1	A	A	G	T	T	A	C	G	G	C

Figure 20.1 Sixteen Reads From the Two Haplotypes, H_0 and H_1, of an Interval in Some Chromosome. Each row displays a single read, aligned to its known location on the chromosome. The matrix of the reads is denoted R. Each read in this example is of length four – dashes in a row indicate positions that are outside of the read. The reads span an interval of length $m = 10$. Each read comes from one of the unknown haplotypes H_0 or H_1, but we *don't* know which one, and a read generally *will contain* errors. A solution to this instance of the HI problem produced the two deduced sequences H'_0 and H'_1, shown at the bottom. These differ from H_0 and H_1 at three sites. The solution assigns Reads {1,3,5,7,9,10,11,14} to H'_0, and assigns the other eight reads to H'_1. There are seven differences between the reads and the deduced sequences, where the reads are placed. Note that the originating sequence pair (H_0, H_1) is not an optimal solution to the haplotype assembly problem for this problem instance. This is due to the errors in some of the reads. Without errors in the reads, (H_0, H_1) would necessarily be an optimal solution to the haplotype assembly problem.

A Simplifying Assumption We will first assume that every column in R contains *at most two* distinct DNA characters, as is the case in Figure 20.1. This is the standard assumption for *SNP* (single nucleotide polymorphism) sites, but it is not difficult to relax this assumption. We leave that to the reader. Note that we are *not* assuming that only two distinct DNA characters appear in all of R – we are only assuming that at most two distinct DNA character appear in any *column* of R. Because of this assumption, we can *recode* the characters in any column as "0" and "1." For concreteness, reading top to bottom in a column, we recode all occurrences of the first DNA character encountered to be "0," and recode all occurrences of the other DNA character to be 1. For example, in Figure 20.1 the second column of R contains the DNA characters "GAG," and is recoded to "010." The recoding makes R a binary matrix, so the optimal H'_0 and H'_1 will be binary sequences.[2]

Some Notation We assume the reads are numbered (arbitrarily) for unique identification. Since we know the location of each read, relative to the (unknown) sequences H_0 and H_1, we can define the character in a read that will be placed at a given position, say k, of H'_0 or H'_1. Assuming that Read j spans position k, we use $c(j,k)$ to denote the character, 0 or 1, in Read j that will be aligned with position k in H'_0, or H'_1.

Note that $c(j,k)$ is *not* a variable, since Read j is given as part of the input R. In Figure 20.1, Read 16 contains the DNA sequence "CGGC" (recoded to 1101) starting at position 7, so $c(16,7)$ is 1; $c(16,8)$ is 1; $c(16,9)$ is 0; and $c(16,10)$ is 1.

The ILP Variables and Objective Function The ILP formulation will use binary variables $H'_0(k)$ and $H'_1(k)$, for k from 1 to m, with the obvious interpretation that the values given to those $2m$ variables define the sequences H'_0 and H'_1 in a solution to the haplotype assembly problem.

For each Read j, the formulation will use the binary variable $A(j)$, with the interpretation that $A(j)$ has value 0 if and only if Read j is assigned to be in sequence H'_0. Equivalently, $A(j)$ has value 1 if and only if Read j is assigned to H'_1. For example, in Figure 20.1, $A(j)$ is 0 for $j \in \{1,3,5,7,9,10,11,14\}$, and $A(j)$ is 1 for all other indices from 1 to 16.

The ILP formulation will also use the binary variable $z_0(j,k)$, which must be set to 1 *if* (but not *only-if*) the value at position k of H'_0 is *unequal* to $c(j,k)$, i.e., the corresponding value in Read j. Similarly, the formulation will use the binary variable $z_1(j,k)$, which must be set to 1 if (but not *only-if*) the value at position k of H'_1 is *unequal* to $c(j,k)$. For example, in the ILP solution that produced sequence H'_0 shown in Figure 20.1, $z_0(13,7)$ must have been set to 1, because $c(13,7) = 1$ (recoded from "C") and $H'_0(7) = 0$ (recoded from "T").

Next, for each Read j and position k covered by Read j, the formulation will use the binary variable $z(j,k)$, with the following interpretation:

> *If* Read j has been assigned to H'_0 *and* $z_0(j,k)$ has value 1 (i.e., $c(j,k) \neq H'_0(k)$), then $z(j,k)$ must be set to 1. Similarly, *if* Read j has been assigned to H'_1 and $z_1(j,k)$ has value 1, then $z(j,k)$ must be set to 1.

For example, in the ILP solution that produced sequence H'_1 shown in Figure 20.1, $z_1(15,8)$ must have been set to 1, because $c(15,8) = 0$ (recoded from 'C') and $H'_1(8) =$ 1 (recoded from "G"). Then, since Read 15 is assigned to H'_1 in that ILP solution, so $z(15,8)$ must have been set to 1.

Finally, for each j from 1 to n, we define the variable $z(j)$ with the inequality

$$z(j) = \sum_{k=1}^{k=m} z(j,k),$$

[2] It is easy to see that for any column k, the character aligned with position k of the optimal H'_0 (respectively, H'_1) must be a character in column k of R. For example, in Figure 20.1, the two characters in column two of R are "A" and "G." Consistent with that, sequence H'_0 has character "A" in position two, and H'_1 has character "G" in position two. Note however, that it is still possible for the same character to appear at a given position k in both H'_0 and H'_1, even if there are two distinct characters in column k of R. An example is in position 6 in Figure 20.1.

so $z(j)$ is the *total* number of errors contributed by Read j in a solution to the haplotype assembly problem. Therefore, the objective function for the ILP formulation is:

$$\text{Minimize} \sum_{j=1}^{j=n} z(j). \tag{20.1}$$

The ILP Inequalities Above, we defined and used the ILP variables $z_0(j,k)$, $z_1(j,k)$, and $z(j,k)$, but we did not develop the linear inequalities used to set their values. Here, we first detail the inequalities used to set variables $z_0(j,k)$ and $z_1(j,k)$. Then we will detail the inequalities used to set the value of $z(j,k)$.

To set the value of $z_0(j,k)$, given H'_0, the ILP needs to determine if $c(j,k)$ is *unequal* to $H'_0(k)$. For that, we can use the NOT-EQUAL idiom for two binary *variables*, discussed in Section 12.3, even though $c(j,k)$ is a constant (0 or 1, given in the input) and not a variable.

Exercise 20.2.1 *Explain why we can use the NOT-EQUAL idiom here, even though $c(j,k)$ is a constant, not a variable.*

Applying the inequalities for the NOT-EQUAL idiom detailed in (12.14), we get:

$$\begin{aligned} z_0(j,k) &\geq c(j,k) - H'_0(k), \\ z_0(j,k) &\geq H'_0(k) - c(j,k). \end{aligned} \tag{20.2}$$

The inequalities for $z_1(j,k)$ are identical, with the replacement of "0" by "1."

Inequalities for $z(j,k)$ For each Read j and position k covered by Read j, the ILP formulation uses the following logic to set $z(j,k)$:

$$\begin{aligned} &a) \textit{ If } (z_0(j,k) = 1) \textit{ AND } (A(j) = 0) \textit{ then } z(j,k) \textit{ must be set to } 1. \\ &b) \textit{ If } (z_1(j,k) = 1) \textit{ AND } (A(j) = 1) \textit{ then } z(j,k) \textit{ must be set to } 1. \end{aligned} \tag{20.3}$$

The inequality that implements the logic for (20.3 a) is:

$$z_0(j,k) - A(j) - z(j,k) \leq 0. \tag{20.4}$$

Exercise 20.2.2 *Write out the linear inequality needed to implement (20.3b). It is analogous to the inequality for (20.3a), but involves more than the replacement of "0" with "1." The solution is shown in Figure 20.2, but don't peek – try to figure it out yourself.*

20.2.2 Summary

Putting together all of the inequalities, the full abstract ILP formulation for the haplotype assembly problem is shown in Figure 20.2.

Exercise 20.2.3 *Another way to think about variable $z(j,k)$ is with the statements:*

$$\begin{aligned} |c(j,k) - H'_0(k)| - A(j) - z(j,k) &\leq 0, \\ |c(j,k) - H'_1(k)| + A(j) - z(j,k) &\leq 1. \end{aligned} \tag{20.5}$$

Explain why this is correct. Also, when we implement the absolute value function with the linear inequalities in (15.13), *discussed in Section 15.3.1, does the resulting ILP formulation for the haplotype assembly problem become the same as the formulation in Figure 20.2?*

$$\text{Minimize} \sum_{j=1}^{j=n} z(j)$$

Such that:

For each *Read* j from 1 to n, and each position k from 1 to m:

$$z_0(j,k) \geq c(j,k) - H'_0(k)$$
$$z_0(j,k) \geq H'_0(k) - c(j,k)$$
$$z_1(j,k) \geq c(j,k) - H'_1(k)$$
$$z_1(j,k) \geq H'_1(k) - c(j,k)$$
$$z_0(j,k) - A(j) - z(j,k) \leq 0$$
$$z_1(j,k) + A(j) - z(j,k) \leq 1$$

For each *Read* j:

$$z(j) = \sum_{k=1}^{k=m} z(j,k)$$

All variables are binary.

Figure 20.2 The Full Abstract ILP Formulation for the Haplotype Assembly Problem.

Exercise 20.2.4 *The ILP formulation sets binary variable $z_0(j,k)$ to 1 if (but not only-if) the value at position k of H'_0 is unequal to the corresponding value in Read j. Explain why the only-if direction is not needed. Then, answer the similar question about the variable $z(j)$.*

Exercise 20.2.5 *Explain how to modify the ILP formulation to the haplotype assembly problem, when a column could contain more than two characters. What changes are needed in the ILP if the reads can be of unequal length?*

Looking forwared to the introduction of new sequencing technologies, the authors in [149] use integer linear programming to solve the haplotype assembly problem when read lengths increase, but with higher error rates.

20.3 THE POPULATION VARIANT OF THE HI PROBLEM

In this section we discuss the *population* variant of the HI problem, where the data does *not* come from just a *single* individual, but from *many, unrelated* individuals. This variant is very common in genomic scans looking for the locations of DNA contributing to genetic diseases. As in the case of the individual HI problem, we want to determine haplotypes from information that is a *mixture* of haplotype information. But the form of the mixture is different from the case of the individual HI variant. Also, in the population variant, we do not assume that a reference sequence is available.

Some Terminology Recall that at any single site c on a chromosome, the two DNA characters in the two homologs form the *genotype* at site c. More generally, the collection of genotypes at the sites in the two homologs of a chromosome is called a *chromosome genotype*, even if haplotype sites do not span a complete chromosome.

If the alleles at site c are P and Q, then we use '$P|Q$' to denote the genotype at site c. For example, if an individual has haplotypes

$$AGGCCA$$

and

$$AAGCCT$$

at six sites, then they will have genotypes $A|A$, $G|A$, $G|G$, $C|C$, $C|C$, $A|T$ at the six sites. Because of the second and sixth sites, it is *not* possible to *uniquely* deduce the two *original* haplotypes, from the genotypes. This is because, in addition to the above pair of haplotypes, the haplotype pair

$$AAGCCA$$

and

$$AGGCCT$$

is also consistent with the genotype data.

Most of the genomic data collected today is *SNP* data, so at most two of the four characters in the DNA alphabet are observed at any site. Hence it is possible to recode the two characters observed *at that site* as binary characters, 0 and 1 (see Figure 20.3).

	1	2	3	4	5	6
individual 1: haplotype 1	0	0	0	0	0	0
individual 1: haplotype 2	0	1	0	0	0	1
individual 1: genotype	0	2	0	0	0	2
individual 2: haplotype 1	0	0	0	0	0	1
individual 2: haplotype 2	0	0	0	0	0	0
individual 2: genotype	0	0	0	0	0	2
individual 3: haplotype 1	0	0	1	1	1	0
individual 3: haplotype 2	0	0	1	1	1	0
individual 3: genotype	0	0	1	1	1	0
individual 4: haplotype 1	1	0	1	1	0	0
individual 4: haplotype 2	0	0	1	1	1	0
individual 4: genotype	2	0	1	1	2	0

Figure 20.3 Four Pairs of SNP Haplotypes and Their Derived Genotypes from Four Individuals. Given only the genotypes, we would know the haplotypes for individual 3, and correctly deduce the haplotypes for individual 2, but can not uniquely deduce the originating haplotype pairs for individuals 1 and 4.

20.3.1 Genotypes and Haplotype Inference

Haplotype Inference For technological reasons, it has been (and still continues to be) much cheaper and easier to obtain genotype data than haplotype data, while generally, haplotype data is more biologically informative: biological and chemical activity tends to take place on each of the two chromosome homologs separately. Further, there are genetic diseases and traits that involve *both* homologs individually, so understanding these traits requires observing the both of the haplotypes [191]. So, a critical technical problem that has received great attention in the last three decades is the problem of inferring the underlying haplotypes (two per individual), from genotypes (one per individual). This is called *phasing* the genotype.

We assume that in each haplotype there are m sites, where each site can have one of two states (alleles), 0 and 1. The sampled population consists of n individuals. For each individual, we know the genotype data possessed by the individual, but do not know the two originating haplotypes creating that genotype. This leads to input information that is *ternary* rather than binary, i.e., consisting of entries from the set $\{0,1,2\}$, as we explain next.

20.3.2 Genotypes from Haplotypes

Two haplotypes of length m define a single *genotype*, g, of length m under the following rules:

> A site in genotype g has a value of 0 or 1, *if* the two corresponding sites in the underlying haplotypes *both* have *that* value. A site in the genotype has value of 2 if the corresponding values in the haplotypes do *not* agree with each other, i.e., one is 0 and one is 1 (see Figure 20.3).

In the *inverse, phasing* problem, one is given a chromosome genotype, g, and tries to *phase* (i.e., reconstruct) the original haplotype pair that gave rise to the genotype.

The HI Problem in Populations Abstractly, input to the haplotype inference (HI) problem in populations consists of a set of n *chromosome genotypes* $\mathcal{G}$, each of length m, where each site in the genotype has value either $0, 1$, or 2. Then, a *solution* to the HI problem is a phasing of each of the n genotypes. So, a solution consists of n *pairs* of binary sequences, which we represent in the matrix $M(\mathcal{G})$. For any genotype $g \in \mathcal{G}$, the associated binary sequences h_1, h_2 in $M(\mathcal{G})$ must both have value 0 (or 1) at any site where g has value 0 (or 1); but for any site where g has value 2, *exactly one* of h_1, h_2 must have value 0, and the other must have value 1. See Figure 20.4 for an instance of the HI problem and two solutions.

Many Solutions For an individual whose genotype vector has k sites with value 2, there are 2^{k-1} haplotype pairs that could appear in a solution to the HI problem. That is, there are 2^{k-1} ways to phase the genotype. For example, if the observed genotype g is 0212, then the pair of binary sequences 0110, 0011 is one feasible phasing, out of two feasible phasings (see Figure 20.4).

Of course, we want to find the explanation that actually gave rise to genotype g, and to all of the genotypes in $\mathcal{G}$. However, without additional biological insight one cannot know which of the many possible phasings is the likely to be the correct one.

	1	2
person 1	2	2
person 2	0	2
person 3	1	0

(a) Matrix $\mathcal{G}$ with three genotypes, each with two sites

	1	2
1	0	0
2	1	1
3	0	0
4	0	1
5	1	0
6	1	0

(b) First HI solution (phasing)

	1	2
1	0	1
2	1	0
3	0	0
4	0	1
5	1	0
6	1	0

(c) Second HI solution (phasing)

Figure 20.4 A Set of Genotypes, $\mathcal{G}$, and Two HI Solutions, i.e., Two Ways to *Phase* the Genotypes. There are three individuals and two sites, shown in panel (a). The first phasing, shown in panel (b), cannot be derived on a perfect phylogeny with all zero ancestral sequence since it violates the perfect-phylogeny theorem (stated in Section 3.2.2). The second phasing, shown in panel (c), can be derived on a perfect phylogeny, and is a solution to the PPH problem.

20.3.3 The Need for a Genetic Model

Algorithm-based haplotype inference would be impossible without the implicit or explicit use of some *genetic model* of *haplotype evolution*, either to assess the *biological fidelity* of proposed phasing, or to guide the algorithm in constructing a phasing. Many of the underlying models that have been articulated are *stochastically* based (often using a *coalescent* model [197] of haplotype evolution), but several are more *combinatorially* based, and one can sometimes view a model in either light.

Combinatorial Methods Combinatorial (or optimization) methods often state an explicit *objective function* that indirectly reflects an evolutionary model for the creation of unknown haplotypes, and hence for the creation of the given genotypes. One then tries to *optimize* the objective in order to solve the HI problem. There are several combinatorial models that have been proposed and extensively studied for the HI problem. Two of these are the *perfect-phylogeny haplotyping (PPH)* model [84], and the *pure-parsimony* model [85], which have both been addressed using integer linear programming.

20.4 PERFECT PHYLOGENY HAPLOTYPING

The Perfect-Phylogeny Haplotype (PPH) Problem:

Given a set of n genotypes, $\mathcal{G}$, determine if there is a phasing, $M(\mathcal{G})$, of $\mathcal{G}$, such that $M(\mathcal{G})$ can be generated by a *perfect phylogeny* $T(\mathcal{G})$, with the all-zero ancestral sequence (see Section 3.2.2 for a definition and discussion of perfect phylogeny).

In addition to specifying the perfect phylogeny $T(\mathcal{G})$, a solution must also assign each genotype in $\mathcal{G}$ to a *pair* of haplotypes derived on $T(\mathcal{G})$, in order to explicitly phase the genotypes in $\mathcal{G}$ (see Figure 20.4 for a simple example). In that example, there are two solutions to the HI problem, but only one solves the PPH problem. That is, there are two phasings of $\mathcal{G}$, but only one of those can be derived on a perfect phylogeny. In general, a PPH problem instance might not have a solution, i.e., when there is no phasing of $\mathcal{G}$ that produces haplotypes that can be generated on a perfect phylogeny (with all-zero ancestral sequence).

A PPH Solution Is Biologically Appealing A solution, $M(\mathcal{G})$, to the PPH problem (if there is one) is a solution to the HI problem. But it is highly constrained by the requirement that the haplotypes in $M(\mathcal{G})$ be derived on a perfect phylogeny with the all-zero ancestral sequence. It is this constraint that makes the PPH problem nontrivial to solve, and, in the right context, makes it more likely to capture the historically correct biological events. That is, when the evolutionary history of the underlying haplotypes, from which the observed genotypes are derived, is representable by a perfect phylogeny, then it is appropriate to require that any HI solution also be representable on a perfect phylogeny. The justification for this model is the same as the general justification for the perfect phylogeny model of SNP evolution, partly discussed in Section 3.2.2. See [86] for a more complete justification of the perfect phylogeny model.

20.4.1 An ILP Formulation for the PPH Problem

An ILP formulation for the PPH problem can be obtained by modifying the ILP formulation for phylogenetic problem IMM (or IM), discussed in Section 14.1.1, as follows:

> Given genotype input $\mathcal{G}$ in matrix form, duplicate each row of $\mathcal{G}$, and change each "2" to a "?," creating a matrix M. Then, create the concrete ILP formulation for Problem IMM with input M. The result is that for each genotype vector $g \in \mathcal{G}$, the ILP formulation will contain two binary variables, $Y(2g-1,j)$ and $Y(2g,j)$, for each site j in vector g.
>
> Next, for each cell (g,j) where $\mathcal{G}(g,j)$ has value 2, add the inequality
>
> $$Y(2g-1,j) + Y(2g,j) = 1,$$
>
> to the ILP formulation. This inequality *forces* exactly one of those two binary variables to have value 1; and the other have value 0.

The objective function for problem PPH is the same as for problem IMM:

$$\text{minimize} \sum_{(p,q)} I(p,q),$$

where the variable $I(p,q)$ was defined in Section 14.1.1.

We claim that for any input $\mathcal{G}$, the optimal solution to the concrete ILP formulation will have value 0, if and only if the genotypes $\mathcal{G}$ can be phased so that the resulting haplotypes can be generated on a perfect phylogeny with all-zero ancestral sequence.

Exercise 20.4.1 *Fully detail and explain the correctness of the above claim, i.e., the correctness of the abstract ILP formulation to problem PPH.*

Exercise 20.4.2 *We can also modify the ILP formulation for problem IMCR (discussed in Section 14.2) to obtain an ILP formulation for the following*

> **Problem PPH-IMCR:** *Remove the minimum number of columns in the input $\mathcal{G}$, so that missing values in the remaining data can be filled in to create a perfect phylogeny.*

Fully detail the abstract ILP formulation for problem PPH-IMCR.

20.4.2 The PPH Problem with Maximum Parsimony

As stated above, not every instance of the PPH problem has a solution. However, when there is a solution, there may be many solutions. In those cases, it is desirable to use additional criteria to select one of the PPH solutions from among the multiple solutions. An attractive *secondary* criterion is to find a PPH solution that uses the *fewest* number of *distinct* haplotypes, over all PPH solutions.

20.4.2.1 Pure Parsimony

The combinatorial model for the HI problem that has been most widely studied is the *pure-parsimony* model (sometimes called the *maximum-parsimony* model), which suggests that a biologically valid solution to the HI problem should use a *small* number of *distinct* haplotypes. The pure-parsimony model is supported by the empirical observation that the number of distinct haplotypes seen in a population is generally small, and vastly smaller than the number of haplotypes possible, given the observed genotypes. For example, in a study of haplotypes associated with asthma [60], only 10 distinct haplotypes were found in a region with 13 SNP sites (which have the potential for $2^{13} = 8,192$ distinct haplotypes). The standard explanation for the small number of distinct haplotypes seen in humans is the rapid expansion of the human population, which has not allowed enough time for the establishment of highly diverse haplotypes.

Here we mix the pure-parsimony model with the PPH model, leading to the

> **MP-PPH** Problem: Given $\mathcal{G}$, phase the genotypes so that the resulting haplotypes can be generated on a perfect phylogeny; *and* over all such phasings, find one that uses the fewest number of distinct (different) haplotypes.

An ILP Formulation for the *MP-PPH* **Problem:** Since the MP-PPH problem is a refinement of the PPH problem, it should not be a surprise that the ILP formulation to solve it is based on the ILP formulation for the PPH problem.

In the ILP formulation of the PPH problem, the objective function is

$$\text{minimize} \sum_{(p,q)} I(p,q),$$

and we saw earlier that there is a feasible solution to the PPH problem, if and only if the ILP formulation has an optimal value of zero. Since the PM-PPH problem assumes that there is a PPH solution for the given input, we can *remove* the objective function and add the inequality:

$$\sum_{(p,q)} I(p,q) = 0, \tag{20.6}$$

to force any feasible ILP solution to produce haplotypes with no incompatibilities, i.e., that can be generated on a perfect phylogeny.

With a perfect phylogeny for $\mathcal{G}$ assured, and no ILP objective function, we now want to add inequalities and a new objective function to the ILP formulation, so that the perfect phylogeny produced from $\mathcal{G}$ has the *fewest* number of distinct haplotypes, over all perfect phylogenies for $\mathcal{G}$. The ILP formulation we will develop for the MP-PPH problem follows ideas in [29, 30, 119] for an ILP formulation solving the *pure-parsimony* problem. A different ILP approach to the pure-parsimony problem appears in [85].

The New Variables Let k be a row in matrix M, where k ranges from 1 to $2n$.

Let k and $k' < k$ be any two indices between 1 and $2n$, i.e., indices for the $2n$ haplotypes produced by a feasible ILP solution to the MP-PPH problem. We will refer to "haplotype k" as a shortcut for "the haplotype indexed by k, created by the ILP solution." For each (k', k) pair, we create the ILP variable $d(k', k)$, which we want to be set to 1 *if* (but not only if) haplotype k' is *distinct* (different) from haplotype k. This is accomplished by creating the following inequalities for (k', k):

For each column j from 1 to m:

$$\begin{aligned} d(k',k) &\geq Y(k,j) - Y(k',j), \\ d(k',k) &\geq Y(k',j) - Y(k,j). \end{aligned} \tag{20.7}$$

Clearly, $d(k', k)$ will be set to 1 if (but not only if) haplotypes k and k' are distinct, since they will be distinct if they differ at least in one column j.

We next introduce an ILP variable $X(k)$, for each haplotype k from 1 to $2n$, which we want to be set to 1 in any feasible ILP solution, *if* (but not only if) haplotype k is distinct from *all* of the haplotypes numbered less than k, i.e., every haplotype $k' < k$. This is achieved with the following inequality:

$$\sum_{k'=1}^{k-1} d(k',k) - k + 2 \leq X(k). \tag{20.8}$$

Finally, the objective function for this ILP formulation is

$$\mathit{Minimize} \sum_{k=1}^{2n} X(k).$$

Why Is This Correct? The correctness of this ILP formulation is a bit subtle. It helps to imagine that each variable $d(k', k)$ will be set to 1 if and *only if* haplotypes k' and k are different. If that were true, then

$$\sum_{k'=1}^{k-1} d(k',k),$$

would be exactly $k-1$, and so

$$\sum_{k'=1}^{k-1} d(k',k) - k + 2,$$

would be exactly 1, *if and and only if* haplotype k is distinct from *every* haplotype $k' < k$. Then,

$$\sum_{k=1}^{2n} X(k),$$

would *exactly* count the number of distinct haplotypes in the $2n$ haplotypes.

However, the inequalities in (20.7) are only guaranteed to set $d(k',k)$ to 1 *if* (but not only-if) haplotypes k and k' differ. Similarly, inequality (20.8) is only guaranteed to set $X(k)$ to 1 *if* (but not only-if) haplotype k differs from every haplotype $k' < k$. Therefore, $\sum_{k=1}^{2n} X(k)$ might *overcount* the number of distinct haplotypes. But, since the objective function is to *minimize* $\sum_{k=1}^{2n} X(k)$, any variable $X(k)$ will be set to value 0 in the optimal solution, unless it is *forced* to be 1, because some variable $d(k',k)$ is forced to be set to value 1. So, in an arbitrary feasible solution, $\sum_{k=1}^{2n} X(k)$ might be an overcount, but in an *optimal* solution, it *exactly* counts the number of distinct haplotypes. Therefore, the ILP formulation we have developed for the MP-PPH problem is correct. Further, as shown in [87], this ILP formulation is efficient for solving instances of the MP-PPH problem sizes of current interest.

Exercise 20.4.3 *As a function of n and m, how many variables and inequalities are used in the above ILP formulation for the MP-PPH problem?*

Exercise 20.4.4 *Consider the pure-parsimony haplotyping problem: Given n genotypes $\mathcal{G}$, find a phasing of $\mathcal{G}$ that minimizes the number of distinct haplotypes used. That is, remove the requirement from problem MP-PPH that the haplotypes be generated on a perfect phylogeny. Write out a full ILP formulation for the pure-parsimony haplotyping problem. Estimate the number of variables and inequalities used as a function of n and m.*

For problems of modest size, a practical ILP formulation for the pure-parsimony haplotyping problem was suggested and studied in [85], and a very large body of literature on the pure-parsimony problem has developed since then. The ILP formulation in [85] solves quickly, but the size of the formulation grows rapidly as a function of n and m, and so is not practical for all problem sizes of current interest. In contrast, the size of the ILP formulation for the pure-parsimony haplotyping problem developed in [29, 30, 119] (and what is intended as an answer in the previous exercise) grows slowly, but it does not solve fast enough for practical use.

The ILP formulation for the MP-PPH problem sits in a *sweet spot* – it's size grows slowly, *and* it solves quickly. Apparently, the requirement in the MP-PPH problem that the optimal HI solution must minimize the number of distinct haplotypes over all *PPH solutions*, rather than over *all HI solutions*, constrains the ILP sufficiently, so that it can be solved efficiently in practice. This illustrates a statement made early in the book, that not all ILP formulations are effective, and related problems can have related ILP formulations that behave differently in practice. Still, enough ILP formulations in computational biology are effective, which is after all the motivating premise of this book.

20.5 SOFTWARE FOR THE MP-PPH HAPLOTYPING PROBLEM

The Perl program *bbminpph.pl* can be downloaded from the book website. That program takes in a matrix, $\mathcal{G}$, of n genotypes, and if $\mathcal{G}$ can be phased so that the $2n$ haplotypes have no incompatibilities (i.e., they can be derived on a perfect phylogeny), then the program will solve the MP-PPH problem for $\mathcal{G}$. If no perfect phylogeny solution is possible, then the ILP will be infeasible. Call the program on a command line in a terminal window as:

```
perl bbminpph.pl genotype-matrix output-file.lp
```

An example genotype matrix is in the file: *mpph.txt* After you solve a concrete ILP formulation and save the solution to a file, you can construct the haplotype matrix with the $2n$ haplotypes found by solving the concrete ILP formulation. The values for the haplotype matrix are given by the values of the Y variables. For example, the entry "Y7,4 1" means that the seventh haplotype has a 1 at position 4. It would be easy for an experienced programmer to write a script to actually create a matrix of the $2n$ haplotypes from what is in the resultfile.

21

More Extended Exercises

In this chapter we explore several topics through an exposition that relies more on exercises for the reader, than do most of the other topics in the book. These exercises are intended to help the reader develop skills in both biological and ILP modeling.

21.1 VISUALIZING HIERARCHICAL CLUSTERING

Here we introduce a problem that arose in the computational biology literature [15, 16], with many possible similar applications. The goal of the section is for the reader, with a few hints, to develop an abstract ILP formulation for the problem, addressing Tasks (A) and (B) enumerated in the Introduction. Then, the basic problem will be modified, and the reader will be asked to modify and/or extend their ILP formulation. All of the concepts needed for this exercise have been discussed in previous parts of the book.

The ILP problem concerns how to organize and *visualize* a binary tree that is created by a process called *hierarchical clustering*. So, before we get to the ILP problem, we briefly discuss hierarchical clustering. Input to the problem consists of n elements, and a measure of the similarity of each *pair* of elements. The following somewhat rewrites (for clarity) a quote, from [16], describing the generic clustering procedure:

> Hierarchical clustering starts by computing an n by n similarity matrix S, containing the similarity measure of all pairs of input elements to be clustered. Initially, each input element corresponds to a single cluster. Using the similarity values, we combine at each step the two most similar remaining clusters, and form a new cluster which contains both clusters. After combining these two clusters (and removing the two individual clusters just combined) we compute the similarity of the new cluster to all the remaining clusters. For n genes, this step is repeated $n-1$ times until we are left with a single cluster the contains all genes.

Clearly, as the clustering method creates clusters – by successively combining two clusters into a single cluster – we can represent those clusters by a *binary tree*, where each leaf represents one of the input elements, and each internal node of the tree represents a new cluster created by the method. When two clusters α and β are joined into a new single cluster γ, we create a new internal node for γ, and extend a directed edge from γ to the nodes representing clusters α and β. So, the result of a hierarchical clustering of n input elements is a rooted binary tree with n leaves and $n - 1$ non-leaves.

Specific hierarchical clustering methods differ in how they define and measure the similarity of a pair of elements, and how they define and measure similarity between clusters. Although the application in [16] concerns gene expression data, hierarchical clustering is used in a *huge* range of applications, both inside and outside of biology. Furthermore, the general term "hierarchical clustering" includes many different clustering methods. What is *common* to all of these methods is that

> They each produce a rooted binary tree with one leaf for each of the input elements; and each cluster created by the clustering algorithm is represented by a subtree containing, at the leaves, all of the elements in the cluster.

21.1.1 The Problem Context

The following is again a quote,[1] from [15]:

> *Hierarchical clustering* is one of the most popular methods for clustering gene expression data. Hierarchical clustering assembles input elements into a single tree, and subtrees represent different clusters. Thus, using hierarchical clustering one can analyze and visualize relationships in scales that range from large groups (clusters) to single genes. ... However ... the *ordering* of the leaves, which plays an important role in analyzing and visualizing hierarchical clustering results, is not defined by the clustering algorithm. Thus, for a binary tree, any one of the $2^n - 1$ orderings is a possible outcome.

21.1.2 The Binary Clustering-Layout Problem

Given a hierarchical clustering, in the form of a rooted *binary* tree, T, with n labeled leaves, we want to *order* the leaves in T so that the tree has a *planar* drawing,[2] and so that local *similarities* of the objects represented by the leaves, are *best* visualized. In particular, we want to create a planar drawing of T to *minimize the total change* seen between *adjacent* leaves. Figure 21.1 shows part of a figure from [15]. The input items consist of the columns shown beneath the tree. Each column represents a gene, and each row represents an individual. The shading in a cell in a row r and column c represents that frequency that individual r expresses gene c. In creating the clustering T, the value $S(i,j)$ is a measure of the similarity of the columns i and j. Tree T shows how the columns are organized into successively larger clusters, but for the best display, we want a planar drawing of T that emphasizes gradual change between columns that are adjacent in the displayed T. This problem was first addressed using dynamic programming in [16], but here we will develop an ILP solution (which, we assert, is much easier to develop).

More Detail Each interior node of T has exactly two children. A planar drawing of T is defined by choosing, at each interior node, which of the two children are placed on the left, and which are placed on the *right* (this assumes a vertical display of T,

[1] I have removed some of the quote, and italicized some words, and inserted the word "clustering" for clarity.

[2] The definition of a planar drawing was given in Section 8.2.1.

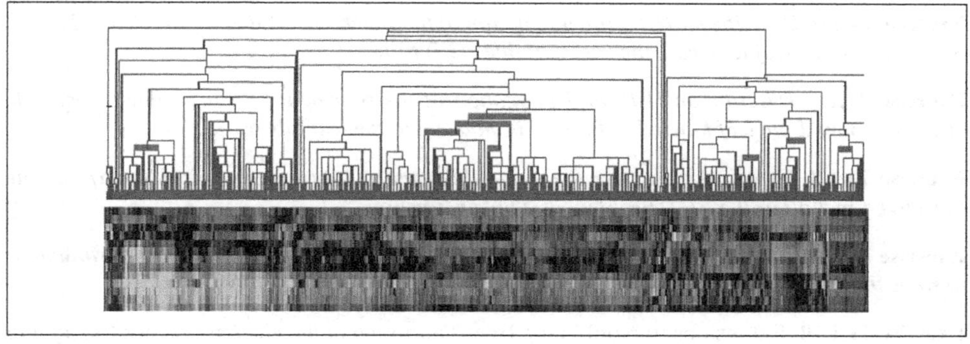

Figure 21.1 A Planar Drawing of the Clustering Tree, Ordering the Leaves of the Given Tree to Minimize the Sum of the Distances between Adjacent Leaves. Figure extracted from Figure 6 in [15].

where the root is at the *top*, and the edges of T are directed away from the root, toward the bottom of the page). That is, in a planar drawing, *no* edges *cross*. Given a specific planar drawing of T, the *value* of the drawing is defined as $\sum S(i,j)$, where the sum is taken over all pairs (i,j) of leaves that are *adjacent* in the drawing of T.

For an informal definition of *adjacency*, imagine that T represents a (strange) building where the root of T is the only entrance and exit; each edge in T is a hallway of the building; and a leaf in T is a place where the hallway is closed. Next consider entering the building at the root, putting your right hand on the wall on the right in the hallway, and walking around the building, always keeping your right hand on the wall. That walk is called a *depth-first* traversal of T. Then, the order that the walk reaches the leaves of T, imposes an order on the leaves. Two leaves are *adjacent* if they are adjacent in that order.

21.1.3 Exercises to Develop An ILP Formulation for the BCLP

Clearly, the *binary clustering-layout problem (BCLP)* is related to the two tanglegram problems discussed in Chapters 8 and 15. The task of creating an abstract ILP formulation for BCLP requires that you fully understand the formulations for the tanglegram problems. Some parts of those formulations can be used for the BCLP, some parts must be removed, and some new parts must be created. This will be detailed through the following exercises.

Exercise 21.1.1 *After reviewing the logic of the solution to the tanglegram problem and for the interleaf-distance tanglegram problem, describe the high-level logic to solve the BCLP. This is Task (A).*

Exercise 21.1.2 *Describe in words how the logic for the BCL problem differs from the logic for the tanglegram problem, and for the interleaf-distance tanglegram problem.*

Exercise 21.1.3 *Examine the abstract ILP formulation for the interleaf tanglegram distance problem in Figure 15.1 (on page 280), and select the part(s) of it that can be used for an abstract ILP formulation to solve the BCLP.*

Exercise 21.1.4 *What part(s) of your answer to Exercise 21.1.1 are not implemented in the solution you describe for Exercise 21.1.3?*

Exercise 21.1.5 *Using the P variables, create a linear function L that is true if and only if the leaf labeled i and the leaf labeled j are adjacent in tree T.*

Exercise 21.1.6 *Describe an ILP implementation of the statement "*If *the leaf labeled i of tree T is adjacent to the leaf labeled j, then set variable z(i, j) to 1."*

Exercise 21.1.7 *Describe an ILP implementation of the statement "Set the value of variable z(i, j) to 1* only if *the leaf labeled i of tree T is adjacent to the leaf labeled j."*

Exercise 21.1.8 *Explain how to use the z variables and the set of values in S to express the objective function or the BCL problem as a linear function.*

Exercise 21.1.9 *Putting together all of your answers, write out the full, abstract ILP formulation to solve the BCL problem.*

Exercise 21.1.10 Extensions to Nonbinary Trees *Suppose that the given tree is* not binary, *but rather, an internal node can have* more than *two children. Again we want a planar layout of the tree to minimize the total sum of the distances between adjacent leaves. Give an ILP formulation to solve this problem, and describe how both the problem and the solution differ from the case of a binary tree.*

21.2 CLUSTERING TO PREDICT *IN VIVO* TOXICITIES FROM *IN VITRO* EXPERIMENTS

In [55], the authors address the question of how to use limited *in vivo* data (i.e., data from *live* organisms) about chemical toxicities, to be be able to *predict* chemical toxicities from *in vitro* data (i.e., data from *nonliving* laboratory systems). The motivation, of course, is that *in vivo* studies are difficult and expensive, while *in vitro* studies are much simpler and cheaper. The downside to *in vitro* studies is that the results might not be relevant or reliable for living organisms – chemicals that are found to be toxic, or nontoxic, in a laboratory system might behave differently in a real, living organism. So, what can one do to make *in vitro* studies more predictive for living organisms? That is the problem addressed in [55].

The main approach to this problem in [55] is to *cluster* the chemicals into a few groups, where the chemicals in each group exhibit similar *in vitro* experimental results. Then, instead of investigating the *in vivo* toxicity of every chemical, the suggestion is to investigate only a *single* chemical in each cluster. If the chemicals in a cluster behave similarly enough, perhaps because they affect the same molecular pathway, or cell, or organ, etc., then this *thinning* of the data will reduce the time and costs of the *in vivo* investigations, increasing their ultimate impact and relevance. Further, after the clusters have been defined, when a new chemical is to be investigated, we can do the *in vitro* experiments on the new chemical, and if the results are consistent with one of the clusters, plausibly *conjecture* that the *in vivo* behavior of the new chemical will be similar to what is known about the *in vivo* behavior of the other chemicals in the cluster.

In more detail, the authors studied about 300 chemicals, using about 600 types of *in vitro* biochemical and cell-based assays. For example, an experiment might add *benzene* (the chemical) to a healthy, laboratory-grown colony of rat liver cells (the assay), to see if the chemical has a toxic effect on the cells.[3] After all of the chemical and assay combinations are examined, the results, either *quantitative*, reflecting the *amount* of toxicity, or *binary*, reflecting *whether or not* any toxicity was demonstrated, are organized into a matrix M. The rows of M represent the chemicals; the columns

[3] I would certainly expect benzene to be an unhealthy food additive.

represent the assays; and a cell $M(i,j)$ represents the result of introducing chemical i to assay j.

Then, for each pair of rows, (p,q), a *similarity score*, $s(p,q)$, is determined, reflecting how similar or different are the data for chemicals p and q. The specific details of such a similarity score are extremely important for the effectiveness of this study, but are a matter of biology, toxicology, and chemistry, and are outside the scope of this book. Still, given these scores, the *rows* of matrix M are then *ordered* (permuted) to *maximize* the sum of the similarity scores of *adjacent* rows. Finally, after the row ordering is fixed, *breakpoints* in the row ordering are determined, which then define *clusters* of the chemicals, which will be used as described above.

Exercise 21.2.1 *How can the problem of ordering the rows of M be addressed and solved by the use of ILP?*

Exercise 21.2.2 *There are many conceptual ways to determine where to place breakpoints in the row ordering to achieve the goals of the clustering. Try to define several ways and discuss the positive and negative properties of each. Consider both the computational problem of determining the breakpoints, and any issues of biological modeling. Consider the case when the* number *of clusters is fixed in advance, and the case that the number is determined during the computation.*

Since clusters *partition* the rows, the problem of choosing breakpoints is *related* to the problems of *community detection* and of *K cuts* discussed in Chapter 13. We can consider each row as a node in a graph G, with an edge between every pair of nodes (p,q), with edge weight $s(p,q)$. Then, we can use community detection or cuts to define a partition of the nodes, which then divides the rows into clusters. However, the community detection problem is now more constrained, because each cluster *must* consist of an *interval* of rows in the row ordering. How do we implement that requirement?

Exercise 21.2.3 *Define a binary ILP variable $X(p,q)$, where p is before q in the row ordering, to mean that row q is in the same cluster as row p. Then, write the inequalities to implement the logic that*

> *$X(p,q)$ has value 1, if and only if every row r that is between p and q (including q) in the ordering, is in the same cluster as row p.*

Exercise 21.2.4 *Use your answer to Exercise 21.2.3 to modify the ILP formulations for community detection, discussed in Chapter 13. Is this a sensible way to choose breakpoints in the row ordering?*

Another approach is to use your answer to Exercise 21.2.3, but set the number of desired breakpoints, k-1, and solve an instance of the k-cut problem. Do you want to maximize or minimize the value in the k cut? Write out the abstract ILP formulation for this approach.

Exercise 21.2.5 *An alternative to first ordering the rows, is to simply compute the similarity scores for each pair of rows, and then use community detection or solve a k-cut problem, to partition the rows into clusters. This skips the problem of finding the best row ordering. Can you see a benefit to first ordering the rows, as described in the exercises here?*

Similar Problems in Other Biological Settings Essentially the same computational problem addressed in the previous exercise on clustering toxicity tests, arises in other biological contexts. The most common one is when we compile information about whether, or how much, a gene is *expressed* in differing conditions. For example, whether a gene is expressed when the temperature is high or low, or the amount of food available to an organism is varied, or during mitosis, or in cancer cells, or in the liver, etc. That kind of data is called *gene-expression data*, which is again organized into a matrix, where the rows represent genes, and the columns represent conditions. An entry in matrix cell (i,j) shows whether (in the binary case) or how much (in the quantitative case) gene i is expressed under condition j.

In the case of gene-expression data, the purpose of clustering (or permuting) the rows based on their similarity, is to make it visually easier to examine the data, and to identify genes that are likely to be expressed together in the same biological process.

21.3 BICLIQUES AND BICLUSTERING IN BIOLOGICAL DATA

21.3.1 The Biological and Computational Problem

The clustering problems and methods that we have considered so far, group taxa (items) on the basis of *all* of their attributes. For example, in the toxicity problem discussed in Section 21.2, the rows of the data matrix, M, were clustered (ordered) based on their similarity over *all* of the experimental conditions tested. However, in many biological applications, and particularly in genomics (and other areas where large sets of high-throughput data are obtained), a *subset*, $\mathcal{S}$, of the taxa have very similar behaviors on a *subset*, $\mathcal{B}$, of experimental conditions,[4] and behave very differently on conditions outside of $\mathcal{B}$. The clustering approaches discussed so far may have difficulty finding such *pairs* of highly correlated subsets $(\mathcal{S}, \mathcal{B})$, and hence may be less biologically informative than alternative methods specifically designed to find them.[5] The following quote, based on a specific application, amplifies this general statement.

> Conventional molecular approaches for the study of organismal response to toxicant exposures or diseases involve the study of one gene or a few genes at a time, whereas biological response is driven by a group of genes. Thus, when normal function of a specific biological process is perturbed, alterations and enrichment in the expression of a subset of co-functioning genes associated with that biological process are observed...The fact that genes interact with each other and are expressed in functionally relevant patterns implies that gene-expression data can be grouped into functionally meaningful gene sets across a subset of conditions...However, derivation of meaningful and relevant gene sets from the thousands of genes showing expression changes following exposure to toxicants is challenging...[201]

So, we want methods that *simultaneously reorder* the rows and columns of a data matrix M, in order to identify a subgroup, $\mathcal{S}$, of taxa and subgroup, $\mathcal{B}$, of conditions, where the behavior of the taxa in $\mathcal{S}$ is highly correlated with the conditions in $\mathcal{B}$. If we permute the rows and columns of M to group together the rows in $\mathcal{S}$, and the columns in $\mathcal{B}$, then data for those taxa and conditions will lie in a rectangle in M (see Figure 21.2). Of course, the techniques we will discuss here are based on integer linear programming.

[4] Different "experimental conditions" could actually be different samples or individuals.

[5] Of course, one approach is to first reorder (cluster) the taxa based on *all* the experimental conditions, and then starting from the reordered data matrix, reorder the conditions based on *all* the taxa. The end result might, as a side consequence, expose the desired pairs of subsets, but it is easy to find examples where this fails, and a more targeted approach is more likely to succeed.

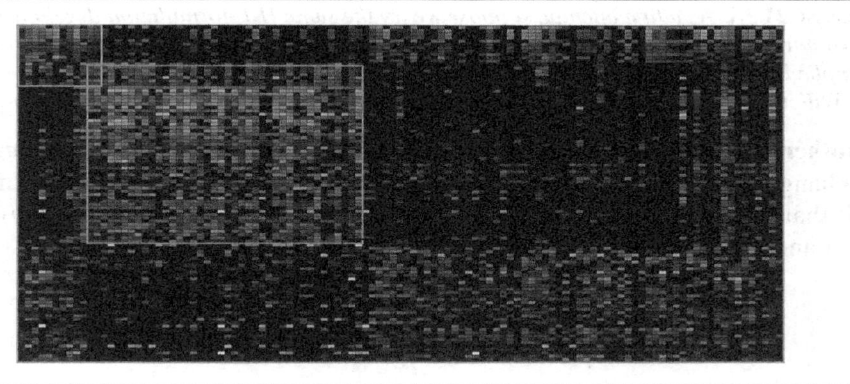

Figure 21.2 The Matrix Shows the Result of Biclustering of Gene Expression Data (Permuting Both Rows and Columns) in Individuals with Lung Cancer. The 100 rows correspond to genes, and the 111 columns correspond to individuals, of whom 53 have adenocarcinoma (AD), and 58 have squamous cell carcinoma (SQ). The bicluster algorithm was given no information on which individuals have AD or SQ cancer. Light cells in the matrix indicate gene expression levels that are higher than normal, and dark cells indicate gene expression levels that are lower than normal. The degree of the lightness or darkness indicates the extent of the deviation from normal. A total of 32 biclusters were identified. Two overlapping biclusters are marked. The biclusters collectively separate most of the individuals with AD cancer from most of the individuals with SQ cancer, showing that different gene are expressed in the two types of lung cancer. The existence of multiple biclusters suggests that the classification of the lung cancers into two subtypes can be further refined into several, narrower subtypes. This figure is extracted from figure 3 in [40].

21.3.2 Bicliques

We start the discussion of biclustering with the narrower, more precisely stated problem – the *maximum biclique* problem.

Recall that a bipartite graph, G, is a graph with two sets of nodes, A and B, where every edge in G has one end-node in set A and one end-node in set B. That is, no edge in G runs between two nodes in A, or two nodes in B (see Figure 7.6b).

A *biclique* in a bipartite graph G is a subgraph of G specified by a subset of nodes $A' \subseteq A$, and a subset of nodes $B' \subseteq B$, such that for any node $a' \in A'$ and any node $b' \in B'$, G contains the edge (a', b').

Just as a single edge in a graph is a clique, a single edge in a bipartite graph is a biclique. But neither of these are likely to have much biological importance. What is of biological interest are *large* bicliques. Here, through a series of exercises, we develop an ILP formulation to find a *maximum-size* biclique in a bipartite graph, and then extend the discussion to *near bicliques, high-density, bicliques*, and *node-balanced* bicliques.

21.3.2.1 Maximum Bicliques

We consider first the problem of finding a maximum-size biclique in a bipartite graph G. We want to develop an ILP formulation to solve that problem.

Exercise 21.3.1 *As a first attempt, suppose we try the same ILP formulation developed for the maximum-clique problem, shown in Figure 2.11. Perhaps when that abstract ILP formulation is applied to a bipartite graph, the result will be a maximum biclique.*

Will it? If not, give an example of where it fails, and explain in detail why it fails. Hint: It fails.

Another Attempt Another suggestion is to modify the formulation in Figure 2.11, by changing the inequality in (2.2) so that instead of applying to *every* pair of nodes (i,j) that have no edge between them, it would *only* apply to node pairs (i,j) where $i \in A$ and $j \in B$. Then, the formulation would have:

$$C(i) + C(j) \leq 1, \text{ for each pair of nodes } (i,j) \text{ in G, where } i \in A,\ j \in B, \text{ and there is } \textit{no} \text{ edge between them.} \tag{21.1}$$

Exercise 21.3.2 *Will this new formulation work? Hint: No. Why not? Give an explicit example where it fails, and fully explain the failure.*

Another Idea How about if we use the inequalities in (21.1), and put a *positive weight* on each edge in the bipartite graph. Then, we also change the objective function so that instead of maximizing the number of *nodes* that are chosen, we maximize the *total weight* of all of the edges in the subgraph induced by the chosen nodes.

Exercise 21.3.3 *Write out the abstract ILP formulation to implement the above idea. The key is to use a binary variable for each edge in the bipartite graph, which will be set to 1 only if both end-nodes of that edge are chosen to be in the biclique.*

Will this ILP formulation solve the problem of finding the maximum size biclique in a bipartite graph? Explain.

Yet Another Idea Another idea is to use the inequalities in (21.1), and *not* use any edge weights, but include inequalities that say that *at least one* node from A must be chosen, and at least one node from B must be chosen.

Exercise 21.3.4 *Will this idea work to solve the maximum-size biclique problem? Fully explain.*

Balanced Bicliques There are many applications where the number of nodes from A that are chosen must be *approximately* equal to the number of nodes from B that are chosen. The resulting biclique is called a *balanced* biclique.

Exercise 21.3.5 *Define a balanced biclique in a way that you think is meaningful, and write out an abstract ILP formulation to find a largest balanced biclique.*

21.3.2.2 Near Bicliques

Because of course!

Exercise 21.3.6 *Reexamine all of the ways that the maximum-clique problem was generalized and modified in earlier chapters. For example, all the ways that near cliques or high-density subgraphs were defined. For each of those definitions, state an analogous definition and problem statement for bipartite graphs and near bicliques. Then write out an abstract ILP formulation to solve those problems.*

Near bicliques are also called *quasi-bicliques* in both the biological and the graph theory literature. As one example, see [38].

21.3.3 Methods for Biclustering

We started the discussion of bicliques and biclustering by motivating *biclustering*, i.e., the discovery of many pairs of (taxa, condition) subsets, which form submatrices in M. Then we looked at the precise problems of finding a maximum-size biclique or near biclique. Now we return to the more general problem of *biclustering*.

The goal of biclustering is to find every biologically informative near biclique in M. These near bicliques can overlap, either sharing nodes representing taxa or nodes representing conditions, or both.

> Unlike classical clustering techniques, biclusters can overlap with each other. This is ideal for mining functionally related gene sets as genes can be associated with more than one biological process. [201]

How to Do It?

Unfortunately, *how* to do biclustering is much less clear than how to find a largest biclique, or near biclique. For one thing, the goal of biclustering, while intuitive, has been much harder to state in a precise, formal way. Further, there is less agreement on what a good bicluster is. And even when a clear, formal biclustering problem has been stated, effective solution methods have been harder to find and justify. There have been many suggested methods, but they feel (to me) very heuristic and *ad hoc*, making it difficult to evaluate their effectiveness. Still, I think there are several things we can observe.

First Biclustering methods can be built on *iterative* uses of methods for finding large *bicliques* or *near bicliques*. If we have a meaningful definition of a near biclique, then a biclustering, with *no* overlapping biclusters, can be obtained by finding a largest near biclique in the bipartite graph, removing the nodes in that biclique, and iterating this procedure until no further appropriate bicliques can be found.

Exercise 21.3.7 *Analogous to what was done for cliques in Section (2.2.7) and (12.3), state abstract ILP formulations to find a* second largest *biclique, and a* second-largest *biclique in a bipartite graph. Consider both the case that the two bicliques share no nodes, and the case that they can share nodes, as long as the second-largest biclique is not a subgraph of the largest biclique.*

Things are more difficult in general, when the biclusters are allowed to overlap. Then, some heuristic ideas have to be used to determine which (meaningful) subset of nodes in the first biclique or near biclique should be removed. Perhaps the nodes with the smallest degrees in the near clique, or the nodes with a small number of edges to nodes outside of the biclique?

Second In the case of nonoverlapping biclusters, it may be useful to view the biclustering problem as a *community finding* problem, or a *multi-cut* partitioning problem. That is, biclustering generalizes the problem of finding a single, large near biclique in a bipartite graph, in the same way that community finding generalizes the problem of finding a single large near clique. Perhaps the definitions and partitioning methods of *modularity* or *k cuts* can be modified for bipartite graphs and biclusters.

Exercise 21.3.8 *Generalize the notion of modularity to apply to bipartite graphs and biclusters. Then state an abstract ILP formulation for the generalized modularity problem.*

Do the analogous thing for k cuts and their related definitions from Section 13.2.

21.4 COMBINATORIAL DRUG THERAPY

> Biological networks are complex, and effective therapies may require combinations of drugs to overcome redundancies, feedback mechanisms, or drug resistance. ...screening is challenging because of the multiplicity of combinations to test. [10]

Drug therapies for many diseases require the *combined* action of several different drugs, from different drug classes. In fact, when only one drug is taken, it may eliminate one contributor to the disease, but also create a more inviting environment for other contributors. So, it is necessary to use a *combination* of drugs to eliminate all the players at the same time.

A recent example was discussed in the journal *Nature* [164]. In colon cancer, tumor progression in the *center* of a tumor is highly influenced by a signaling pathway called NOTCH, while tumor progression on the *exterior* of the tumor is influenced by a different pathway, called MAPK. There are drugs that suppress the activity of NOTCH, and other drugs that suppress the activity of MAPK. But targeting only one pathway is not highly successful in killing the tumor. In fact, suppressing only one pathway can make matters worse, because

> ... drugs that block MAPK signaling caused tumour cells with high NOTCH activity to flourish. And in reverse, drugs that blocked NOTCH allowed cancer cells with high MAPK activity to thrive. [164]

To be effective in slowing the growth of colon tumors, drugs that suppress NOTCH must be given *together* with drugs that suppress MAPK.

Combinatorial drug therapy has also been central to the success of combating HIV. Quoting from Wikipedia: [222]

> Combinations of antiretrovirals create multiple obstacles to HIV replication ... If a mutation that conveys resistance to one of the drugs being taken arises, the other drugs continue to suppress reproduction of that mutation. With rare exceptions, no *individual* antiretroviral drug has been demonstrated to suppress an HIV infection for long; these agents must be taken in *combinations* in order to have a lasting effect. As a result, the standard of care is to use combinations of antiretroviral drugs. Combinations usually consist of three drugs from at least two different classes. This three drug combination is commonly known as a triple cocktail. Combinations of antiretrovirals are subject to positive and negative synergies, which limits the number of useful combinations.

21.4.1 Drug Mixtures and Cocktail Therapies

Combinatorial drug therapy is often successful, and sometimes essential, but finding good combinations of drugs is a complex problem. Trial and error on human subjects is slow, expensive and sometimes immoral. Finding effective cocktails by using *computation* is then clearly desirable.

> Computational models ... can be used ... to rationally identify chemotherapy-potentiating drug combinations. [5]

21.4.2 Exercises: Biological Models and ILP Formulations

We start with the most *biologically naive* drug/protein model and problem statement, and then successively add in more biological reality.[6]

(a) The Basic Drug/Protein Model is that we have a set of proteins, $\mathcal{P}$, and a set of drugs, $\mathcal{D}$, and information about which drugs affect (e.g., inactivate) which proteins. That is, for each protein $p \in \mathcal{P}$, there is a subset $\mathcal{D}(p) \subseteq \mathcal{D}$ of drugs that affect protein p. We will assume that every drug in $\mathcal{D}$ can affect some protein in $\mathcal{P}$ or else there is no point in having d in $\mathcal{D}$. For now, we also assume that any individual drug in $\mathcal{D}(p)$ can *fully inactivate* the function of protein p.

The same information can be reorganized so that for any drug $d \in \mathcal{D}$, there is a subset $\mathcal{P}(d)$ of proteins such that d can inactivate any protein p in $\mathcal{P}(d)$. These are two different, but equivalent, ways to express the basic information. We can repurpose Figure 7.6 (on p. 139) to be an illustration of these relationships, interpreting each node on the $\mathcal{D}^2$-side of B as representing a single drug, and each node on the $\mathcal{P}^2$-side of B as a single protein, and an edge between a node representing drug d and a protein p if and only if p is in $\mathcal{P}(d)$ (equivalently, d is in $\mathcal{D}(p)$).

We define a subset Q of $\mathcal{D}$ to be a *drug cover of* $\mathcal{P}$ if the drugs in Q collectively inactivate *all* of the proteins in $\mathcal{P}$. More formally, Q is a *drug cover* if for every protein $p \in \mathcal{P}$, there is at least *one* drug $d \in Q$ that is in $\mathcal{D}(p)$.

The Basic Drug Cover Problem: Given the input discussed above, select the *smallest* drug cover, $D^* \subseteq \mathcal{D}$, of $\mathcal{P}$.

In more general settings, the basic drug cover problem is another instance of the classic *set cover* problem that was introduced in Section 7.3.1.2.

Exercise 21.4.1 An Abstract ILP Formulation *For each drug $d \in \mathcal{D}$, let $X(d)$ be a binary ILP variable that will be given value 1 if d is chosen to be in D^*, and will be given value 0 otherwise. Using these variables only, create and explain an abstract ILP formulation for the basic drug cover problem.*

(b) Adding in a Bit More Reality to the drug/protein model, each drug $d \in \mathcal{D}$ has a *cost*, $C(d)$, which could be a price, or could be a number that reflects adverse side effects caused by the drug. We want a drug cover of $\mathcal{P}$ that has the lowest total cost.

Exercise 21.4.2 *Modify the ILP formulation for the basic drug cover problem to solve the version of the problem with a cost, $C(d)$ for each drug d. This is an easy extension.*

(c) The Cost/Benefit/Budget Variant Now, we add into the drug/protein model the concept of *benefit* as well as cost. For any particular protein, $p \in \mathcal{P}$, there is a *benefit*,[7] $B(p)$, from inactivating p. This leads to many different computational problems. One is:

The Maximum Benefit Drug Cover Problem Given the basic information about $\mathcal{D}$ and $\mathcal{P}$ (including costs and benefits), and given a budget b, find a subset of drugs D^* that have a total cost at most b, *maximizing* the total benefit of the proteins that are covered, i.e., inactivated. More formally, find Q^* to

[6] The material in this section is loosely based on the paper "Combinatorial therapy discovery using mixed integer linear programming" [146].

[7] How a benefit is measured is a often a thorny issue, so here we only say that it is a positive number somehow reflecting the positive value of inactivating a given protein.

$$\text{maximize} \sum_{p \in \mathcal{P}(d), d \in Q^*} B(p),$$

so that

$$\sum_{d \in Q^*} C(d) \leq b.$$

A budget can also be used to limit the number of drugs chosen, i.e., to put a bound on $|Q^*|$.

Exercise 21.4.3 *Modify the ILP formulation developed for Exercise 21.4.2 to solve the maximum benefit drug cover problem.*

Hint: As before, the critical variables are the $X(d)$ variables, indicating whether or not drug d will be selected. These are used to limit the total cost. But, the tricky part of the formulation is how to incorporate the benefits. One approach is to use a binary ILP variable $Y(p)$ for each protein p, where $Y(p)$ gets set to value 1 if and only if *there is a drug d chosen for Q^*, where $p \in \mathcal{D}(p)$. These $Y(p)$ variables can then be used to incorporate benefit into the objective function.*

Exercise 21.4.4 *Can the* "if and only if" *in the previous hint be changed to* "only if," *or just to* "if"*? That is, would the ILP formulation that results still be correct with either of those changes? Explain.*

(d) Protein Complexities So far, we have examined the problem of finding the smallest, or least expensive, set of drugs to inactivate every protein in $\mathcal{P}$; and the problem of finding a subset of drugs to maximize the benefit obtained from them, while keeping the cost below a given budget, b. In those problems, we assumed that any drug in $\mathcal{D}(p)$ could inactivate protein p *by itself*. In reality, it often takes a *combination* of drugs to inactivate a single protein. Here, we follow that reality, starting with the simplest case and then adding in greater complexity.

Exercise 21.4.5 *Suppose that to inactivate a protein p in $\mathcal{P}$,* two *different drugs from $\mathcal{D}(p)$ must be chosen. To start, assume that* any two *drugs in $\mathcal{D}(p)$ are sufficient. Modify the ILP formulations developed so far, to incorporate this change.*

More realistically, specific *combinations* of drugs may be needed in order to inactivate a particular protein, p. So now, for any protein p, $\mathcal{D}(p)$ will be a *set* of *subsets* of $\mathcal{D}$ that can inactivate protein p. One of those subsets of drugs must be chosen in order to inactivate p.

Also, the meaning of $\mathcal{P}(d)$ changes from "the set of proteins that can be inactivated by d" to "the set of proteins that can be inactivated by a subset of drugs containing d."

Exercise 21.4.6 *Given the new information and constraints on drug combinations, modify the prior ILP formulations to incorporate those changes. Hint: OR and AND idioms may be very helpful here.*

(e) Drug Menus In some cases, the combinations of drugs that can inactivate a particular protein have a "menu-like" structure, where certain dishes are grouped together as "appetizers," or "soups," or "salads," or "main entree," or "desserts." A meal consists of choosing one dish from the appetizers, one from either the soups or the salads, one main entree, and then one dessert. For now, we will assume that no dish is in more than one group.

In a similar way, the drugs in $\mathcal{D}$ might be divided into several groups, and in order to inactivate a particular protein p, one drug from each group must be chosen.

Making matters more complex, it may be that the way the drugs are divided into groups is different for different proteins that we want to inactivate.

Exercise 21.4.7 *Incorporating the new information about drug menus, modify the ILP formulations for the problem variants we have considered (e.g., minimizing the number or cost of the drugs chosen, so that all protein are inactivated; or the cost/benefit/budget variants).*

(f) Grouping Proteins Now we turn to the way that proteins may be grouped, and how that affects the ILP formulations. The main reason to inactivate a protein is that is it part of a *disease-causing pathway*. For now, assume that a pathway is just a single series of proteins, so that inactivating any protein in the series inactivates the pathway. The goal is not to inactivate every protein in $\mathcal{P}$, but to inactivate every *pathway* in a given set of pathways, denoted $\mathcal{N}$. A given protein p may be in several pathways in $\mathcal{N}$, and typically is.

Exercise 21.4.8 *To tackle pathway problems, we first go back to the assumption that each drug d can fully inactivate every protein in $\mathcal{P}(d)$. Modify the ILP formulation to solve the problem of finding the smallest set of drugs so that every pathway is inactivated. More formally, find the smallest set $Q^* \subset \mathcal{D}$ such that for each pathway W in $\mathcal{N}$, there is at least one drug $d \in Q^*$, which is in $\mathcal{D}(p)$ for some protein p in pathway W.*

Next, assuming each drug has a given cost, modify the ILP formulation to find the least expensive set of drugs to inactivate every pathway. Also, given a benefit for inactivating a pathway, modify the ILP formulation for the cost/benefit/budget variant of the problem. Finally, tackle the cases where diseases are associated with protein pathways as described in paragraph (f), and to inactivate a given protein p, drugs must be used in combinations, as described in paragraph (e). Consider both the case that we want to inactivate all the pathways, and the case that we want to maximize the benefit of inactivating pathways, within a given cost budget.

(g) The Bad Side Effects of Drugs All drugs have some undesirable side effects that occur with varying probabilities. The possible bad side effects of an *individual* drug can be modeled as part of its "cost," so the ILP formulations considered so far essentially address the issue of side effects of individual drugs. However, another reality is *drug interactions*, where certain *combinations* of drugs can have very harmful effects if they are taken *together*. In fact, bad drug interactions can occur with just two drugs, and few drug interaction studies have gone beyond combinations of two drugs. How can interaction issues be incorporated into the drug/protein models and problems?

Exercise 21.4.9 *For each of the prior drug/protein problems considered, and the abstract ILP formulations developed, incorporate constraints to avoid, or limit the costs of, choosing drugs with known bad interactions. That is, create abstract ILP formulations that can find the optimal set of drugs, Q^*, satisfying the requirements of that model, while taking into account what is known about side effects of drug interactions.*

(h) Off-Target Effects Now we add into the drug/protein model a very important consideration that makes real problems more difficult. Until now, we have only considered the set of proteins, $\mathcal{P}$, that we wanted to inactivate. But a drug can have an undesired effect on a protein that is *not* in *P*. These are called *off-target* proteins. In one way, this has been modeled indirectly in the cost of choosing a drug, but often we need to be certain that the damage is limited, so we need to be more explicit.

Given $\mathcal{D}$, the set of drugs that affect proteins in $\mathcal{P}$, we let $\overline{\mathcal{P}}$ be the set of proteins *not* in $\mathcal{P}$ that are affected by some drug in $\mathcal{D}$. Then, there are several computational problems that result. We consider one in the next exercise.

Exercise 21.4.10 *Develop an abstract ILP formulation to try to select a set of drugs, Q^*, so that every protein in $\mathcal{P}$ is inactivated, and so that for every protein $\bar{p} \in \overline{\mathcal{P}}$, no drug that affects $\bar{p}$ is selected for Q^*. If there is such a set of drugs, find a set, Q^*, with the smallest size or cost, or a set with the greatest benefit within a given cost benefit.*

Exercise 21.4.11 *For more reality, the inactivation of a protein in $\overline{\mathcal{P}}$ may require a known combination of drugs from $\mathcal{D}$, similar to the case of a protein in $\mathcal{P}$, discussed in paragraphs (d) and (e). We want to inactivate every protein in $\mathcal{P}$ without inactivating any protein in $\overline{\mathcal{P}}$, if possible. Create an abstract ILP for this problem, and for all the variants of the drug/protein model considered above.*

21.4.3 Numbers

The drug/protein models developed in [146] are somewhat different from the ones developed here, but are in the same *spirit*. To get a sense for the sizes of real problem instances, we discuss some real numbers.

The authors extracted drug/protein information from existing databases, and then removed any drug not known to affect a protein involved with human disease. That left 4,233 drugs, 2,058 target proteins, and 9,669 drug/protein interactions. Six diseases were studied in detail. The largest set $\mathcal{P}$ has size 41, and the largest set $\mathcal{D}$ has size 77. The sizes of $\overline{\mathcal{P}}$ were not reported, but illustrations of the drug/protein interaction networks show $\overline{P}$ to be significantly larger than $\mathcal{P}$. As one case, for the disease of *hypertension*, there were 26 proteins known to be involved in creating hypertension (so $|\mathcal{P}| = 26$), and 77 known drugs that affect one or more of those 26 proteins (so $|\mathcal{D}| = 77$). However, only 19 of the 26 proteins were known to be affected by any of the drugs. So, the best result would be a small set of drugs, Q^*, that inactive those 19 proteins, and do not affect any proteins uninvolved in hypertension. When limited to at most five drugs, the best solution they found affected 10 of the 19 proteins, but also affected one protein not involved in hypertension, i.e., in $\overline{\mathcal{P}}$.

21.5 RETURN TO THE CAPE OF SOUTH AFRICA

At the end of our discussion of the *threatened-species-protection problem* in Chapter 1, we said that enhancements to the problem would be introduced after the appropriate ILP techniques had been studied. Well, that time has come.

Encouraging Bordering Regions In the way that the threatened-species-protection problem was defined in Chapter 1, the regions that could be preserved were treated *independently*. It did not matter whether two regions shared a common border, or where located far apart. However, both from a preservation, ecological standpoint, and from a financial and management standpoint, it is better to preserve a large, *connected* set of regions, rather than the same number of *isolated* regions. So, in the ILP formulation of the species-protection problem, we would like to encourage the preservation of *adjacent, connected* regions whenever it is economical to do so.

To be more precise, suppose that the stated *cost*, $C(A)$ in Chapter 1, for preserving a region A includes the cost of *fencing in* the entire region, or posting signs around the border of the region. So, $C(A)$ can be expressed as $P(A) + F(A)$, where $P(A)$ is the *purchase* price for region A, and $F(A)$ is the cost to put a *fence* along the entire border of region A. We are implicitly assuming here that either an entire region is preserved, or none of it is.

Now, if two *preserved* regions, say A and B, share a border, then *no* fence is required along that border. This means that some amount of cost should be *subtracted* from $F(A)+F(B)$. Let's call that subtracted amount $F(A, B)$, and assume that $F(A, B)$ is specified as part of the input to the problem. More generally, for each pair of regions, (P, Q), that *could* be preserved, and that *share* a border, the problem input will contain the value of $F(P, Q)$, i.e., the reduction in the fencing cost *if both* regions are preserved.

Exercise 21.5.1 *Suppose regions A and B share a border. Write the linear inequalities to set binary variable $Z(A, B)$ to value 1 if and only if both regions A and B are preserved.*

Now suppose that, other than A and B, no regions that could be preserved, share any of their borders. How is variable $Z(A, B)$ used in the objective function to implement the reduction in total cost by $F(A, B)$, if and only if both regions A and B are preserved?

Generalizing from the case of just two bordering regions, suppose that *many* of the regions that could be preserved share borders. In the extreme case, suppose that a region A is completely surrounded by other regions that could be preserved. Then, if all of those surrounding regions are preserved, no fencing is needed for region A. In a less extreme case, some fencing is needed for region A if one or more of its borders is *unshared* with any other preserved region.

Exercise 21.5.2 *How is the abstract ILP formulation for the species-protection problem modified (both in the inequalities and in the objective function) to handle all possible border conditions, and choices of which regions are preserved?*

Hint: At first exposure, this looks like it might have a very messy *answer. For example, suppose region A shares borders with ten other regions that could be preserved. The obvious generalization from the case of just two regions discussed above, would create a Z-type binary variable for* each *of the* 2^{10} subsets *of regions that border A. One of these Z variables would be set to value 1 if and only if the subset of regions associated with it were exactly the regions bordering A that were chosen to be preserved, along with region A. That approach would create an unruly zoo of variables and inequalities. But, there is instead a very* simple *solution that avoids the mess. Play with some examples of regions bordering regions to discover it, and then answer the question.*

Fixed and Varying Costs So far, we have assumed that the *purchase price* for a region is a known and unchangeable quantity. But suppose it partly depends on the *number* or on the *total list prices* of regions that are chosen to be preserved. For example, perhaps the total purchase price will be reduced by 10% if more than five regions are preserved; or reduced by a percentage that increases with increasing total purchase price. For example, if the total list price of the regions chosen is more than 1 million dollars, then there will be a 5% discount, etc. Modify the ILP formulation for the species-protection problem to incorporate these kinds of realistic possibilities.

Another kind of cost is called a *fixed* cost, which must be paid for a region, if any *nonzero* amount of it is chosen for preservation. There are several specific ways that fixed costs can occur. One way is when a *fractional* amount of a region can be preserved, but there are things that must be done, for example building a road to the region, if *any* amount of the region is preserved. Then, the total cost for preserving part of a region A is $I(A) + C(A) \times X(A)$, where $X(A)$ is the *fraction* of region A that will be preserved, $C(A)$ is the cost for preserving all of region A, and $I(A)$ is the *fixed* (or initial) cost that must be paid if any nonzero amount of region A is chosen for preservation.

Exercise 21.5.3 *Modify the abstract ILP formulation for the species-protection problem to incorporate fixed costs, as defined above.*

In the above exercise, fractional amounts of a region could be preserved, and that was key to a sensible scenario. But fixed costs can arise in cases where *all or none* of a region will be protected. For example, suppose there are several regions where certain threatened species is actively being *poached*. Preservation of those regions requires increasing security, perhaps through the purchase of a drone. The cost of the drone must be paid if at least one region is chosen for preservation, no matter how many regions are preserved (assuming that amount is within the monitoring ability of the drone).

Exercise 21.5.4 *Show how to handle this kind of fixed cost in the abstract ILP formulation.*

Exercise 21.5.5 *In a more open-ended challenge, think of realistic-seeming scenarios and variations to the species-protection problem, and show how the abstract ILP formulation can be changed to incorporate those changing ecological models.*

21.6 COMPARING THREE-DIMENSIONAL PROTEIN STRUCTURE WITH CONTACT MAPS

> Protein structure comparison is a fundamental problem for structural genomics, with applications to drug design, fold prediction, protein clustering, and evolutionary studies. [35]

Essentially, protein structure comparison is intended to provide a *measure of the similarity* of the three-dimensional structures of two proteins whose structures are each known. Significant structural similarity is possible even with large differences in the amino acid sequences of two proteins. The key difficulties in developing a structure comparison method are first to define a *quantitative measure* that reflects important biological features of protein structure; and second to show how the measure can be effectively computed from protein structure data. The first goal has been addressed through *protein contact maps* and the *contact map overlap (CMO) measure* [78]. The second goal requires a practical method to compute the CMO measure on protein data.

In this exercise I will define the CMO measure and outline an abstract ILP formulation, developed in [35], that computes the measure. You will fill in the details of the ILP formulation.

A Protein Contact Map The three-dimensional structure of a protein can be specified by listing the X, Y, Z coordinates of every atom in the protein, but such a mass of detail makes it hard to compare the structures of two different proteins. A simpler approach builds on *less* detail, while maintaining important information that allows meaningful protein structure comparisons.[8]

Given a protein $\mathcal{P}$ and its three-dimensional structure, we build an undirected *graph* called a *contact map*, $CG(\mathcal{P})$. Graph $CG(\mathcal{P})$ has one node for each amino-acid residue in $\mathcal{P}$. We display the graph by placing the nodes on a horizontal line, using the same order that their corresponding amino acids occur on $\mathcal{P}$. Then, considering the three-dimensional structure of $\mathcal{P}$, we connect a pair of nodes, (v_i, v_j) in $CG(\mathcal{P})$,

[8] See again the quote on the role of models in studying protein structure, on page 129.

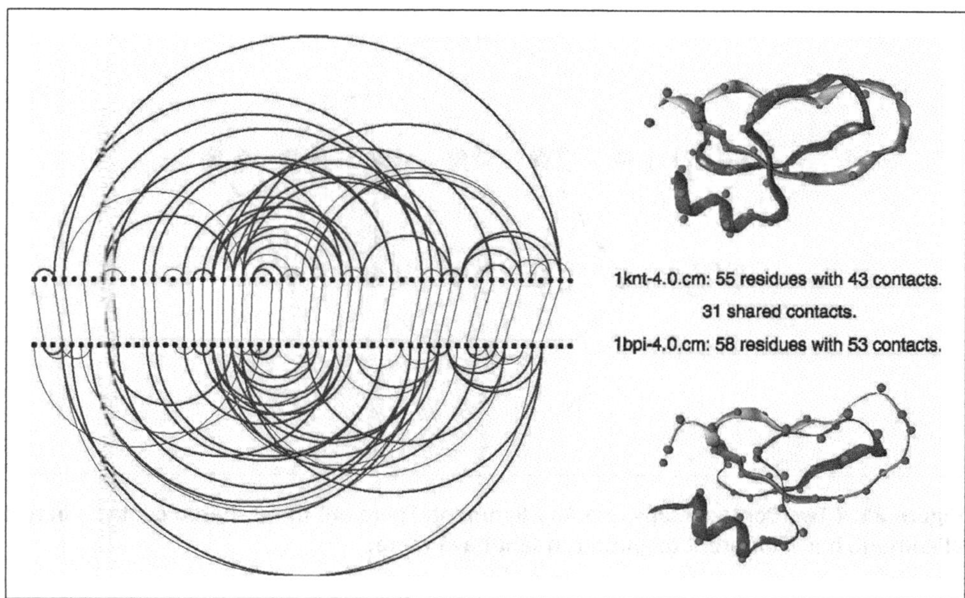

Figure 21.3 Two Contact Maps and An Alignment. The top row of 55 nodes, with 43 curves, is the contact map for the protein shown to its right; and the bottom row of 58 nodes, with 53 curves, is the contact map for the protein shown to its right. Each curve in a contact map connects two nodes representing two amino acids that are "in contact" (are close to each other) in the three-dimensional structure of the protein shown to the right of the contact map. The alignment of a node in the top contact map with a node in the bottom contact map is shown by a straight (vertical-ish) line between the two nodes. The alignment is *non-crossing*, meaning that no pair of these lines cross. This figure is figure 1 in [35]. Printed under license from Mary Ann Liebert, Inc.

with an undirected edge, if (and only if) the i'th and j'th amino acids in $\mathcal{P}$ are located "close" to each other in the given *three-dimensional* structure of $\mathcal{P}$ (see Figure 21.3, where those edges are drawn as *curves*). How close the two amino acids need to be to be considered "close" is a chemical-modeling issue, and the level is set by the user. Certainly, even though the graph is called a *contact* graph, actual physical contact is not usually required. The contact map $CG(\mathcal{P})$ represents certain features of the full three-dimensional structure of $\mathcal{P}$, and these features are considered [35, 78] sufficient for meaningful protein structure comparison.

Aligning Two Contact Maps How do we measure the similarity of two contact maps, $CG(\mathcal{P}_1)$ and $CG(\mathcal{P}_2)$, for proteins $\mathcal{P}_1$ and $\mathcal{P}_2$? We start by defining an *alignment* of the two maps. An alignment of contact maps, $CG(\mathcal{P}_1)$ and $CG(\mathcal{P}_2)$, consists of a set of *pairs* of nodes, where each pair contains one node from $CG(\mathcal{P}_1)$ and one node from $CG(\mathcal{P}_2)$, such that:

(a) No node is in more than one pair.

(b) By numbering the nodes consecutively, left to right, starting at number 1, if a node i in $CG(\mathcal{P}_1)$ is aligned with a node j in $CG(\mathcal{P}_2)$, and a node $i' > i$ in $CG(\mathcal{P}_1)$ is aligned with a node j' in $CG(\mathcal{P}_2)$, then $j' > j$. That is, the aligned pairs do *not cross*

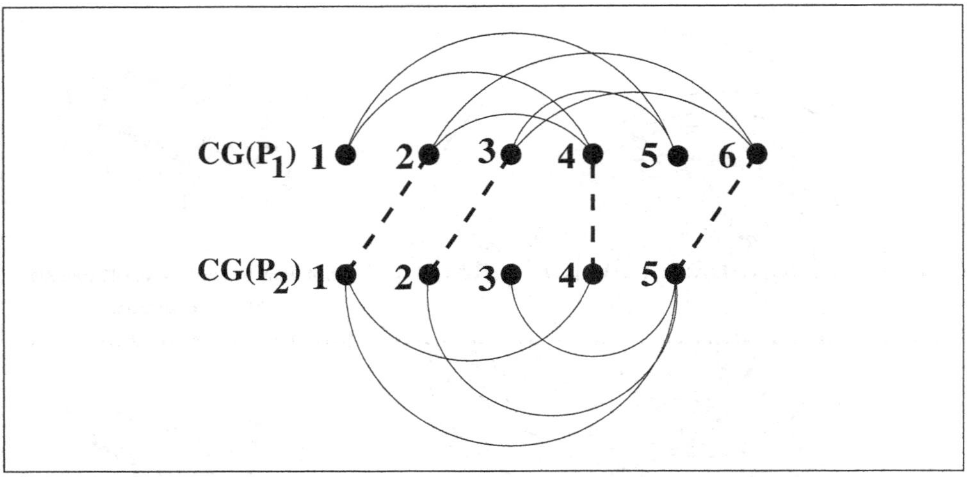

Figure 21.4 Two Contact Maps and An Alignment. There are three shared contacts in this alignment, but a different alignment might have more.

(see Figures 21.3 and 21.4, where each aligned pair is connected by a dashed, straight [vertical-ish] line).

These conditions imply that number of nodes of $CG(\mathcal{P}_1)$ that are in an alignment must be equal to the number of nodes of $CG(\mathcal{P}_2)$ that are in the alignment. Also, the fact that the alignment does not cross is equivalent to saying that the associated lines do not cross, or share a node. This should be familiar to the reader from the discussion of *simple RNA folding* in Chapter 6.

Shared Contacts and Alignment Scores Given an alignment, $\mathcal{A}$, of two contact maps, $CG(\mathcal{P}_1)$ and $CG(\mathcal{P}_2)$, how do we measure the *quality* of the alignment? The simplest approach is the following: We say that two node pairs, $(i,j), (i',j')$, both in alignment $\mathcal{A}$, form a *shared contact* if there is an edge (curve) in $CG(\mathcal{P}_1)$ between nodes i and i', and also an edge in $CG(\mathcal{P}_2)$ between j and j'. For example, in Figure 21.4, the pairs $(2,1), (6,5)$ form a shared contact because both of the node pairs are in the alignment, and there is a curve between nodes 2 and 6 in $CG(\mathcal{P}_1)$, and a curve between nodes 1 and 5 in $CG(\mathcal{P}_2)$.

Given this definition, the *alignment score* of an alignment, $\mathcal{A}$, is equal to the number of shared contacts it has. For example, the alignment in Figure 21.4 has three shared contacts. This leads naturally to

> **The CMO Problem:** Given a pair of contact maps, $CG(\mathcal{P}_1)$ and $CG(\mathcal{P}_2)$, find an alignment of $CG(\mathcal{P}_1)$ and $CG(\mathcal{P}_2)$, *maximizing* the alignment score over all alignments of $CG(\mathcal{P}_1)$ with $CG(\mathcal{P}_2)$.

Exercise 21.6.1 *What are the shared contacts in the alignment shown in Figure 21.4? Can you find an alignment between the contact maps in Figure 21.4, which has more than three shared contacts?*

Exercise 21.6.2 *What is the biological/structural justification for requiring an alignment to be non-crossing? What is the biological/structural justification for using the* number *of shared contacts as the score of an alignment, i.e., where a larger score implies a higher level of three-dimensional similarity compared to a lower score.*

21.6.1 An Abstract ILP Formulation for the CMO Problem

Since the key decision is which alignment to form, the ILP formulation will need variables whose values specify an alignment, and inequalities that force the satisfaction of the constraints, (*a*) and (*b*), in the definition of an alignment. For each pair of nodes, $i \in CG(\mathcal{P}_1)$ and $j \in CG(\mathcal{P}_2)$, let $P(i,j)$ be a binary variable indicating whether nodes *i* and *j* will be aligned to each other, or not.

Exercise 21.6.3 *What are the differences in the definitions of an* alignment *in the CMO problem, and a non-crossing pairing in the RNA-folding problem?*

Exercise 21.6.4 *Write out inequalities that force the satisfaction of constraints* (*a*) *and* (*b*) *in the definition of an alignment. Model these inequalities after the inequalities described in Section 6.1 for a nested pairing in an RNA sequence, but note that the problem of RNA folding only involves one sequence, while here we have two contact maps.*

Next, the ILP formulation needs variables and inequalities to determine if two node pairs form a shared contact. Define $C(i,j,i',j')$ as a binary variable, which will be set to 1 *only if*:

(i) (i,j) is an aligned pair.
(ii) (i',j') is an aligned pair.
(iii) There is an edge in $CG(\mathcal{P}_1)$ between i and i'.
(iv) There is an edge in $CG(\mathcal{P}_2)$ between j and j'.

Exercise 21.6.5 *Write out* only-if *inequalities that implement the setting of variable* $C(i,j,i',j')$ *(as specified above), for each ordered pair of nodes* $i < i'$ *in* $CG(\mathcal{P}_1)$*, and each ordered pair of nodes* $j < j'$ *in* $CG(\mathcal{P}_2)$.

Exercise 21.6.6 *Using the C variables defined above, write out the objective function for the ILP formulation of the CMO problem.*

Exercise 21.6.7 *Why doesn't the ILP formulation need inequalities that set* $C(i,j,i',j')$ *to 1* if *conditions (i) through (iv) are satisfied?*

Exercise 21.6.8 *Suppose that the two contact maps have n nodes each. As described above, the number of* $C(i,j,i',j')$ *variables is* $(\frac{n(n-1)}{2})^2$*, which is about* $n^4/4$*. This number grows very quickly as a function of n. We would like to reduce that growth.*

Let E_1 *denote the number of pairs of nodes in* $CG(\mathcal{P}_1)$ *that are connected by an edge, each edge representing a pair of amino acids in protein* $\mathcal{P}_1$ *that are "close" in the three-dimensional structure of* $\mathcal{P}_1$*. Similarly, let* E_2 *denote the number of pairs of nodes in* $CG(\mathcal{P}_2)$ *that are connected by an edge. Generally,* E_1 *and* E_2 *will each be much smaller than* $n^2/2$*. Explain how to reduce the number of C variables in the ILP formulation to just* $E_1 \times E_2$.

Exercise 21.6.9 *Using the idea in the previous exercise, write out the concrete ILP formulation to solve the CMO problem for the pair of contact maps shown in Figure 21.4. Of course, ignore the alignment shown in that figure. Then use Gurobi to solve that instance of the CMO problem.*

Extending the Scoring Scheme Until now, the score of an alignment has been the *number* of shared contacts induced by the alignment. But more biologically informed scores are possible, and more meaningful scoring should lead to more meaningful alignments. Certainly, each edge (i,j) in a contact map might be given a weight to indicate *how close* amino acid i is to amino acid j in the known structure; or the weight may be a reflection of the *reliability* of that contact information. Also, we could give a weight to each edge (p,q) between two contact maps. Such a weight might reflect the similarly of the two amino acids p and q (e.g., are they the same amino acid; or in the same amino acid class).[9] In the extreme case, two amino acids in two CMO maps would be allowed to pair only if they were the same type of amino acid.

Exercise 21.6.10 *Discuss how to modify the CMO problem to incorporate edge weights, and show how to change the ILP formulation to handle such weights. For example, what is an appropriate objective function when edge weights are used?*

Exercise 21.6.11 *Given the relationship of the simple RNA-folding problem to the maximum-clique problem (as discussed in Section 6.1.4), and the relationship of the CMO problem to the RNA-folding problem, formulate the CMO problem as a maximum clique problem on a graph G. Be sure to explicitly define the nodes and edges in G. What would be weighted in G (i.e., nodes or edges) if we consider the weighted CMO problem discussed in the previous exercise.*

21.6.2 The Protein Threading Problem

The CMO problem gives us a way to *compare* the *known* structures of two amino acid sequences, and measure their structural similarity. However, the same machinery can be used to help *determine* the three-dimensional structure of amino acid sequence, $\mathcal{P}_1$, when the structure of a closely related sequence, $\mathcal{P}_2$, is already known. This general approach to protein structure prediction is called *protein threading*. It seeks the best *superposition* of $\mathcal{P}_1$ onto $\mathcal{P}_2$, using the *known* structure of P_2 to help deduce the *unknown* structure of P_1. Protein threading has been a successful way to *predict* protein structure, because families of proteins have similar three-dimensional structures due to common evolutionary history, and/or common biochemical constraints.

How to Thread For protein threading, we change the CMO problem so that now only *one* real contact map, $CG(\mathcal{P}_2)$ (for $\mathcal{P}_2$), is given at input. The only information given about $\mathcal{P}_1$ is its amino acid *sequence*, so we don't have a real contact map for $\mathcal{P}_1$. But, for ease of exposition, we say that the input $CG(\mathcal{P}_1)$ is just an ordered row of nodes, one for each amino acid in $\mathcal{P}_1$.

A *solution* to the threading problem consists of a (*non-crossing*) alignment, $\mathcal{A}$, of nodes in $CG(\mathcal{P}_1)$ with nodes in $CG(\mathcal{P}_2)$; and a *pairing* of some nodes of $CG(\mathcal{P}_1)$, where each pair in the pairing then creates a *shared contact* with nodes in $CG(\mathcal{P}_2)$. That is, two nodes i and i' in $CG(\mathcal{P}_1)$ are permitted to pair, if and only if i and i' are respectively aligned in $\mathcal{A}$ to some nodes j, j' in $\mathcal{P}_2$, and there is an edge (curve) between j and j' in $CG(\mathcal{P}_2)$. The pairing of some nodes in $CG(\mathcal{P}_1)$ creates a *proposed* contact map $CG(\mathcal{P}_1)$ for $\mathcal{P}_1$, which mirrors the contact map $CG(\mathcal{P}_2)$ as best possible. Each pair (i, i') in $CG(\mathcal{P}_1)$ indicates that amino acids i, i' are *predicted* to be "close" to each other in the three-dimensional structure of $\mathcal{P}_1$.

The score of a threading of P_1 to $CG(P_2)$ is given by a combination of a score for the alignment between $\mathcal{P}_1$ and $\mathcal{P}_2$, and a score for the pairing in $\mathcal{P}_1$. A structurally meaningful scoring scheme, balancing the contributions from the alignment with the contributions from the chosen pairing in $\mathcal{P}_1$, is vital to the success of protein threading. However, a discussion of specific scoring schemes is outside the scope of this book.

[9] Several ways to classify and divide amino acids into classes have been developed, and are extensively used.

Exercise 21.6.12 *In the protein-threading problem, why is it* essential *that each edge used in the alignment of node in $\mathcal{P}_1$ and $\mathcal{P}_2$ be weighted to reflect the biological relationship between the amino acids at the end points of the edge?*

Exercise 21.6.13 *Develop a full, abstract ILP formulation for the protein threading problem.*

See [206] for a more in-depth discussion of use of ILP in protein threading.

21.6.3 From Protein to RNA

The general goal of protein threading is to use a known three-dimensional structure of one protein as a *template* for the structure of a *related* protein whose structure is unknown but expected to be similar to the known one. The same idea is useful in the *RNA folding* problem, even in two dimensions.

In the RNA threading problem, the sequence and the two-dimensional fold of an RNA molecule, R_2, is given, along with the sequence (only) of a *related* RNA molecule, R_1, whose structure is expected to similar to R_2. As in protein threading, we want to find a (non-crossing) *alignment* of nucleotides in R_1 to nucleotides in R_2, and use the known fold information from R_2 to deduce a folding of R_1. And, as in protein threading, we need scores to evaluate the goodness of the alignment and of the induced fold of R_1.

The *score* of the alignment should *positively* weight the alignment of *identical* nucleotides, and *negatively* weight alignment of *nonidentical* nucleotides, and possibly distinguish between the types of mismatches (e.g., between nonidentical but *complementary* nucleotides or between nonidentical *noncomplementary* nucleotides). The score of an alignment might also reflect how well *runs* of nucleotides in R_2 are aligned to runs of nucleotides in R_1, in order to model *stacks* and *stacking*.

With an alignment, the known fold of R_2 is used to *induce* a fold of R_1. In particular, if nucleotides in positions i and i' in R_1 are aligned to nucleotides in positions j and j', respectively in R_2, and positions j and j' are paired in the fold of R_2, then we pair nucleotides i and i' in the predicted fold of R_1. The quality of the overall predicted fold of R_1 should be scored, for example, giving a positive score for each pair of complementary nucleotides, and a score for stacks, etc, as discussed in detail in Chapter 6.

Exercise 21.6.14 *Above we have only sketched the broad outline of RNA threading. Now, playing the role of an RNA modeler, complete the RNA-threading model; write out an explicit statement of the RNA-threading problem, and an explicit objective function; and detail an abstract ILP formulation to solve your stated RNA-threading problem.*

21.7 RECONSTRUCTING SIBLING RELATIONS

The following problem arises in the wildlife conservation where we have genetic information about members of a population who are alive today, or who lived at the same time in the past, but we don't have information from preceding generations. The main task is to determine which of the individuals are *full siblings*, i.e., individuals who had the *same two parents*. Information from the parental generation is generally not available or is very limited.

As one example, in addressing the current pandemic of *chronic wasting disease* that is affecting deer, elk, and moose in Colorado and neighboring regions, remains of deer that have clearly died from the disease are collected in the field, and genomic information is extracted from their remains. Then, in order to try to understand how wasting disease *spreads* in a population, researchers use the genomic information to reconstruct family relationships among the dead deer.[10] An ILP solution to this problem was developed in [17], and we discuss here a modified version of that solution.

21.7.1 Problem Details

Input to the problem consists of n individuals, each with a pair of specified haplotypes from a single identified chromosome (for example, chromosome 14). In this version of the problem we will assume that the n individuals come from several, *unknown* families, so that each pair of individuals are either *full siblings*, or share *no* parents. That is, there are *no half-siblings* among the n individuals, and no parents of those individuals are in the population. Also, contrary to the case in the pedigree reconstruction problem in Chapter 19, we assume that for the chromosome of interest, with its two haplotypes, a parent transmits to a child an identical copy of one of their two haplotypes. That is, (for now) there is no recombination between the chromosome homologs in this model. It's complicated enough.

The Sibling Reconstruction Problem (SRP):

> Partition the n individuals into the *smallest* number of "families" of full-siblings, obeying the genetic assumptions given above.

To better understand this, we need to more fully state what is required of a family.

What Is a family? Consider k individuals who are in the same family, so they have the same mother and same father. The mother has two haplotypes, that we call *maternal* haplotypes, and the father has two haplotypes (called *paternal*), so the mother and father have at most *four distinct* haplotypes between them. They could have fewer (1, 2, or 3) because identical haplotypes are possible, and in fact are common. Certainly there can only be those same (at most) four distinct haplotypes between the k children. And, each of the k children in the family must have one of the maternal haplotypes, and one of the paternal haplotypes. A complete description of the family genetics would explicitly specify for each of the k individuals, which of their two haplotypes is maternal (coming from their mother), and which is paternal (coming from their father).

Again, we don't have any information about the parents of the n individuals. We only know the two haplotypes that each the n individuals posses, and we don't know for any individual which of their two haplotypes are maternal, and which are paternal.

Exercise 21.7.1 *If k individuals have the same mother and father, then they can have at most four distinct haplotypes between them. That is necessary. Show by example, that it is* not *sufficient. That is, give an example where the k individuals only have four distinct haplotypes between them, but they cannot possibly have the same mother and father. Hint: Gender.*

[10] Another realistic example is trying to sort out the sibling relations of the children on Peter Pan's island, long after the children have forgotten them.

What Is a Solution? Formally, a solution to the SRP is a partition of the n individuals into families, so that in each declared family, with say k members, the following hold:

(a) The k individuals in a family have at most four distinct haplotypes between them.

(b) For each individual, one of their haplotypes is declared to be maternal, and the other is declared to be paternal.

(c) Over all the k members of the family, at most two distinct haplotypes are declared to be maternal, and at most two are declared to be paternal.

These are just the conditions of basic Mendelian genetics in diploid species, where each individual has two "copies" (homologs) of any chromosome, and a child inherits one haplotype from their mother and one from their father.

We Want a Parsimonious Solution Note that declaring each of the n individuals as a family by themselves satisfies the three conditions of a solution. But, as is common in many inference problems justified by biological parsimony arguments, we want a solution with the *fewest number* of families possible. Here we will outline the logic of an ILP solution to this problem. You will fill in the details.

21.7.2 An ILP Formulation

We start with some definitions and notation. Over the n individuals, let H denote the set of *distinct* haplotypes they collectively posses. In worst case, H could be as large as $2n$, but in real populations the size of H will likely be a small fraction of that. We will describe a proposed parent by the two haplotypes that they posses. Certainly a proposed parent who only has one haplotype in H can be replaced by one who has both haplotypes in H, without changing the number of proposed families.

Exercise 21.7.2 *We claim that there is an optimal solution where no proposed parent has a haplotype not in H. Why is this claim correct? Explain.*

Given the claim, each proposed parent can be described by *two* haplotypes in H. Now, consider an individual, I, and suppose I has haplotype pair (h_1, h_2). Then, one parent of I must be in the set $\{(h_1, h) : h \in H\}$ and the other parent of I must be in $\{(h_2, h') : h' \in H\}$. We denote the union of those two sets as $P(I)$, which is of size at most $2|H|$.

So, there is a well-defined, finite set of possible parents for each of the n individuals, and for the n individuals overall. It follows that there is a finite set of possible parent *pairs*, and these parent pairs can be mechanically generated from the input.

Exercise 21.7.3 *Given the discussion above, one ILP solution for the* sibling reconstruction problem *is based on the ILP formulation for the* set-cover *problem (defined in Section 7.3.1.2 and also used in section 21.4, earlier in this chapter). Figure out and write the full details of this* set cover *approach to the sibling reconstruction problem.*

21.7.2.1 Practicality

Next, we want to examine if the set-cover approach is practical. For a single individual, I, the size of parent set, $P(I)$, is at most $2|H|$, so the number of possible *parent pairs* of I is at most $4|H|^2$. Then since there are n individuals, the number of possible parent pairs overall is bounded by $n \times 4|H|^2$. Further, $|H|$ is bounded by $2n$ (if each of the n individuals has distinct haplotypes), so the number of possible parent pairs, over the n individuals, is at most $8n^3$. But these estimates are unrealistically large. The actual number of possible parents of I is $|\bigcup_I P_I|$,[11] and for real data, the number of possible parent pairs will be much less than $8n^3$. The next exercises look at this issue in more detail.

Exercise 21.7.4 *The ILP approach was tested in [17] using several real problem instances where the family structure was known, so a comparison could be made between the "truth" and what the ILP produced. In the case of a dataset of* flies, *there were 190 individuals that came from six families. Hence, for this data $|H| \leq 24$. Why? That is, explain where the number 24 comes from.*

Using $n = 190$ and $|H| = 24$, compute the two upper bounds derived above. Notice how large the gap is between them. What is the take-home lesson here?

Exercise 21.7.5 *To reduce the size and complexity of a problem instance, we claim first we may remove any input individual who is identical to some other input individual. So, the ILP formulation needs only be created for a subset of the input, where each individual is distinct from the others. Explain this.*

Second, explain why a parental pair *can have at most* four distinct *children, where a child is described by its two haplotypes. So, if a parental pair has more than four children, two or more must be identical (with respect to their two haplotypes). Because of this, after identical individuals are removed, no ILP formulation can have a parental pair with more than four* potential *children. This simplifies the structure of any problem instance. Can you see a way to use this fact to help the ILP formulation be solved faster?*

Exercise 21.7.6 *So far in the discussion of the sibling reconstruction problem we have assumed that none of the n individuals are* half-siblings, *i.e., sharing only one parent with some other individual. Now, assume we allow half-siblings. We still want to minimize the number of parental pairs (now with some parent possibly in more than one pair) that together could have parented all of the n given individuals.*

Show by example, that when half-siblings are allowed, a reconstruction of parental pairs must assign genders *to some of the parents in order to be sure that each parental pair has one male and one female. Why was this not necessary when only full-siblings were allowed?*

Exercise 21.7.7 *Show how to modify the ILP formulation to handle the case of allowing* half-siblings. *Be sure to address the gender issue from the previous exercise.*

When half-siblings are allowed, how will the number of parental pairs in the optimal solution compare to the number of parental pairs in the optimal solution when no *half-siblings are allowed? Do you think that generally the ILP formulation will solve faster when half-siblings are allowed, or when they are not allowed? Explain.*

More Variations There are many variations and extensions of the sibling reconstruction problem that can be imagined. We have already seen the variation of allowing half-siblings. Another interesting, and real, variation is in the study of male and female *oak trees* (did you know oak trees have gender?) In a single female oak tree, each of its acorns must be pollinated by a male oak tree. So, the acorns are the n input individuals. And, because all the acorns have the same mother, whose two haplotypes are known, the minimum number of parental pairs is actually the minimum number of male oak trees needed to pollinate the sampled acorns. According to the authors

[11] The symbol "$\bigcup$" denotes the *union* of several sets, which are identified in the subscript of $\bigcup$. This is in contrast to symbol "$\cup$," which denotes the union of just two sets.

of [17], this computational study has "provided additional supporting evidence that oak pollen disperses much farther than previously thought."[12]

Exercise 21.7.8 *Show how to modify the ILP formulation for the sibling reconstruction problem to solve the oak tree variant of the problem.*

[12] An acorn might not fall far from the tree, but pollen does.

22

What's Next?

If you want to learn more about the topics covered in this book, there are two natural directions to go: *biology* and *technology*. The first direction is to learn more about the wide variety of *biological areas* where integer programming has been, or could be, of value. The second direction is to learn more about techniques to *formulate* and *solve* integer linear programs.

Biology In the biological direction, the opportunities are vast. Although this book has discussed problems from a wide range of biology, there are many other (sub)areas in biology not discussed, where integer programming is being used, or could be. Of course, even within the biological areas that were covered, there are many specific problems, publications, and ILP formulations that were not discussed. And, I have no doubt that the range of biology where ILP can be helpful will greatly expand in the (near) future. Hopefully, this book, and its readers will contribute to that expansion.

Technology In the technical direction, there are two subdirections: creating ILP formulations and solving them.

Creating ILP Formulations On the topic of *creating* ILP formulations, I think the book has covered a large fraction of the *types* of models, and *styles* of ILP formulations that have been used in computational and systems biology.[1] Of course, there are additional idioms that were not discussed, and I know of a few styles of formulations that were omitted. But on whole, I think the book has done a good job of introducing and discussing the kinds of ILP formulations that are currently in use. No doubt, new ILP formulation styles will emerge as new biological problems and areas are addressed through the use of ILP.

Solving ILP Formulations On the topic of *solving* ILP formulations, the book has barely scratched the surface. Here again, there are two subdirections: *using* ILP solvers, and *understanding and mastering* advanced *solution techniques* that are not currently implemented in the solvers. There is a huge literature for both of these subdirections, well beyond what I covered in the book.

[1] By "style," I mean clique-like formulations, or TSP-like formulations, or Steiner-Tree-like formulations, or set-cover-like formulations, etc.

Using Solvers I mentioned earlier in the book that there are many parameters in Gurobi that can be set, but we only discussed a few of them. More significant, the book uses Gurobi in only the *command-line* mode, which is the simplest way to learn about, and to use, Gurobi. But Gurobi has two other modes, the *interactive* mode, and the *API* (application programming interface) mode, which gives the user more features, and more control over the solver.

The interactive mode allows the user to issue successive commands, interacting with the solver. Moreover, while in interactive mode, the system remembers all the prior commands and parameter settings. This allows the user more control and the use of a richer set of functions. For example, in interactive mode, the user can ask Gurobi to find all of the optimal solutions and output the variable values for each.[2] The interactive mode does not require knowing any computer programming language.

The API mode allows one to *program* the use of Gurobi so, for example, the program can supervise Gurobi's progress in solving an ILP formulation; or pause the solver to make changes in the ILP formulation as the execution proceeds, etc. This is the way, for example, that the relaxation/separation approach for TSP works in the Python program *tsp.py*, discussed in Section 9.8.4.1: running, supervising, and pausing the ILP execution to check if any of the subtour inequalities are violated, and adding them to the formulation, if needed.

The API mode gives the user the most control over the behavior of Gurobi, but its use requires knowing one of several computer programming languages. Python is (implicitly at least) the language that is most recommended by Gurobi. The way that I use a computer programming language to create concrete ILP formulations is related, but (as practiced in this book) I don't use the Gurobi API mode.

Another use of Gurobi that was not mentioned in the book, is its ability to run on parallel and distributed computer systems with many processors, and in the cloud. These approaches allow much greater computational resources to be thrown at solving a problem instance, greatly speeding up its solution compared to running on my MacBook Pro laptop. Some of the problems we examined in the book, where the ILP executions were not practical when running on my MacBook Pro, might be practical on those more industrial-strength systems.

Solution Techniques On the subdirection of advanced solution techniques, there are several well-developed techniques that are only partially implemented in ILP solvers. The basic approach of ILP solvers is *branch-and-bound*, using linear programming to provide bounds, with the added use of some cutting planes to exclude fractional solutions, and heuristics find better feasible solutions. When a branch-and-bound approach is used together with the application of cutting planes, the overall method is called *branch-and-cut*.[3] However, there other solution techniques that are not easily implemented in a general ILP solver, because each application of the technique requires specialized understanding of the application. The names of some of these methods are *column generation*, *cutting-plane methods*, *branch-and-*

2 Recall that in Section 2.2.7.2, we discussed how to use Gurobi in command-line mode, to find the *number* of distinct optimal solutions to a problem, but Gurobi will only output values for the variables in for one of the optimal solutions.

3 These words will probably not be meaningful to you, and I am only planting the word seeds in your brains in case you do go deeper into ILP solving techniques after this book.

cut, *branch-and-price*, *lift and project*, *relaxation/separation*, and *Lagrange relaxation*. Advanced solution methods are discussed in [18, 47, 121, 203].

Learning LP and ILP Theory Another area that I have not even mentioned in the book is the mathematical theory of linear programming and integer linear programming.[4] But, the theory of linear programming is truly *beautiful*, both mathematically and in the way that it touches many topics outside of linear programming itself (for example, convex geometry, game theory, economics, combinatorics, computational complexity, etc.). Learning linear programming theory requires a bit of knowledge about matrix theory and linear algebra, but if you have that background I highly recommend LP theory for the mathematically inclined. There are many excellent books that discuss that topic.

The theory of *integer* linear programming is more difficult, and less beautiful (to me), and I have to admit I have never mastered much of it. The educational material on ILP theory is also much more limited than for LP theory.

Learning to Program Finally, if you want to completely master the "work flow" of using integer linear programming, a basic competency in computer programming is needed, in some programming language. This is true whether you use the API mode, or write computer programs to create concrete ILP formulations, the way I do it. The favored language of Gurobi (and others) is currently Python, which is a relatively easy language to learn. For the briefest of introductions, tailored around two of the programs that accompany this book, see the tutorial on Python posted on the book website.

[4] I promised that there would be no math in the book, so a discussion of theory was out of the question, and out of keeping with what I wanted the book to accomplish.

23

Epilogue: Some Very Opinionated Comments for Advanced Readers

Ok, I have written a long book on practical issues that I hope will have a positive impact on biology and the future of the world. Maybe someone who would otherwise not know about ILP will learn about it and use it to get a better solution to a hard computational problem, and that will lead to a better analysis and understanding of some data, and that will lead to a better understanding of disease, and that will lead to better understanding of cancer, and that will lead to a better treatment, which might help someone. If so, I get some karma credit for my labors. Spending that credit forward, I now take the liberty of writing a few opinionated pages that (hopefully) some people will find debatable. I invite all readers of this book to read these comments, but unlike the rest of the book, they are aimed more at readers with background in computer science theory.

23.1 ON THE POWER OF LINEARITY AND LINEAR MODELS

There is a general belief that *linear phenomena* are rare, and therefore *linear models* are unable to represent more than a small slice of the "real world." The following quote (which itself contains a quotation), from a physicist, is a typical reflection of this belief:

> There is a well known saying: "Dividing the universe into things that are linear and things that are non-linear is very much like dividing the universe into things that are bananas and things that are not." ... Non-linearity is a hallmark of the real world. It occurs anytime outputs of a system cannot be expressed in terms of a sum of inputs, each multiplied by a simple constant – a rare occurrence in the grand scheme of things. ... we have a tendency to try and view the world in terms of linear models much for the same reason that looking for lost keys under a lamppost might make sense: because that is where the light is. [22]

The implication is that linear phenomena are so rare, that tools and methods for working with linear systems will not be of great value, no matter how well they work. Even though it may be easier to search where the light is, you would be fooling yourself to think that the search is likely to be productive.

A related story is told in [48], about the lecture in 1948 where George Dantzig announced the simplex algorithm for linear programming. After the lecture,

> Harold Hotelling, great in both academic stature and physical size, rose from his seat, stated simply "But, we all know that the world is non-linear," and sat down.

The book relates how John von Neumann defended Dantzig, and goes on to argue for the importance of linear models and linear programming:

> Fortunately for the world, many of its complexities can in fact be described in sufficient detail by linear models ... The scope of the use of linear programming in industry is breathtaking ... Take that Professor Hotelling. [48]

And, If Linear Is Rare Let me conjecture what Hotelling would have said about *integer* linear programming, and linear models where all the variables are required to have integer values:

> "But, we all know that the world is non-linear, and measured in *fractional* quantities."

If linear phenomena are rare, integer linear phenomena must be phenomenally rare – and if linear phenomena are like bananas, then *integer* linear phenomena must be like *Musa Troglodytarum* (Google it).

Yes, But Yes, its true that most real-world phenomena are nonlinear and measured in fractions.[1] And, yes its true that the behavior of most natural systems "cannot be expressed in terms of a sum of inputs, each multiplied by a simple constant." And, the reply that linear phenomena are of sufficient importance that linear models and linear programming have vast application, is also true. BUT, these objections to linear models, and this reply to those objections, miss something huge that computer science and mathematics came to understand in the 1970s.

> There may be very few *natural* phenomena that *behave* according to an integer linear function (model), but we may still be able *describe* the behavior of the system using a set of integer *linear* inequalities.[2] Natural expression is one thing, but mathematical expressability is another.

And this expressability is *not* achieved by some *linearization* or a *linear approximation* of the nonlinear phenomena. Instead, this fact comes from the realization that integer linear systems have great *expressive* power. Integer linear programming is NP hard, so any instance of a problem in NP can be *reduced* (in polynomial time as a function of the input size) to an integer linear program. The ILP may *look nothing* like the *natural description* of the problem, but the solution of the ILP will identify a solution to the original problem.

And it gets better. For many problems in NP, the corresponding ILP instance *exactly* captures the *solution space* of the original problem, so that the solutions to the problem and feasible solutions to the ILP are in one-to-one correspondence. Then, the ILP exactly captures all of the possible behaviors of the "system." [3]

[1] Excluding quantum phenomena.

[2] See Section 1.2 for a definition of an integer linear function and inequality.

[3] Let me state this a bit more formally and precisely. A "language," L, is just a set of objects with some stated property. For example the set of graphs that each contain a clique with at least half the nodes of the graph, forms a language. NP is a set of languages (with very particular properties, but we don't need those here). So each language in NP is a set of objects. For each language L in NP, and each object x which might be in L, we can constructively create, in time that is bounded by a polynomial function in the number of bits needed to describe x, a system of linear inequalities, $\mathcal{I}$, such that $x \in L$ if and only if there is an assignment of *integer* values to the variables in $\mathcal{I}$ that satisfy all of the

Further, if we add a linear integer objective function to the set of inequalities, we get an ILP formulation whose solution identifies an *optimal* solution (i.e., an object that is best according to the criteria in the objective function). So, any optimization problem that is stated with a linear objective function over a set in NP, can be efficiently reduced to an ILP formulation. This is huge.

23.1.1 Let's Make This Concrete

Consider the *maximum clique* problem. Clearly, there is a function, $F(G)$, which takes in any graph G, and returns the size of the maximum-size clique in G. But is it a *linear* function? Equivalently, we can describe a system that takes in a graph G and outputs the integer $F(G)$. Is the behavior of this system linear? Can its behavior "be expressed in terms of a sum of inputs, each multiplied by a simple constant"?

The question of whether $F(G)$ is a linear function (or the system behaves linearly) is almost nonsense. What does it mean to be linear in this context? Linear in what variables? The idea that $F(G)$ is a linear function not only seems impossible, it seems *imparsable*.

But here's the thing: Given a graph G, we can write a *small*[4] set of *linear inequalities* whose *integer* feasible solutions exactly specify what it is to be a clique in G. We did that in Chapter 2. The solution space of those integer linear inequalities is in one-to-one correspondence with the set of cliques of G. And, when we add an integer linear objective function, maximizing the number of variables whose value is set to one, the solution value will be exactly $F(G)$.

The phenomenon of being a clique in a graph does not sound at all like a linear phenomenon (it doesn't even sound like a banana), and is not *naturally* described as one. Yet, the cliques in G can be put into one-to-one correspondence with the solutions of an integer linear system of inequalities, and the maximum clique problem can be *expressed* in an ILP formulation.

23.1.2 And, We See It Throughout This Book

The problems discussed in this book are anchored in biological phenomena. Almost none of those particular phenomena have a *natural* description as a linear phenomenon. And even the simplified, formalized models that are created from those biological phenomena are not *naturally* linear. But, in each case, the behavior of the model was expressed precisely in terms a system of integer linear inequalities; where a "best" behavior was identified by adding an integer linear objective function. The mechanics of expressing a behavior, phenomenon, or problem in terms of integer linear inequalities and an integer linear objective function, is called a *reduction* to

inequalities in $\mathcal{I}$. And, except for the problems known to be in P (which are not relevant for this discussion, and are nice anyway), the time for this construction is much less than for any known method to determine if $x \in L$. So, we can't cheat by first figuring out if x is in L, and then creating a trivial linear inequality based on the answer. The construction is really doing something substantive.

The importance of all this comes from the fact that there are a huge number of interesting, real problems (languages) in NP, and for most of them there is *no known* polynomial-time algorithm that determines membership in the language. Hence the utility of a reduction to ILP, and the importance of a good ILP solver.

[4] Whose size is bounded by a polynomial function of the size of the graph G.

ILP. In this light, we can now see that most of what we have done in this book is to develop and describe reductions to ILP.

23.1.3 But Any NP-Hard Problem Should Work

Now, everything I have said about the expressability of integer programming holds for *any* NP-hard problem. So what is special about integer linear programming? Two words: Gurobi and Cplex. These are commercial companies that are heavily invested in integer linear programming. They continually work to improve their products, and provide extensive documentation, education, online user groups, and customer service (for paying customers). Plus, there is a large body of work in computational biology that is based on the use of Gurobi and Cplex. This all gives them the Big-Mo.

There are academic solvers for other NP-hard problems, and serious devotees who work on them. But, that is still a cottage industry in comparison to real industry. The most highly developed solvers for an alternative NP-hard problem are for *Boolean satisfiability* in conjunctive normal form (CNF). These are called *SAT-solvers*. How does the ease of formulating an ILP compare to the ease of creating a CNF SAT formula, and how do the SAT-solvers compare to ILP-solvers?

I don't have a well-defined answer for either question – I (and an undergraduate student, Chase Maguire) have played with one SAT-solver and tried solving one combinatorial problem with it: protein folding under the HP model, discussed in Chapter 7. From that limited experience, I can say that I find creating and using SAT formulas very painful. SAT has very little structure compared to systems of linear inequalities. Relations and statements that are trivial to implement with linear inequalities seem torturous to implement with SAT formulas. And, of course, optimization is very difficult with SAT formulas (one has to do binary search over the possible solution values). Testing if the optimal objective value is larger (or smaller) than some constant number, *can sometimes* be implemented with a SAT formula but its not obvious how to do it [110]. Certainly, there are idioms in SAT programming, and maybe the reason I find SAT programming painful is that I am not yet facile enough in using the idioms – but I don't think so.

And, paraphrasing Don Knuth (as best I remember it, and far from a verbatim quote): "Several hundred years of solving systems of linear equations, over a hundred years of linear algebra, and decades of numerical linear algebra gives ILP a huge headstart"[5]. Can SAT, or some other NP-hard problem, catch up to ILP with more research and time, or is there some *fundamental* reason that modeling with linear inequalities is more natural? I vote for the latter, but I don't really know.

23.2 WHY IS ILP IN COMPUTATIONAL BIOLOGY DIFFERENT FROM TRADITIONAL ILP?

Most of the uses of ILP in computational and systems biology involve developing (polynomial time) *reductions* of problems (languages) in NP (and mostly NP-complete ones) to ILP formulations. The end result is often an ILP that seems weird and disconnected from the original problem. And that is what is different from traditional uses of ILP, where many more of the problems are *naturally* stated in terms of

[5] Don Knuth. Personal communication.

integer linear inequalities. Then, the creation of an ILP formulation mostly involves translating the native description of the problem into symbols and inequalities in the format needed for an ILP solver.

As an example, recall the *threatened-species-protection problem*, discussed in Chapter 1. The native description of the problem involved a linear objective function, and costs and requirements and constraints that were directly encoded into linear inequalities. The restriction of the problem to integer values just restricted the values of the variables, so even the integer version of the problem has a direct translation into an ILP formulation. Thus, the version of the protection problem in Chapter 1 is an example of a fairly typical problem in traditional LP and ILP modeling. But, it is a nontypical problem in computational and systems biology that come from reductions.

The differences between traditional ILP formulations and those implementing reductions are not just conceptual – they can also be practical. For example, traditional ILP formulations typically have more variables than inequalities, while the opposite is true for most of the formulations discussed in this book.

23.3 BLACK BOXES VERSUS CLEAR SEMANTICS AND ILP

Integer programming in computational biology illustrates another strength of modeling with optimization, compared to several competing modeling and computational methods that are often implemented as *black boxes*. Those methods include traditional statistical inference (using either maximum-likelihood or Bayesian approaches), and the more recent approaches involving machine learning and deep neural nets.

23.3.1 What Is a Black Box and What's Wrong with It?

There are many computer programs whose internal workings are not understood by anyone, even their designers and programmers. This sometimes happens because of the accumulation of numerous changes and compromises in the program (perhaps contributed over time by multiple individuals); or because the program implements a high-level computational approach that is only vaguely understood; or because the behavior of the program depends on specific parameters and values that are only obtained empirically (i.e., by some form of trial and error experimentation), without any deeper guiding principles. Programs of this type are called *black boxes*.

There are black boxes whose performance is amazing, but it is unclear why. The problem of explaining the workings of black boxes is the focus of a significant research effort, called *explainable A.I.*, supported by the U.S. Department of Defense. The impetus for this project is the belief that if people don't understand how a computer program comes to a decision, or obtains a result, they will not trust or follow it. More generally, this is the "interpretability problem in machine learning: the difficulty in explaining how a particular AI [program] has come to its conclusions" [97].

23.3.2 Semantics: What versus How

Ideally, even if a user does not know *how* a computer program works, they should know and understand *what* it does, i.e., what its workings *achieve*. For a computer program to have understandable *semantics*, there must be a *simple, short* statement of "what it does." And, the statement has to be complete and correct. It is not enough to have a simple statement of "what we want it to do," or "what we designed it to do." For example, if I know that a program takes in an integer x and outputs the number of prime numbers less than x, that is a clear, simple statement of *what* the program *does*, without any hint of *how* it does it (in fact, I would like to know how it does it). And, that is very different from having a program where all we can say is that "It tries to find the number of primes less than x. It seems to work well, so far."

Typically, black boxes lack clear semantics. Often, the only true statement about what the black box achieves, is that it executes the program running on it. Then you have to understand the "how" to understand the "what," and there may not be a simple, correct statement of what the "how" is.

In contrast, when a problem is formulated as an integer program, there is a precise statement of what the ILP does: It optimizes the given objective function over the set of feasible solutions specified by the inequalities. And, in the context of biological models, there is usually a precise and meaningful statement of what the feasible solutions are, and what the objective function means in that context. There is no guarantee of this, but the results of optimization methods are generally very *interpretable*. For that conclusion, it doesn't matter if we don't understand *how* an ILP solver works (that can be considered a black box, especially since the best ILP solvers are proprietary). And, for this conclusion, it doesn't matter if the solver is efficient – there still is a clear, simple statement of *what* the solver produces.

23.3.3 A Specific Case

Recall the haplotyping problem in populations discussed in Chapter 20, and the *pure parsimony* ILP approach discussed in Section 20.4.2.1. I claim that the pure parsimony ILP approach has clear semantics. The inequalities guarantee that the set of feasible solutions to the ILP formulation are exactly the ways that the n input genotypes could be phased. That is, a feasible solution to the ILP is a set of n haplotype pairs, such that the i'th genotype would be generated by the i'th pair of haplotypes. Conversely, any such set of n haplotype pairs define a feasible solution to the ILP. The objective function calls for a feasible solution with the *minimum number of distinct* (different) haplotypes used in the n pairs. That is a simple, clean and correct statement of *what* the ILP *does* (ok, assuming that the solver solves – but that is different from the issue of interpretability).

Now consider the computer program called *Phase* [188], which was the gold-standard haplotyping program in its day (it is no longer the gold standard because it did not scale well to the larger problem instances that came later). Its accuracy was amazing. Comparing the haplotype solutions it produced to the (few) ground-truth data sets that were known, its accuracy was typically in the 95–100% range (for a particular error measure we don't need to discuss). And, the initial inspiration for Phase came from clearly articulated ideas and statistical-genetic models. The high-level game plan of Phase is similar to many statistical-inference approaches: Phase derives from a stochastic model of haplotype evolution starting from a small founder set of haplotypes, to the haplotypes observed today. The general stochastic model is called *the coalescent model*, [197] and the evolutionary forces in that model are site mutations and meiotic recombinations. The coalescent model uses the *infinite sites* assumption, which is another clear piece of the evolutionary model underlying Phase. Then, given n observed genotypes, the problem Phase tries to solve is to find n pairs of haplotypes that maximize the probability (under the articulated stochastic model) of observing the n given genotypes. This is a classic maximum-likelihood approach, of the kind discussed in Chapter 10.

The problem is that the model has several parameters, and the methods of finding a maximum-likelihood solution all involve taking a long, semi-random walk through the solution space, aided by some heuristics and parameter tuning obtained by experimentation, compromises in the model, etc. The particular method used in Phase to search the solution space is called *Gibbs sampling*, which is a form of the Markov-Chain Monte-Carlo (MCMC) method.

In the end, the Phase program, no matter how well-founded it was in principle, became a highly engineered black box – a very high-performing black box, but one where it was no longer clear *how* it worked, or *what* it did. Yes, it produced a set of haplotypes that explain the genotypes, and that is what the designers *wanted* it to do. But, since Phase can't guarantee it will produce the maximum-likelihood solution, and its internals are so obscured, it lacks clear, simple semantics.

In fact, one of the designers of Phase conjectured (in different words) that maybe what Phase is doing is solving what we now call the pure-parsimony problem [58]. And when I compared the solutions obtained from the pure-parsimony ILP to the solutions obtained from Phase, it was clear that the conjecture is very close to the truth [85]. In reality, the number of distinct haplotypes that Phase creates is very close to the minimum necessary, which is vastly smaller than the number of haplotypes possible. Further, the haplotyping solutions produced by the ILP solutions to the pure-parsimony problem, were also amazingly accurate – a bit less than for Phase, but still amazing in comparison to the size of the solution space. So, in this case, some clear semantics for Phase became possible, but only because of an ILP formulation whose semantics are absolutely clear and correct.

23.3.4 The Bottom Line

Optimization formulations (and ILP in particular), typically use simple objective functions, and precise constraints, which naturally lead to clear semantics. In contrast, statistical-inference methods, and in more extreme ways, machine learning and deep neural nets, often lack clear semantics. It is not clear what we learn from the success of a black box (except to use it when appropriate). Worse, what can we learn from the *failure* of a black box, if we can't state *what* the box does? What would we have learned from Phase if it had been a failure? In contrast, if the pure-parsimony ILP produced highly inaccurate haplotyping solutions, we at least would have learned that nature and evolution is doing something other than minimizing the number of distinct haplotypes. That would be scientific progress, even if it was not an engineering success.

Bibliography

[1] G. Agarwal and D. Kempe. Modularity-maximizing graph communities via mathematical programming. *The European Physical Journal B*, 66:409–418, 2008.

[2] R. Agarwala, D. L. Applegate, D. Maglott et al. A fast and scalable radiation hybrid map construction and integration strategy. *Genome Research*, 10:350–364, 2000.

[3] R. K. Ahuja, T. L. Magnanti, and J. B. Orlin. *Network Flows: Theory, Algorithms, and Applications*. Prentice Hall, 1993.

[4] F. Alizadeh, R. M. Karp, D. Weisser, and G. Zweig. Physical mapping of chromosomes using unique probes. *Journal of Computational Biology*, 2:159–184, 1995.

[5] O. Alkan, B. Schoeberl, M. Shah et al. Modeling chemotherapy-induced stress to identify rational combination therapies in the DNA damage response pathway. *Science Signaling*, 11(540), 2018.

[6] U. Alon. *An Introduction to Systems Biology: Design Principles and Biological Circuits*. Chapman and Hall, 2006.

[7] E. Althaus, G. W. Klau, O. Kohlbacher, H. P. Lenhof, and K. Reinert. Integer linear programming in computational biology. In *Festschrift Mehlhorn, LNCS 5760*, pages 199–218. Springer, 2009.

[8] E. Alvarez-Miranda, I. Ljubic, and P. Mutzel. The maximum weight connected subgraph problem. In M. Junger and G. Reinelt, editors, *Facets of Combinatorial Optimization*, pages 245–270. Springer, 2013.

[9] D. Applegate, R. Bixby, V. Chvatal, and W. Cook. The concorde TSP solver, 2003. Retrieved from www.math.uwaterloo.ca/tsp/concorde/index.html (last accessed January 24, 2019).

[10] C. Ash and J. Smith. In other journals: Nanoscreening for drug combinations. *Science*, 361:39–40, 2018.

[11] N. Atias and R. Sharan. iPoint: An integer programming based algorithm for inferring protein subnetworks. *Molecular BioSystems*, 9:1662–1669, 2013.

[12] C. Backes, A. Rurainski, G. W. Klau et al. An integer linear programming approach for finding deregulated subgraphs in regulatory networks. *Nucleic Acids Research*, 40:e43, 2012.

[13] V. Bafna and P. Pevzner. Sorting by reversals: Genome rearrangements in plant organelles and evolutionary history of X chromosome. *Journal of Molecular Biology and Evolution*, 12:239–246, 1995.

[14] H. J. Bandelt, P. Foster, B. Sykes et al. Mitochondrial portaits of human populations using median networks. *Genetics*, 141:743–753, 1995.

[15] Z. Bar-Joseph, E. D. Demaine, D. K. Gifford et al. K-ary clustering with optimal leaf ordering for gene expression data. *Bioinformatics*, 19:1070–1078, 2003.

[16] Z. Bar-Joseph, D. K. Gifford, and T. S. Jaakkola. Fast optimal leaf ordering for hierarchical clustering. *Bioinformatics*, 17(suppl_1): S22–S29, 2001.

[17] T. Berger-Wolf, S. Sheikh, B. DasGupta et al. Reconstructing sibling relationships in wild populations. *Bioinformtics*, 23:i49–i56, 2007.

[18] D. Bertsimis and R. Weismantel. *Optimization Over Integers*. Dynamic Ideas, 2005.

[19] M. Blanchette, G. Bourque, and D. Sankoff. Breakpoint phylogenies. In S. Miyano and T. Takagi, editors, *Genome Informatics*, pages 25–34. University Academy Press, 1997.

[20] V. Blondel, J.L Guillaume, R. Lambiotte et al. Fast unfolding of communities in large networks. *Journal of Statistical Mechanics: Theory and Experiment*, 10:10008–10020, 2008.

[21] C. Blum and P. Festa. *Metaheuristics for String Problems in Bio-informatics*. Wiley, 2016.

[22] G. Boccaletti. Scale analysis. In J. Brockman, editor, *This Will Make You Smarter. The Edge Question for 2011*, pages 184–187. Harper Collins, 2012.

[23] P. Bonizzoni, C. Brghin, R. Dondi et al. The binary perfect phylogeny with persistent characters. *Theoretical Computer Science*, 454:51–63, 2012.

[24] P. Bonizzoni, A. P. Carrieri, G. D. Vedova et al. Algorithms for the constrained perfect phylogeny with consistent characters, 2014. arXiv:1405.7497v1.

[25] P. Bonizzoni, A. P. Carrieri, G. D. Vedova et al. Explaining evolution via constrained persistent perfect phylogeny. *BMC Genomics*, 15(Suppl 6):S10, 2014.

[26] P. Bonizzoni, A.P. Carrieri, G. Della Vedova et al. When and how the perfect phylogeny model explains evolution. In N. Jonoska and M. Saito, editors *Discrete and Topological Models in Molecular Biology*, Natural Computing Series, chapter 4. Springer, 2013.

[27] P. Bonizzoni, S. Ciccolella, G. Della Vedova et al. Beyond perfect phylogeny: Multisample phylogeny reconstruction via ILP. *Proceedings of ACM-BCB Conference 2017*, 2017.

[28] R. Bosch and M. Trick. Integer programming. In E. K. Burke and G. Kendall, editors, *Search Methodologies: Introductory Tutorials in Optimization and Decision Support Techniques*, pages 67–92. Springer US, Boston, MA, 2014.

[29] D. Brown and I. Harrower. A new integer programming formulation for the pure parsimony problem in haplotype analysis. In *WABI, Workshop on Algorithms in Bioinformatics*, volume 3240, pages 254–265. LNCS, Springer, 2004.

[30] D. Brown and I.M. Harrower. Integer programming approaches to haplotype inference by pure parsimony. *IEEE/ACM Transactions on Computational Biology and Bioinformatics*, 3(2):141–154, 2006.

[31] S. Bruckner, F. Huffner, R. M. Karp et al. Topology-free querying of protein interaction networks. *Journal of Computational Biology*, 17:237–252, 2010.

[32] A. Burt and R. Trivers. *Genes in Conflict*. Belknap Press, 2006.

[33] S. Canzar and M. El-Kebir. A mathematical programming approach to marker-assisted gene pyramiding. In *WABI, Workshop on Algorithms in Bioinformatics*, volume 6833 of *Lecture Notes in Computer Science*, pages 26–38. Springer, 2011.

[34] A. Caprara. Sorting by reversals is difficult. In *Proceedings of RECOMB 97: The First International Conference on Computational Molecular Biology*, pages 75–83. ACM Press, 1997.

[35] A. Caprara, R. Carr, S. Istrail et al. One thousand and one PDB structure alignments: Integer programming methods for finding the maximum contact map overlap. *Journal of Computational Biology*, 11:27–52, 2004.

[36] D. Catanzaro, S. E. Shackney, A. Schäffer et al. Classifying the progression of ductal carcinoma from single-cell sampled data via integer linear programming: A case study. *IEEE/ACM Transactions on Computational Biology and Bioinformatics*, 13:643–655, 2016.

[37] V. Chandru, M. R. Rao, and G. Swaminathan. Protein folding on lattices: An integer programming approach. In M. Grotschel, editor, *The Sharpest Cut: The Impact of Manfred Padberg and His Work*. SIAM Press, 2004.

[38] W. C. Chang, S. Vakati, R. Krause et al. Exploring biological interaction networks with tailored weighted quasi-bicliques. *BMC Bioinformatics*, (Suppl 10):S16, 2012.

[39] C. Chauve and E. Tannier. A methodological framework for the reconstruction of contiguous regions of ancestral genomes and its application to mammalian genomes. *PLOS Computational Biology*, 4(11), 2008.

[40] H-C Chen, W. Zou, T-P Lu et al. A composite model for subgroup identification and prediction via bicluster analysis. *PLoS ONE*, 9:e111318, 2014.

[41] Z. Z. Chen, F. Deng, and L. Wang. Exact algorithms for haplotype assembly from whole-genome sequence data. *Bioinformatics*, 29:1938–1945, 2013.

[42] O. Chernomor, B. Q. Minh, F. Forest et al. Split diversity in constrained conservation prioritization using integer linear programming. *Methods in Ecology and Evolution*, 6:83–91, 2015.

[43] M. Chimani, S. Rahmann, and S. Bocker. Exact ILP solutions for phylogenetic minimum flip problems. In *Proceedings of the First ACM-BCB Conference*, pages 147–153, 2010.

[44] C. Chothia. One thousand families for the molecular biologist. *Nature*, 357:543–544, 1992.

[45] D. Clayton, S. Bush, and K. Johnson. *Coevolution of Life on Hosts*. University of Chicago Press, 2016.

[46] K.D. Cocks and I.A. Baird. Using mathematical programming to address the multiple reserve selection problem: An example from the Eyre Peninsula, south australia. *Biological Conservation*, 49:113–130, 1989.

[47] M. Conforti, G. Corneujels, and G. Zanbelli. *Integer Programming*. Springer, 2014.

[48] W. J. Cook. *In Pursuit of the Traveling Salesman*. Princeton University Press, 2012.

[49] T. Cormen, C. Leiserson, R. Rivest et al. *Introduction to Algorithms, 3rd edition*. MIT Press, 2009.

[50] P. Csermely, T. Korcsmaros, H. Kiss et al. Structure and dynamics of molecular networks: A novel paradigm of drug discovery: A comprehensive review. *Pharmacology and Therapeutics*, 138:333–408, 2013.

[51] J. Cussens, M. Bartlett, E. Jones et al. Maximum likelihood pedigree reconstruction using integer linear programming. *Genetic Epidemiology*, 37(1):69–83, 2013.

[52] G. B. Dantzig, D. R. Fulkerson, and S. M. Johnson. Solution of a large-scale travelling-salesman problem. *Operations Reseach*, 2:393–410, 1954.

[53] B. Dilkina and C.P. Gomes. Solving connected subgraph problems in wildlife conservation. In *CPAIOR10: Proceedings of the 7th International Conference on Integration of AI and OR Techniques in Constraint Programming*, 2010.

[54] K. A. Dill, S. Bromberg, K. Yue et al. Principles of protein folding – A perspective from simple exact models. *Protein Science*, 4:561–602, 1995.

[55] P. DiMaggio, A. Subramani, R. Judson et al. A novel framework for predicting *in vivo* toxicities from *in vitro* data using optimal methods for dense and sparse matrix reordering and logistic regression. *Toxicological Sciences*, 118:251–265, 2010.

[56] M. Dittrich, G.W. Klau, A. Rosenwald et al. Identifying functional modules in protein-protein interaction networks: An integrated exact approach. *Bioinformatics*, 24:i223i231, 2008.

[57] L. Dollo. Le lois de l'évolution. *Bulletin de la Societé Belge de Géologie de Paléontologie et d'Hydrologie*, 7:164–167, 1893.

[58] P. Donnelly. Comments made in a lecture given at the DIMACS Conference on Computational Methods for SNPs and Haplotype Inference, November 2002.

[59] R. F. Doolittle. What we have learned and will learn from sequence databases. In G. Bell and T. Marr, editors, *Computers and DNA*, pages 21–31. Addison-Wesley, 1990.

[60] C. Drysdale, D. W. McGraw, C. B. Stack et al. Complex promoter and coding region β2-adrenergic receptor haplotypes alter receptor expression and predict *in vivo* responsiveness. *Proceedings of the National Academy of Sciences (USA)*, 97:10483–10488, 2000.

[61] R. L. Dunbrack and M. A. Karplus. A backbone dependent rotamer library for proteins: application to sidechain prediction. *Journal of Molecular Biology*, 230:543–571, 1993.

[62] I.M. Ehrenreich, Y. Hanzawa, L. Chou et al. Candidate gene association mapping of arabidopsis flowering time. *Genetics*, 183:325–335, 2009.

[63] M. El-Kebir, L. Oesper, H. Acheson-Field et al. Reconstruction of clonal trees and tumor composition from multi-sample sequencing data. *Bioinformatics*, 31:i62i70, 2015.

[64] L. Elbroch, M. Levy, M. Lubell et al. Adaptive social strategies in a solitary carnivore. *Science Advances*, 3:e1701218, 2017.

[65] E. Estrada and P. Knight. *A First Course in Network Theory*. Oxford Press, 2015.

[66] J. Felsenstein. *Inferring Phylogenies*. Sinauer, 2004.

[67] J. Ferrell, J. Pomerening, E. M. Machleder et al. Simple, realistic models of complex biological processes: Positive feedback and bistability in a cell fate switch and a cell cycle oscillator. *FEBS Letters*, 583:3999–4005, 2009.

[68] M. Fessenden. Protein maps chart the causes of disease. *Nature*, 549:293–295, 2017.

[69] R. Forrester and H. J. Greenberg. Quadratic binary programming models in computational biology. *Algorithmic Operations Research*, 3:110129, 2008.

[70] L. R. Foulds and R. L. Graham. The Steiner tree problem in phylogeny is NP-complete. *Advances in Applied Math*, 3, 1982.

[71] K. Fox, B. Gavish, and S. Graves. An n-constraint formulation of the (time-dependent) traveling salesman problem. *Operations Research*, 28:101821, 1980.

[72] J. Frumkin, B. N. Patra, A. Sevold et al. The interplay between chromosome stability and cell cycle control explored through gene-gene interaction and computational simulation. *Nucleic Acids Research*, 44:8073–8085, 2016.

[73] J. Gao, L. Li, and C. Reidys. Inverse folding of RNA pseudoknot structures. *Algorithms for Molecular Biology*, 20105:27, 2010.

[74] M. Garey and D. Johnson. *Computers and Intractability*. W.H. Freeman, 1979.

[75] B. Gavish and S. Graves. The travelling salesman problem and related problems. Working Paper OR 078-78. Technical report, MIT, Operations Research Center, 1978.

[76] L. R. Gerber and M. C. Runge. Endangered species recovery: A resource allocation problem. *Science*, 362:284–286, 2018.

[77] M. R. Goddard and A. Burt. Recurrent invasion and extinction of a selfish gene. *Proceedings of the National Academy of Sciences (USA)*, 96:13880–13885, 1999.

[78] A. Godzik and J. Skolnick. Flexible algorithm for direct multiple alignment of protein structures and sequences. *Computer Applications in the BioSciences*, 10:587–596, 1994.

[79] Y. Guan, D. Gorenshteyn, M. Burmeister et al. Tissue-specific functional networks for prioritizing phenotype and disease genes. *PLoS Computational Biology*, 8(9):e1002694, 2012.

[80] K. S. Guimares, R. Jothi, E. Zotenka et al. Predicting domain-domain interactions using a parsimony approach. *Genome Biology*, 7:R104, 2006.

[81] K. S. Guimares and T. M. Przytycka. Interrogating domain-domain interactions with parsimony based approaches. *BMC Bioinformatics*, 9:171, 2008.

[82] D. Gusfield. Integer linear programming in computational biology: Overview of ILP, and new results for traveling salesman problems in biology. In T. Warnow, editor, *Bioinformatics and Phylogenetics*. Springer Nature Switzerland AG, 2019.

[83] D. Gusfield. *Algorithms on Strings, Trees and Sequences: Computer Science and Computational Biology*. Cambridge University Press, 1997.

[84] D. Gusfield. Haplotyping as perfect phylogeny: Conceptual framework and efficient solutions. In *RECOMB, The Annual International Conference on Research in Computational Molecular Biology*, pages 166–175. ACM Press, 2002.

[85] D. Gusfield. Haplotype inference by pure parsimony. In R. Baeza-Yates, E. Chavez, and M. Chrochemore, editors, *Proceedings of the Annual Symposium on Combinatorial Pattern Matching*, volume 2676, pages 144–155. LNCS, Springer, 2003.

[86] D. Gusfield. *ReCombinatorics: The Algorithmics of Ancestral Recombination Graphs and Explicit Phylogenetic Networks*. MIT Press, 2014.

[87] D. Gusfield, Y. Frid, and D. Brown. Integer programming formulations and computations solving phylogenetic and population genetic problems with missing or genotypic data. In *Proceedings of 13th Annual International Conference on Combinatorics and Computing*, pages 51–64. LNCS 4598, Springer, 2007.

[88] S. Hannenhalli and P. Pevzner. Transforming cabbage into turnip: Polynomial algorithm for sorting signed permutations by reversals. *Proceedings of the 27th ACM Symposium on the Theory of Computing*, pages 178–189, 1995.

[89] S. Hannenhalli and P. Pevzner. Transforming mice into men: Polynomial algorithm for genomic distance problem. *Proceedings of the 36'th IEEE Symposium on Foundations of Comp. Sci.*, pages 581–592, 1995.

[90] V. Hatzimanikatis, C. Floudas, and J. Bailey. Analysis and design of metabolic reaction networks via mixed-integer linear optimization. *AIChE Journal*, 42:1277–1292, 1996.

[91] D. M. Hillis. SINEs of the perfect character. *Proceedings of the National Academy of Sciences (USA)*, 96:9979–9981, 1999.

[92] C. Hitte, T. D. Lorentzen, R. Guyon et al. Comparison of MultiMap and TSP/CONCORDE for constructing radiation hybrid maps. *Journal of Heredity*, 94:9–13, 2003.

[93] B. Holland, H. Spencer, T. Worthy et al. Identifying cliques of convergent characters: Concerted evolution in the cormorants and shags. *Systematic Biology*, 59:433–445, 2010.

[94] F. Honti, S. Meader, and C. Webber. Unbiased functional clustering of gene variants with a phenotypic-linkage network. *PLoS Computational Biology*, 10(8):e1003815, 2014.

[95] S. M. H. Hosseini, F. Hoeft, and S. R. Kesler. GAT: A graph-theoretical analysis toolbox for analyzing between-group differences in large-scale structural and functional brain networks. *PLos ONE*, 7(7):e40709, 2012.

[96] T. J. P. Hubbard, A. M. Lesk, and A. Tramontano. Gathering in the fold. *Nature Structural Biology*, 4:313–313, April 1996.

[97] M. Hutson. Has artificial intelligence become alchemy? *Science*, 360:478–478, 2018.

[98] E. Huttlin, L. Ting, R. J. Bruckner et al. The bioplex network: A systematic exploration of the human interactome. *Cell*, 162:425–440, 2015.

[99] H. Jabbari, I. Wark, and C. Montemagno. RNA secondary structure prediction with pseudoknots: Contribution of algorithm versus energy model. *PLoS ONE*, 13(4), 2018.

[100] H. JH, R Lyngso, and J. Gorodkin. The FOLDALIGN web server for pairwise structural RNA alignment and mutual motif search. *Nucleic Acids Res.*, 33 (July 1), 2005.

[101] O. Johnson and J. Liu. A traveling salesman approach for predicting protein functions. *Source Code for Biology and Medicine*, 1, 2006.

[102] Y. Kato, K. Sato, M. Hamada et al. Ractip: Fast and accurate prediction of RNA–RNA interaction using integer programming. *Bioinformatics*, 26(18), 2010.

[103] J. D. Kececioglu and D. Sankoff. Efficient bounds for oriented chromosome inversion distance. *Proceedings 5'th Symposium on Combinatorial Pattern Matching. Springer LNCS 807*, pages 307–325, 1994.

[104] J. D. Kececioglu and D. Sankoff. Exact and approximation algorithms for sorting by reversal. *Algorithmica*, 13:180–210, 1995.

[105] A. Kehagias and L. Pitsoulis. Bad communities with high modularity. *The European Physiscal Journal B*, 86:330, 2013.

[106] C. L. Kingsford, B. Chazelle, and M. Singh. Solving and analyzing side-chain positioning problems using linear and integer programming. *Bioinformatics*, 21:1028–1036, 2005.

[107] S. Kinreich, N. Intrator, and T. Hendler. Functional cliques in the amygdala and related brain networks driven by fear assessment acquired during movie viewing. *Brain Connectivity*, 1:484–495, 2011.

[108] G. Klau, S. Rahmann, A. Schliep et al. Optimal robust non-unique probe selection using integer linear programming. *Bioinformatics*, 20 Suppl. 1:i186–i193, 2004.

[109] J. Kleinberg and E. Tardos. *Algorithm Design*. Addison-Wesley Longman, 2005.

[110] D. Knuth. *The Art of Computer Programming, Volume 4, Fascicle 6: Satisfiability*. New York: Addison-Wesley, 2015.

[111] C. Korostensky and G. Gonnet. Near optimal multiple sequence alignments using a traveling salesman problem approach. *Proceedings of String Processing and Information Retrieval Symposium*, 1999.

[112] C. Korostensky and G. Gonnet. Using traveling salesman problem algorithms for evolutionary tree construction. *Bioinformatics*, 16:619–627, 2000.

[113] A. Kreimer, E. Borenstein, U. Gophna et al. The evolution of modularity in bacterial metabolic networks. *Proceedings of the National Academy of Sciences (USA)*, 105(19):6976–6981, 2008.

[114] S. Krivov and M. Karplus. Hidden complexity of free energy surfaces for peptide (protein) folding. *Proceedings of the National Academy of Sciences (USA)*, 101:14766–14770, 2004.

[115] J. Kuipers, K. Jahn, and N. Beerenwinkel. Advances in understanding tumour evolution through single-cell sequencing. *Biochimica et Biophysica Acta*, 1867:27–138, 2017.

[116] J. Kuipers, K. Jahn, B. Raphael et al. Single-cell sequencing data reveal widespread recurrence and loss of mutational hits in the life histories of tumors. *Genome Research*, 27:1885–1894, 2017.

[117] G. Lancia. Integer programming models for computational biology problems. *Journal of Computer Science and Technology*, 19:6077, 2004.

[118] G. Lancia. Mathematical programming in computational biology: An annotated bibliography. *Algorithms*, 1:100129, 2008.

[119] G. Lancia, C. Pinotti, and R. Rizzi. Haplotyping populations by pure parsimony: Complexity, exact and approximation algorithms. *INFORMS Journal on Computing, Special Issue on Computational Biology*, 16:348–359, 2004.

[120] G. Lancia, F. Rinaldi, and P. Serafini. A unified integer programming model for genome rearrangement problems. In F. Ortuño and I. Rojas, editors, *Bioinformatics and Biomedical Engineering: Third International Conference, IWBBIO 2015, LNCS Vol. 9043*, pages 491–502. Springer, LNCS, 2015.

[121] G. Lancia and P. Serafini. *Compact Extended Linear Programming Models*. Springer, 2018.

[122] J. K. Lanctot, M. Li, B. Ma et al. Distinguishing string selection problems. *Information and Computation*, 185:41–55, 2003.

[123] E. L. Lawler. *Combinatorial Optimization: Networks and Matroids*. Holt, Rinehart and Winston, New York, 1976.

[124] S. Lee, C. Phalakornkule, M. Domach et al. Recursive MILP model for finding all the alternate optima in LP models for metabolic networks. *Computers and Chemical Engineering*, 24:711–716, 2000.

[125] G. Li, D. D. Serba, M. C. Saha et al. Genetic linkage mapping and transmission ratio distortion in a three-generation four-founder population of Panicum virgatum (L.). *G3: Genes—Genomes—Genetics*, 4:913–923, 2014.

[126] X. Li, P. Chavali, and M. Babu. Capturing dynamic protein interactions. *Science*, 359:1105–1106, 2018.

[127] Z. Li, R-S Wang, and X-S Zhang. Mass flow model and essentiality of enzymes in metabolic networks. In *Second International Symposium on Optimization and Systems Biology*, volume 9 of *Lecture Notes in Operations Research*, pages 182–190, 2008.

[128] E. Lorenzo, K. Camacho-Caceres, A. J. Ropelewski et. al. An optimization-driven analysis pipeline to uncover biomarkers and signaling paths: Cervix cancer. *Microarrays*, 4:287–310, 2015.

[129] W. Lu, T. Tamura, J. Song et al. Integer programming-based method for designing synthetic metabolic networks by minimum reaction insertion in a boolean model. *PLoS ONE*, 9:e92637, 2014.

[130] C. Y. Ma, P. Chen, B. Berger et al. Identification of protein complexes by integrating multiple alignment of protein interaction networks. *Bioinformatics*, 33:1681–1688, 2017.

[131] S. Malikic, A. McPherson, N. Donmez et al. Clonality inference in multiple tumor samples using phylogeny. *Bioinformatics*, 31:1349–1356, 2015.

[132] T. Marschall, I. Costa, S. Canzar et al. CLEVER: Clique-enumerating variant finder. *Bioinformatics*, 28:2875–2882, 2012.

[133] A. Mazza, K. Klockmeier and E. Wanker. An integer programming framework for inferring disease complexes from network data. *Bioinformatics*, 32:i271–i277, 2016.

[134] C. Miller, R. Tucker, and R. Zemlin. Integer programming formulation of traveling salesman problems. *Journal of the Association for Computing Machinery*, pages 326–329, 1960.

[135] W. Miller, S. J. Wright, Y. Zhang et al. Optimization methods for selecting founder individuals for captive breeding or reintroduction of endangered species. *Pacific Symposium on Biocomputing*, 15:43–53, 2010.

[136] S. Miyazawa and R. L. Jernigan. Residue–residue potentials with a favorable contact pair term and an unfavorable high packing density term, for simulation and threading. *Journal of Molecular Biology*, 256:623–644, 1996.

[137] B. Moret, D. A. Bader, and T. Warnow. High-performance algorithm engineering for computational phylogenetics. *The Journal of Supercomputing*, 22:99–111, 2002.

[138] M. Mutsuddi, D. Morriss, S. Waggoner et al. Analysis of high-resolution HapMap of DTNBP1 (dysbindin) suggests no consistency between reported common variant associations and schizophrenia. *American Journal of Human Genetics*, 79:903–909, 2006.

[139] E. Navieva, K. Jim, A. Agarwal et al. Whole-proteome prediction of protein function via graph-theoretic analysis of interaction maps. *Bioinformatics*, 21,(Supp. 1):i302–i310, 2005.

[140] M. Newman. Modularity and community structure in networks. *Proceedings of the National Academy of Sciences (USA)*, 103:8577–8582, 2006.

[141] L. Nunes, L. Galvao, H. Lopes et al. An integer programming model for protein structure prediction using the 3D-HP side chain model. *Discrete Applied Mathematics*, 198:206–214, 2016.

[142] J. Orth, I. Thiele, and B. Palsson. What is flux balance analysis? *Nature Biotechnology*, 28:245–248, 2010.

[143] R. Page. *Tangled Trees: Cospeciation and Coevolution*. University of Chicago Press, Chicago, 2002.

[144] G Palla, I. Derenyi, I. Farkas et al. Uncovering the overlapping community structure of complex networks in nature and society. *Nature*, 435:814–818, 2005.

[145] J. Palmer and L. Herbon. Plant mitochondrial DNA evolves rapidly in structure, but slowly in sequence. *Journal of Molecular Evolution*, 27:87–97, 1988.

[146] K. Pang, Y. W. Wan, W. T. Choi et al. Combinatorial therapy discovery using mixed integer linear programming. *Bioinformatics*, 30:1456–1463, 2014.

[147] G. Pataki. The bad and the good-and-ugly. Technical report, Columbia University, IEOR, 2000.

[148] G. Pataki. Teaching integer programming formulations using the traveling salesman problem. *SIAM Review*, 65:116–123, 2003.

[149] M. Patterson, T. Marschall, N. Pisanti. WhatsHap: Weighted haplotype assembly for future-generation sequencing reads. *Journal of Computational Biology* 22:498–509, 2015.

[150] E. Pennisi and M. Price. Research news: Molecular ‘barcodes’ reveal lost whale hunts. *Science*, 361:119, July 2018.

[151] A. Platzer, P. Perco, A. Lukas et al. Characterization of protein-interaction networks in tumors. *BMC Bioinformatics*, 8:224, 2007.

[152] U. Poolsap, Y. Kato, and T. Akutsu. Prediction of RNA secondary structure with pseudoknots using integer programming. *BMC Bioinformatics*, 10(Suppl 1):S38, 2009.

[153] M. Pradhan, K. Nagulapalli, and M. Palakal. Cliques for the identification of gene signatures for colorectal cancer across populations. *BMC Systems Biology*, 6(Suppl 3):S17, 2012.

[154] J. Pritchard and N. Rosenberg. Use of unlinked genetic markers to detect population stratification in association studies. *The American Journal of Human Genetics*, 65:220–228, 1999.

[155] T. Przytycka. Stability of characters and construction of phylogenetic trees. *Journal of Computational Biology*, 14:539–549, 2007.

[156] T. Przytycka, G. Davis, N. Song et al. Graph theoretical insights into evolution of multidomain proteins. *Journal of Computational Biology*, 13:351–363, 2006.

[157] A. Pujol, R. Mosca, J. Farres et al. Unveiling the role of network and systems biology in drug discovery. *Trends in Pharmacological Sciences*, 31:115–123, 2010.

[158] D. Quammen. *The Tangled Tree: A Radical New History of Life*. Simon and Schuster, 2018.

[159] Y. Qui, H. Jiang, W.K. Ching et al. Discovery of Boolean metabolic networks: Integer linear programming based approach. *BMC Systems Biology*, 12(Supp. 1):7, 2018.

[160] S. Rash and D. Gusfield. String barcoding: Uncovering optimal virus signatures. In *Proceedings of RECOMB 2002: The Sixth Annual International Conference on Computational Biology*, pages 254–261, 2002.

[161] D. A. Ray, J. Xing, A-H. Salem et al. SINEs of the *nearly* perfect character. *Systematic Biology*, 55:928–935, 2006.

[162] G. Reinelt. TSPLIB A traveling salesman problem library. *ORSA Journal on Computing*, 3:376–384, 1991.

[163] J. Reiter, A. Makohon-Moore, J. Gerold et al. Reconstructing metastatic seeding patterns of human cancers. *Nature Communications, article 14114*, 8, 2017.

[164] Research Highlight. A double-pronged attack on colon tumours succeeds where one doesn't. *Nature*, 557, 2018.

[165] J. De Las Rivas and C. Fontanillo. Protein–protein interaction essentials: Key concepts to building and analyzing interactome networks. *PLoS Computational Biology*, 6(6):e1000807, 2010.

[166] I. B. Rogozin, Y. I. Wolf, V. N. Babenko et al. Dollo parsimony and the reconstruction of genome evolution. In V. A. Albert, editor, *Parsimony, Phylogeny, and Genomics*. Oxford University Press, 2006.

[167] R. Salari, S. S. Saleh, D. Kashef-Haghighi et al. Inference of tumor phylogenies with improved somatic mutation discovery. Journal of Computational Biology, 20(110):933–944, 2013.

[168] D. Sankoff and M. Blanchette. Multiple genome rearrangement and breakpoint phylogeny. *Journal of Computational Biology*, 5:555–570, 1998.

[169] K. Sato, Y. Kato, M. Hamada et al. Ipknot: Fast and accurate prediction of RNA secondary structures with pseudoknots using integer programming. *Bioinformatics [ISMB/ECCB]*, 27(13):85–93, 2011.

[170] T. Sawik. A note on the Miller-Tucker-Zemlin model for the asymmetric traveling salesman problem. *Bulletin of the Polish Academy of Sciences, Technical Sciences*, 64:517–520, 2016.

[171] S. T. Schmidt, S. Zimmerman, J. Wang et al. Quantitative analysis of synthetic cell lineage tracing using nuclease barcoding. *ACS Synthetic Biology*, 6:936–942, 2017.

[172] R. Schwartz and A. Schäffer. The evolution of tumour phylogenetics: Principles and practice. *Nature Reviews: Genetics*, 18:213–229, 2017.

[173] F. Seersholm, T. Cole, A. Grealy et al. Subsistence practices, past biodiversity, and anthropogenic impacts revealed by New Zealand-wide ancient DNA survey. *Proceedings of the National Academy of Sciences (USA)*, 115(30):7771–7776, 2018.

[174] C. Semple and M. Steel. *Phylogenetics*. Oxford University Press, 2003.

[175] M. Shao, Y. Lin, and B. M.E. Moret. An exact algorithm to compute the DCJ distance for genomes with duplicate genes. *Journal of Computational Biology*, 22(5):425–435, 2015.

[176] M. Shao and B. M. E. Moret. Comparing genomes with rearrangements and segmental duplications. *Bioinformatics*, 31(12):i329–i338, 2015.

[177] M. Shao and B. M. E. Moret. A fast and exact algorithm for the exemplar breakpoint distance. *Journal of Computational Biology*, 23(5):337–346, 2016.

[178] M. Shao and B. M. E. Moret. On computing breakpoint distances for genomes with duplicate genes. *Journal of Computational Biology*, 24(6):571–580, 2017.

[179] R. Sharan, R. Karp, T. Ideker et al. Conserved patterns of protein interaction in multiple species. *Proceedings of the National Academy of Sciences (USA)*, 102(6):1974–1979, 2005.

[180] V. Sharma, T. Lehmann, H Stuckas et al. Loss of RXFP2 and INSL3 genes in Afrotheria shows that testicular descent is the ancestral condition in placental mammals. *PLoS Biology*, 16(6):e2005293, 2018.

[181] Z. Shi, C. Derow, and B. Zhang. Co-expression module analysis reveals biological processes, genomic gain, and regulatory mechanisms associated with breast cancer progression. *BMC Systems Biology*, 4, 74, 2010.

[182] T. Shlomi, M. Cabili, M. Herrgard et al. Network-based prediction of human tissue-specific metabolism. *Nature Biotechnology*, 26:1003–1010, 2008.

[183] R. Shrestha, E. Hodzic, J. Yeung et al. HITnDRIVE: Multi-driver gene prioritization, based on hitting time. In R. Sharan, editor, *Proceedings of RECOMB 2014*, volume 8394 of *Lecture Notes in Bioinformatics*, pages 293–306, 2014.

[184] R. Shrestha, E. Hodzic, T. Sauerwald et al. HITnDRIVE: Patient-specific multidriver gene prioritization for precision oncology. *Genome Research*, 27:1573–1588, 2017.

[185] W. A. Smith, K. Oakeson, K. Johnson et al. Phylogenetic analysis of symbionts in feather-feeding lice of the genus columbicola: Evidence for repeated symbiont replacements. *BMC Evolutionary Biology*, 13:109, 2013.

[186] S. Sridhar, F. Lam, G. E. Blelloch et al. Mixed integer linear programming for maximum-parsimony phylogeny inference. *IEEE/ACM Transactions on Computational Biology and Bioinformatics*, 5:323–331, 2008.

[187] S. Sridhar, S. Rao, and E. Halperin. An efficient and accurate graph-based approach to detect population substructure. In T. Speed and H. Huang, editors, *Research in Computational Molecular Biology: 11th Annual International Conference, RECOMB 2007*, pages 503–517, 2007.

[188] M. Stephens, N. Smith, and P. Donnelly. A new statistical method for haplotype reconstruction from population data. *American Journal of Human Genetics*, 68:978–989, 2001.

[189] T. Tamura, W. Lu, and T. Akutsu. Computational methods for modification of metabolic networks. *Computational and Structural Biotechnology Journal*, 13:376–381, 2015.

[190] T. Tamura, K. Takemoto, and T. Akutsu. Finding minimum reaction cuts of metabolic networks under a Boolean model using integer linear programming and feedback vertex sets. *International Journal of Knowledge Discovery in Bioinformatics (IJKDB)*, 1:14–31, 2010.

[191] R. Tewhey, V. Bansal, A. Torkamani et al. The importance of phase information in human genomics. *Nature Reviews Genetics*, 12:215–223, 2011.

[192] N. Tuncbag, F. Salman, O. Keskin et al. Analysis and network representation of hotspots in protein interfaces using minimum cut trees. *Proteins*, 78:2283–2294, 2010.

[193] K. E. van Rens, V. Makinen, and A. Tomescu. SNV-PPILP: Refined SNV calling for tumor data using perfect phylogenies and ILP. *Bioinformatics*, 31:1133–1135, 2015.

[194] B. Venkatachalam, J. Apple and K. St. John. Untangling tanglegrams: Comparing trees by their drawings. *IEEE/ACM Transactions on Computational Biology and Bioinformatics*, 7(4):588–597, 2010.

[195] B. Venkatachalam and D. Gusfield. Generalizing tanglegrams, 2018. Technical Report, UC Davis, College of Engineering. Retrieved from https://escholarship.org/uc/item/0bg5p8ch (last accessed January 24, 2019).

[196] A. von Kamp and S. Klamt. Enumeration of smallest intervention strategies in genome-scale metabolic networks. *PLoS Computational Biology*, 10(1):e1003378, 2014.

[197] J. Wakeley. *Coalescent Theory*. Roberts and Co., 2009.

[198] W. Wang, Z. Zack, M. Costanzo et al. Pathway-based discovery of genetic interactions in breast cancer. *PLOS Genetics*, 13, 2017.

[199] J.S. Waters and J.H. Fewell. Information processing in social insect networks. *PLoS ONE*, 7(7):e40337, 2012.

[200] Ask Well. Science Times, *The New York Times*, October 9, 2018.

[201] A. Williams and S. Halappanavar. Application of biclustering of gene expression data and gene set enrichment analysis methods to identify potentially disease causing nanomaterials. *Beilstein Jounal of Nanotechnology*, 6:2438–2448, 2015.

[202] E. O. Wilson. A consistency test for phylogenies based on contemporaneous species. *Systematic Zoology*, 14:214–220, 1965.

[203] L. Wolsey. *Integer Programming*. John Wiley, 1998.

[204] Y. Wu. A practical method for exact computation of subtree prune and regraft distance. *Bioinformatics*, 25(2):190–196, 2009.

[205] S. Wuchty, Z. N. Oltvai, and A. Barabási. Evolutionary conservation of motif constituents in the yeast protein interaction network. *Nature Genetics*, 35:176–179, 2003.

[206] J. Xu, M. Li, D. Kim et al. Raptor: Optimal protein threading by linear programming. *Journal of Bioinformatics and Computational Biology*, pages 95–117, 2003.

[207] Y. Xu, D. Xu, and H. Gabow. Protein domain decomposition using a graph-theoretic approach. *Bioinformatics*, 16:1091–1104, 2000.

[208] N. Yanev, M. Traykov, P. Milanov et al. Protein folding prediction in a cubic lattice in hydrophobic-polar model. *Journal of Computational Biology*, 24:412–421, 2017.

[209] N. Yosef, E. Zalckvar, A. D. Rubenstein et al. ANAT: A tool for constructing and analyzing functional protein networks. *Science Signaling*, 4:pl1, 2011.

[210] H. Yu, A. Paccanaro, V. Trifonov et al. Predicting interactions in protein networks by completing defective cliques. *Bioinformatics*, 22:823–829, 2006.

[211] K. Yue, K. Fieberg, P. Thomas. A test of lattice protein folding algorithms. *Proceedings of the National Academy of Sciences (USA)*, 92:325–329, 1995.

[212] W. Zachary. An information flow model for conflict and fission in small groups. *Journal of Anthropological Research*, 33:452–473, 1977.

[213] G. Zhang, B. B. Beck, W. Luo et al. Development of a phylogenetic tree model to investigate the role of genetic mutations in endometrial tumors. *Oncology Reports*, 25:1447–1454, 2011.

[214] J. Zheng, I. B. Rogozin, E. V. Koonin et al. Support for the *coelomatc* clade of animals from a rigorous analysis of the pattern of intron conservation. *Molecular Biology and Evolution*, 24:2583–2592, 2007.

[215] X. Zhu, M. Gerstein, and M. Snyder. Getting connected: Analysis and principles of biological networks. *Genes and Development*, 21:10101024, 2007.

[216] Wikipedia. Genus, 2018. Retrieved from https://en.wikipedia.org/wiki/Genus (last accessed December 28, 2018).

[217] Wikipedia. Metabolism, 2018. Retrieved from https://en.wikipedia.org/wiki/Metabolism (last accessed December 28, 2018).

[218] Wikipedia. Factorial, 2018. Retrieved from https://en.wikipedia.org/wiki/Factorial (last accessed December 28, 2018).

[219] Wikipedia. The Traveling Salesman Problem, 2018. Retrieved from https://en.wikipedia.org/wiki/Travelling_salesman_problem (last accessed December 28, 2018).

[220] Wikipedia. Metabolic Engineering, 2018. Retrieved from https://en.wikipedia.org/wiki/Metabolic_engineering (last accessed December 28, 2018).

[221] Wikipedia. Chromosome, 2018. Retrieved from https://en.wikipedia org/wiki/Chromosome (last accessed January 23, 2019).

[222] Wikipedia. Management of HIV/AIDS, 2018. Retrieved from https://en.wikipedia.org/wiki/Management_of_HIV/AIDS (last accessed January 23, 2019).

Index